Technological Innovation Driving Sustainable Entrepreneurial Growth in Developing Nations

Nana Yaw Asabere
Accra Technical University, Ghana

Glenn Kwabena Gyimah
Accra Technical University, Ghana

Amevi Acakpovi
Accra Technical University, Ghana

Florence Plockey
Accra Technical University, Ghana

A volume in the Advances in Logistics, Operations, and Management Science (ALOMS) Book Series

IGI Global
PUBLISHER of TIMELY KNOWLEDGE

Published in the United States of America by
IGI Global
Business Science Reference (an imprint of IGI Global)
701 E. Chocolate Avenue
Hershey PA, USA 17033
Tel: 717-533-8845
Fax: 717-533-8661
E-mail: cust@igi-global.com
Web site: http://www.igi-global.com

Library of Congress Cataloging-in-Publication Data

Names: Asabere, Nana, 1981- editor.
Title: Technological innovation driving sustainable entrepreneurial growth in developing nations / edited by Nana Yaw Asabere, Glenn Gyimah, Amevi Acakpovi, Florence Plockey.
Description: Hershey, PA : Business Science Reference, [2024] | Includes bibliographical references and index. | Summary: "The main objective of this publication is to provide a bridge for technological innovations through applied research thereby driving sustainable entrepreneurial growth in developing nations. This objective will be achieved through discussion of ideas, results, generalizations, hypotheses, and plans among researchers representing institutions from academia and industry"-- Provided by publisher.
Identifiers: LCCN 2023013642 (print) | LCCN 2023013643 (ebook) | ISBN 9781668498439 (h/c) | ISBN 9781668498446 (s/c) | ISBN 9781668498453 (eISBN)
Subjects: LCSH: Technological innovations--Developing countries. | Sustainable development--Developing countries.
Classification: LCC T173.5.D48 T43 2024 (print) | LCC T173.5.D48 (ebook) | DDC 303.48/3091724--dc23/eng/20230808
LC record available at https://lccn.loc.gov/2023013642
LC ebook record available at https://lccn.loc.gov/2023013643

This book is published in the IGI Global book series Advances in Logistics, Operations, and Management Science (ALOMS) (ISSN: 2327-350X; eISSN: 2327-3518)

British Cataloguing in Publication Data
A Cataloguing in Publication record for this book is available from the British Library.

For electronic access to this publication, please contact: eresources@igi-global.com.

Advances in Logistics, Operations, and Management Science (ALOMS) Book Series

ISSN:2327-350X
EISSN:2327-3518

Editor-in-Chief: John Wang, Montclair State University, USA

MISSION

Operations research and management science continue to influence business processes, administration, and management information systems, particularly in covering the application methods for decision-making processes. New case studies and applications on management science, operations management, social sciences, and other behavioral sciences have been incorporated into business and organizations real-world objectives.

The **Advances in Logistics, Operations, and Management Science** (ALOMS) Book Series provides a collection of reference publications on the current trends, applications, theories, and practices in the management science field. Providing relevant and current research, this series and its individual publications would be useful for academics, researchers, scholars, and practitioners interested in improving decision making models and business functions.

COVERAGE

- Computing and information technologies
- Decision analysis and decision support
- Networks
- Marketing engineering
- Organizational Behavior
- Information Management
- Production Management
- Services management
- Operations Management
- Political Science

IGI Global is currently accepting manuscripts for publication within this series. To submit a proposal for a volume in this series, please contact our Acquisition Editors at Acquisitions@igi-global.com or visit: http://www.igi-global.com/publish/.

The Advances in Logistics, Operations, and Management Science (ALOMS) Book Series (ISSN 2327 350X) is published by IGI Global, 701 E. Chocolate Avenue, Hershey, PA 17033-1240, USA, www.igi-global.com. This series is composed of titles available for purchase individually; each title is edited to be contextually exclusive from any other title within the series. For pricing and ordering information please visit http://www.igi-global.com/book-series/advances-logistics-operations-management-science/37170. Postmaster: Send all address changes to above address.

Titles in this Series

For a list of additional titles in this series, please visit:
http://www.igi-global.com/book-series/advances-logistics-operations-management-science/37170

Black Women's Formal and Informal Ways of Leadership Actualizing the Vision of a More Equitable Workplace
Audrey D. Bowser (Arkansas State University, USA) Kimberley M. Davis (University of Arkansas at Pine Bluff, USA) and Natalie A. Johnson-Leslie (Arkansas State University, USA)
Information Science Reference • copyright 2023 • 340pp • H/C (ISBN: 9781668438275) • US $215.00 (our price)

Multidisciplinary Approaches to Organizational Governance During Health Crises
Carla Sofia Vicente Negrão (CeBER, University of Coimbra, Portugal) Isabel Guerreiro Pimentel Maia (Centre for Business and Economics Research, University of Coimbra, Portugal) and João António Furtado Brito (University of Cape Verde, Cape Verde)
Business Science Reference • copyright 2023 • 300pp • H/C (ISBN: 9781799892137) • US $250.00 (our price)

New Perspectives and Possibilities in Strategic Management in the 21st Century Between Tradition and Modernity
Javier Martínez-Falcó (University of Alicante, Spain & Stellenbosch University, South Africa) Bartolomé Marco-Lajara (University of Alicante, Spain) Eduardo Sánchez-García (University of Alicante, Spain) and Luis A. Millan-Tudela (University of Alicante, Spain)
Business Science Reference • copyright 2023 • 514pp • H/C (ISBN: 9781668492611) • US $315.00 (our price)

Government Impact on Sustainable and Responsible Supply Chain Management
Atour Taghipour (Normandy University, France)
Information Science Reference • copyright 2023 • 354pp • H/C (ISBN: 9781668490624) • US $240.00 (our price)

For an entire list of titles in this series, please visit:
http://www.igi-global.com/book-series/advances-logistics-operations-management-science/37170

IGI Global
PUBLISHER of TIMELY KNOWLEDGE

701 East Chocolate Avenue, Hershey, PA 17033, USA
Tel: 717-533-8845 x100 • Fax: 717-533-8661
E-Mail: cust@igi-global.com • www.igi-global.com

Table of Contents

Detailed Table of Contents

health institutions and governments plan ahead to forestall COVID-19 deaths and implement policies toward preventive care.

In this chapter, a model of spatial sustainable entrepreneurship is presented by integrating sustainable development with entrepreneurship and spatial data. The model contributes to the three dimensions of sustainable development. Spatial data can promote economic sustainability by identifying opportunities, optimizing supply chains, and creating jobs. It can contribute to environmental sustainability through resource management and renewable energy projects. Additionally, it can enable social sustainability by addressing social needs and promoting inclusion. Policymakers, entrepreneurs, and researchers can collaborate to support training programs, open data policies, and collaborations to maximize the use of spatial data in sustainable entrepreneurship. Future work can involve advanced analytics, privacy considerations, stakeholder engagement, impact assessment, infrastructure development, cross-disciplinary research, and policy frameworks.

Statistically, to retain the originality of data, appropriate probability distribution must be employed rather than transforming the existing dataset. Classical Distributions (CD) lack the ability to model and describe some important real-life events. In this study, weighted Weibull distribution (WWD) is modified using Exponentiated Weighted Weibull distribution (EWWD). The shapes of the probability density function and hazard rate functions are investigated together with some structural statistical properties of the distribution. The performances of the maximum likelihood estimation of the unknown parameters of the distribution were done by implementing the Monte Carlo simulation method. We provide applications to two datasets to prove the flexibility of the model. The processes were performed using R-software. The EWWD is considered to be more tractable and capable than the WWD. The Exponentiated weighted Weibull distribution assumes a superior performance among the compared distributions as evidenced by the Kolmogorov-Smirnov statistics and Akaike Information Criteria values.

The use of antibiotics is widespread throughout the world, and Ghana is not an exception. They are one of the most frequently prescribed medications. Antibiotic residue concentrations are rising in the environment and are found in practically all water bodies, including lakes and rivers, because of inadequate management of residential and medical wastes. Contrarily, there is a substantial chance of medication toxicity and resistance. Therefore, removal of these pollutants from water bodies is essential, yet conventional wastewater or water treatment methods frequently fall short of efficiently removing these chemicals. Antibiotics will be used as a case study in this chapter to discuss the concentration of pharmaceutical pollutants in the environment, specifically in Ghanaian waterbodies, and risk evaluations of these pollutants. Additionally, this chapter will look into the possible removal technologies report and establish the present investigation on removing these pollutants with microplastics.

The aim of this study was to investigate antioxidant and anticancer potentials of natural cocoa extracts. Antioxidant properties of the extracts were evaluated by DPPH assay, total phenolic content by Folin-Ciocalteau and reduced glutathione content by O-phthalaldehyde assays. Antiproliferative activity was also evaluated on leukaemia and prostate cancer cell lines using MTT assay. The hydroethanolic extracts of cocoa roots and beans and aqueous cocoa leaves showed high total antioxidant activities with the root extract having the strongest scavenging effect on DPPH and the highest total phenolics content. The hydroethanolic cocoa bean crude extract showed the highest anticancer activity against LNCaP cells, whiles the ethyl acetate fraction of hydroethanolic cocoa leaf exhibited the highest anticancer against PC3. Jurkat and PC3 cells had similar susceptibilities to the natural cocoa extracts. Thus, cocoa has antioxidant and anticancer activity, and the specific bioactive compounds responsible for these activities need to be elucidated in future research.

The aim of this study was to apply ultrasound technology to extract bioactive compounds from Moringa oleifera Lam. leaves. Bringing modern trends of extraction into perspective, ultrasound-assisted extraction (UAE) of polysaccharides of Moringa leaves (MLP) was established engaging response surface methodology (RSM). The experimental yield of 9.24% of crude polysaccharide was close to the predicted 9.57%. Three graded ethanol precipitation of the UAE polysaccharides with corresponding molecular weights (55,250, 60,330 and 34,260 g/mol) was obtained. Despite similar Fourier transform infrared spectroscopy among fractions there was different molar ratios of monosaccharides composition and different surface morphology under scanning electron microscopy. All fractions were not of helical structures as demonstrated by the Congo red assay. MLP precipitated at 80% was more viscous (G″ loss modulus) and had the highest radical scavenging ability, thus can prevent damage of biomolecules. The polysaccharides extract obtained may be incorporated in pharmaceutical and nutraceutical products.

Surfaces and devices in public institutions are likely to be contaminated with various microorganisms as people congregate there for various reasons. Swab samples from devices like ATMs, seats, teller counters, door handles, pens, writing desks, toilet flush handles, and tap heads were obtained from banks and churches. 60 samples in all

Although sizing is usually established through the development of a size chart, the real elements of fit can only be ascertained after the creation of a garment pattern, based on the measurements of the size chart. Therefore, one of most fundamental stages in the garment manufacturing business is patternmaking. This study developed basic pattern blocks for garment manufacture using anthropometric data of Ghanaian girls (7 to 12 years), measured, and processed into a size chart in a previous study. First of all, the measurements of the size chart were extracted, after which one extra size (GH 6) was created based on the intervals observed between the sizes. Basic pattern blocks were developed and validated by producing trial samples and testing them on a sub-section of the target population. The major contribution of this study is the availability of a variation of ready-made standard garment blocks for use by garment manufacturers, and an enhanced fit of garments produced for the children's wear sector.

The study was conducted to design and construct a drying chamber for millinery and accessories production to reduce the risks involved in the old ways of drying hat. The design of the study is descriptive while the population includes fashion schools. The sample size for the study was 100 respondents while a researcher's self-developed questionnaire was used as the instrument for data collection. The blocked hat used the sun drying method, rather than the drying chamber, and considered the drying chamber more effective than sun drying method. The dryer constructed could dry the hats effectively and efficiently.

This chapter focused on sculptural fabricated prosthetic socket for a transtibial (below-knee) amputee. As part of this study, an in-depth examination of the special needs for lower limb prostheses with quality and attractiveness was conducted in Sekondi-Takoradi Metropolis. Qualitative inquiry employing studio-based and descriptive research designs was drawn from collecting data from ten (10) purposively selected

Preface

In an era defined by the rapid evolution of technology and the imperative of sustainable growth, the confluence of innovation and entrepreneurship emerges as a beacon of hope for developing nations. Against the backdrop of a world reshaped by the COVID-19 pandemic, the need for ingenious solutions to complex challenges has never been more pronounced. This reference book, *Technological Innovation Driving Sustainable Entrepreneurial Growth in Developing Nations,* stands as a testament to the collaborative spirit and intellectual vigor of scholars and visionaries from Accra Technical University, Ghana.

Nana Yaw Asabere, Glenn Kwabena Gyimah, Amevi Acakpovi, and Florence Plockey, distinguished voices from Accra Technical University, spearheaded this endeavor with a profound understanding of the pivotal role technology and entrepreneurship play in Africa's journey toward sustainable development. This collection of chapters, born from the vibrant discussions at the 2nd Applied Research Conference of Technical Universities in Ghana held in Accra Technical University (2022), crystallizes their commitment to harnessing knowledge for tangible impact.

Africa, home to remarkable diversity and untapped potential, is at a crossroads demanding innovative interventions. As the world grapples with the Sustainable Development Goals outlined by the United Nations, this book assumes the role of a guidepost, illuminating pathways through which research and practical insights can coalesce to effect change. The overarching objective is clear: to cultivate an ecosystem where technological advancements are seamlessly woven into the fabric of sustainable entrepreneurial growth, reshaping the trajectory of nations on the rise.

The chapters presented herein traverse a rich spectrum of themes, each underlining the interconnectedness of technological innovation, entrepreneurship, and sustainable development. "Subtheme 1: Information and Communication Technology (ICT)," beckons us to harness the power of the digital age, bridging gaps and fostering inclusivity. "Subtheme 2: Entrepreneurship, Industry and University Linkages" underscores the symbiotic relationship between academia and industry, a synergy that propels progress. "Subtheme 3: Innovations in Science and Health" reflects the urgency of enhancing well-being through pioneering scientific solutions. Lastly,

"Subtheme 4: Engineering" resonates with the transformative potential of engineering feats in shaping our shared future.

Our intended readership extends to the corridors of academia and the realms of industry, for it is in their collaboration that the seeds of change germinate. Educators, researchers, and industry leaders will find in these pages a wellspring of ideas, a reservoir of data, and a catalyst for action. As editors, we firmly believe that this book serves as a cornerstone for dialogue, an enabler of innovation, and a compass guiding the ship of progress through uncharted waters.

We extend our deepest gratitude to the contributors, whose unwavering dedication to advancing knowledge fuels the engine of change. The journey from research to application is rarely linear, and this book stands as a tribute to their perseverance.

ORGANIZATION OF BOOK

Chapter 1

This chapter explores the innovative application of laser shock peening (LSP) to extend the fatigue life of metals. The study delves into the effects of pulse energy on the fatigue lifespan of AA 2024-T351 alloy. Through experimental and finite element analysis, the chapter confirms that LSP induces higher residual compressive stress, preventing superficial failures and enhancing fatigue life. The correlation between experimental and FE results, coupled with micro-scale fatigue striations, underscores the potential of LSP in increasing the durability of alloys.

Chapter 2

Addressing environmental pollution, this chapter investigates the potential of recycled polyethylene to reinforce cement-less composites for civil engineering applications. By skillfully blending laterite and polyethylene, the study produces composites with exceptional mechanical properties. The composite with a specific weight percentage combination exhibits impressive compressive and flexural strength, paving the way for a sustainable material alternative in construction.

Chapter 3

Focused on technology's role in healthcare, this chapter employs IBM's Auto AI Experiment to predict COVID-19 mortality in the USA. Leveraging extensive datasets, the research identifies pipelines with superior prediction outcomes, contributing valuable insights for health institutions and governments to plan ahead for preventive care strategies.

Chapter 4

This chapter proposes an innovative model that integrates spatial data with sustainable development and entrepreneurship. Encompassing economic, environmental, and social dimensions, the model underscores the power of spatial data in promoting sustainable entrepreneurship. Recommendations for collaboration and policy frameworks underline the potential for driving inclusive and impactful entrepreneurship.

Chapter 5

Exploring the realm of statistical distributions, this chapter introduces the Exponentiated Weighted Weibull Distribution (EWWD). The study compares its performance with classical distributions, highlighting the flexibility and tractability of EWWD. Through simulations and applications, the chapter presents a superior distribution model for real-life event modeling.

Chapter 6

Focusing on environmental challenges, this chapter examines the presence of antibiotic residues in water bodies, particularly in Ghana. The study evaluates the potential risks and toxicity associated with inadequate wastewater treatment. Additionally, the chapter delves into removal technologies, emphasizing the urgency of addressing pharmaceutical pollutants for ecological and human health.

Chapter 7

Centered on natural resources, this chapter explores the antioxidant and anticancer properties of cocoa extracts. Through comprehensive analyses, the study identifies potent hydroethanolic cocoa extracts with remarkable antioxidant and anticancer activity. The findings highlight cocoa's potential as a valuable source of bioactive compounds for pharmaceutical and nutraceutical applications.

Chapter 8

Leveraging modern techniques, this chapter showcases the application of ultrasound technology to extract bioactive compounds from Moringa oleifera leaves. The research introduces a new method to enhance the extraction process, producing polysaccharides with diverse properties. The study's results underline the potential for integrating ultrasound technology into pharmaceutical and nutraceutical products.

Chapter 9

Examining public health concerns, this chapter investigates the microbial contamination present on commonly used public surfaces. Through comprehensive sampling and analysis, the study highlights the potential sources of infections and emphasizes the importance of proper hand hygiene, particularly in the midst of the ongoing pandemic.

Chapter 10

Addressing water pollution, this chapter explores the trend of using Moringa oleifera seeds for water and wastewater treatment. Highlighting their effectiveness in comparison to synthetic chemicals, the study underscores the potential of these natural resources to mitigate water pollution and offer cost-effective solutions for water treatment in developing countries.

Chapter 11

In pursuit of enhanced garment fit and manufacturing precision, this chapter develops basic pattern blocks tailored to the anthropometric data of Ghanaian girls. Through meticulous measurement and analysis, the chapter establishes a valuable resource for garment manufacturers, offering standardized pattern blocks that cater to the unique needs of the children's wear sector.

Chapter 12

In the context of garment manufacturing, this chapter presents an innovative drying chamber design for millinery and accessories production. The research introduces an alternative approach to traditional drying methods, enhancing the efficiency and effectiveness of the drying process for various headwear.

Chapter 13

Centered on healthcare and design, this chapter explores the development of prosthetic sockets for transtibial amputees. By incorporating anthropometric data and qualitative inquiry, the study addresses the unique needs of amputees and offers insights into enhancing the fit and functionality of prosthetic devices.

Chapter 14

Focusing on micro and small enterprises (MSEs), this chapter investigates the impact of digital technologies on their survival post-COVID-19 era. The study delves into the most effective digital tools for MSEs' performance and survival. Through quantitative analysis, the chapter provides valuable insights for MSEs in adopting digital resources for sustainability.

Chapter 15

Exploring the intersection of regulations, stakeholder orientation, and sustainability, this chapter examines the influence of state regulations on the relationship between stakeholder market orientation and sustainability performance. Through empirical analysis, the study provides insights into the complex dynamics of regulatory frameworks and business practices.

Chapter 16

This chapter delves into sustainable practices in the construction industry, particularly in Ghana. Through a survey-based approach, the study evaluates the level of environmental awareness among construction contractors. The results provide valuable insights into contractors' grasp of sustainability concepts and practices.

Chapter 17

Centered on nutritional awareness, this chapter investigates the nutritional content of makuea poo-ung, a seasonal crop. The research highlights the potential of this crop to contribute to dietary allowances, showcasing its viability as a natural food supplement.

In closing, *Technological Innovation Driving Sustainable Entrepreneurial Growth in Developing Nations* encapsulates not only the aspirations of a continent but also the universal spirit of human ingenuity. May its pages inspire, inform, and ignite a movement that amplifies the impact of innovation and entrepreneurship in shaping a more equitable and prosperous world.

Nana Yaw Asabere
Accra Technical University, Ghana

Glenn Kwabena Gyimah
Accra Technical University, Ghana

Amevi Acakpovi
Accra Technical University, Ghana

Florence Plockey
Accra Technical University, Ghana

Chapter 1
The Impact of 3 J Laser Pulse Energy to Enhance the Fatigue Lifespan of AA2024-T351 Induced via LSP

Enoch Asuako Larson
https://orcid.org/0000-0001-9224-080X
University for Development Studies, Ghana

Salifu Tahiru Azeko
https://orcid.org/0000-0003-3860-147X
Tamale Technical University, Ghana

Eric Akowuah
University of Energy and Natural Resources, Ghana

Prince Owusu Ansah
Kumasi Technical University, Ghana

Samuel Adu-Gyamfi
Sunyani Technical University, Ghana

Augustine Kwame Milku
https://orcid.org/0000-0002-1319-0291
Bolgatanga Technical University, Ghana

Jamal-Deen Kukurah
Tamale Technical University, Ghana

Philip Yamba
Tamale Technical University, Ghana

Jacob Kofi Mensah
https://orcid.org/0000-0003-1997-2291
Tamale Technical University, Ghana

Anthony Akayeti
Tamale Technical University, Ghana

ABSTRACT

Laser shock peening is a new method employed to increase the fatigue life of metals by inducing in-depth Compressive Residual Stress to about 1 mm. This helps prevent superficial failures, stress corrosion cracking and elongate the fatigue life and this

DOI: 10.4018/978-1-6684-9843-9.ch001

has provided varying applications in the aviation industry. This study investigates the effect of pulse energy on fatigue lifespan of AA 2024-T351 and the results validated by the finite element method. When compared with the as-received specimen, the result confirms that LSP can significantly produce higher residual compressive stress, and elongate fatigue life of alloys marginally at the superficial layer while decreasing in-depth. The FE analysis of the laser peened specimen produced a good correlation with the experimental work. The fracture morphologies also show visible signs of fatigue striations in the micro-scale. The layout between each striation space is large for the as-received specimen, demonstrating a rapid fatigue growth rate whereas relatively small in the LSP-treated specimen indicating a stable fatigue growth.

1.INTRODUCTION TO LASER SHOCK PEENING (LSP)

1.1 Overview of LSP Process

Contemporary laser shock peening (LSP) is a remarkable surface modification process (M. Abeens et al., 2019; Dhakal & Swaroop, 2018; Luo et al., 2013) that produces CRS on the superficial layers of alloys or metals. This enhances their mechanical properties and resist corrosion (Kumar et al., 2023), increases fatigue life (H. Zhang et al., 2020), wear resistance (Maleki et al., 2021), SCC (Lu et al., 2012) and surface integrity. LSP has experienced several applications in diverse industries including the aerospace and aviation industry, nuclear plant, automobile manufacturing sector (Luo et al., 2013) and high-speed rail industry (Hu et al., 2020).

Above all, the LSP treatment uses a mechanical process rather than a thermal one to prevent thermal impacts (Ding & Ye, 2006). And so, to prevent thermal damage a black coating paint is mostly put on the top surface of the material to protect it from heat deformation occurring from the LSP treatment. The top surface of the metal or alloy is exposed to intense laser pulse energy, which raises the plasma pressure in the water confinement zone (Berthe et al., 1997; Wu & Shin, 2005). During this time, the wave induces deep CRS that impacts the surface layer and sends shockwaves into the metallic alloy (Braisted & Brockman, 1999), this essential enhances the metal or alloy's wear resistance and increases the fatigue properties (Cao et al., 2010).

Research shows that finite element simulation of LSP induces residual stress (Braisted, W. and Brockman, 1999; Ding, 2003). In order to simulate short interval wave development and the resulting residual pressures on the material, the Abaqus Explicit/Standard software package was used. (Peyre et al., 2003; Peyre et al. 2007) used the J-C model to characterize the dynamic performance of materials using laser shock peening. (Adu-Gyamfi et al., 2018)) employed finite element simulation and experiential work on different scanning patterns by LSP on AA2024 aluminium alloy to generate RS to enhance fatigue life via the Johnson Cook model. Kim & Lee (2013) and (Ding & Ye, 2006), and Ding (Ding & Ye, 2006) established an axisymmetric two-dimensional (2D) finite element model (FEM) to simulate multiple shots in a single impact site.

It has been reported by many researchers that laser shock peening though induces residual stress in metallic material, the laser energy plays vital role in the distribution. Mao-Zhong Ge et al. studied the effect of laser energy on microstructure of Mg-3Al-1Zn alloy treated by laser shock peening (Ge et al., 2018). They saw in their studies that the microstructural feature, residual stress distribution and bending degree of Mg-3Al-1Zn alloy subjected to LSP under different laser pulse energy levels showed some changes. The grain size on the top surface decreased as the energy was reduced from 6.5 J to 8.5 J, and the depth of plastic deformation layer increased from 0.8 mm for 6.5 J to 0.95 μm for 10.5 J. Furthermore, residual stress and arc height of all the laser shocked samples increased with laser pulse energy increasing. Also, Gao et al. (2021) in their study on the effects of LSP on nickel aluminium bronze alloy with emphasis on residual stress distribution and surface roughness on the fatigue life of the material. The results indicated that the high-cycle fatigue life variation trend of the nickel aluminium bronze specimens under different laser pulse energies agreed well with the fatigue stress concentration factor, and the beneficial surface roughness and compressive residual stress induced by LSP were considered the main factors that contributed to the improvement of the fatigue properties.

Consequently, Qin et al. (2022) came out with observations after the investigation of the surface integrity and high cycle fatigue (HCF) and very high cycle fatigue (VHCF) properties of 2024-T351 aluminum alloy. The surface roughness, surface topography, microstructural, microhardness and residual stress of specimens treated with two different LSP pulse energy levels, 10 J and 20 J were analyzed. Their results showed that surface roughness Ra value and microhardness of LSP treated specimens increased appreciably comparing to the untreated specimen with both energies. X-ray diffraction analysis also showed that sub-grain sizes were refined to the nanoscale. Furthermore, high-level compressive residual stresses were introduced to the surface layer of the specimens. The S-N curve results showed that LSP treatment reduced the fatigue life of specimens, and the higher the laser pulse energy, the more obvious the effect of reduction.

A novel technique based on the FE model of the material and laser shock peening experiment is provided to accurately estimate the plasma shock wave pressure. In this study, AA2024-T351 aluminium alloy was chosen as the experimental material, and a 20 x 20 x 3 (mm^3) experimental sample size was treated with multiple LSP impacts. The $Sin^2\psi$ was used to calculate the CRS of the experimental sample following LSP treatment using an X-ray residual stress instrument. This study can help researchers acquire the correct FEM of materials and provide knowledge of the theoretical aspect of LSP.

Abaqus software was employed to create a three-dimensional (3D) model to investigate the effect of pulse energy effect on RS at the superficial via in-depth direction. The experimental results were compared to the simulation results of the CRS field. The impact of pulse energy on residual compressive stress distribution and fatigue life was investigated before and after the LSP process.

1.2 Historical Development of LSP

When laser shock peening (LSP) was first developed in the 1960s, research primarily concentrated on the fundamental process development, understanding of mechanisms, utilization of high laser power density to achieve high pulse pressures (Fairand et al., 1972), and creation of physical models to describe laser peening processes (Peyre et al., 1996).

The LSP process since its invention at Battelle Memorial Institute in 1972, has been investigated and studied by several scholars worldwide (Adu-Gyamfi, 2018; Larson et al., 2023; Larson et al., 2019).

From 1986, a number of scholars have also helped improve and expanded upon this technique by addressing effects of altered laser shape, characteristics of the shock-wave and their propagation, including modeling the induced mechanical properties (Ballard et al., 1991; Berthe et al., 1997) and their earlier applications performed in countries like China (Guo et al., 1999), Japan (Sano et al., 1997) and France (Peyre & Fabbro, 1995). In the investigations, particular focus was placed on several important LSP-related variables, such as laser parameters, and thermos-protected coatings, confined overlays, which have a big impact on the mechanical reactions of metallic materials.

A surface treatment method called laser shock peening was created to replace shot peening. In order to prevent fracture propagation and enhance the lifespan of metallic components, the procedure involves introducing higher levels of compressive residual stresses into the target material.

The procedure is primarily employed in the aerospace sector to repair jet engine turbine blades, which are vulnerable to damage from foreign objects during takeoff and landing because rocks and ice are sucked into the engine, damaging the blades and shortening their operational life.

LSP was not widely used in industry because of the exorbitant expense of the lasers that could produce such large-sized beam spots with several intensities of GW/cm^2. However, the development of micro-scale laser shock peening (μLSP) was spurred on by recent developments in the production of micro electromechanical systems (MEMS) and other micron-sized devices, as well as the utilization of metallic components inside them.

1.3 Importance of LSP in Enhancing Material Fatigue Resistance

When a metal component weakens as a result of cyclic loading, metal fatigue develops (Fatemi et al., 2019; Liao et al., 2020). The amount and depth of generated compressive residual stresses present in a component determine how much stronger it is against fatigue. Higher in-depth compressive residual stresses offer improved protection against fracture propagation and failure(Gu et al., 2023; van Aswegen & Polese, 2021). Comparing laser shock peening to other surface enhancing techniques, compressive residual stresses are often produced up to 10 times deeper (Luo et al., 2021).

2. LITERATURE REVIEW

2.1 Previous Studies on LSP for Enhancing Material Fatigue Resistance

Wang et al. (2022) studied Ti-6Al-4V that had been damaged by foreign objects and was examined for fatigue strength enhancement using a combination of laser shock peening and shot peening. It was discovered that samples treated with both LSP and SP simultaneously had samples with harder surfaces than the as-received material. When it came to fatigue testing, combined LSP and SP specimens that were shot by 1 mm steel balls at high cycle fatigue (HCF) with 10^7 cycles showed the greatest improvement, with a strength that was around 35% greater than that of the untreated samples.

In Hu et al. (2022), laser shock peening technology was used to repair a titanium alloy's double-sided argon arc welded joint. In comparison to the untreated specimens, there was a noticeable improvement in the joints' hardness, residual stress, fatigue life, and mechanism of fatigue crack nucleation. Their findings demonstrated that the welded joint's hardness had improved, and that the site of the fatigue crack at the initiation had shifted from the surface defect with the highest concentration of stress in-depth. The high stress magnitude 590 MPa was enhanced by around 4.6 times with stress ratio of 0.1, while the medium stress magnitude 330 MPa was enhanced by 9.9 times.

2.2 Effects of Laser Pulse Energy on LSP

Numerous researches have revealed that although LSP causes residual stress in metallic materials, the laser energy play a crucial in the distribution.

Ge et al. (2018) and colleagues examined how the microstructure of an alloy made of magnesium, aluminum, and zinc after laser shock peening was affected by laser energy. They discovered in their research that the Mg-3Al-1Zn alloy exposed to LSP at various laser pulse energy levels displayed certain changes in its microstructural characteristic, residual stress distribution, and degree of bending. As the energy was raised from 6.5 J to 8.5 J, the grain size in the top surface reduced while the depth of the plastic deformation layer enhanced from 0.8 mm for 6.5 J to 0.95 m for 10.5 J.

Additionally, all of the laser shocked specimen also saw an increase in residual stress and arc height as laser pulse intensity increased.

Additionally, Gao et al. (2021) in their research on the impact of LSP on nickel-aluminum bronze alloy with focus on residual stress distribution and surface roughness on the material's fatigue life.

The results showed that the nickel aluminum bronze specimens high-cycle fatigue life variation under various laser pulse energies correlated with the fatigue stress concentration factor, and the advantageous surface roughness and residual stress induced by LSP were found to be the main factors that resulted in the enhancement of the fatigue properties.

Following their examination into the surface integrity, high cycle fatigue (HCF), and very high cycle fatigue (VHCF), features of the aluminum alloy 2024-T351, Zhi Qin et al. (Qin et al., 2022) published their findings. Their findings demonstrated that LSP treated specimens surface roughness Ra values and microhardness enhanced significantly in comparison to the untreated material with both energies. Additionally, X-ray diffraction research revealed that subgrain sizes were optimized to the nanoscale.

Also demonstrated by X-ray diffraction studies was the nanoscale refinement of subgrain sizes. In addition, the specimens surface layer received high-level compressive residual stresses. The LSP treatment shortened the specimens fatigue life, and the S-N curve findings revealed that this impact became more pronounced and greater than the laser pulse energy used.

Prochazka et al. (2023) studied the impact of strain-assisted tempering on the fatigue characteristics of 54SiCr6 steel, which is utilized for spring steel wires in a range of automotive applications. The strain-assisted tempering material fatigue resistance was also improved by using LSP procedure and the static and fatigue characteristics compared to commercial steel treatment was significantly improved, according to their research.

3. MATERIAL AND METHODS

3.1 Description of AA 2024-T351 Alloy and Setup

In this current study, the fatigue specimen used was made from AA2024-T351 and utilized for the fatigue specimen. Dog-bone specimen thickness used was 3 mm. Table 1 (Dorman et al., 2012) and Table 2 (Correa et al., 2015) respectively, show the aluminum alloy chemical and mechanical properties. The material under study weight-to-strength ratio is high, meaning it possess good corrosion resistance and ductile at low temperatures. Figure 1 shows the LSP setup, specimen and scanning direction.

Table 1. Chemical composition in wt% of base metal AA2024-T351(Dorman et al., 2012).

Element	Cu	Mg	Si	Fe	Mn	Zn	Cr	Ti	Other			Al
				Max		Max	Max	Max	Max. (each)	Max. (total)	Max.	
wt%	90.7-94.7	3.8-4.9	0.46	0.5	1.2-1.8	0.5	0.15	0.25	0.1	0.05	0.15	Bal.

Table 2. Mechanical properties of AA2024-T351 Aluminium Alloy (Correa et al., 2015).

Material	Young modulus (GPa)	Yield stress (MPa)	Ultimate stress (MPa)	Elongation at fracture (%)
Al 2024-T351	72	360	481	19

Figure 1. Experimental setup: (a) LSP Setup (b) Specimen and laser scanning direction

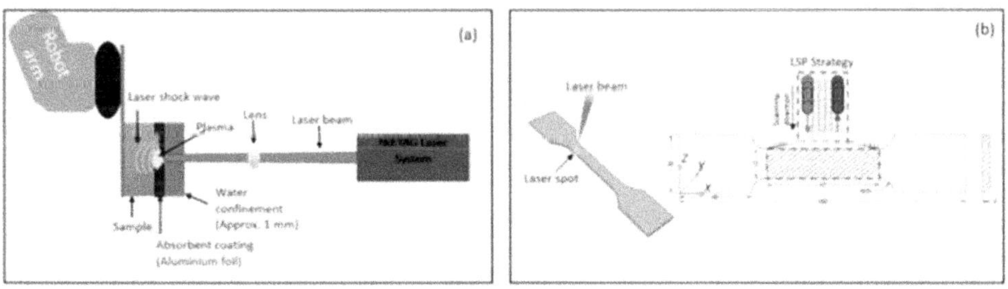

3.2 Numerical Simulation of LSP

The numerical simulation of LSP includes procedures for determining material parameters, type of study, loading condition, and mesh sensitivity analysis. In this study, a three-dimensional (3-D) FE model is created with ABAQUS to analyze the LSP-generated residual stress patterns in a predefined geometrical model 20 × 20 × 3 mm^3 (length × width × thickness) dimensions.

The mesh densities have a significant impact on the FE simulation results because the RS in-depth direction changes substantially. For this model, 98,496 elements mesh C3D8R 8-node linear brick were generated, reduced integration, and hourglass control for the entire model. To compute the laser peak pressure of the induced plasma shockwave, the pressure model used is provided (Fabbro et al., 1990).

$$P(GPa) = 0.01 \times \sqrt{\frac{\alpha}{2\alpha + 3}} \times \sqrt{Z}\left(I_0\right)^{1/2} \quad (1)$$

$$\frac{2}{Z} = \frac{1}{Z_{water}} + \frac{1}{Z_{target}} \quad (2)$$

$$I_o = \frac{4E}{\pi d^2 \tau} \quad (3)$$

Where P denotes the peak pressure of the generated plasma shockwave, α denotes the fraction of the internal energy devoted to the thermal energy, I_o represents power density. Z represent synthetic acoustic of target material and the acoustic impedance of the constrain layer whiles Z_{water} and Z_{target} represents the acoustic impedance of specimen and the constraining layer $\left(1.420 \times 10^6 g / cm^{-2} s^{-1}\right)$ and $\left(0.165 \times 10^6 g / cm^{-2} s^{-1}\right)$, respectively. d, τ and E are the spot diameter of the laser (mm), laser pulse time (ns), and pulse energy (J), respectively. The shockwave pressure curve is shown in Figure 2.

However, the acoustic impedance of the specimen is calculated by the expression,

$$Z_i = \rho_i \times D_i \quad (4)$$

Figure 2. Shockwave pressure

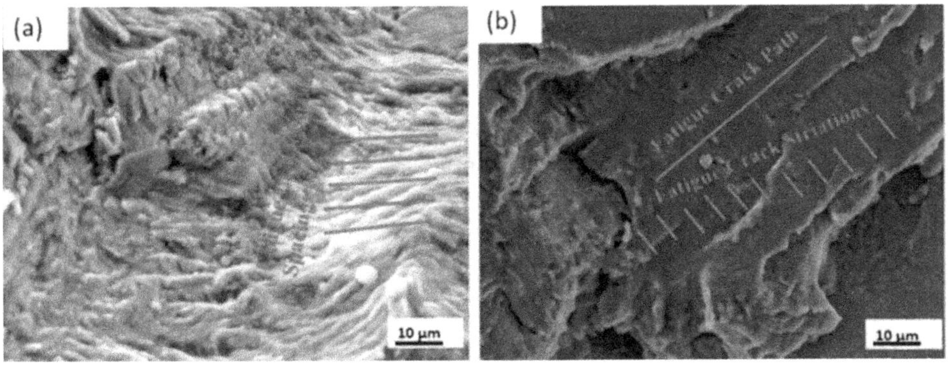

Figure 3. Abaqus FEM of LSP for AA2024-T351 Aluminium alloy

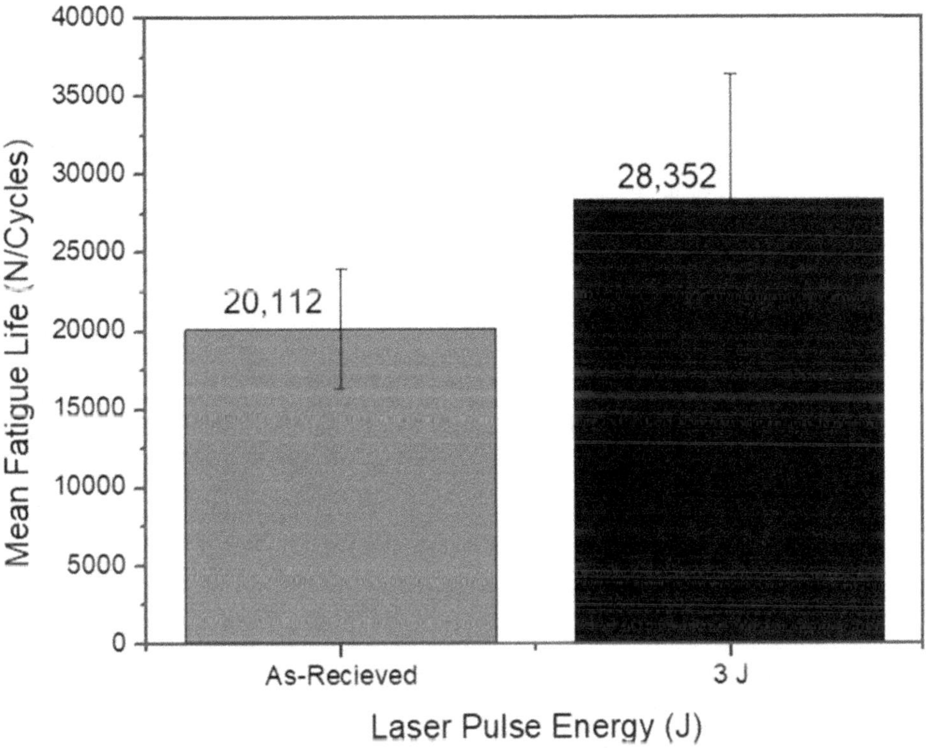

Where ρ_i denotes the mass density of the specimen, and D_i denotes the sound velocity of the specimen.

The strain rate at the target's surface exceeds 10^6 s^{-1}, and Hugoniot Elastic Limit (HEL) (Y. Hu et al., 2009) can be used to calculate the elastic limit stress in the shock wave propagation area. When the peak pressure exceeds the target's HEL, plastic deformation occurs. Figure 3 shows the finite element model of the AA2024-T351 Aluminium alloy specimen treated with LSP.

3.3 The JC model

The Johnson-Cook (JC) model is an employed constitutive material modeling used in this study to represent the impacts of the FE parametric including strain rate, temperature, among others on the material's dynamic yield stress. In addition, the Johnson-Cook model is established as follows (Li, 2013).

Where $T^* = \left(T - T_o\right) / \left(T_m - T_o\right)$, denotes non-dimensional temperature, T_o denotes ambient temperature, and T_m denotes material's melting temperature. The measures A, B, and n indicate the material's strain hardening, C denotes strain rates effect on properties of the material, and m denotes temperature softening effect. Where $\varepsilon^* = \varepsilon / \varepsilon_0$ is the non-dimensional plastic strain rate and represents the plastic strain. J-C parameters are listed in Table 3.

Table 3. J-C model (AA2024-T351)

Initial yield of the material (*A*, MPa)	Strain Hardening constant (*B*, MPa)	Strain Hardening coefficient (*n*)	Strengthening coefficient of strain rate (*C*)	Thermal Softening parameter (*m*)	Plastic strain rate (μ^*)
369	684	0.73	0.0083	1.70	1

4. RESULTS AND DISCUSSION

4.1 The Effect of the Laser Pulse on CRS

Dynamic deformation with ultra-high strain rates will occur on the superficial of the material due to the laser-induced plasma waves, thus resulting in an in-depth residual stress level. This result will significantly enhance metallic materials' fatigue properties. Figure 4 show RS distribution through depth variations of as-received specimen and the LSP specimen with the 3 J laser energy on AA2024-T351.

Residual stress on the superficial layer of the LSPed specimen reached a near-optimal level, then gradually diminish as the depth of the specimen drifts away from the top surface. The compressive residual stress at the top surface of the untreated was 50 MPa. Nonetheless, after laser peening, the top surface stress state transformed from tensile to compressive, and the specimen maximum surface RS was -225 MPa. The residual stress measurement for the FE simulation after LSP was -267 MPa. This results is attributed to the severe plastic deformed area which is because of the induced RS layer in the LSP region. In addition, the depth of the RS was around 0.8 mm.

The specimen surface without LSP had positive residual stress reading, indicating that the as-received specimen surface is filled with tensile residual stress. However, after LSP, the residual stress values became negative, indicating that laser shock peening produces compressive residual stress on the specimen surface.

After LSP, the stress state of the specimen was altered from tensile stress to a high residual stress level as shown in Figure 5(b). For the laser pulse energy level, the CRS distribution indicates a similar pattern in the depth direction as the simulated results for Von mises stress (s11) in Figure 5.

Figure 4. Experimental and FEA residual stress

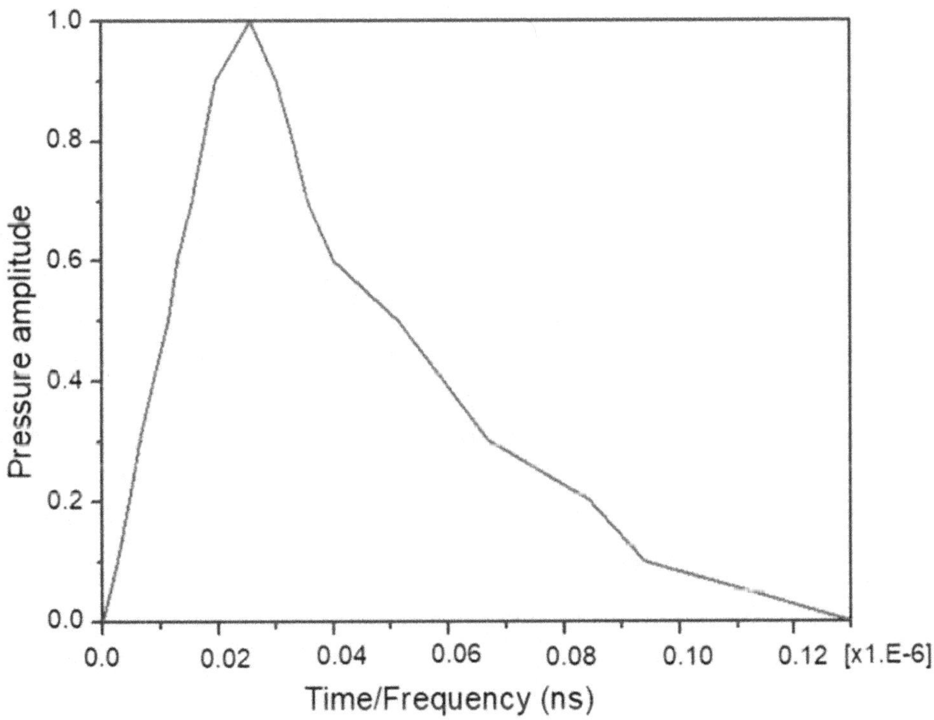

Figure 5. FEA simulated compressive residual stress (s11); (a) After LSP (b) Section (X-X)

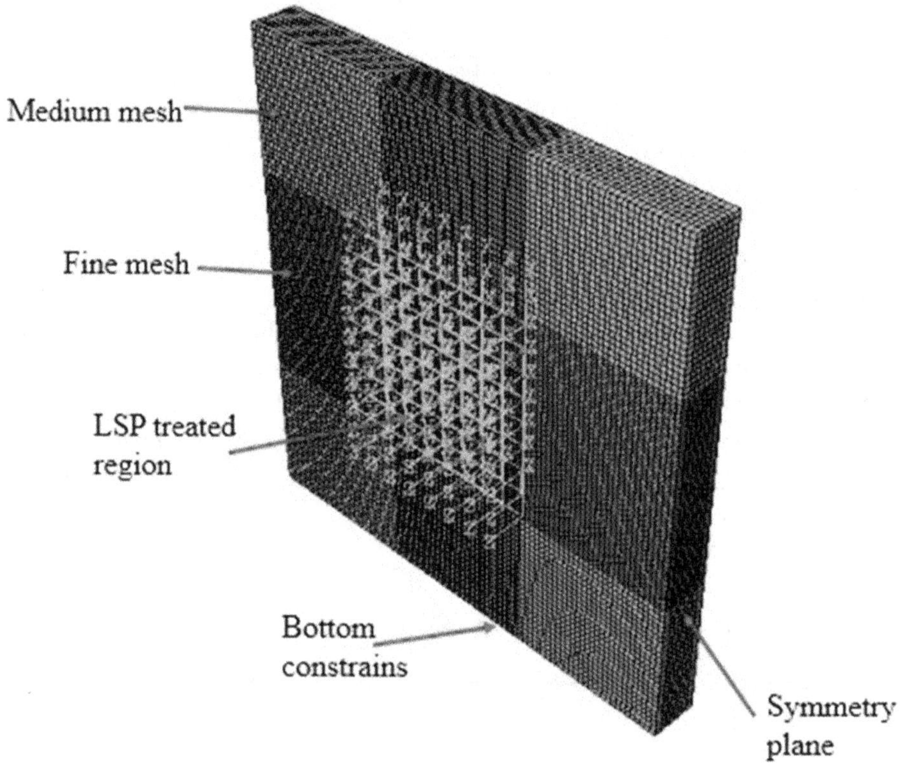

4.2 Fatigue Fractography Analysis

The fatigue fractography of both as-received and LSPed specimen was examined using SEM following the fatigue test. The crack initiation originated at the weakest portion of the specimen, where there is stress or strain concentration. In addition, an initial micro-crack was observed, which steadily enlarged until it fractured abruptly. The fatigue life of the specimen under varying cyclic loading can be estimated by the initiation and propagation of the crack. Many metallic materials, when subjected to cyclic loading, go through a stage known as stable growth, during which crack formation is rapid. The features of FCI regions for both LSPed and as-received specimen are shown in Figures 6 (a) and (b). The region at the arc angle part of the opposite end is where the untreated specimen fracture first originated (Tang et al., 2020). However, the fatigue fracture origin of the LSP-treated specimen was distant from the site of the sharp angle region, indicating the presence of residual stress after

Figure 6. Fatigue crack origin (a) As-received (b) LSPed specimen

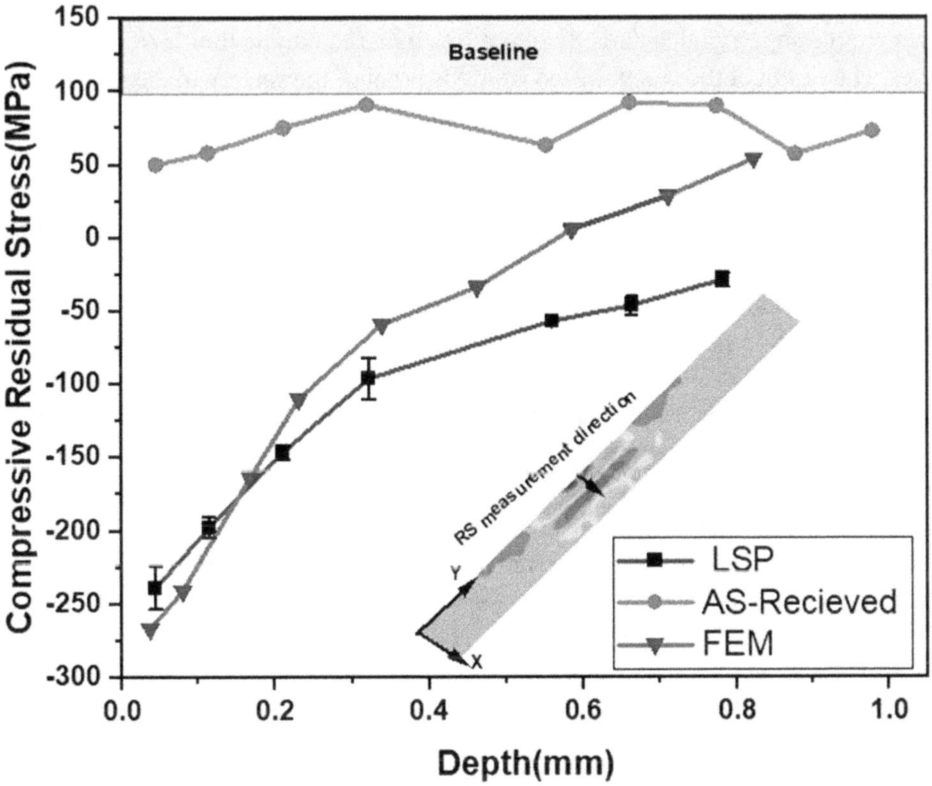

Figure 7. Fracture morphology; (a) Untreated, (b) 3 J

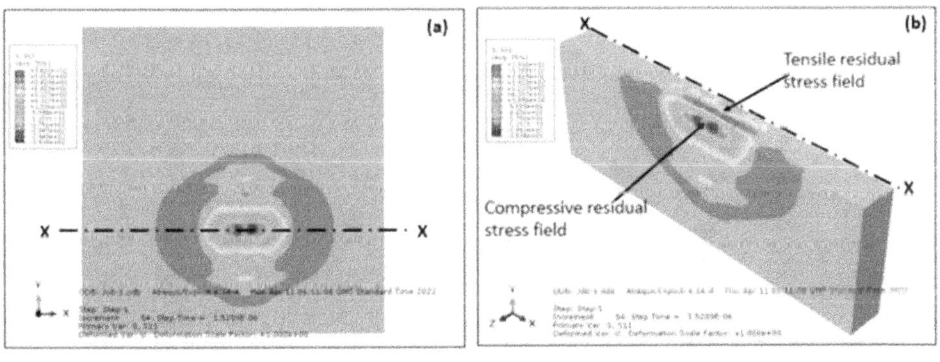

LSP treatment in Figure 6 (b). At the arc angle where the tensile stress concentration was high, the residual stress layer reduces the tensile stress concentration. The total stress of the superficial layer is therefore less than the fatigue limit's strengthening layer. The origin of the crack shifted from the point of the surface to the underneath surface, where residual stress is low approximately 0.1 mm in-depth.

Figure 7 show the microstructure of fractured surface starting at the fracture origin and progressing steadily via the thickness of the as-received and LSPed specimen. The morphology of the material reveals dimples with scattered inclusions on the fatigue crack scan portions. Once the induced stress is far greater than the metallic material yield strength plastic deformation is introduced. During this time, dislocations pile up resulting in stress concentration and microcavities. As strain increases, tiny spaces within the material enlarge, resulting in inclusions beneath the dimples and isometric dimples on the cracked surface (Pantelakis et al., 2007).

Figures 8 (a and b) show SEM fatigue surface fracture of untreated and LSPed-treated specimen, respectively. Figure 8 (a) show that the crack initiation was developed at the weak location showing that stress concentration is low. The crack propagation rate was steady until a fracture occurred due to the presence of compressive residual stress at the crack propagation (CP) zone as indicated in Figures 9 (a) and (b).

Figure 8(b) show fracture of the surface magnified microscopic nature, with internal cracks present in the specimen mid-thickness area. This causes the fatigue striations distance to increase as the crack length increases. Fatigue crack develops from these internal cracks, propagates and reaches the point of fracture. As the crack length increase, the space between the fatigue striation increases. Fatigue cracks develop from these cracks and propagate until they reach a point of fracture.

Figure 8. Fatigue fracture morphology; (a) untreated specimen (b) Magnified internal crack

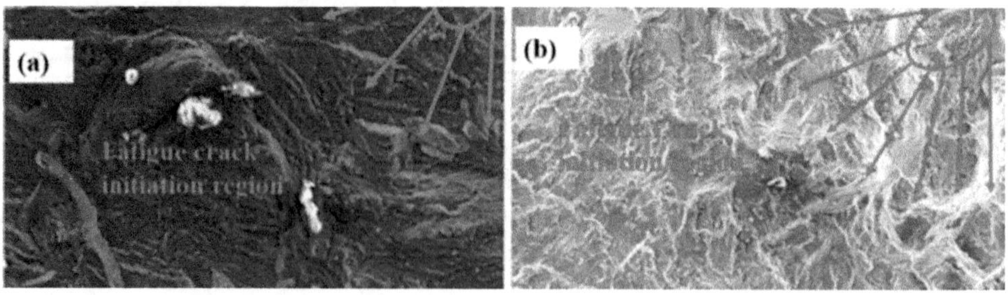

Figure 9. Fatigue fracture morphology; (c) 3 J specimen and (d) Magnified internal cracks

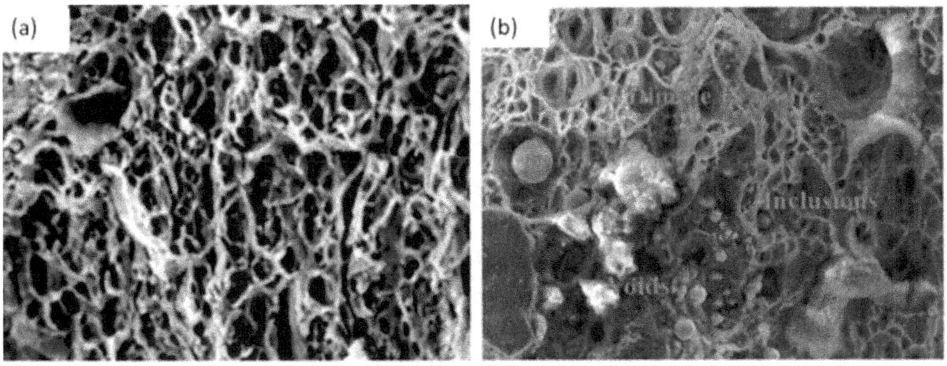

4.3 Fatigue Crack Behavior and Microstructure on the Fractured Surfaces

Generally, fatigue fractography morphology is a direct outcome of metallic material failure (Zhou et al., 2012), and the striation spacing could reciprocate the change in fatigue crack development rate da / dN . The fatigue behavior and microstructure on the crack surface in-depth is about 0.5 mm as shown in Figure 10(a) using the thickness of the as-received and LSPed specimen. Fatigue striations are observed on the specimen fractured surface. Fatigue striation spacing of the untreated dog-bone specimen is approximately 0.77 m, but it decreases gradually to 0.38 m after 3 J as shown in Fig. 10(b). Because of the induced RS via LSP, the fatigue striation

Figure 10. Fatigue striations of; (a) Untreated, and (b) 3 J specimen

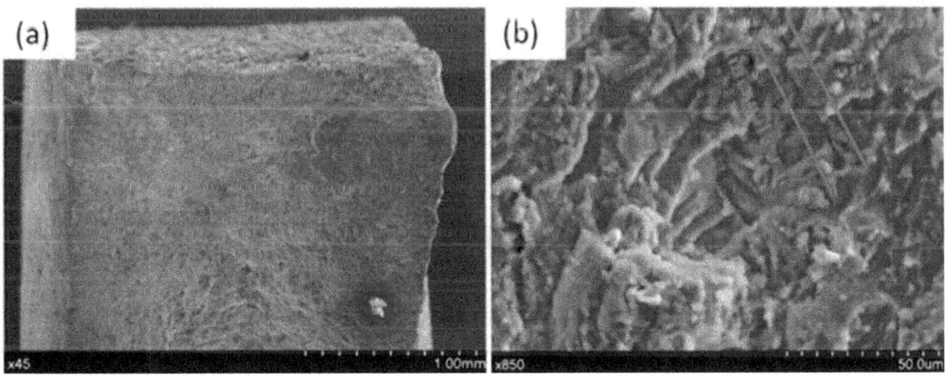

spacing (FSS) decrease slowly resulting in a gradual decrease in fatigue crack growth rate.

Metallurgical factors like grain orientation and stress intensity factor might affect striation spacing (Ye et al., 2010). By inducing CRS at the top surface layer by laser processing, the stress intensity factor (SIF), as well as fatigue crack growth surrounding the fracture point may be reduced (Ye et al., 2010; Zhou et al., 2012).

4.4 Fatigue Life

Figure 11 represent the fatigue life of the as-received and LSPed specimen against laser pulse energy. The fatigue behavior of AA2024-T351 was improved after the LSP treatment of 3 J laser pulse energy as per the experimental results. For the untreated LSP specimen, the stress amplitude increases after 20,112 (2×10^5 cycles). The fatigue strength of the LSPed specimen exhibited fatigue life of 28,352 (2×10^5 cycles), with a stress amplitude of 180 MPa, using a pulse energy of 3 J. It was discovered that the stress amplitude increased the specimen fatigue life.

The RS produced by LSP can reduce the stress intensity factor (K), and is used to predict fatigue life, and this played a vital role in improving the fatigue strength of the metallic material. As a result, the fatigue resistance and dislocations density in the metallic specimen was found to have a good correlation. The improvement in fatigue life was evaluated between the as-received and LSPed specimen.

Figure 11. Fatigue life performance

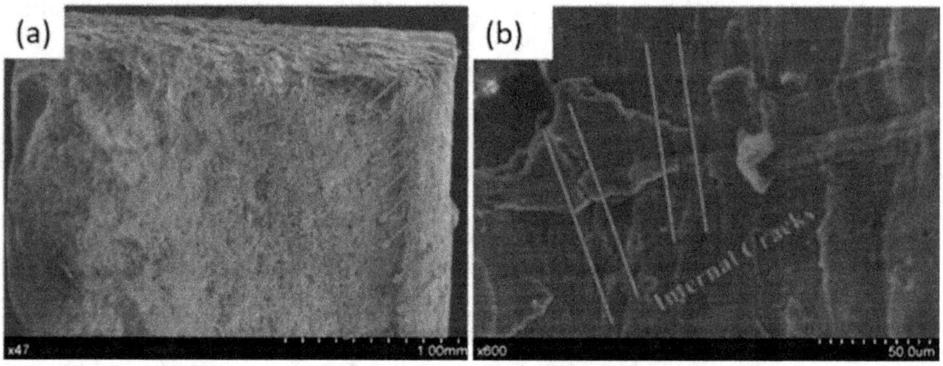

5. APPLICATIONS AND FUTURE DIRECTIONS

In recent times, LSP has receive several industrial applications due to its potential to increase fatigue life and enhance metallic properties across fields. Notwithstanding, additive manufacturing (AM) has garnered a lot of interest in the LSP scientific community due to its capability to create components with complicated geometries. However, there are myriad of challenges with AM metals, such as increased porosity and the presence of tensile residual stresses resulting in poor fatigue behavior (Zhang et al., 2021). In order to solve this issue, (Hackel et al., 2018) treated AM specimens using LSP and SP to address this problem, and they found that the fatigue behavior was significantly enhanced. It is worth mentioning that, LSP has a stronger ability to enhance the fatigue behavior of AM specimens than shot peening. This is due to the fact that LSP can produce more compressive residual stress and a deeper impacted layer than SP. These findings show that laser shock peening is a successful technique for treating AM metallic materials.

Ceramics, which has exceptionally weak plasticity, like SiC and Al2O3, fail prematurely when subjected to external loads due to their high cracking sensitivity and low fracture toughness. And so, application of LSP on ceramic material is hampered because of its intrinsic brittle nature of the material and its property of high melting point, hardness, wear resistance, and low density but makes its application on structural and functional materials useful.

LSP can produce compressive residual stresses (CRS) on the superficial layer of metallic materials, as described in earlier articles. These stresses may partially counteract the external load, raising the stress threshold for crack development. Additionally, CRS have the potential to lessen in effect the stress intensity factor (SIF) near the crack tip and hence prevent crack progression. Few research work on LSP-treated ceramics have shown that the advantageous compressive residual stresses can also be created in ceramics (Zhang et al., 2021).

LSP on metallic glasses makes it have numerous potential applications due to their outstanding qualities, such as high strength, high hardness, and good resistance to corrosion and wear (Zhu et al., 2015). Notwithstanding, when exposed to external stress at room temperature, metallic glasses have low plasticity and will break with almost no plastic deformation. When the temperature is close to or above the glass transition point, uniform viscous flow is used to achieve plastic deformation of metallic glasses however, when the temperature is significantly below the glass transition point, shear bands are used instead (Zhang et al., 2021). When the shear bands are increased, it will cause a corresponding increase in the interatomic distance and decrease the bonding force among atoms, hence improving the plasticity of the metallic glass, since the number of shear bands is correlated with the free volume of the material. As a result, the LSP process' free volumes created by impact wave can enhance the metallic glass plasticity as reported by ((Fu et al., 2014; Fu et al., 2015; Fu et al., 2016; Wang et al., 2015) on the treatment of zirconium-based metallic glass via LSP process.

5.1 Limitations and Challenges of LSP

The use of LSP has had its corresponding limitation and challenges as well. Firstly, the high cost of laser system startup and complicated laser equipment may limit the advancement for the LSP processing facility. Laser equipment are expensive to acquire and require high initial investment cost to start up especially when research and academic institution are not adequately resourced financially.

Secondly, the usage of laser round spot rather than laser square spot, limit the exact overlap coverage area, which makes it expensive to laser Peen larger surface area and more precise.

Finally, the need to split the laser beam in half would allow for concurrent treatment of the specimen's front and back sides, thus decreasing the asymmetry of the residual stress field and the resulting physical bending effect of the specimens.

5.2 Potential Applications of LSP in Various Industries

Laser shock peening has a great potential in a wide range of industries including the aerospace, biomedical and automotive industries. In the aerospace industry, especially during takeoff or landing, hard items (including stone, sand, or birds) are drawn into the engine at high velocity and cause unrepairable dent to the engine compressors or fan blades (Zhao et al., 2020). When a fast-speed impact and high magnitude of applied stress concurrently occur at the same time, the edges of damage are vulnerable to developing microcracks. Due to this LSP has been used tremendously to induced in-depth compressive residual stress in aerospace engines to enhance the fatigue life and microstructure evolution of aerospace materials (Zabeen et al., 2015).

In addition, several research applications have been performed on biomedical implant materials (Cao et al., 2015; Ruixia, 2018; Zhang et al., 2018) on biocompatibility, hardness, ion release rate and wear resistance among others via LSP. In the field of automotive, research have been conducted using LSP on leaf springs, gears, crankcase and pistons.

5.3 Future Directions for Research and Development of LSP Technology

The future direction of LSP can be confronted with a lot of difficulties even though it is an established technique in the field of laser surface enhancement technique. For instance, the traditional LSP treatment still calls for the application of black paint or tape as absorbing layer. Hence, the overall process efficiency of LSP is significantly decreased by the time required for applying and removing an absorbent layer.

Therefore, an efficient way to increase the LSP efficiency will be to create an automatic method for utilizing the absorbent layer. Additionally, LSP technique will require an online monitoring technology to allow for the processing parameters to be able to control it in real time.

6. CONCLUSION

To simulate one-sided LSP impact processing, a FEA method founded on Abaqus's Explicit and experimental work was explained. This novel method was employed to model the shock wave growth and to predict the CRS distribution in AA2024-T351 aluminium alloy metallic material induced by LSP. The experimental and FE analysis findings revealed that after the LSP process, the presence of RS and severe plastic deformation was induced in-depth from the superficial layer, and subsequently to a tensile stress field of the material. The study further revealed a good correlation between experimental work and FE simulation after the residual stress field was computed. The laser pulse energy significantly enhanced the fatigue life of metallic material after LSP demonstrating a uniform residual stress field. The fatigue life for the as-received specimen was 20,112 (2×10^5) cycles however, after LSP the fatigue life was enhanced to 28, 352 (2×10^5) cycles. Fatigue striations spacing was observed both on the as-received and the LSPed treated specimen. When compared to the original specimen, the ductile striations spacing decreased because of the presence of the CRS effect.

6.1 Summary of Findings

Novel laser shock peening is a surface enhancement technique that was used to produce deeper levels of compressive residual stress in metallic materials to enhance the fatigue lifespan, microstructure evolution and corrosion resistance such as on AA2024 aluminum alloy application on aircraft lug.

The following findings can be drawn from the study.

1. When compared with the as-received specimen, the result confirms that LSP can significantly produce higher residual compressive stress, and elongate fatigue life of the alloy material marginally at the superficial layer while decreasing in-depth.
2. The FE analysis of the laser peened specimen produced a good correlation with the experimental work.
3. The fracture morphologies also show visible signs of fatigue striations in the micro-scale.

4. The layout between each striation space is large for the as-received specimen, demonstrating a rapid fatigue growth rate whereas relatively small in the LSP-treated specimen indicating a stable fatigue growth.

6.2 Implications for Materials Science and Engineering

Laser shock peening process is a science whose underlying application is buttress from the physics of material science and engineering. This has allowed several metallic materials to be understudied by various researchers (Q. Abeens et al., 2021; W. Fu et al., 2022; Maleki et al., 2022; Pathak et al., 2022; J. Zhou et al., 2022)

6.3 Directions for Future Research

Researchers have previously conducted few simulations on simple geometries and experimental study into the impact of LSP parameters on fatigue life. Hence, the need for an industrial application, to determine the optimal process parameters for a real-world 3D problem will be quite helpful. This arguably can be done by utilizing surrogate models as in this study and can further be studied to solve practicable problems related to aero-plane lug.

This experimental and FEA study will help with the creation of an optimization method for real-world material surface improvement and fatigue life enhancement problems as well as give further validation for the simulation work.

REFERENCES

Abeens, M., Muruganandhan, R., & Thirumavalavan, K. (2019). Comparative analysis of different surface modification processes on AA 7075 T651. *IOP Conference Series. Materials Science and Engineering*, *574*(1), 012018. Advance online publication. doi:10.1088/1757-899X/574/1/012018

Abeens, Q., Rajasekaran, E., Premnath, M., Sarukasan, D., Muruganandhan, R., & Thirumavalavan, K. (2021). Experimental and numerical investigation of residual stresses in low pulsed laser shock peening on AA7075 T651. *IOP Conference Series. Materials Science and Engineering*, *1126*(1), 012086. doi:10.1088/1757-899X/1126/1/012086

Adu-Gyamfi, S. (2018). Effects of Laser Shock Peening on Mechanical Properties and Surface Morphology of AA2024 Alloy. *American Journal of Mechanical and Materials Engineering*, *2*(2), 15. doi:10.11648/j.ajmme.20180202.11

Adu-Gyamfi, S., Ren, X. D., Larson, E. A., Ren, Y., & Tong, Z. (2018). The effects of laser shock peening scanning patterns on residual stress distribution and fatigue life of AA2024 aluminium alloy. *Optics & Laser Technology, 108*, 177–185. doi:10.1016/j.optlastec.2018.06.036

Ballard, P., Fournier, J., Fabbro, R., & Frelat, J. (1991). Residual stresses induced by laser-shocks. *Journal de Physique. IV, 1*(C3), 1. doi:10.1051/jp4:1991369

Berthe, L., Fabbro, R., Peyre, P., Tollier, L., & Bartnicki, E. (1997). Shock waves from a water-confined laser-generated plasma. *Journal of Applied Physics, 82*(6), 2826–2832. doi:10.1063/1.366113

Braisted, W., & Brockman, R. (1999). Finite Element Simulation of Laser Shock Peening. *International Journal of Fatigue, 21*(7), 719–724. doi:10.1016/S0142-1123(99)00035-3

Braisted, W., & Brockman, R. (1999). Finite element simulation of laser shock peening. *International Journal of Fatigue, 21*(7), 719–724. https://doi.org/10.1016/S0142-1123(99)00035-3. doi:10.1016/S0142-1123(99)00035-3

Cao, Y., Shin, Y. C., & Wu, B. (2010). Parametric Study on Single Shot and Overlapping Laser Shock Peening on Various Metals via Modeling and Experiments. *Journal of Manufacturing Science and Engineering, 132*(6), 061010. doi:10.1115/1.4002850

Cao, Y., Xie, X., Antonaglia, J., Winiarski, B., Wang, G., Shin, Y. C., Withers, P. J., Dahmen, K. A., & Liaw, P. K. (2015). Laser shock peening on Zr-based bulk metallic glass and its effects on plasticity: Experiment and modeling. *Scientific Reports, 5*(1), 10789. doi:10.1038rep10789 PMID:25991412

Correa, C., Peral, D., Porro, J. A., Díaz, M., Ruiz de Lara, L., García-Beltrán, A., & Ocaña, J. L. (2015). Random-type scanning patterns in laser shock peening without absorbing coating in 2024-T351 Al alloy: A solution to reduce residual stress anisotropy. *Optics & Laser Technology, 73*, 179–187. doi:10.1016/j.optlastec.2015.04.027

Dhakal, B., & Swaroop, S. (2018). Review: Laser shock peening as post welding treatment technique. In Journal of Manufacturing Processes, 32. doi:10.1016/j.jmapro.2018.04.006

Ding, K. (2003). FEM simulation of two sided laser shock peening thin sections of Ti-6Al-4V alloy. *Surface Engineering, 19*(2), 127–133. doi:10.1179/026708403225002568

Ding, K., & Ye, L. (2006). Simulation of multiple laser shock peening of a 35CD4 steel alloy. *Journal of Materials Processing Technology*, *178*(1–3), 162–169. doi:10.1016/j.jmatprotec.2006.03.170

Dorman, M., Toparli, M. B., Smyth, N., Cini, A., Fitzpatrick, M. E., & Irving, P. E. (2012). Effect of laser shock peening on residual stress and fatigue life of clad 2024 aluminium sheet containing scribe defects. *Materials Science and Engineering A*, *548*, 142–151. doi:10.1016/j.msea.2012.04.002

Fabbro, R., Fournier, J., Ballard, P., Devaux, D., & Virmont, J. (1990). Physical study of laser-produced plasma in confined geometry. *Journal of Applied Physics*, *68*(2), 775–784. doi:10.1063/1.346783

Fairand, B. P., Wilcox, B. A., Gallagher, W. J., & Williams, D. N. (1972). Laser shock-induced microstructural and mechanical property changes in 7075 aluminum. *Journal of Applied Physics*, *43*(9), 3893–3895. doi:10.1063/1.1661837

Fatemi, A., Molaei, R., Simsiriwong, J., Sanaei, N., Pegues, J., Torries, B., Phan, N., & Shamsaei, N. (2019). Fatigue behaviour of additive manufactured materials: An overview of some recent experimental studies on Ti-6Al-4V considering various processing and loading direction effects. In Fatigue and Fracture of Engineering Materials and Structures, 42(5). doi:10.1111/ffe.13000

Fu, J., Zhu, Y., Zheng, C., Liu, R., & Ji, Z. (2014). Effect of laser shock peening on mechanical properties of Zr-based bulk metallic glass. *Applied Surface Science*, *313*, 692–697. doi:10.1016/j.apsusc.2014.06.056

Fu, J., Zhu, Y., Zheng, C., Liu, R., & Ji, Z. (2015). Evaluate the effect of laser shock peening on plasticity of Zr-based bulk metallic glass. *Optics & Laser Technology*, *73*, 94–100. doi:10.1016/j.optlastec.2015.04.009

Fu, J., Zhu, Y., Zheng, C., Liu, R., & Ji, Z. (2016). Enhanced plasticity of bulk metallic glass in different aspect ratios via laser shock peening with multiple impacts. *Optics & Laser Technology*, *83*, 43–48. https://doi.org/https://doi.org/10.1016/j.optlastec.2016.03.028. doi:10.1016/j.optlastec.2016.03.028

Fu, W., Huang, Y., Sun, J., & Ngan, A. H. W. (2022). Strengthening CrFeCoNiMn0.75Cu0.25 high entropy alloy via laser shock peening. *International Journal of Plasticity*, *154*, 103296. doi:10.1016/j.ijplas.2022.103296

Gao, Y., Yang, W., Huang, Z., & Lu, Z. (2021). Effects of residual stress and surface roughness on the fatigue life of nickel aluminium bronze alloy under laser shock peening. *Engineering Fracture Mechanics*, *244*, 107524. Advance online publication. doi:10.1016/j.engfracmech.2021.107524

Ge, M. Z., Xiang, J. Y., Fan, Z., Lu, Y. L., & Lei, W. N. (2018). Effect of laser energy on microstructure of Mg-3Al-1Zn alloy treated by LSP. *Journal of Alloys and Compounds*, *734*, 266–274. doi:10.1016/j.jallcom.2017.11.032

Gu, H., Yan, P., Jiao, L., Chen, S., Song, Y., Zou, S., & Wang, X. (2023). Effect of laser shock peening on boring hole surface integrity and conformal contact fretting fatigue life of Ti-6Al-4 V alloy. *International Journal of Fatigue*, *166*, 107241. doi:10.1016/j.ijfatigue.2022.107241

Guo, D., Wu, H., Wang, S., Xin, H., Wang, J., Dai, Y., Xia, X., Zhang, Y., Zhang, H., Hong, Y., & Yu, C. (1999). Study of mechanism of strengthening materials with laser shock processing. *Science in China, Series E. Technological Sciences*, *42*(3), 288–293. Advance online publication. doi:10.1007/BF02916775

Hackel, L., Rankin, J. R., Rubenchik, A., King, W. E., & Matthews, M. (2018). Laser peening: A tool for additive manufacturing post-processing. *Additive Manufacturing*, *24*, 67–75. doi:10.1016/j.addma.2018.09.013

Hu, X., Wu, J., Zhao, J., Yang, Y., & Qiao, H. (2020). Numerical simulation of the surface morphology and residual stress field of IN718 alloy by Gaussian mode laser shock. *Optik (Stuttgart)*, *207*, 164441. doi:10.1016/j.ijleo.2020.164441

Hu, X., Zhao, J., Teng, X., Nie, X., Jiang, Y., & Zhang, Y. (2022). Fatigue resistance improvement on double-sided welded joints of a titanium alloy treated by laser shock peening. *Journal of Materials Engineering and Performance*, *31*(12), 10304–10313. doi:10.100711665-022-07006-w

Hu, Y., Gong, C., Yao, Z., & Hu, J. (2009). Investigation on the non-homogeneity of residual stress field induced by laser shock peening. *Surface and Coatings Technology*, *203*(23), 3503–3508. doi:10.1016/j.surfcoat.2009.04.029

Kim, J. H., & Lee, J. W. (2013). Effects of simulation parameters on residual stresses in 3D finite element laser shock peening analysis. *Global Journal of Researches in Engineering Mechanical and Mechanics Engineering*, *13*(9). https://globaljournals.org/GJRE_Volume13/1-Effects-of-Simulation.pdf

Larson, E. A., Adu-Gyamfi, S., Augustine, M., Yamba, P., Kukurah, J., Abdulai, K., Sekyi-Ansah, J., Abdul-Razak, O., Akurugu, E., & Kuzmin, A. (2023). Optimize multiple peening effects on surface integrity and microhardness of aluminum alloy induced by LSP. *Materials Sciences and Applications*, *14*(3), 208–221. doi:10.4236/msa.2023.143012

Larson, E. A., Ren, X., Adu-Gyamfi, S., Zhang, H., & Ren, Y. (2019). Effects of scanning path gradient on the residual stress distribution and fatigue life of AA2024-T351 aluminium alloy induced by LSP. *Results in Physics*, *13*(February), 102123. doi:10.1016/j.rinp.2019.02.059

Li, Y. H. (2013). *Theory and technology of laser shock processing*. Science Press.

Liao, D., Zhu, S. P., Correia, J. A. F. O., De Jesus, A. M. P., & Berto, F. (2020). Recent advances on notch effects in metal fatigue: A review. *Fatigue & Fracture of Engineering Materials & Structures*, *43*(4), 637–659. Advance online publication. doi:10.1111/ffe.13195

Lu, J. Z., Luo, K. Y., Yang, D. K., Cheng, X. N., Hu, J. L., Dai, F. Z., Qi, H., Zhang, L., Zhong, J. S., Wang, Q. W., & Zhang, Y. K. (2012). Effects of laser peening on stress corrosion cracking (SCC) of ANSI 304 austenitic stainless steel. *Corrosion Science*, *60*, 145–152. doi:10.1016/j.corsci.2012.03.044

Luo, K. Y., Lu, J. Z., Wang, Q. W., Luo, M., Qi, H., & Zhou, J. Z. (2013). Residual stress distribution of Ti-6Al-4V alloy under different ns-LSP processing parameters. *Applied Surface Science*, *285*(PARTB), 607–615. doi:10.1016/j.apsusc.2013.08.100

Luo, X., Dang, N., & Wang, X. (2021). The effect of laser shock peening, shot peening and their combination on the microstructure and fatigue properties of Ti-6Al-4V titanium alloy. *International Journal of Fatigue*, *153*, 106465. doi:10.1016/j.ijfatigue.2021.106465

Maleki, E., Bagherifard, S., Unal, O., Bandini, M., & Guagliano, M. (2022). On the effects of laser shock peening on fatigue behavior of V-notched AlSi10Mg manufactured by laser powder bed fusion. *International Journal of Fatigue*, *163*, 107035. doi:10.1016/j.ijfatigue.2022.107035

Maleki, E., Unal, O., Guagliano, M., & Bagherifard, S. (2021). The effects of shot peening, laser shock peening and ultrasonic nanocrystal surface modification on the fatigue strength of Inconel 718. *Materials Science and Engineering A*, *810*, 141029. doi:10.1016/j.msea.2021.141029

Pantelakis, S. G., Petroyiannis, P. V., Bouzakis, K. D., & Mirisidis, I. (2007). Surface hardness increase of 2024 aluminum alloy subjected to cyclic loading. *Theoretical and Applied Fracture Mechanics*, *48*(1), 68–81. doi:10.1016/j.tafmec.2007.04.002

Pathak, S., Zulic, S., Kaufman, J., Kopecek, J., Stránsky, O., Böhm, M., Brajer, J., Beránek, L., Shukla, A., Ackermann, M., Vêle, F., & Mocek, T. (2022). Post-processing of selective laser melting manufactured SS-304L by laser shock peening. *Journal of Materials Research and Technology*, *19*, 4787–4792. doi:10.1016/j.jmrt.2022.07.014

Peyre, P., Chaieb, I., & Braham, C. (2007). FEM calculation of residual stresses induced by laser shock processing in stainless steels. *Modelling and Simulation in Materials Science and Engineering*, *15*(3), 205–221. doi:10.1088/0965-0393/15/3/002

Peyre, P., & Fabbro, R. (1995). Electromagnetic Gauge study of laser-induced shock waves in aluminum alloys. *Journal de Physique. III*, *5*(12), 1953–1964. doi:10.1051/jp3:1995241

Peyre, P., Fabbro, R., Merrien, P., & Lieurade, H. P. P. (1996). Laser shock processing of aluminium alloys. Application to high cycle fatigue behaviour. *Materials Science and Engineering A*, *210*(1–2), 102–113. doi:10.1016/0921-5093(95)10084-9

Peyre, P., Sollier, A., Chaieb, I., Berthe, L., Bartnicki, E., Braham, C., & Fabbro, R. (2003). FEM simulation of residual stresses induced by laser peening. *The European Physical Journal Applied Physics*, *23*(2), 83–88. doi:10.1051/epjap:2003037

Procházka, R., Stehlík, A., Kotous, J., Salvetr, P., Bucki, T., Stránský, O., & Zulić, S. (2023). Fatigue properties of spring steels after advanced processing. *Materials (Basel)*, *16*(9), 3327. doi:10.3390/ma16093327 PMID:37176208

Qin, Z., Li, B., Huang, X., Zhang, H., Chen, R., Adeel, M., & Xue, H. (2022). The effect of laser shock peening on surface integrity and high and very high cycle fatigue properties of 2024-T351 aluminum alloy. *Optics & Laser Technology*, *149*, 107897. doi:10.1016/j.optlastec.2022.107897

Renjith Kumar, R., Devasena, T., & Abeens, M. (2023). Improved mechanical properties of nano structured 316 L stainless steel by statistically optimized low pulsed laser shock peening process for aeronautical application. *Optik (Stuttgart)*, *276*, 170654. doi:10.1016/j.ijleo.2023.170654

Ruixia, Z. (2018). Effects of laser shock peening on the corrosion behavior and biocompatibility of nickel-titanium alloy. *Journal of Biomedical Materials Research. Part B, Applied Biomaterials*, *107*(6), 1854–1863. PMID:30550636

Sano, Y., Mukai, N., Okazaki, K., & Obata, M. (1997). Residual stress improvement in metal surface by underwater laser irradiation. *Nuclear Instruments & Methods in Physics Research. Section B, Beam Interactions with Materials and Atoms*, *121*(1–4), 432–436. doi:10.1016/S0168-583X(96)00551-4

Tang, Y., Ge, M. Z., Zhang, Y., Wang, T., & Zhou, W. (2020). Improvement of fatigue life of GH3039 superalloy by laser shock peening. *Materials (Basel), 13*(17), 3849. doi:10.3390/ma13173849 PMID:32878304

van Aswegen, D. C., & Polese, C. (2021). Experimental and analytical investigation of the effects of laser shock peening processing strategy on fatigue crack growth in thin 2024 aluminium alloy panels. *International Journal of Fatigue, 142*, 105969. doi:10.1016/j.ijfatigue.2020.105969

Wang, L., Wang, L., Xue, Y., Zhang, H., & Fu, H. (2015). Nanoindentation response of laser shock peened Ti-based bulk metallic glass. *AIP Advances, 5*(5), 57156. doi:10.1063/1.4921898

Wang, L., Zhou, L., Liu, L., He, W., Pan, X., Nie, X., & Luo, S. (2022). Fatigue strength improvement in Ti-6Al-4V subjected to foreign object damage by combined treatment of laser shock peening and shot peening. *International Journal of Fatigue, 155*, 106581. doi:10.1016/j.ijfatigue.2021.106581

Wu, B., & Shin, Y. C. (2005). A self-closed thermal model for laser shock peening under the water confinement regime configuration and comparisons to experiments. *Journal of Applied Physics, 97*(11), 113517. doi:10.1063/1.1915537

Ye, C., Liao, Y., & Cheng, G. J. (2010). Warm laser shock peening driven nanostructures and their effects on fatigue performance in Aluminum Alloy 6160. *Advanced Engineering Materials, 12*(4), 291–297. doi:10.1002/adem.200900290

Zabeen, S., Preuss, M., & Withers, P. J. (2015). Evolution of a laser shock peened residual stress field locally with foreign object damage and subsequent fatigue crack growth. *Acta Materialia, 83*, 216–226. doi:10.1016/j.actamat.2014.09.032

Zhang, C., Dong, Y., & Ye, C. (2021). Recent developments and novel applications of laser shock peening: A review. In Advanced Engineering Materials, 23(7). doi:10.1002/adem.202001216

Zhang, H., Cai, Z., Wan, Z., Peng, P., Zhang, H., Sun, R., Che, Z., Guo, C., Li, B., & Guo, W. (2020). Microstructure and mechanical properties of laser shock peened 38CrSi steel. *Materials Science and Engineering A, 788*, 139486. doi:10.1016/j.msea.2020.139486

Zhang, R., Zhou, X., Gao, H., Mankoci, S., Liu, Y., Sang, X., Qin, H., Hou, X., Ren, Z., Doll, G. L., Martini, A., Dong, Y., Sahai, N., & Ye, C. (2018). The effects of laser shock peening on the mechanical properties and biomedical behavior of AZ31B magnesium alloy. *Surface and Coatings Technology, 339*, 48–56. doi:10.1016/j.surfcoat.2018.02.009

Zhao, Z., Wang, L., Lu, K., Li, Y., Chen, W., & Liu, L. (2020). Effect of foreign object damage on high-cycle fatigue strength of titanium alloy for aero-engine blade. *Engineering Failure Analysis*, *118*, 104842. doi:10.1016/j.engfailanal.2020.104842

Zhou, J., Zhou, X., Li, H., Hu, J., Han, X., & Liu, S. (2022). In-situ laser shock peening for improved surface quality and mechanical properties of laser-directed energy-deposited AlSi10Mg alloy. *Additive Manufacturing*, *60*, 103177. Advance online publication. doi:10.1016/j.addma.2022.103177

Zhou, J. Z., Huang, S., Sheng, J., Lu, J. Z., Wang, C. D., Chen, K. M., Ruan, H. Y., & Chen, H. S. (2012). Effect of repeated impacts on mechanical properties and fatigue fracture morphologies of 6061-T6 aluminum subject to laser peening. *Materials Science and Engineering A*, *539*, 360–368. doi:10.1016/j.msea.2012.01.125

Zhu, Y., Fu, J., Zheng, C., & Ji, Z. (2015). Effect of laser shock peening without absorbent coating on the mechanical properties of Zr-based bulk metallic glass. *Optics & Laser Technology*, *75*, 157–163. doi:10.1016/j.optlastec.2015.06.029

Chapter 2
Production of Cement-Less Composite Reinforced With Polyethylene for Mechanical and Civil Engineering Applications

Salifu Tahiru Azeko
https://orcid.org/0000-0003-3860-147X
Tamale Technical University, Ghana

Joseph Asiamah Boadu
Tamale Technical University, Ghana

Moses Kingsford Flomo
Cuttington University, Liberia

Enoch Asuako Larson
https://orcid.org/0000-0001-9224-080X
University for Development Studies, Ghana

Jacob Kofi Mensah
https://orcid.org/0000-0003-1997-2291
Tamale Technical University, Ghana

Philip Yamba
Tamale Technical University, Ghana

Anthony Akayeti
Tamale Technical University, Ghana

ABSTRACT

The accumulation of polyethylene (PE) and the production of cement from industrial activities causes environmental pollution, thus posing severe threats to the lives of human beings and other living creatures. This work, therefore, presents the results of both experimental and theoretical modeling of cement-less composite reinforced with recycled polyethylene for mechanical and civil engineering applications. The composites are produced with different weight percentages of laterite and PE, and

DOI: 10.4018/978-1-6684-9843-9.ch002

homogeneously mixed to produce unique composites with outstanding mechanical properties. The composite with 40 wt.% laterite and 60 wt.% PE exhibited the highest compressive strength (~25 MPa), flexural strength (~7.3 MPa) and fracture toughness (~0.6 MPa√m). The SEM images reveal evidences of bonding and ligament bridging in the cement-less composite reinforced with 40 wt.% laterite and 60 wt.% PE. This study shows that the mixture of molten polyethylene fibers with laterite can be used as a sustainable material for the manufacturing of pavement and building blocks/bricks.

1. INTRODUCTION

Polyethylene (PE), one of the wastes generated from industries has become very necessary in recent times. Globally, the high demand for low density polyethylene (LDPE), medium density polyethylene (MDPE) and high-density polyethylene (HDPE) causes environmental pollution (Krueger et al. 2015; Azeko et al. 2018; Flomo et al. 2021). Several researches have shown that PE wastes can averagely take more than fifty (50) years for complete degradation when left on the landfill without engineered microbial activities (Azeko et al. 2015; Fotopoulou and Karapanagioti 2017; Urbanek et al. 2018; Mohanan et al. 2020; Yadav et al. 2020).

The difficulties of PE and other plastics to completely degrade under natural environmental settings causes the accumulation of these industrial waste on the environment. In trying to reduce environmental pollution caused by these plastics, few researchers (Azeko et al. 2016; Gu and Ozbakkaloglu 2016; Aldahdooh et al. 2018; Azeko et al. 2022) have found novel ways of re-using these plastics into cement-based structures for civil and mechanical engineering applications. Cement-based structures are not cost effective and environmentally friendly and contribute to the global rise of greenhouse gas emission.

Inasmuch as few researchers succeeded (Azeko et al. 2016; Gu and Ozbakkaloglu 2016; Aldahdooh et al. 2018; Azeko et al. 2022) in integrating reprocessed PE into cement-based structures for applications in the mechanical and civil engineering industries, they fail to eliminate the use of cement, that serves as an industrial pollutant. This work will therefore, address the issue of the cement being a pollutant by eliminating it to produce a cement-less composite materials reinforced with PE. The novel composites are manufactured by mechanically mixing different weight percentages of laterite and PE to produce unique laterite/PE composites with outstanding mechanical properties comparable to cement-based structures (Al-Fakih et al. 2021; Zahra et al. 2022).

2. SHEAR LAG THEORY

Ductile and/or unidirectional fiber reinforcement form the basis for the shear lag theory that are applicable in composite materials (Tevatia and Srivastava 2015; Peng et al. 2020). This theory can be used in predicting the behavior of cement-based structures that are reinforced with whiskers or short fibers (Wei et al. 2015; Peng et al. 2020). The basic assumption of this model is that, the applied load is transferred from the matrix material to the fibers through shearing.

The Cox model assumes that the displacement gradient in the fiber is proportional to the difference between the displacement in the fiber and the matrix (Peng et al. 2020):

$$\frac{d\sigma_f}{dt} = h(u - v) \tag{1}$$

Where $\frac{d\sigma_f}{dt}$ represent the displacement gradient in the fiber, u and v are respectively, the axial fiber and axial matrix displacements, while h is a proportionality constant.

Applying simple mathematical force balance for matrix-fiber interaction (Kuhn and Wehler 2020):

$$\frac{1}{4}\sigma_{fi}\pi d_f^{\,2} = \tau_m \pi d_f (dx) \tag{2}$$

Where, σ_{fi} represents the stress of the fiber, d_f denote the fiber diameter, and τ_m represents the shear stress of the matrix.

By solving for σ_f in eq. 2 gives:

$$\sigma_f = \frac{-4\tau_m}{d} \tag{3}$$

By substituting eq. (3) into eq. (1), we obtain:

$$\sigma_f = \frac{4h(u - v)}{d} \tag{4}$$

Differentiating eq. (4) with respect to z yields:

$$\frac{d^2\sigma_f}{dz^2} - h\frac{du}{dz} - h\frac{dv}{dz} = h\left(\frac{\sigma_f}{E_f} - e\right) \tag{5}$$

The solution to eq. (7) is given by:

$$\sigma_f = E_f\varepsilon_f + K_1\cosh(\beta z) + K_2\sinh(\beta z) \tag{6}$$

Where, β *is described as the shear lag parameter*, C_1 *and* C_2 are constants, Cosh (βz) denote the hyperbolic cosine function of βz, sinh (βz) represents the hyperbolic sine function of βz, E_f represents the elastic/Young's modulus of the fiber, and ε_f is the strain of the fiber.

Where $\beta = \sqrt{\frac{h}{E_f}}$ and $h = \frac{UG_m \ln\left(\frac{D}{d}\right)}{d^2}$, and K_1 *and* K_2 are constants.

Therefore, $$\beta = \sqrt{\frac{8G_m}{d^2E_f}}\ln\left(\frac{D}{d}\right) = \frac{2\sqrt{2}}{d}\sqrt{\frac{G_m}{E_f}}\ln\left(\frac{D}{d}\right) \tag{7}$$

Where G_m is the matrix shear modulus, E_f is the fiber Young's modulus, d is the fiber diameter, and D is the diameter of the unit cell.

3. MATERIALS AND METHODS

3.1 Processing of Polyethylene-Laterite Composite

Waste 'Pure water sachets' were collected at the Tamale Technical University campus. This nuisance was washed with detergents, then watered and air dried for six (6) hours at radian temperature. The dried 'pure water sachets' classified as low-density polyethylene (LDPE) (Azeko et al. 2015) were measured in different weight percentages (40-70 wt.%). These LDPE were melted above 130 °C in fume hood and mechanically mixed with different weight percentages of laterite (30-60%) until complete homogeneity using the rule of mixture strength model. The resulting polyethylene-laterite composites were then allowed to air dry at room temperature (30 °C) for 24 hours. The as-ready samples were then prepared for mechanical testing using the Universal Testing Machine to determine the compressive, flexural and fracture toughness values.

3.2 Compression And Flexural Testing

The compressive strength and flexural strength measurements were caried out using the Universal Mechanical Testing Machine (TIRAtest Model 2810, Schalkau, Thuringia, Germany). This was done under displacement control with a 0.03 mm/s displacement rate a 0.02/s strain rate. A monotonic loading was carried out for all the samples until the samples were separated into two or more pieces due to failure. In carrying out the actual testing, precise dimensions of all samples were measured using a pair of digital vernier calipers.

3.3 Fracture Toughness Testing

Thirty (30) samples were prepared for fracture toughness test using the Universal Mechanical Testing Machine (TIRAtest Model 2810, Schalkau, Thuringia, Germany) to carry out the actual measurement. The tests with a notch width of ~ 3mm were carried out using Single Edge Notch (SEN) specimens of dimensions 50 mm× 13 mm× 6.5 mm. This experiment was carried out under load control with 0.01 MPa $\sqrt{m}s^{-1}$ stress intensity factor (K). These samples were loaded by increasing the load gradually until fracture.

4. DISCUSSION OF FINDINGS AND RESULTS

4.1 Compressive Strength

The results for the compressive strength of the polyethylene-laterite composite without the inclusion of cement is illustrated in Figures 1a-1d. Critically examining Figures 1a, 1c and 1d, the composites reached their ultimate compressive strength, then after the onset of necking was visible. Unlike Figure 1b, although the composite reached its ultimate compressive strength, the onset of necking was negligible, but the composite abruptly failed. Figure 1d recorded the lowest compressive strength of 7.5 MPa. The composite in Figure 1c had the highest compressive strength of 25 MPa, which is higher than previous work on polyethylene-cement composites (Azeko et al. 2016; Gu and Ozbakkaloglu 2016; Aldahdooh et al. 2018; Azeko et al. 2022). The reason behind this tremendous jump is that, the melted polyethylene served as an effective glue/binder, due to proper binding and interactions of polyethylene-laterite interphase.

Figure 1. Compressive Strength of Cementless Polyethylene-Laterite Composites with: (a) 60% L+40% PE (b) 50% L+50% PE (c) 40% L+60% PE (d) 30% L+70%PE
Note: L=Laterite, PE=Polyethylene

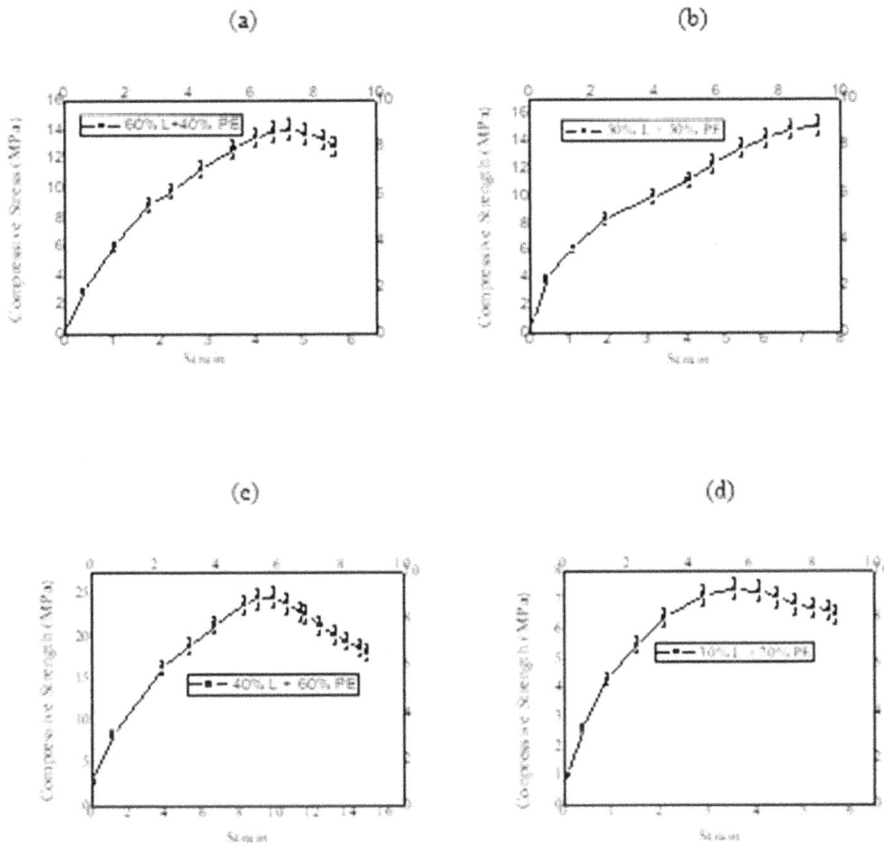

4.2 Flexural Strength

The results for the flexural strength of the polyethylene-laterite composite without the inclusion of cement is illustrated in Figures 2a-2d. Critically examining Figures 2(a, c and d), the composites reached their ultimate flexural strength, then after the onset of necking was visible.

Examining Figure 2b, the composite reached its ultimate flexural strength without visible occurrence of necking, however, the composite failed abruptly without major evidence of plastic deformation. Figure 2d recorded the lowest flexural strength of ~1.8 MPa. The composite in Figure 2c had the highest flexural strength of ~7.3 MPa. The reason behind this increase is that, the melted polyethylene served as an

Figure 2. Flexural Strength of Cementless Polyethylene-Laterite Composites with: (a) 60% L+40% PE (b) 50% L+50% PE (c) 40% L+60% PE (d) 30% L+70%PE
Note: L=Laterite, PE=Polyethylene

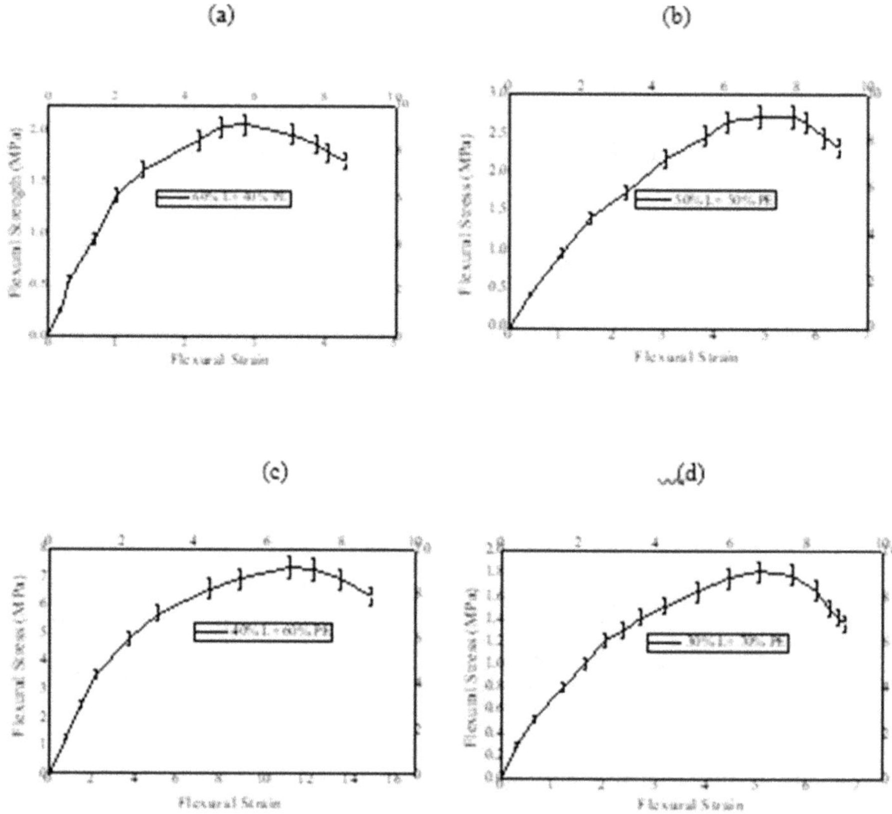

effective glue/binder, due to proper binding and interactions of polyethylene-laterite interphase.

4.3 Fracture Toughness

The results for the fracture toughness of the polyethylene-laterite composite without the inclusion of cement is illustrated in Figure 3. It is shown that the polyethylene-laterite composite (without inclusion of cement) with 60 wt.% PE and 40 wt.% laterite had the highest fracture toughness of ~ $0.6MPa\sqrt{m}$, with the least fracture toughness of ~ $0.1MPa\sqrt{m}$ being the composite with 30 wt.% laterite and 70 wt.% PE. The increased in fracture toughness of this composite is due to the PE serving as impediment for dislocation and propagation of cracks as well as filling voids in

Figure 3. Fracture Toughness of Cementless Polyethylene-Laterite Composites
Note: L=Laterite, PE=Polyethylene

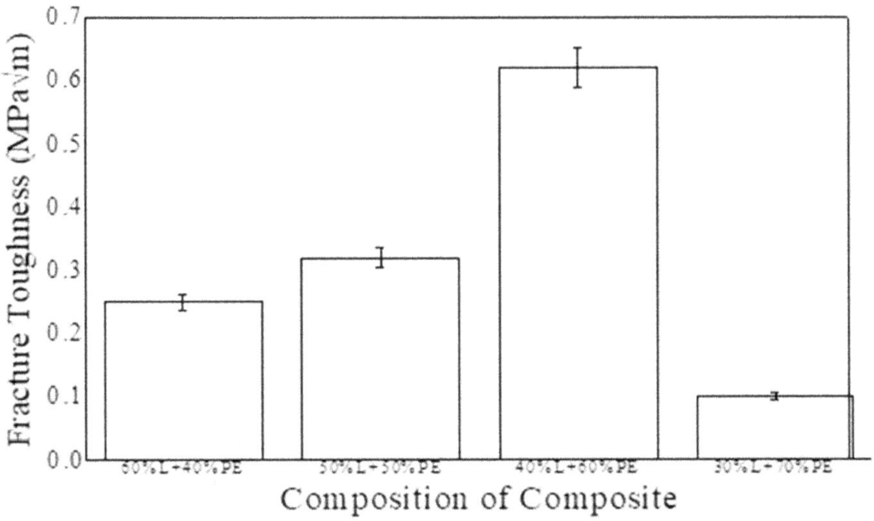

cracks and micro-cracks. In this case, when cracks are initiated, they begin to propagate until within the shortest possible time, they reach the PE binder/glue. Then the cracks must continue to propagate but because they have met an obstacle, excess energy is needed for the crack (s) to either glide or climb. This excess energy needed gives the composite its overall strength, since longer period of time is needed to overcome the PE glue/binder.

4.4 Scanning Electron Microscopy (SEM) Analysis

The result for the scanning electron microscopy analysis for the laterite-polyethylene composites without the inclusion of cement is shown in Figure 4. From the results, matrix-fiber bonding, matrix-fiber debonding and appearance of micro-cracks and secondary/multi cracks are visible. The composite with 40 wt.% laterite and 60 wt.% PE show evidences of bonding between the matrix and PE fiber, although some visibility of gully erosions can be seen in the middle of the image. Critically looking at this gully erosion, ligament bridging is evident at certain locations, thus, serving as impediment for crack propagation.

Figure 4. Scanning Electron Microscopy Analysis of Laterite Polyethylene Composites Reinforced with 40% L + 60% PE
Note: L=Laterite, PE=Polyethylene

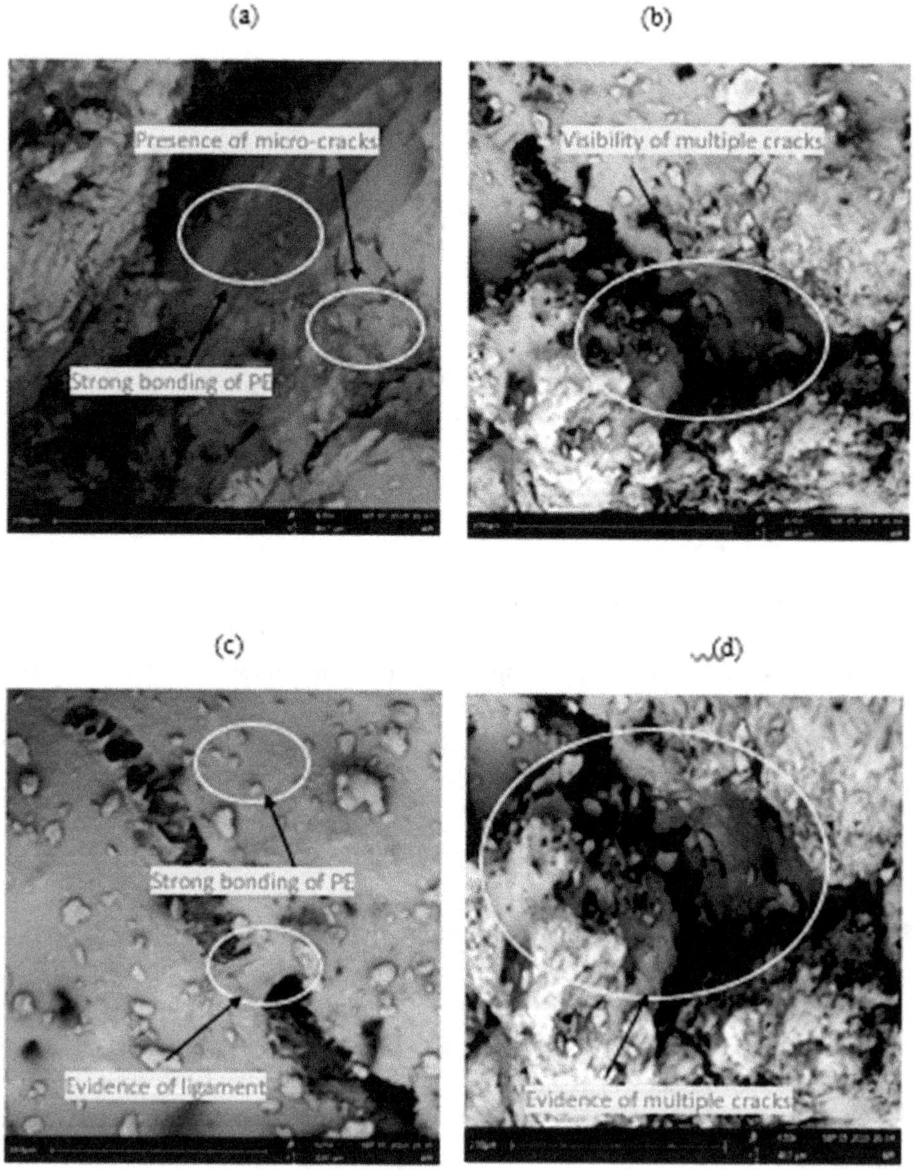

5. CONCLUSION

In this study, we examined the effects of polyethylene (PE) fibers that was homogeneously mixed with laterite in different weight percentages to produce polyethylene-laterite composites without the inclusion of cement. The study shows that polyethylene in its molten state when incorporated with laterite to produce polyethylene-laterite composite possess outstanding compressive strength, flexural strength and fracture toughness that are comparable to cement-based structures. The composites were produced with different weight percentages of laterite and PE and homogeneously mixed with laterite to produce unique composites with outstanding mechanical properties. The composite with 40 wt.% laterite and 60 wt.% PE displayed the highest compressive strength, flexural strength and fracture toughness of 25 MPa, 7.3 MPa and $0.6MPa\sqrt{m}$, respectively. The results obtained in this work can be used for several applications in the mechanical and civil engineering industry for the production of pavement and building blocks/bricks.

6. ACKNOWLEDGMENTS

The authors are grateful to Tamale Technical University for their financial support.

REFERENCES

Al-Fakih, A., Mohammed, B. S., Wahab, M. M. A., Liew, M. S., Amran, Y. M., Alyousef, R., & Alabduljabbar, H. (2020, August). Characteristic compressive strength correlation of rubberized concrete interlocking masonry wall. In *Structures* (Vol. 26, pp. 169–184). Elsevier.

Aldahdooh, M. A. A., Jamrah, A., Alnuaimi, A., Martini, M. I., Ahmed, M. S. R., & Ahmed, A. S. R. (2018). Influence of various plastics-waste aggregates on properties of normal concrete. *Journal of Building Engineering*, *17*, 13–22. doi:10.1016/j.jobe.2018.01.014

Azeko, S. T., Arthur, E. K., Danyuo, Y., & Babagana, M. (2018). Mechanical and physical properties of laterite bricks reinforced with reprocessed polyethylene waste for building applications. *Journal of Materials in Civil Engineering*, *30*(4), 04018039. doi:10.1061/(ASCE)MT.1943-5533.0002205

Azeko, S. T., Arthur, E. K., Minh, D. P., Lyczko, N., Nzihou, A., & Soboyejo, W. O. (2022). Mechanical and Thermal Properties of Sustainable Composite Building Materials Produced by the Reprocessing of Low-Density Polyethylene, Biochar, Calcium Phosphate, and Phosphogypsum Wastes. *Journal of Materials in Civil Engineering*, *34*(2), 04021457. doi:10.1061/(ASCE)MT.1943-5533.0004021

Azeko, S. T., Etuk-Udo, G. A., Odusanya, O. S., Malatesta, K., Anuku, N., & Soboyejo, W. O. (2015). Biodegradation of linear low-density polyethylene by Serratia marcescens subsp. marcescens and its cell free extracts. *Waste and Biomass Valorization*, *6*(6), 1047–1057. doi:10.100712649-015-9421-0

Azeko, S. T., Mustapha, K., Annan, E., Odusanya, O. S., & Soboyejo, W. O. (2016). Recycling of polyethylene into strong and tough earth-based composite building materials. *Journal of Materials in Civil Engineering*, *28*(2), 04015104. doi:10.1061/(ASCE)MT.1943-5533.0001385

Flomo, M. K., Azeko, S. T., Arthur, E. K., Kukurah, J. D., Mustapha, K., Annan, E., & Agyei-Tuffour, B. (2021). Reinforcement of cement mortar with recycled polyethylene waste for construction applications. *Journal of Composite Materials*, *55*(21), 2867–2875. doi:10.1177/00219983211000249

Gu, L., & Ozbakkaloglu, T. (2016). Use of recycled plastics in concrete: A critical review. *Waste Management (New York, N.Y.)*, *51*, 19–42. doi:10.1016/j.wasman.2016.03.005 PMID:26970843

Krueger, M. C., Harms, H., & Schlosser, D. (2015). Prospects for microbiological solutions to environmental pollution with plastics. *Applied Microbiology and Biotechnology*, *99*(21), 8857–8874. doi:10.100700253-015-6879-4 PMID:26318446

Kuhn, C., & Wehler, S. (2020). A Force-Balanced Fiber Retardation Model to Predict Fiber-Matrix-Separation during Polymer Processing. *Journal of Composites Science*, *4*(4), 165. doi:10.3390/jcs4040165

Mohanan, N., Montazer, Z., Sharma, P. K., & Levin, D. B. (2020). Microbial and enzymatic degradation of synthetic plastics. *Frontiers in Microbiology*, *11*, 2837. doi:10.3389/fmicb.2020.580709 PMID:33324366

Peng, Z., Wang, X., & Wu, Z. (2020). A bundle-based shear-lag model for tensile failure prediction of unidirectional fiber-reinforced polymer composites. *Materials & Design*, *196*, 109103. doi:10.1016/j.matdes.2020.109103

Tevatia, A., & Srivastava, S. K. (2015). Modified shear lag theory-based fatigue crack growth life prediction model for short-fiber reinforced metal matrix composites. *International Journal of Fatigue*, *70*, 123–129. doi:10.1016/j.ijfatigue.2014.09.004

Urbanck, A. K., Rymowicz, W., & Mirończuk, A. M. (2018). Degradation of plastics and plastic-degrading bacteria in cold marine habitats. *Applied Microbiology and Biotechnology, 102*(18), 7669–7678. doi:10.100700253-018-9195-y PMID:29992436

Wei, X., Filleter, T., & Espinosa, H. D. (2015). Statistical shear lag model–Unraveling the size effect in hierarchical composites. *Acta Biomaterialia, 18*, 206–212. doi:10.1016/j.actbio.2015.01.040 PMID:25684701

Yadav, V., Sherly, M. A., Ranjan, P., Tinoco, R. O., Boldrin, A., Damgaard, A., & Laurent, A. (2020). Framework for quantifying environmental losses of plastics from landfills. *Resources, Conservation and Recycling, 161*, 104914. doi:10.1016/j.resconrec.2020.104914

Zahra, T., Thamboo, J., & Asad, M. (2021). Compressive strength and deformation characteristics of concrete block masonry made with different mortars, blocks and mortar beddings types. *Journal of Building Engineering, 38*, 102213. doi:10.1016/j.jobe.2021.102213

Chapter 3
Leveraging IBM Watson Auto AI to Predict the Occurrence of Deaths Among COVID-19 Patients in USA

Victor Worlanyo Gbedawo
https://orcid.org/0000-0002-2476-174X
Ho Technical University, Ghana

Andrews Dodzi Kobla Dzikunu
https://orcid.org/0009-0007-3208-7904
Zhejiang Normal University, China

ABSTRACT

Technology and preventive healthcare have become the key solutions to the COVID-19 pandemic. This study explores the COVID-19 dataset to predict the occurrence of deaths among COVID-19 patients in the USA. 1,386,100 observations and 5 variables were selected from the USA COVID-19 Dataset obtained from the online repository of the Center for Systems Science and Engineering (CSSE) on Johns Hopkins University's (JHU) website. IBM's Auto AI Experiment was leveraged to determine the best prediction outcomes. All eight pipelines were trained and tested using the COVID-19 dataset and the results showed that Pipelines 3, 4, 5, 6, 7, 8 outperformed the other two algorithms in terms of accuracy, precision, recall, and the F scores. Pipelines 1 and 2 achieved the worst results among the models with an F score of 0.9915. The research demonstrates the promising performance to help health institutions and governments plan ahead to forestall COVID-19 deaths and implement policies toward preventive care.

DOI: 10.4018/978-1-6684-9843-9.ch003

1. INTRODUCTION

The coronavirus COVID-19 pandemic has become the defining global health crisis of our time and the greatest challenge we have had to face since World War II. We reached the tragic milestone of one million deaths, and the human family has to suffer under an almost intolerable burden of loss (Afewerky, 2020).

The pandemic was much more than a health crisis, it was also an unprecedented socio-economic crisis, stressing every one of the countries it touched, it had the potential to create devastating socio-economic and political effects that left deep and long standing scars (Carnut et al., 2020).

Predicting COVID-19 deaths before they occurred, would have definitely been a breakthrough for Health institutions and Governments. The novelty of this study lies in, leveraging IBM's Watson Auto AI platform to evaluate prediction models using the USA COVID-19 dataset.

1.1. Motivation

The use of Emerging Technologies in solving a global crisis such as the prediction of deaths among the novel coronavirus patients was paramount to the very existence of the human race (Abujaber et al., 2020). Machine Learning as a subset of Artificial intelligence forms part of the current trends in patient healthcare which is capable of providing sustainable and useful medical service (Derakhshan & Yousefi, 2019).

1.2. Problem Statement

If one could have predicted the outbreak and spread of coronavirus some hundred (100) days before it got worse, Health Institutions and Governments could have planned and contained it.

1.3. Research Objectives

This study;

1. Seeks to explore the performance of IBM's Auto AI model in predicting COVID-19 Deaths.
2. Evaluate, improve and validate the developed model, using the USA COVID-19 dataset.

1.4. Research Questions

Owing to the above research objectives, this study aims at addressing the following gaps in questions:

1. What are the existing Literature on IBM Watson and COVID-19 Predictions?
2. How is the performance of IBM Watson Auto AI on COVID 19 Dataset?

2. BACKGROUND

2.1. Artificial Intelligence (AI) and Machine Learning (ML) Concepts

AI is the ability of digital computers or computer controlled robots to solve problems that are normally associated with the higher intellectual processing capabilities of humans .

A particular strength of human intelligence is adaptivity. We are capable of adjusting to various environmental conditions and change our behavior accordingly through learning. Precisely because our learning ability is so vastly superior to that of computers, machine learning is, according to Rich's definition, a central subfield of AI. Machine learning algorithms build a mathematical model based on sample data, known as "training data", in order to make predictions or decisions without being explicitly programmed to do so. Most real-world problems tend to have infinite and mostly unpredictable set of possible outcomes. It is in this light that, machine learning algorithms are employed for better outcomes (GANGULY, 2021).

2.2. Types of Machine Learning

In general, all machine learning algorithms are mainly classified into four categories:

Supervised, Unsupervised, Semi-supervised, and Reinforcement learning algorithms.

2.2.1. Supervised Machine Learning

Supervised machine learning algorithms work by feeding some data into the computer and annotate them to let them know what each data output signifies. Based on their features, the computer decides to form a mathematical model so that it can predict the output of some unknown data.

2.2.2. Unsupervised Machine Learning

Unsupervised machine learning algorithms are mainly used for classification purposes where we don't quite know the class of each data sample. Similar to supervised machine learning algorithms, we feed the data into the computer, but we don't tell the computer what output each data signifies; or in other words, we don't annotate the data to its proper classes. Instead, we let the computer decide what each data sample signifies. The computer itself labels an output to each data sample based on some features and that also the computer itself decides. (GANGULY, 2021)

2.2.3. Semi-Supervised Machine Learning

Semi-supervised learning provides a comfortable middle ground between supervised and unsupervised learning. It employs a smaller labeled data set to aid classification and feature extraction from a larger, unlabeled data set during training. Semi-supervised learning can address the issue of insufficient labeled data for a supervised learning algorithm. It also helps if labeling enough data is prohibitively expensive (Marsland, 2014).

2.2.4. Reinforcement Machine Learning

Reinforcement learning bridges the gap between unsupervised learning, where the algorithm can only use similarities in the data to cluster it, and supervised learning, where the algorithm is trained on the correct responses given in the target data. In the middle, it is explained whether or not the response is accurate but not how to make it better (Marsland, 2014).

2.2.5. Linear Regression

The goal of machine learning is to create a functional equation from a problem that can be resolved by providing the necessary inputs. The equation is easily reduced to:

$$f(x) = x^2$$

The difficulties won't be that straightforward in real-world situations, and it would be close to impossible to deduce a fixed equation from the problem statement. In order to identify the closest straight line that might suit the points, it is best to plot the problem's x and f(x) points, or the parameters and the outcome. The majority of machine learning algorithms are built around the linear regression method of fitting a line. The renowned line equation:

$$y = mx + c$$

x is the co-ordinate value of x-axis
y is the co-ordinate value of y-axis
m refers to the slope of the line
c is the y-intercept

With the two sets of x and y values provided, we may say that we can determine the values of m and c in the equation. Now that we have the line (the m and c values), we can use the x value to determine the value of any y. This idea is what the machine learning algorithm for linear regression is based on. In actual problems, the x is not just one value but rather a collection of values, or a matrix. Therefore, when one y value corresponds to a group of x values, the line we are trying to draw turns into a hyperplane. The equation then changes to look like this:

$$y = a^0 + a^1x^1 + a^2x^2 + a^3x^3 + ... + a^nx^n$$

We collect data for each of x^1, x^2, x^3,..., x^n and their related y values in order to build curve fitting. This will result in the following set of equations:

$$y^1 = a^0 + a^1x^{11} + a^2x^{12} + a^3x^{13} + ... + a^nx^{1n} \quad (1)$$

$$y^2 = a^0 + a^1x^{21} + a^2x^{22} + a^3x^{23} + ... + a^nx^{2n} \quad (2)$$

$$y^3 = a^0 + a^1x^{31} + a^2x^{32} + a^3x^{33} + ... + a^nx^{3n} \quad (3)$$

$$y^n = a^0 + a^1x^{n1} + a^2x^{n2} + a^3x^{n3} + ... + a^nx^{nn} \quad (n)$$

As a result, we must determine the values of all the coefficients of a using a set of values for all the x and their associated y.

2.2.6. Computers Learn

In order for a computer to learn, we must first allow it to choose random coefficient values. Then describe the concept of the loss function. The computer is informed by this function of how many errors it would be committing. Based on the errors, the computer will adjust its course. We shall define the mean squared error loss function as the loss function:

$$MSE = \frac{1}{n}\sum_{i=1}^{n}(y' - y)$$

In this case, *Y'* is the actual value of *Y*, and *Y'* is the value obtained when computing the output value using the coefficients derived at each step. *N* is also the quantity of data samples. A set of data (often 25%) is set aside when as sample of data for learning, while the remaining data are used for training. These random coefficient values and the *X* values are used to calculate the values of *Y* for each equation. These values of *Y* that were obtained can either be predicted or derived. The loss function is used to compute the error made. Based on the inaccuracy, the coefficient matrix's values are adjusted and then recalculated as the anticipated *Y* values using this new set of coefficient values. After doing the maximum number of iterations, the saved test data is taken and used to forecast the values of *Y* using the final values of the coefficients obtained. The computer has at this point supposedly learned the coefficient values. The results of cross validation provides one such significant estimate. All conceivable combinations of hyperparameter value combinations are evaluated using cross validation.

2.2.7. IBM Watson's Auto AI

Without writing a single line of code, IBM Watson's Auto AI solution can choose the optimal model from a variety of models. A dataset is analyzed by auto AI, which then fits the data into several machine learning models (GANGULY, 2021). Hyper-parameter tuning is used to determine the ideal parameter, while hyper-parameter optimization is used to improve the model. AutoAI automatically runs the following tasks to build and evaluate candidate model pipelines:

2.2.7.1. Data Pre-Processing

Standard machine learning algorithms only function with integers and no missing values, despite the fact that the majority of data sets comprise a variety of data formats. In order to evaluate, purify, and prepare raw data for machine learning, AutoAI uses a variety of methods or estimators. Using features like data type—categorical or numerical—this automatically discovers and classifies values. AutoAI employs hyper-parameter optimization to find the ideal mix of missing value imputation, feature encoding, and feature scaling algorithms for your data based on categorization.

2.2.7.2. Automated Model Selection

AutoAI uses automated model selection to identify the best model for any type of data. This novel approach tests potential models against small subsets of the data and ranks them based on accuracy. AutoAI then selects the most promising models and increases the size of the data subset until it identifies the best match. This approach saves time and improves performance by gradually narrowing down the potential models based on accuracy.

2.2.7.3. Automated Feature Engineering

The optimum model for any sort of data is found using automated model selection, which is a feature of AI. This innovative method evaluates alternative models using small data subsets and ranks them according to their correctness. The size of the data subset is then increased while AutoAI continues to choose the most promising models until the best match is found. By gradually reducing the number of viable models depending on accuracy, this method saves time and enhances performance.

2.2.7.4. Hyperparameter Optimization

The top-performing models are improved through hyperparameter tweaking. For some machine learning-related function evaluations, like model training and scoring, AutoAI employs a novel hyperparameter optimization approach. Despite taking a while to evaluate each model, this method swiftly pinpoints the best one.

2.2.7.5. Ensembling and Incremental Learning

creating pipelines using BatchedTreeEnsemble on top of ranking pipelines. If necessary, the ensemble pipelines can be utilized to continue training using the remaining data from a subsampled source by splitting the remaining data into batches. The optimized metric is used to independently score each batch of training.

2.3. Confusion Matrix and Metrics

To be able to establish how well a model performed, there is need to evaluate the model using some kind of metric.

Table 1. Confusion matrix

		Relevant documents in results set	Irrelevant documents in results set
Algorithm result	Retrieved	True positive (Tp)	False positive (Fp)
	Not retrieved	False negative (Fn)	True negative (Tn)

Nonetheless, a confusion matrix, also known as an error matrix in Table 1, is a specific table layout that enables visualization of the performance of an algorithm. Based on the results of an algorithm's perfomance,

The True positive (Tp) value answers the question, how many relevant documents were retrieved by the algorithm?

The False positive (Fp) value answers the question, how many irrelevant documents were retrieved by the algorithm?

The False negative (Fn) value answers the question, how many relevant documents did the algorithm not retrieve?

The True negative (Tn) value answers the question, how many irrelevant documents did the algorithm not retrieve?

The objective is to improve the algorithm to decrease the Fp and Fn values.

Accuracy is as a numeric measure of how good your algorithm is. It calculates how many correct results your solution managed to identify, which is the proportion of true results among the total number of cases that are examined. Accuracy is defined by the following formula, which includes the Tp, Tn, Fp, and Fn metrics:

$$Accuracy = \frac{(Tp + Tn)}{(Tp + Tn + Fp + Fn)}$$

Precision is a numeric measure that represents the fraction of retrieved documents that are relevant. It is defined by the following formula:

$$Precision = \frac{Tp}{(Tp + Fp)}$$

Recall sometimes also called sensitivity or True Positive Rate (TPR). is a numeric measure that represents the fraction of relevant documents that were retrieved. It is defined by the following formula:

$$Recall = \frac{Tp}{(Tp + Fn)}$$

The F-score (also called F-measure) is a measure that enables you to tradeoff precision against recall by approximately averaging the precision and recall values. The higher the F-score value is, the better the algorithm is. The formula for F-score is:

$$F - score = \frac{(2 \times precision \times \mathrm{Re}\,call)}{(precision + \mathrm{Re}\,call)}$$

Another often used technique with binary classifiers is the receiver operating characteristic (ROC) curve. It is very similar to the precision/recall curve, except the ROC curve compares the true positive rate (another term for recall) against the false positive rate (FPR), rather than precision vs recall. The FPR measures the proportion of negative events that are mistakenly labeled as positive. It is equal to the true negative rate (TNR), which is 1.

The area under the curve (AUC) is a metric that can be used to compare classifiers. The ROC AUC, which runs from 0 to 1, will be equal to 1 for a perfect classifier (the lower, the worse). (GANGULY, 2021)

3. METHODOLOGY

The first step was to carry out a systematic literature review on IBM Watson and COVID-19 predictions there by gaining enough leverage to propose a four step process as shown in Figure 1

1. Identification of credible sources and Data collection.
2. Data Analysis and Variable definition of Dataset required for COVID-19 model development and model feature testing processes.
3. Model feature testing to identify the required set of experiments to assess individual features.
4. The Key Performance Indicator (KPI) identification associated with the performance of the prediction model.

Figure 1. Methodology design

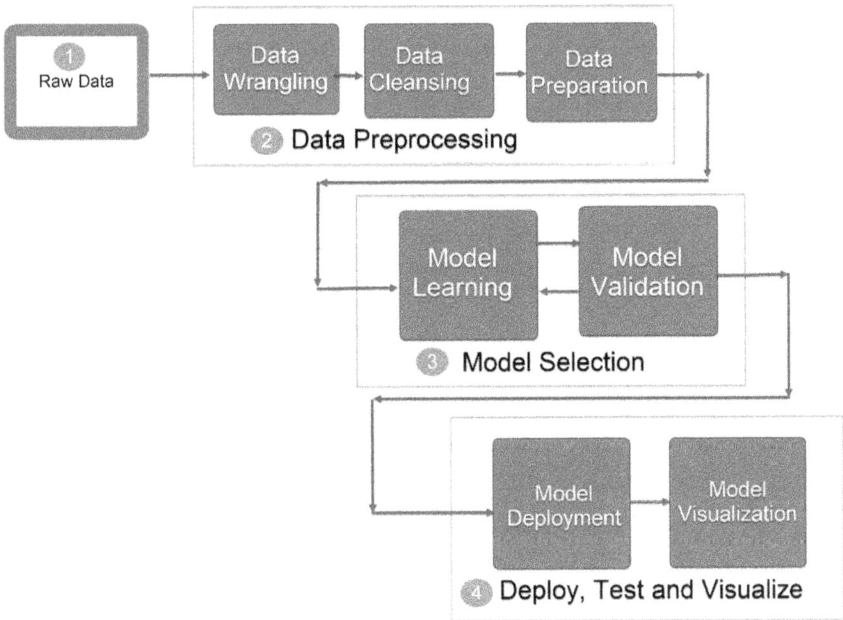

3.1. Systematic Literature Review

A systematic review was carried out until March 26, 2023 and PRISMA guidelines were adhered to in reporting the results of this study (Haddaway et al., 2022). The VOSviewer software and the R studio were used to carry out co-author analysis and bibliographic coupling as captured in Table 2, Figure 2, Table 3, Figure 3, Figure 4 and Figure 5.

3.2. Selection Criteria

A systematic literature review was carried out for existing articles between 2019 to 2023, in the Web of Science database, using the following search terms: ("IBM Watson" OR "COVID-19 predictions" AND PUBYEAR > 2019-2023). PRISMA guidelines were adhered to in reporting the results. The database searches generated a total of 10,658 studies. 3969 studies were excluded because the studies did not meet the criteria after reviewing the abstract. The full text of the remaining 175 studies were reviewed. After applying inclusion criteria and exclusion criteria, a total of 25 studies were identified for inclusion in the review. The systematic Literature review

Table 2. First search results from web of science with search term "COVID-19 predictions"

AUTHOR ID	AUTHOR	h_index	g_index
1	GUEST PAUL C	0	0
2	ADHIKARI MAINAK	2	3
3	BRAUN PETER	4	4
4	MA FENG	3	4
5	ACHARYA U RAJENDRA	0	0
6	CHEN XIN	0	0
7	HUANG DENGSHI	2	3
8	KIM JAE-HO	1	1
9	KIM JANGYOUNG	1	1
10	LI XIANG	2	3
11	LIANG CHAO	2	3
12	LU XINJIE	3	3
13	NANDY SUDARSHAN	2	2
14	SHEN S P	3	3
15	WANG LIN	1	2
16	WEI Y Y	3	3
17	WOODCOCK BARRY G	0	0
18	ZHANG FENG	1	3
19	ABBASI MARYAM	1	1
20	ABBASIFARD MITRA	0	0

provides the current situation regarding IBM Watson and COVID-19 predictions. The SLR discusses research findings and the gaps toward future directions into COVID-19 predictions.

3.3. Review of Existing Literature

For COVID-19 forecasts in India, a deep learning architecture utilizing CNN and layered Bi-GRU has been developed (Ahuja et al., 2022). On COVID-19 datasets, the proposed technique was contrasted with the Gaussian process regression (GPR) model. The simulation results demonstrated that the suggested framework outperformed the GPR model for COVID-19 predictions with greater precision. The lack of a global model for COVID-19 prediction is the paper's main area for further research.

Figure 2. Author impact analysis with search term "COVID-19 predictions"

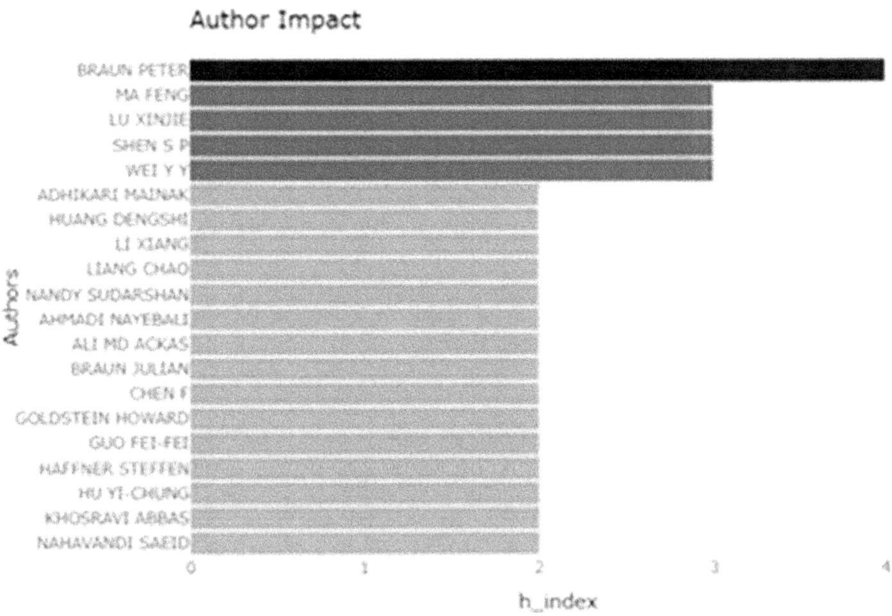

Further, An interactive tool to forecast US hospital needs in the coronavirus

Figure 3. Co-authorship analysis (authors) biggest network eight node connected

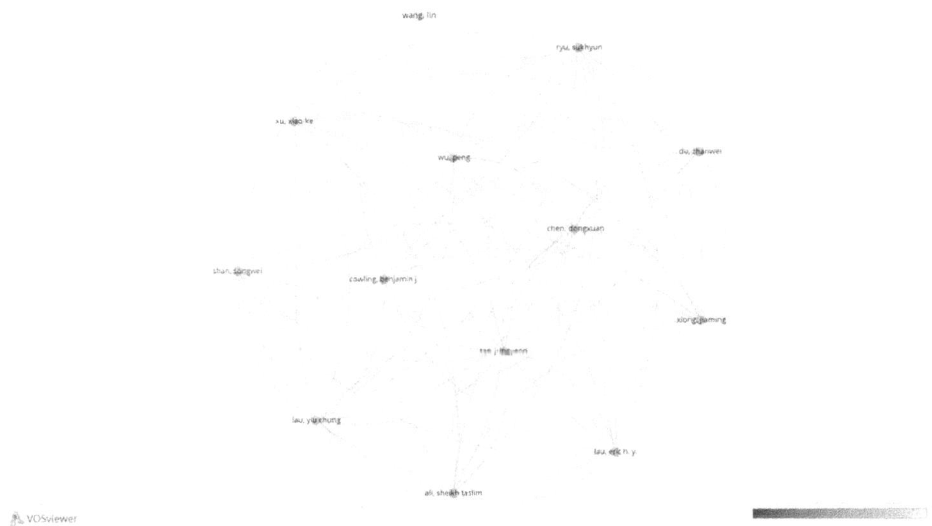

Table 3. Second search results from web of science with search term "IBM watson"

Description	Results
Timespan	2018:2023
Sources (Journals, Books, etc)	130
Documents	165
Average years from publication	2.16
Average citations per documents	6.358
Average citations per year per doc	1.464
References	106
DOCUMENT TYPES	
article	122
journal article	33
meeting abstract	3
research-article	7
DOCUMENT CONTENTS	
Keywords Plus (ID)	11
Author's Keywords (DE)	11
AUTHORS	
Authors	875
Author Appearances	1007
Authors of single-authored documents	7
Authors of multi-authored documents	868
AUTHORS COLLABORATION	
Single-authored documents	7
Documents per Author	0.189
Authors per Document	5.3
Co-Authors per Documents	6.1
Collaboration Index	5.49

2019 pandemic was looked at (Locey et al., 2020). The application uses COVID-19 forecasts in conjunction with models of time lags, patient carryover, and length of stay to forecast hospital visits, admits, discharges, and demands for hospital beds, ventilators, and personal protective equipment. Users could select one of seven COVID-19 models, alter 23 parameters, look at test and hospitalization trend patterns, and obtain forecast data. The application correctly forecasted how COVID-19 would spread throughout the states and territories. Our institution and others frequently

Figure 4. Co-authorship analysis (authors) biggest network 13 node connected

use its hospital level projections. The problem is that more work needs to be done to combine the regional dispersion of COVID-19 with local healthcare requirements, avoid over-reliance on a single model, and give meaningful predictions.

Another literature worth noting, is the Analysis and Forecast of the Number of Deaths, Recovered Cases, and Confirmed Cases From COVID-19 for the Top Four Affected Countries Using Kalman Filter (Ahmadini et al., 2021). For the four most afflicted nations—the United States, India, Brazil, and Russia—the Kalman filter approach (KF) was used. According to the results, the KF technique was almost always able to maintain track of the actual COVID-19 data. The Kalman filters used in the model backdrop gave accurate COVID-19 predictions. The KF method's findings lend assistance to the decision-making process for short-term COVID-19 outbreak management measures. However, the study did not take into account or measure additional factors such as the number of lockdowns, social isolation, and self-isolation as well as the association of those who have had vaccinations and the quantity of daily new instances.

According to predictions from a Monte Carlo stochastic model, confinement time is necessary to prevent a quick rebound of COVID-19 (Girona, 2020). In this instance, local viral transmission was modeled using a Monte Carlo stochastic framework. He stated that improper application renders suppression tactics not only ineffectual but also potentially detrimental because there is a good likelihood that they will increase the duration of effective confinement without decreasing the

Figure 5. PRISMA flow chart
(Haddaway et al., 2022)

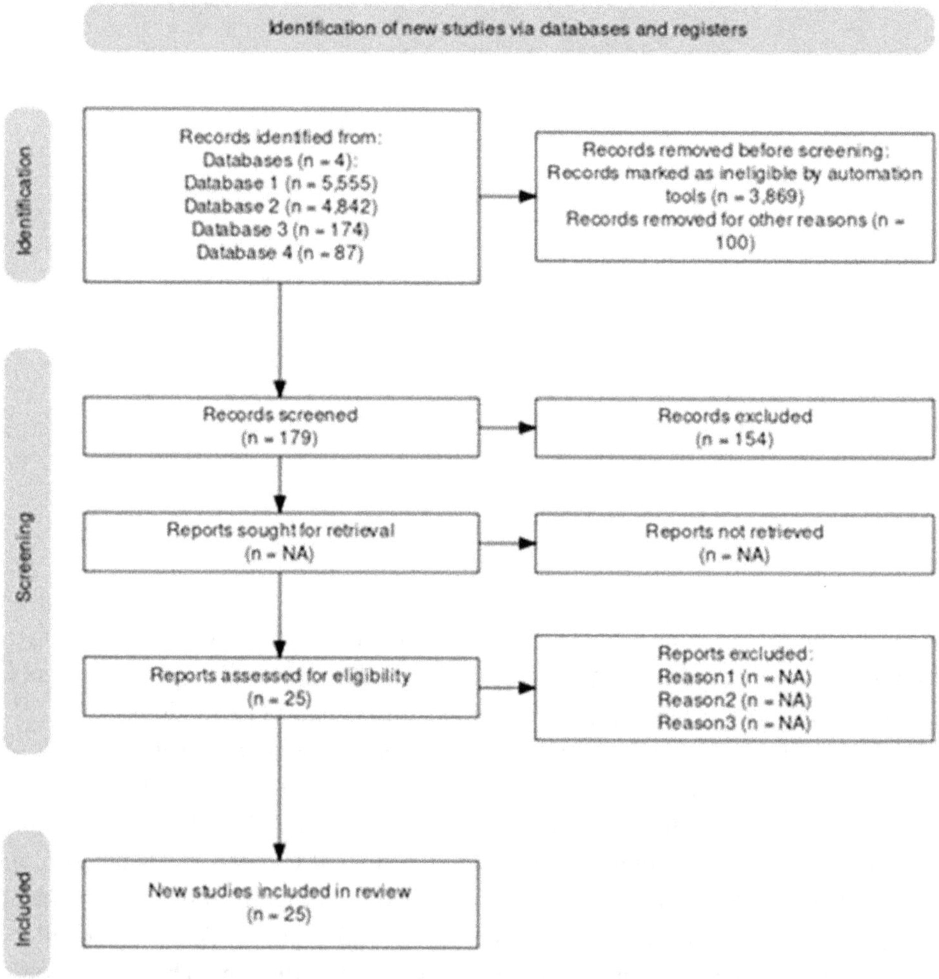

overall number of infections. This is the outcome of a nonlinear interaction between the level of confinement, the length of confinement, and social isolation. Beyond the 95% confidence interval and based on the model assumptions, confinement is effective.

Al-Rashedi & Al-Hagery (2023) published Deep Learning Algorithms for Forecasting COVID-19 Cases in Saudi Arabia. The procedure was as follows:

(1) Data collection: The Saudi Arabian Ministry of Health provided the necessary COVID-19 dataset.

(2) Pre-Analysis: The data underwent some analysis in order to find any hidden patterns.
(3) Processing: Cleaning up missing values or irrelevant variables and processing data.
(4) Time Series Extraction: The process of separating a day, a month, and a year into distinct qualities for the analysis.
(5) Data Scaling: When using models, scaling the data is crucial for achieving optimal performance. The final phase, Model Building, involves creating three models based on the LSTM, CNN, and ARIMA algorithms.
(7) Predicting Results: We forecast the results of several scenarios in cities for each model.
(8) Analysis's findings: Following the application of the models, each algorithm's outcomes are examined independently.

The flaw in this research is that other cutting-edge deep learning and machine learning algorithms weren't investigated, and their effectiveness wasn't compared to the techniques employed. Constructing city-wide forecasting models to foresee the growth of COVID-19 cases in significant Saudi cities like Riyadh and Mecca. To avoid overfitting and underfitting issues that ultimately result in the generalization of the models, take into account alternative feature selection techniques to identify the best possible feature combinations and by applying feature extraction engineering to improve the datasets, we can identify more pertinent features that can improve forecast accuracy

Automating the automatic search for the optimal settings of the DL algorithms' hyperparameters utilizing cutting-edge optimization techniques. A functional time series forecasting approach was put forth, and the daily median absolute errors were given in (Kong et al., 2020), but not the more widely used measure of mean absolute percentage error (MAPE). As a result, it cannot be used as a standard for comparing results in experiments. Using a Kalman filter estimator, a MAPE of 18% to 30% was found.

3.4. Data Collection

Owing to challenges in obtaining reliable data, and for the purposes of achieving the objectives of study, dataset was obtained from the online repository of the Center for Systems Science and Engineering (CSSE) at the Johns Hopkins University (JHU) website. The rationale for using a publicly available dataset was that, it was reliable and can be verified.

3.5. Data Analysis

A study of this kind involves the use of both qualitative and quantitative data analysis mechanisms. According to a post (Brownlee, 2014), Hilary Mason and Chris Wiggins described the OSEMN process as a taxonomy of tasks that a data scientist should feel comfortable working on. In addition to a list of tasks, **OSEMN** can be used as a blueprint for working on data problems using machine learning tools.

1. **O**btain Data: Using APIs to access data, where an API may be public or internal to your organization. Often data is presented in JSON and scripting languages like Python can make data retrieval a lot easier.
2. **S**crub Data: Real data can have inconsistencies, missing values and various other forms of corruption. If it was scraped from a difficult data source, it may require tripping and cleaning up. Even clean data may require post-processing to make it uniform and consistent.
3. **E**xplore Data: Data exploration is useful for getting to know your data, for building an intuition for its form and for getting ideas for data transforms and even predictive models to use later on in the process.
4. **M**odel Data: Model accuracy is often the ultimate goal for a given data problem. This means that the most predictive model is the filter by which a model is chosen. Generally the goal is to use a model predict and interpret. Prediction can be evaluated quantitatively, whereas interpretation is softer and qualitative. A model's predictive accuracy can be evaluated by how well it performs on unseen data. It can be estimated using methods such as cross validation. The algorithms that you try and your biases and reduction on the hypothesis space of possible models that can be constructed for the problem.
5. i**N**terpret Results: The purpose of computing is insight, not numbers. The interpretative power of a model is its abilities to suggest the most interesting experiments to perform next. It gives insights into the problem and the domain.

Data collection refers to a standardised procedure for collecting data pertinent to an experiment. The columns in the dataset are called variables which make up the object of our enquiry. Although this study identifies a set of variables to carry out the task, it is worthwhile to note that the variable set might change depending on the model feature selection. Furthermore, any additional variables will only optimize the performance of the model. The devised approach categorises the variables into two sets, namely, mandatory and optional variables. The mandatory variables which are numerical in nature, comprise COVID-19 Case count variables, COVID-19 Death count, Outcome, Province/State and Date variables. In the case of the outcome variables, a computational assumption was made since COVID-19 meant, not normal

Table 4. Variable description

Mandatory Variables	
Variable type	**Variable**
COVID-19 Case count	Covid_case
COVID-19 Death count	Covid_deaths
Outcome (Predictive)	outcome
Province/State	Provincestate
Date	date
Optional Variables	
Country/Region	countryregion
User	admin2

times, if the death count is less than or equal to 5 (<=5), it was coded as Good News otherwise, Bad News. The optional variables included Country/Region variable and the Administrator variable which were mostly categorical in nature. While mandatory variables influenced the initial order of feature importance, the optional variables just enhanced the dynamics of the feature inferance. It is also worth noting that, the various combinations of variables and variable types influenced the very specific differences between groups and the very specific relationships between variables. Further details about the variable sets are outlined in Table 6.

The data analysis procedure involved an initial statistical analysis of the COVID-19 dataset patterns to describe any existing patterns in their means. This step was crucial in the generation of generalised models and represents a novel process in the proposed methodology. As the mean only gives an indication of data existence, the mean variations needed to be established using tests of statistical significance such as ANalayis Of VAriance (ANOVA).

This technique was used to infer whether, means of two or more groups are significantly different from each other. This technique compared the means of different samples to identify the impact of one or more factors. Furthermore, the ANOVA procedure was enhanced through additional available information that reflected possible correlations with the factors under investigation. For this particular research, the dataset obtained from the John Hopkins webstite, was regarded as the Population and the different samples taken from it, represented the different groups in ANOVA. Each sample constituted one single test. ANOVA was formulated using the following set of equations.

$$Null\ Hypothesis\ H_0 : \mu 1 = \mu 2 = ... = \mu k \tag{1}$$

$$Alternate\ Hypothesis\ H_a : \mu k \neq \mu m \quad (2)$$

where $(\mu 1, \mu 2, ... \mu k)$ represented individual group means, k represents the number of a group and μm represented the group mean when $(\mu 1 = \mu 2 ... = \mu k - 1)$. The alternate hypothesis suggested that there were at least two group means that were statistically and significantly different from each other. ANOVA usually involves the calculation of F-test statistic that represents the ratio of two quantities expected to be roughly equal under the null hypothesis. Large F-test values signify the variability in group means is large compared to the within group variability, which is essential to reject the null hypothesis. F-test values are used in conjunction with p-values to establish the significance. For instance, A large F-test value with a small *p*-value (less than the defined significance levels) represents a rejected null hypothesis. In this case, from the scatter plot in Fig 6, we fail to accept the null hypothesis since there is enough evidence that there is statistical difference between the means at a 95% confidence level, the P-value is 0.0000 which is apparently lower than the Alpha value of 0.05.

3.6. Experimental Results

3.6.1. Dataset

COVID-19 dataset (https://github.com/CSSEGISandData/COVID-19):

The initial results presented in this section are based on the COVID-19 Data Repository by the Center for Systems Science and Engineering (CSSE) at Johns Hopkins University. It is a publicly available database. The details have been provided in Table 7, Table 8 and Table 9.

3.6.2. Implementation Details

The experiment leveraged IBM's Auto AI Platform at a capacity of 16vCPU and 64GB RAM and 1 (one) NIVIDA V100 GPU.

3.6.3. Dataset Preparation

Before importing the COVID-19 dataset obtained from the online repository into IBM watson, the dataset was cleansed of missing values as shown in Table 8 and Table 9. The percentage split between train and test datasets were as follows: 90% of train set and 10% Holdout data as test set.

Figure 6. Scatter interval of COVID-19 cases and deaths

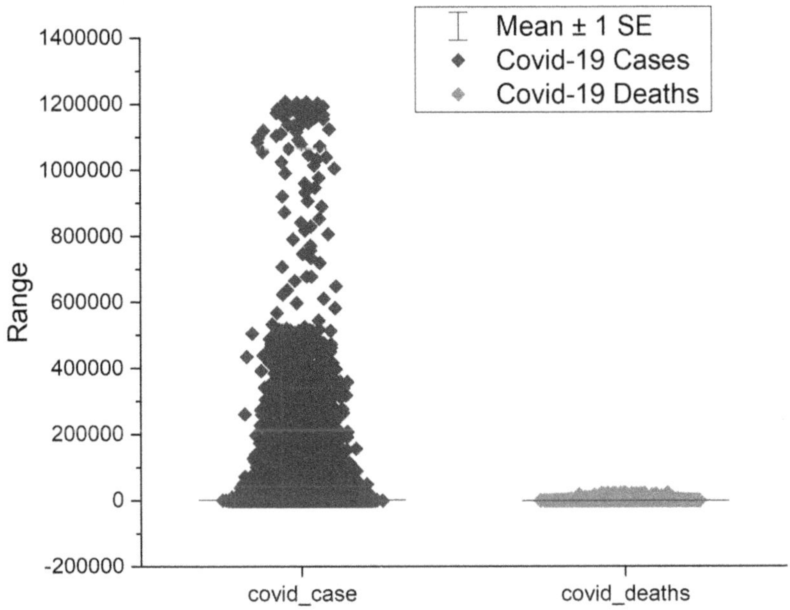

Table 5. Dataset description

Dataset Description	Number of variables	Covid Description	Province/ State	Observations	Size MB
us_confirmed_ updated	5	covid_cases	provincestate	1,386,100	107.07
us_deaths_updated	5	covid_deaths	provincestate	1,386,100	104.43

Table 6. Dataset cleansed

```
covid_case

                type:  numeric (long)

               range:  [0,1208024]                  units:  1
       unique values:  43728                    missing .:  0/1386100

                mean:   2563.22
            std. dev:   14538.7

         percentiles:        10%       25%       50%       75%       90%
                               0         5       175      1181      4342

provincestate

                type:  string (str24)

       unique values:  58                      missing "":  0/1386100

            examples:  "Illinois"
                       "Michigan"
                       "North Carolina"
                       "Tennessee"

             warning:  variable has embedded blanks

date

                type:  string (str10)

       unique values:  415                     missing "":  0/1386100

            examples:  "2020-04-13"
                       "2020-07-05"
                       "2020-09-26"
                       "2020-12-18"
```

4. DISCUSSION

The results of the supervised experiment as shown in Table 10 and Figure 49, shows pipeline 3 as the top ranked algorithm for prediction in this experiment. This means that Pipelines 3, 4, 5, 6, 7, 8 outperformed the other two algorithms in terms of Acuracy, Precision, Recall and the F scores. Pipelines 1 and 2 achieved the worst results among the models with an F score of 0.9915 under Auto AI. Overall, the Auto AI Experiment carried out even though produced very promising results, the resultant values outlook appeared too good to be real, partly because of the automated nature of the tasks.

Table 7. Breakdown of COVID 19 cases in the USA

Province/State	Freq.	Percent	Cum.
Alabama	28,635	2.07	2.07
Alaska	12,865	0.93	2.99
American Samoa	415	0.03	3.02
Arizona	7,055	0.51	3.53
Arkansas	31,955	2.31	5.84
California	24,900	1.80	7.63
Colorado	27,390	1.98	9.61
Connecticut	4,150	0.30	9.91
Delaware	2,075	0.15	10.06
Diamond Princess	415	0.03	10.09
District of Columbia	1,245	0.09	10.18
Florida	28,635	2.07	12.25
Georgia	66,815	4.82	17.07
Grand Princess	415	0.03	17.10
Guam	415	0.03	17.13
Hawaii	2,905	0.21	17.34
Idaho	19,090	1.38	18.71
Illinois	43,160	3.11	21.83
Indiana	39,010	2.81	24.64
Iowa	41,915	3.02	27.66
Kansas	44,405	3.20	30.87
Kentucky	50,630	3.65	34.52
Louisiana	27,390	1.98	36.50
Maine	7,470	0.54	37.04
Maryland	10,790	0.78	37.81
Massachusetts	7,055	0.51	38.32
Michigan	36,105	2.60	40.93
Minnesota	36,935	2.66	43.59
Mississippi	34,860	2.51	46.11
Missouri	48,970	3.53	49.64
Montana	24,070	1.74	51.38
Nebraska	39,425	2.84	54.22
Nevada	7,885	0.57	54.79
New Hampshire	4,980	0.36	55.15
New Jersey	9,545	0.69	55.84
New Mexico	14,525	1.05	56.89
New York	26,560	1.92	58.80
North Carolina	42,330	3.05	61.86
North Dakota	22,825	1.65	63.50
Northern Mariana Islands	415	0.03	63.53
Ohio	37,350	2.69	66.23
Oklahoma	32,785	2.37	68.59
Oregon	15,770	1.14	69.73
Pennsylvania	28,635	2.07	71.80
Puerto Rico	33,200	2.40	74.19
Rhode Island	2,905	0.21	74.40
South Carolina	19,920	1.44	75.84
South Dakota	28,220	2.04	77.87
Tennessee	40,255	2.90	80.78
Texas	106,240	7.66	88.44
Utah	15,355	1.11	89.55
Vermont	6,640	0.48	90.03
Virgin Islands	415	0.03	90.06
Virginia	56,025	4.04	94.10
Washington	17,015	1.23	95.33
West Virginia	23,655	1.71	97.04
Wisconsin	30,710	2.22	99.25
Wyoming	10,375	0.75	100.00
Total	1,386,100	100.00	

Figure 7. IBM Watson studio project

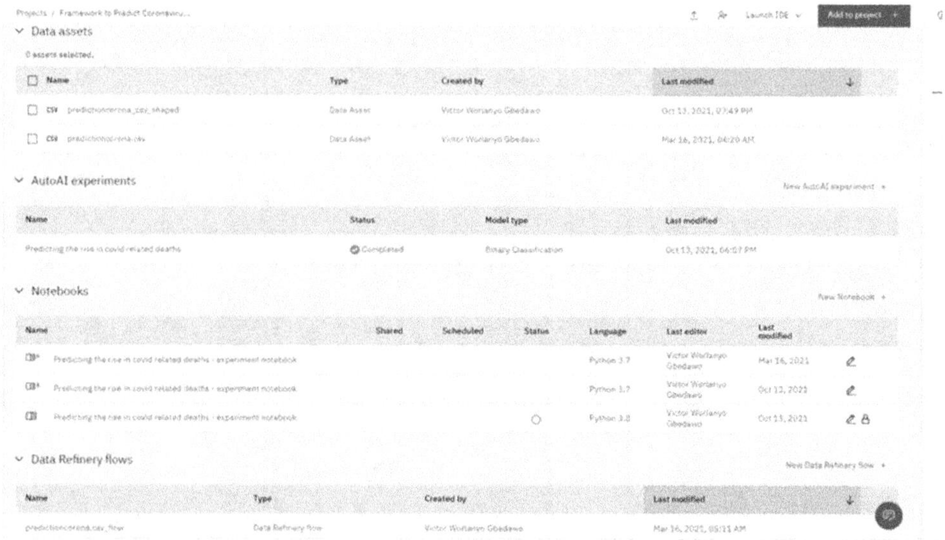

Figure 8. Pipeline 1 model information

Figure 9. Pipeline 1 feature summary

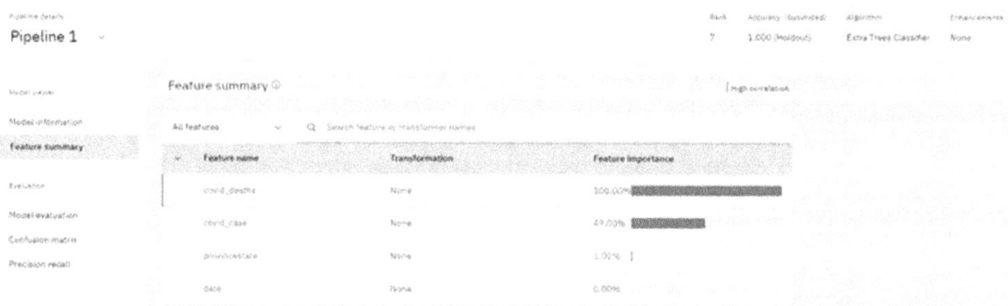

Figure 10. Pipeline 1 receiver operator characteristic (ROC) curve

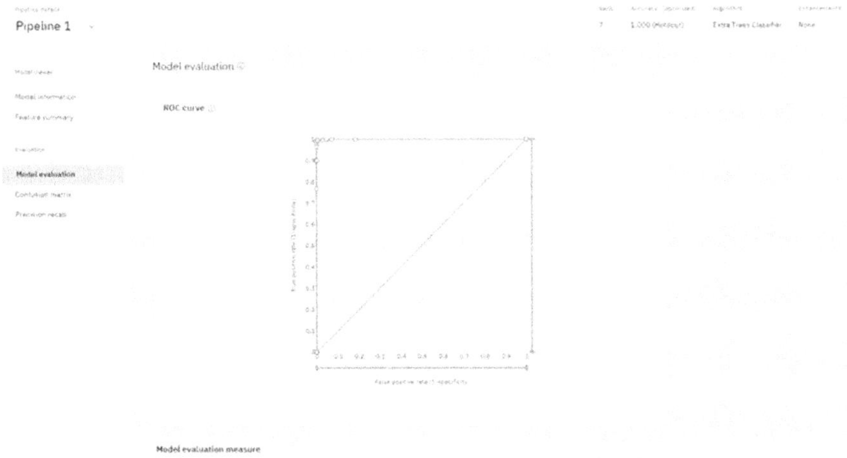

Figure 11. Pipeline 1 confusion matrix

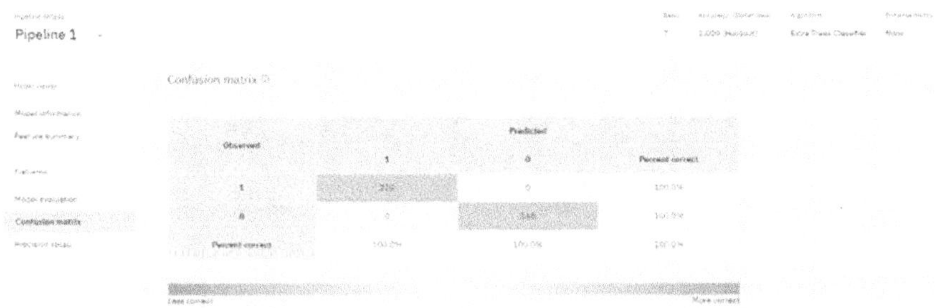

Figure 12. Pipeline 1 precision recall curve

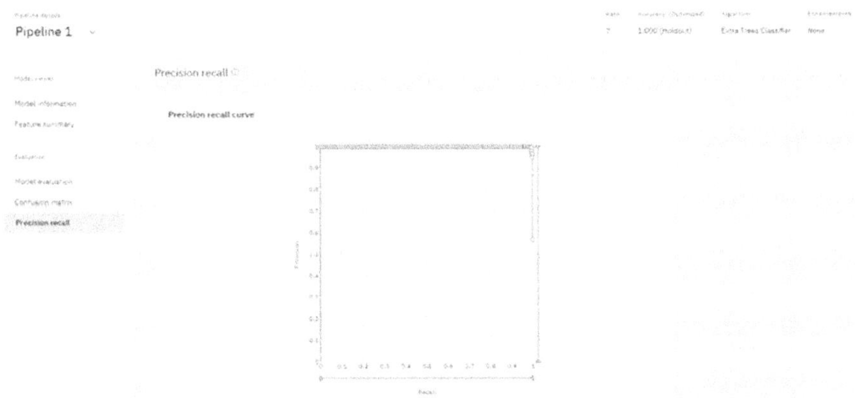

Figure 13. Pipeline 2 model Information

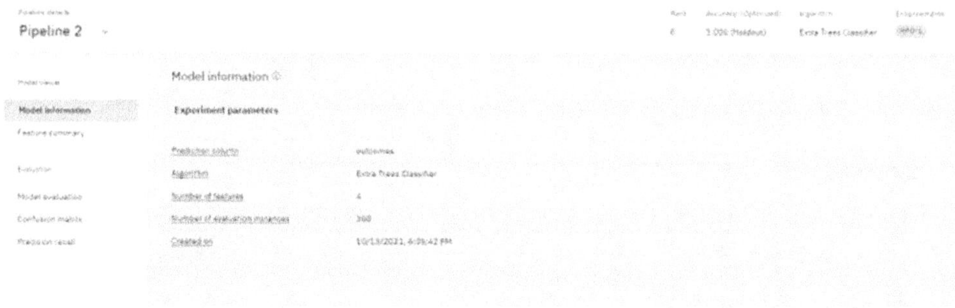

Figure 14. Pipeline 2 feature summary

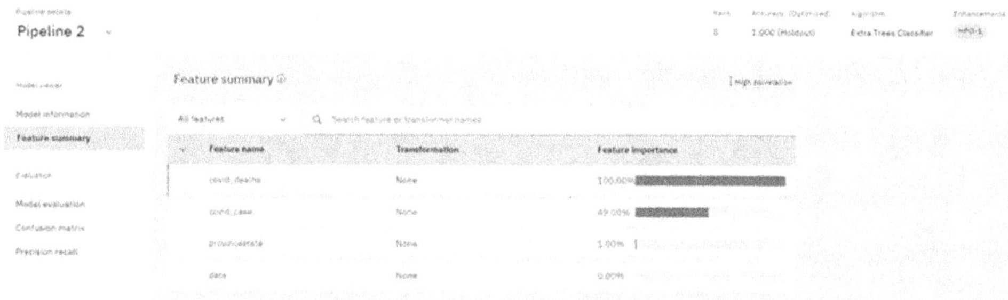

Figure 15. Pipeline 2 receiver operator characteristic (ROC) curve

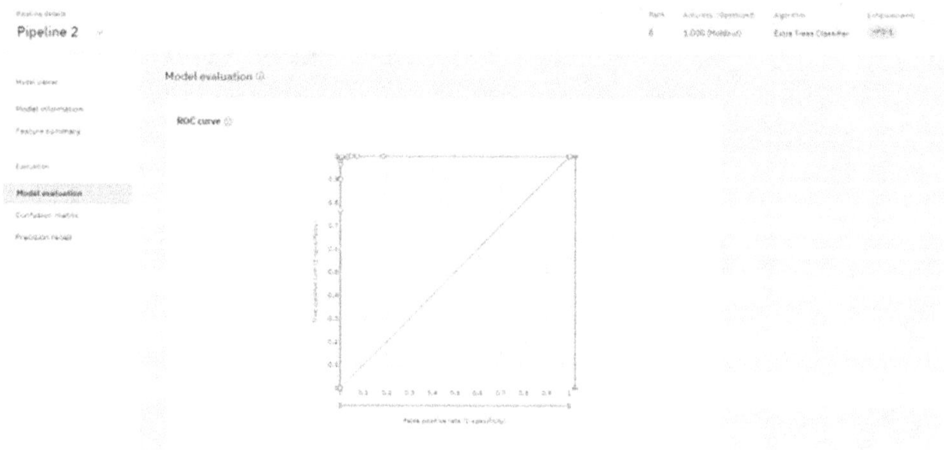

Figure 16. Pipeline 2 confusion matrix

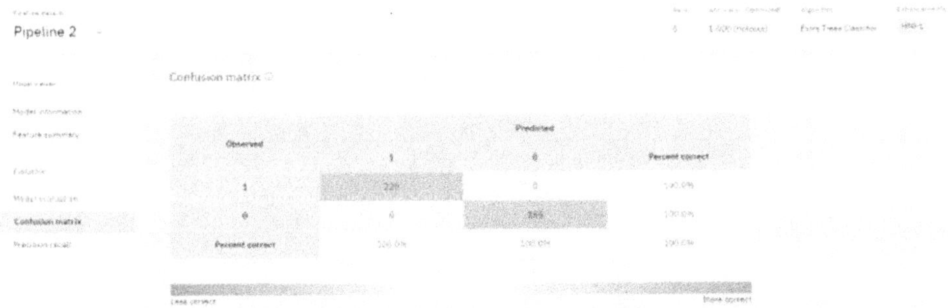

Figure 17. Pipeline 2 confusion matrix

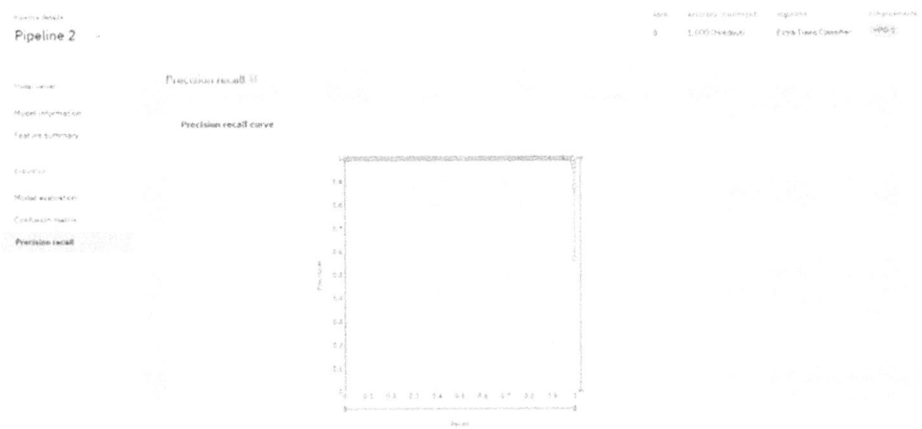

Figure 18. Pipeline 3 model information

Figure 19. Pipeline 3 feature summary

Figure 20. Pipeline 3 receiver operator characteristic (ROC) curve

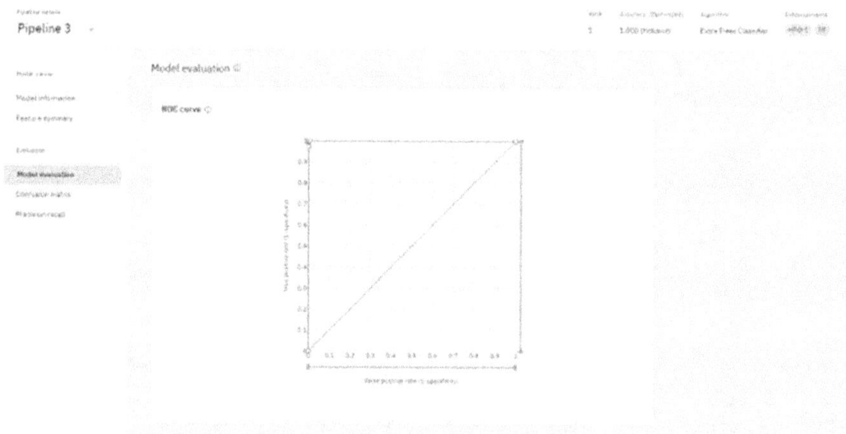

Figure 21. Pipeline 3 confusion matrix

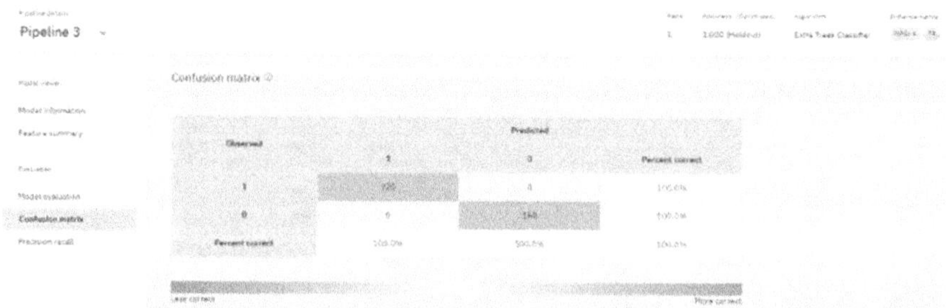

Figure 22. Pipeline 3 precision recall curve

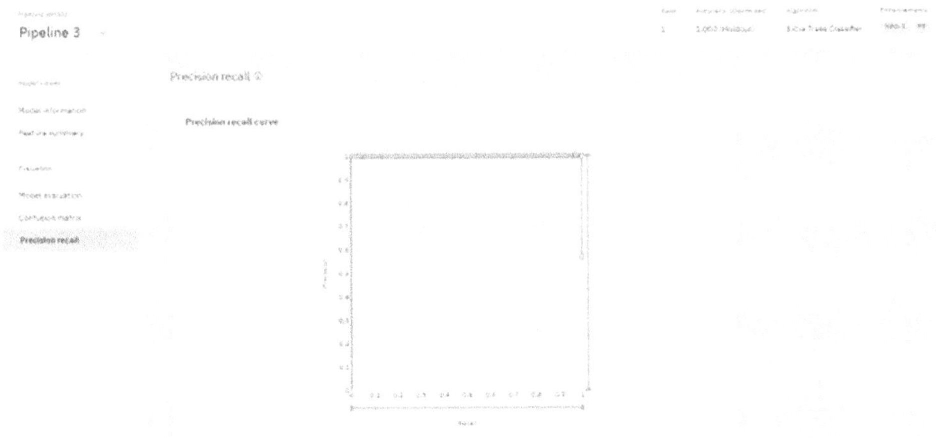

Figure 23. Pipeline 4 model information

Figure 24. Pipeline 4 feature summary

Figure 25. Pipeline 4 receiver operator characteristic (ROC) curve

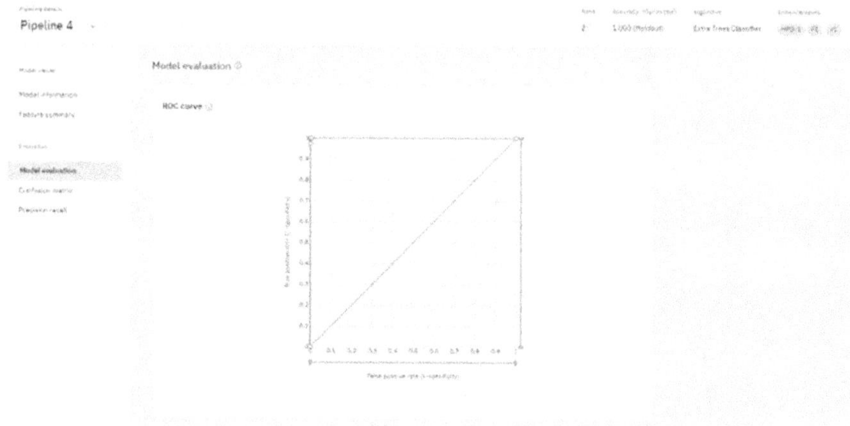

Figure 26. Pipeline 4 confusion matrix

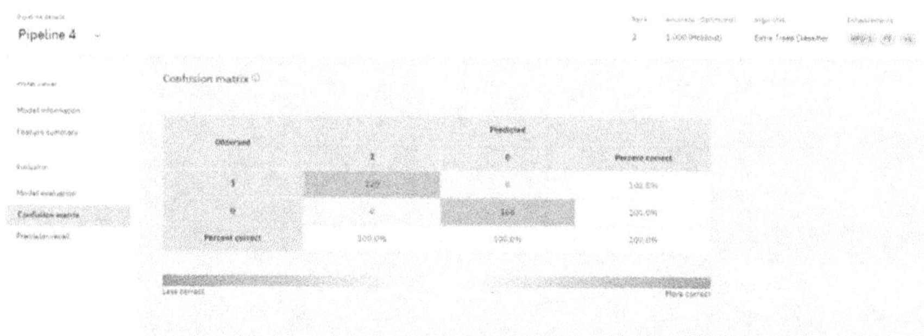

Figure 27. Pipeline 4 precision recall curve

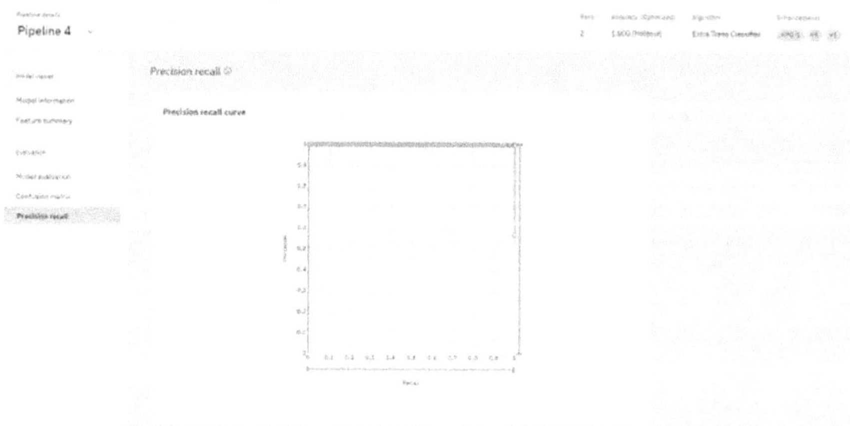

Figure 28. Pipeline 5 model information

Figure 29. Pipeline 5 feature summary

Figure 30. Pipeline 5 receiver operator characteristic (ROC) curve

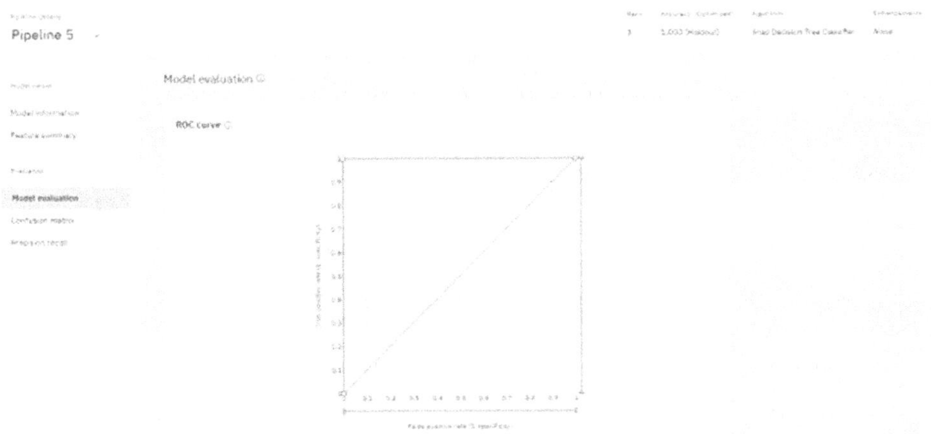

Figure 31. Pipeline 5 confusion matrix

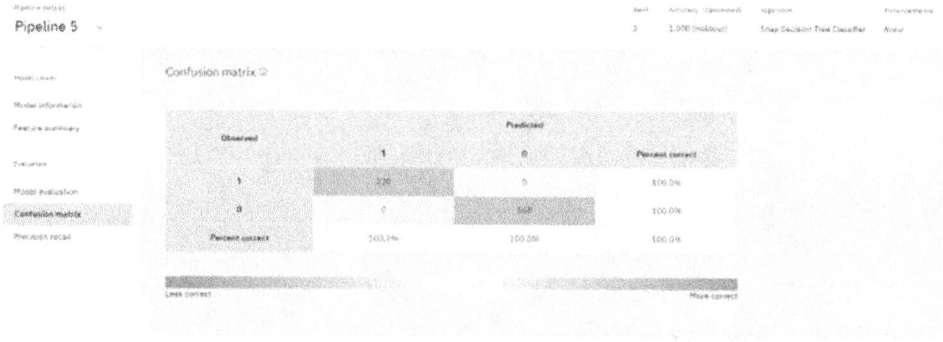

Figure 32. Pipeline 5 precision recall curve

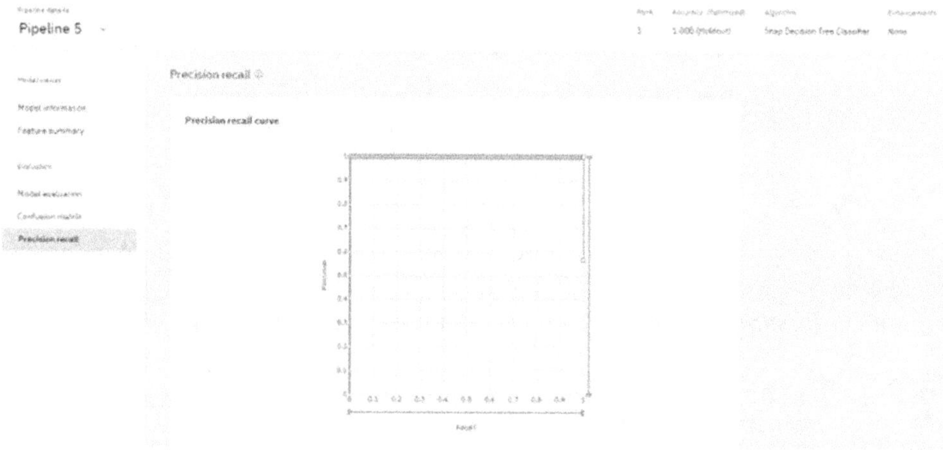

Figure 33. Pipeline 6 model Information

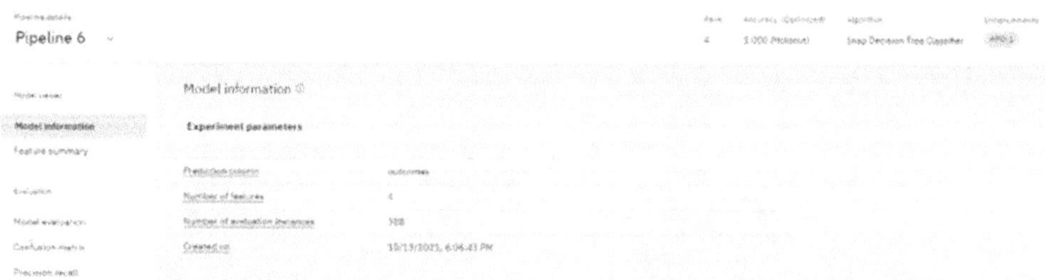

Figure 34. Pipeline 6 feature summary

Figure 35. Pipeline 6 receiver operator characteristic (ROC) curve

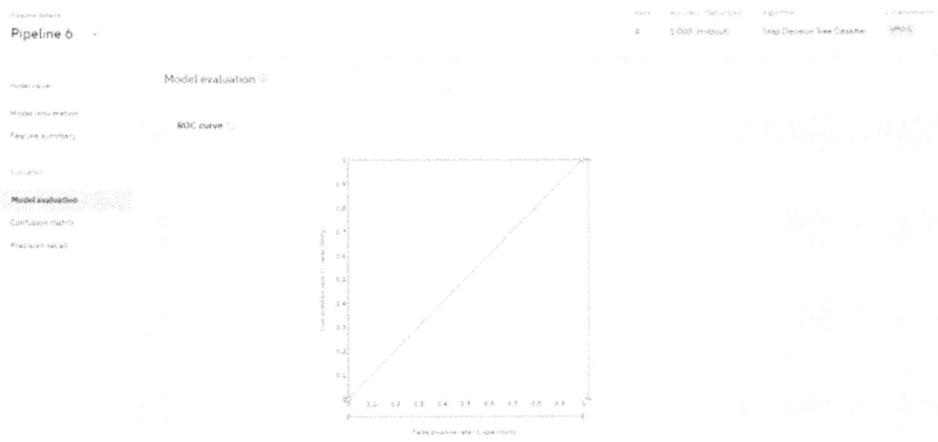

Figure 36. Pipeline 6 confusion matrix

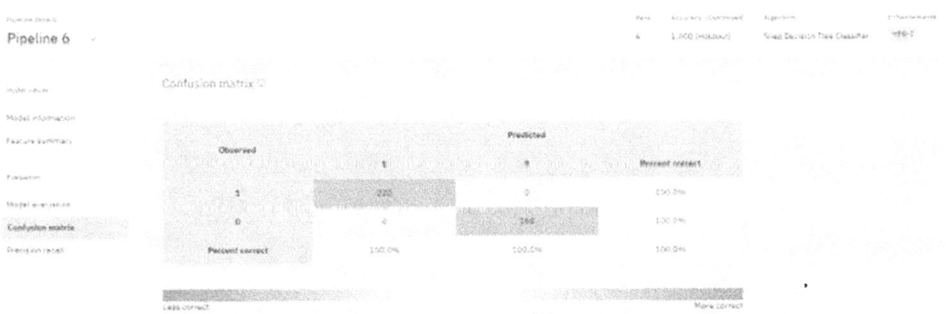

Figure 37. Pipeline 6 precision recall curve

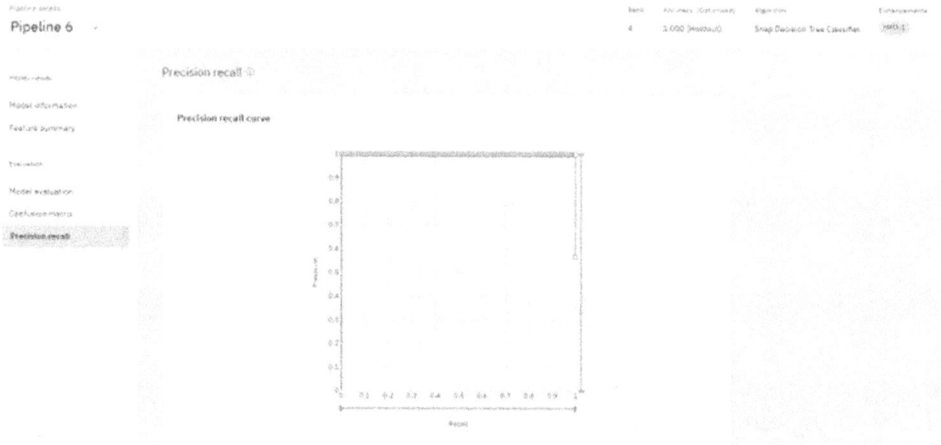

Figure 38. Pipeline 7 model information

Figure 39. Pipeline 7 feature summary

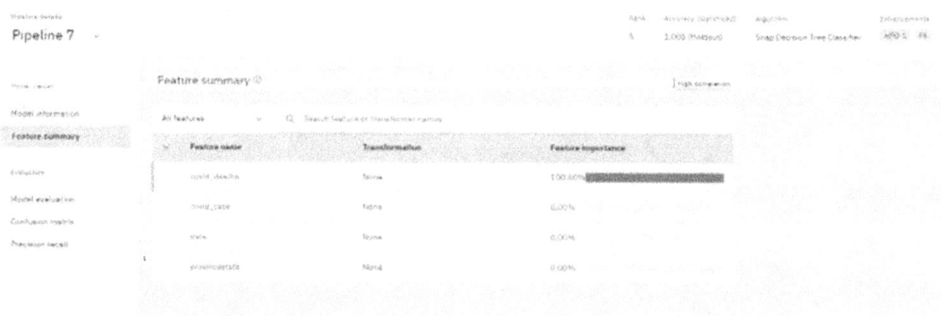

Figure 40. Pipeline 7 receiver operator characteristic (ROC) curve

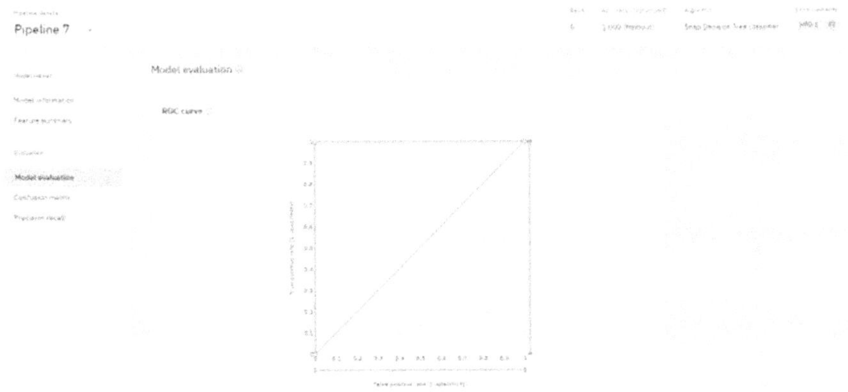

Figure 41. Pipeline 7 confusion matrix

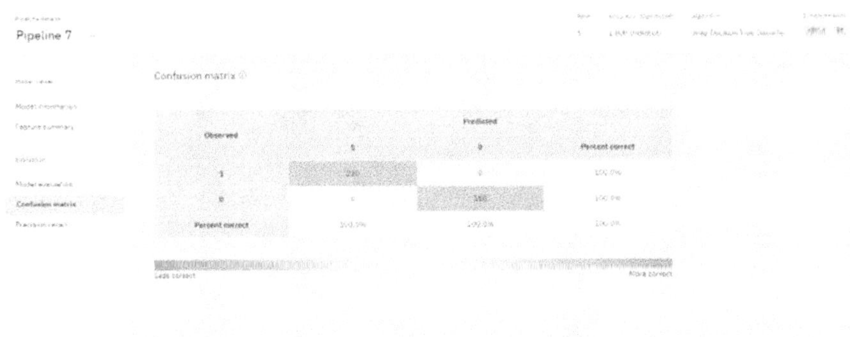

Figure 42. Pipeline 7 precision recall curve

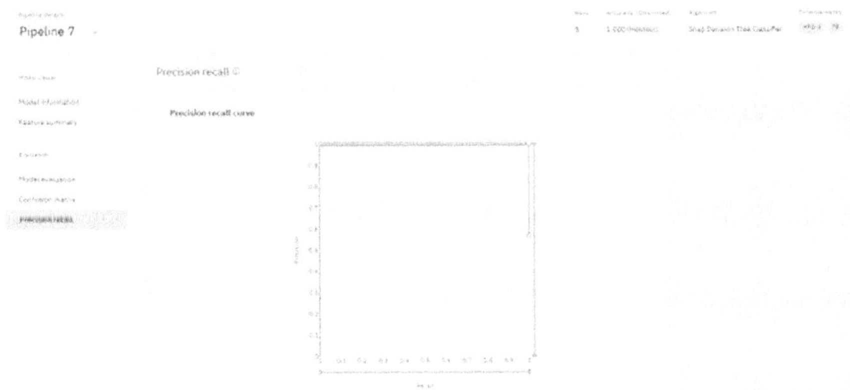

Figure 43. Pipeline 8 model information

Figure 44. Pipeline 8 feature summary

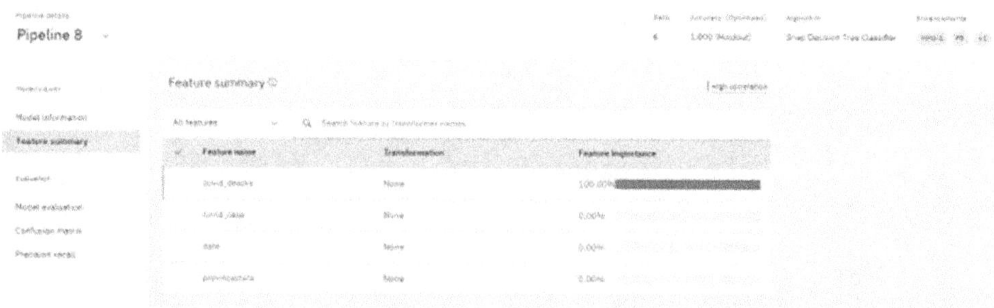

Figure 45. Pipeline 8 receiver operator characteristic (ROC) curve

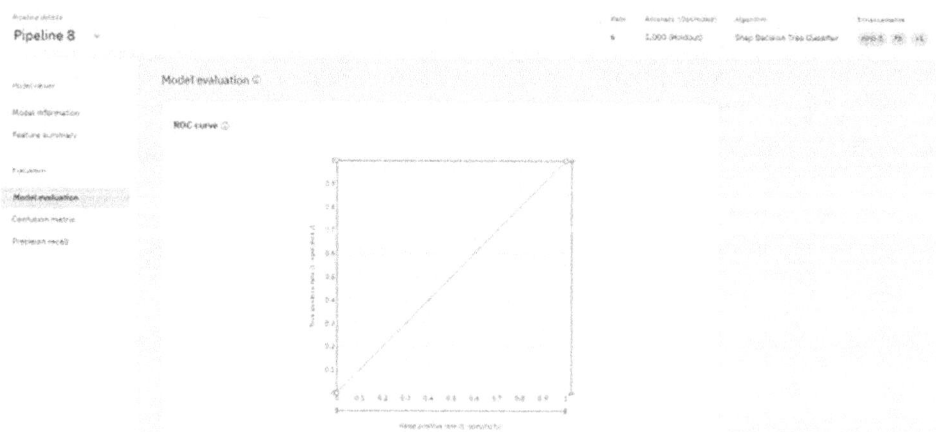

Figure 46. Pipeline 8 confusion matrix

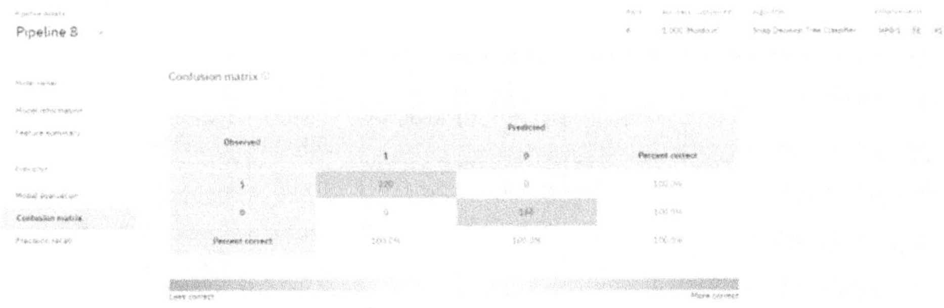

Figure 47. Pipeline 8 precision recall curve

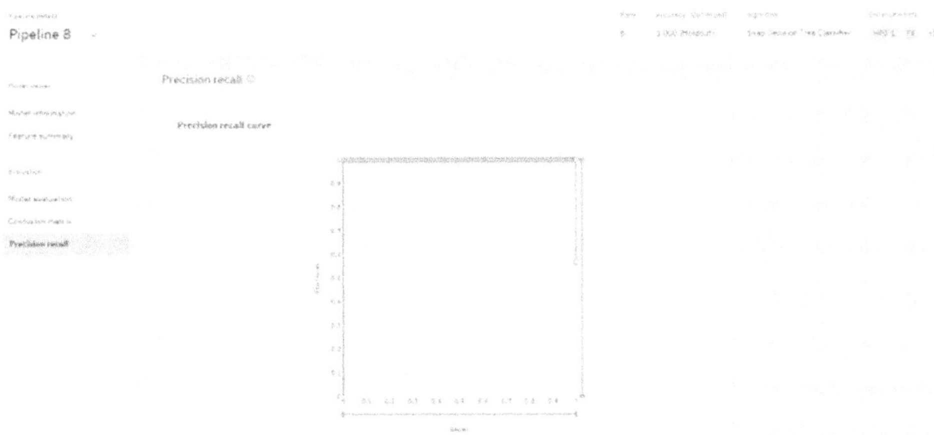

5. CONCLUSION, RECOMMENDATIONS, AND FUTURE WORK

Because of the devastating nature of the epidemic and its quick global spread, COVID-19 has almost brought the planet to an end. In order to forecast the likelihood of mortality among COVID-19 cases in the USA, this report contributes to the body of research on COVID-19 using IBM Watson Auto AI. The effectiveness of the automatic model selection was demonstrated, but due to overfitting, it appeared to be too good to be true. Government officials might utilize IBM's Auto AI to plan and stop upcoming epidemics. In choosing the dataset from the USA for this paper, Long Short-Term Memory neural networks would be used in the future with additional datasets. The findings of the study, will guide development agencies in developing countries, to improve upon the mechanisms used in the aquisition of

Table 8. Ranked pipeline comparison

Pipeline Details	Rank	Algorithm	Number of Features	Model Evaluation Measure-Cross Validation Score				
				Accuracy	Area under ROC	Precision	Recall	F-Score
Pipeline 1	7	Extra trees Classifier	4	0.99	1	0.991	0.991	0.991
Pipeline 2	8	Extra trees Classifier	4	0.99	1	0.991	0.991	0.991
Pipeline 3	1	Extra trees Classifier	8	1	1	1	1	1
Pipeline 4	2	Extra trees Classifier	8	1	1	1	1	1
Pipeline 5	3	Snap Decision Tree Classifier	4	1	1	1	1	1
Pipeline 6	4	Snap Decision Tree Classifier	4	1	1	1	1	1
Pipeline 7	5	Snap Decision Tree Classifier	4	1	1	1	1	1
Pipeline 8	6	Snap Decision Tree Classifier	4	1	1	1	1	1

Figure 48. Progress map of the prediction column

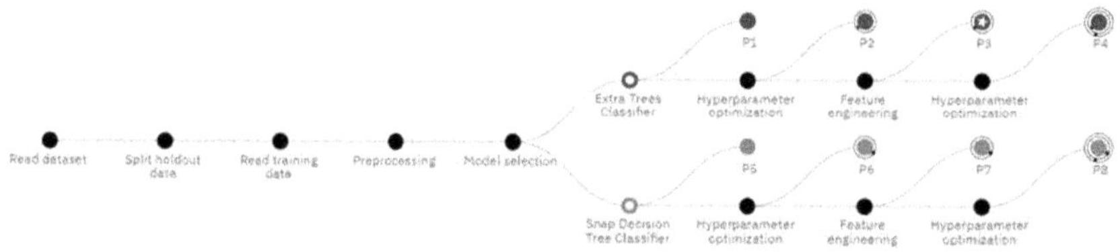

Figure 49. Auto AI pipeline comparison

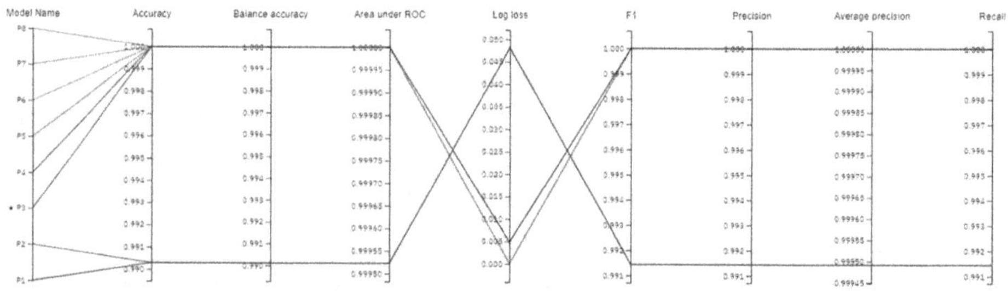

medical datasets. Findings of this study will also contribute to the existing body of knowledge in the fields of Artificial Intelligence and specifically to Medical and Public Health Science. This will to a great extent, improve upon access to preventive healthcare in developing countries. Whilst achieving this, the study will also prompt further investigations into these areas, hence serving as a catalyst for knowledge development. Finally, business oriented organizations, such as consultants and private IT firms, can also take advantage of IBM's Services to develop and build Health related APIs.

REFERENCES

Abujaber, A., Fadlalla, A., Gammoh, D., Abdelrahman, H., Mollazehi, M., & El-Menyar, A. (2020). Prediction of in-hospital mortality in patients with post traumatic brain injury using National Trauma Registry and Machine Learning Approach. *Scandinavian Journal of Trauma, Resuscitation and Emergency Medicine*, *28*(1), 44. doi:10.118613049-020 00738-5 PMID:32460867

Afewerky, H. K. (2020). Pathology and pathogenicity of severe acute respiratory syndrome coronavirus 2 (SARS-CoV-2). *Experimental Biology and Medicine*, *245*(15), 1299–1307. doi:10.1177/1535370220942126 PMID:32635753

Ahmadini, A. A. H., Naeem, M., Aamir, M., Dewan, R., Alshqaq, S. S. A., & Mashwani, W. K. (2021). Analysis and Forecast of the Number of Deaths, Recovered Cases, and Confirmed Cases From COVID-19 for the Top Four Affected Countries Using Kalman Filter. *Frontiers in Physics (Lausanne)*, *9*, 629320. doi:10.3389/fphy.2021.629320

Ahuja, S., Shelke, N. A., & Singh, P. K. (2022). A deep learning framework using CNN and stacked Bi-GRU for COVID-19 predictions in India. *Signal, Image and Video Processing*, *16*(3), 579–586. doi:10.100711760-021-01988-1 PMID:34335985

Al-Rashedi, A., & Al-Hagery, M. A. (2023). Deep Learning Algorithms for Forecasting COVID-19 Cases in Saudi Arabia. *Applied Sciences (Basel, Switzerland)*, *13*(3), 1816. doi:10.3390/app13031816

Brownlee, J. (2014, December 18). How To Work Through A Problem Like A Data Scientist. *Machine Learning Mastery*. https://machinelearningmastery.com/how to work-through-a-pro blem-like-a-data-scientist/

Carnut, L., Mendes, Á., & Guerra, L. (2020). Coronavirus, Capitalism in Crisis and the Perversity of Public Health in Bolsonaro's Brazil. *International Journal of Health Services, 0020731420965137*. doi:10.1177/0020731420965137 PMID:33070682

Derakhshan, F., & Yousefi, S. (2019). A review on the applications of multiagent systems in wireless sensor networks. *International Journal of Distributed Sensor Networks, 15*(5), 1550147719850767. doi:10.1177/1550147719850767

Ganguly, A. (2021). *IBM WATSON SOLUTIONS FOR MACHINE LEARNING achieving successful results across computer... Vision, natural language processing and ai project.* BPB Publications.

Girona, T. (2020). Confinement Time Required to Avoid a Quick Rebound of COVID-19: Predictions From a Monte Carlo Stochastic Model. *Frontiers in Physics (Lausanne), 8*, 186. doi:10.3389/fphy.2020.00186

Haddaway, N. R., Page, M. J., Pritchard, C. C., & McGuinness, L. A. (2022). PRISMA2020: An R package and Shiny app for producing PRISMA 2020-compliant flow diagrams, with interactivity for optimised digital transparency and Open Synthesis. *Campbell Systematic Reviews, 18*(2), e1230. doi:10.1002/cl2.1230 PMID:36911350

Kong, X., Li, C., Zheng, F., & Wang, C. (2020). Improved Deep Belief Network for Short-Term Load Forecasting Considering Demand-Side Management. *IEEE Transactions on Power Systems, 35*(2), 1531–1538. doi:10.1109/TPWRS.2019.2943972

Locey, K. J., Webb, T. A., Khan, J., Antony, A. K., & Hota, B. (2020). An interactive tool to forecast US hospital needs in the coronavirus 2019 pandemic. *JAMIA Open, 3*(4), 506–512. doi:10.1093/jamiaopen/ooaa045 PMID:33619466

Marsland, S. (2014). *Machine Learning: An Algorithmic Perspective* (2nd ed.). Chapman and Hall/CRC. doi:10.1201/b17476

Chapter 4
Spatial Data as a Catalyst to Drive Entrepreneurial Growth and Sustainable Development

Munir Ahmad
https://orcid.org/0000-0003-4836-6151
Survey of Pakistan, Pakistan

ABSTRACT

In this chapter, a model of spatial sustainable entrepreneurship is presented by integrating sustainable development with entrepreneurship and spatial data. The model contributes to the three dimensions of sustainable development. Spatial data can promote economic sustainability by identifying opportunities, optimizing supply chains, and creating jobs. It can contribute to environmental sustainability through resource management and renewable energy projects. Additionally, it can enable social sustainability by addressing social needs and promoting inclusion. Policymakers, entrepreneurs, and researchers can collaborate to support training programs, open data policies, and collaborations to maximize the use of spatial data in sustainable entrepreneurship. Future work can involve advanced analytics, privacy considerations, stakeholder engagement, impact assessment, infrastructure development, cross-disciplinary research, and policy frameworks.

INTRODUCTION

Spatial data denotes information about the physical location and characteristics of objects or phenomena on the Earth's surface (Rajabifard & Williamson, 2001). By evaluating spatial data, entrepreneurs can gain insights into the opportunities and

DOI: 10.4018/978-1-6684-9843-9.ch004

challenges of a particular geographic area and accordingly develop strategies for sustainable growth and entrepreneurship. The use of spatial data has gained momentum due to its potential to offer valuable insights into market trends (Argiolas, 2014; X. Li & Kao, 2022), consumer behavior (Widaningrum et al., 2020), and resource allocation (Shang et al., 2021), as well as its capability to support entrepreneurs to detect new business opportunities, improve their operations, and curtail risks (Cliquet, 2020). The methods for gathering, analyzing, and using spatial data have been significantly improved by technological advancements. Geographic information systems (GIS) development has simplified the process of collecting, storing, analyzing, and visualizing spatial data (D. Li et al., 2016). GIS is a potent tool for spatial analysis in various sectors, including agriculture, health, urban planning, education, and disaster management (Franch-Pardo et al., 2020; Goodchild et al., 1992). In addition, the integration of GIS with other technologies such as remote sensing, artificial intelligence, and machine learning has opened up new possibilities for using spatial data in various industries (Nizeyimana, 2020; Omali & Umoru, 2019; Shabanpour et al., 2022).

Economic development is largely driven by entrepreneurial growth (Prasetyo, 2019), and the use of spatial data can help in this process by enhancing resource allocation, lowering waste, and minimizing environmental impacts. Using spatial data for entrepreneurship has many advantages, one of which is the capacity to explore new company prospects. Entrepreneurs can use spatial data to find undiscovered markets, target particular consumer segments, and create new goods and services (Cliquet, 2020). GIS, for instance, can be used to analyze demographic information and pinpoint locations with a lot of potential clients. Additionally, spatial data can assist business owners in streamlining their operations by lowering transportation costs (Rikalovic et al., 2018), enhancing supply chain management (Karan & Irizarry, 2014), and minimizing waste (Ayoub & Musharavati, 2014).

Another benefit of using spatial data for entrepreneurship is the ability to minimize risks. Entrepreneurs can identify potential risks using spatial data and create mitigation plans for them. GIS, for instance, can be used to analyze natural risks like floods, earthquakes, and wildfires and create backup plans to lessen the effects of these occurrences (Armenakis & Nirupama, 2013; Le Roux et al., 2019). Utilizing spatial data can help in advancing sustainable development. The use of spatial data can support economic growth while also promoting environmental protection by enhancing resource allocation, decreasing waste, and minimizing environmental consequences. GIS, for instance, can be utilized to analyze land use trends and create plans to lessen the effects of urbanization on natural habitats (Ramachandra et al., 2015; Report, 2010).

The use of spatial data for business and sustainable development is not without its difficulties, though. The accessibility and quality of the data present a significant barrier. Collecting spatial data can be time-consuming and expensive, and factors like sensor resolution, weather, and data processing techniques can all have an impact accuracy of data (Basiri et al., 2019; H. Li et al., 2022). Another challenge is the legal and regulatory framework governing the use of spatial data (Ali et al., 2021). Entrepreneurs must comply with a range of laws and regulations governing data privacy, intellectual property rights, and environmental protection.

Against this backdrop, the main objective of this chapter is to provide a comprehensive overview of the role of spatial data in driving entrepreneurial growth for sustainable development. To achieve the objective, a model of spatial sustainable entrepreneurship is developed that demonstrates the role of spatial data as a catalyst in the three dimensions of sustainable entrepreneurship. The model is based on the three pillars of sustainable development i.e., economic, social, and environmental.

The chapter is organized as follows. Section 2 highlights the connection between spatial data and entrepreneurial growth. Section 3 describes the model of spatial sustainable entrepreneurship. Section 4 designates recommendations for policymakers, researchers, and entrepreneurs. The last section serves as the conclusion of the chapter along with future work.

SPATIAL DATA AND ENTREPRENEURIAL GROWTH

Businesses and entrepreneurs are leveraging spatial data to better understand consumer behaviour, streamline processes, and spot untapped market potential. This section will examine the value of spatial data for entrepreneurship, with a focus on how it might enhance market intelligence and decision-making. Let's examine spatial data's significance in decision-making and market intelligence in more detail to go deeper.

Importance of Spatial Data in Decision-Making and Market Intelligence

By offering valuable insights into the physical environment in which businesses function, spatial data can influence decision-making and market intelligence. Entrepreneurs can leverage spatial data to impact decisions on where to place their companies, which markets to focus on, and how to best run their operations. One of the main advantages of spatial data is its capacity to give business owners a thorough understanding of the competitive landscape. Entrepreneurs may learn where potential clients are located as well as information about their demographics and behaviour by analyzing spatial data. Spatial data, for instance, can be utilized to locate high-density population areas, comprehend client commute patterns, and evaluate the intensity of competition in a specific area (Birkin, 2019).

Entrepreneurs can use spatial data to guide their decision-making capabilities. Entrepreneurs can determine regions with strong demand for their goods or services, analyze the availability of resources like water and electricity, and assess the accessibility of transportation infrastructure by analyzing spatial data (Krohn et al., 2022; Wan et al., 2021). Furthermore, by enhancing productivity, lowering costs, and improving logistics, geographical data can be utilized to optimize corporate processes. For instance, supply chains can be made more effective by using spatial data to pinpoint the best delivery and transit routes. Entrepreneurs may optimize inventory levels and reduce waste by using spatial data to predict demand trends (Ciampi et al., 2020).

Spatial data may help business owners monitor and manage their operations in real time in addition to offering insights into the market and the physical environment. Spatial data, for instance, can be used to monitor the whereabouts and movement of assets like machinery and vehicles. Entrepreneurs can take proactive steps to reduce risks by using spatial data to monitor environmental factors like temperature and humidity (Bunce et al., 2008). For entrepreneurs, spatial data is an essential part of market intelligence and decision-making. Spatial data may assist entrepreneurs in making knowledgeable decisions, lowering costs, and increasing efficiency, promoting economic growth and sustainable development. It does this by offering insights into the market landscape, optimizing business operations, and enabling real-time monitoring and management. Table 1 shows some potential spatial data applications for decision-making and market intelligence.

Table 1. Role of spatial data for entrepreneurship

Spatial Data Application	Relevance with Entrepreneurship
Site selection for new business	Spatial data can help to identify optimal business locations in combination with different factors such as demographic, economic, and geographic data.
Customer profiling and segmentation	Spatial data can assist businesses in comprehending their customer base and to target marketing strategies more effectively.
Supply chain optimization	Spatial data can help businesses to optimize their supply chain by identifying the most efficient transportation routes and distribution centers.
Real estate market analysis	Spatial data can help real estate businesses to realize market trends and to develop informed decisions.
Environmental impact assessment	Spatial data can support businesses in recognizing the potential environmental effects of their operations and observing regulatory laws

MODEL OF SPATIAL SUSTAINABLE ENTREPRENEURSHIP

This section describes the process of developing a model that integrates spatial data with sustainable entrepreneurship. The development of the model involves four distinct steps. Subsequent sub-sections provided detailed explanations of each step.

Components of Sustainable Development

The first critical step in developing the model of spatial sustainable entrepreneurship is to define the components of sustainable development. These components are derived from the three pillars of sustainable development: economic, environmental, and social (Ranjbari et al., 2019), as shown in Figure 1.

Figure 1 demonstrates the three dimensions of sustainable development that will be utilized for the development of a spatial sustainable entrepreneurship model. The first dimension, economic sustainability, addresses the public's demand for achieving sustainable economic growth. The second component, environmental sustainability, emphasizes utilizing natural resources without causing harm to the environment. Fulfilling the people's fundamental needs, including social welfare, healthcare, and education, is referred to as "social sustainability".

Figure 1. Components of sustainable development

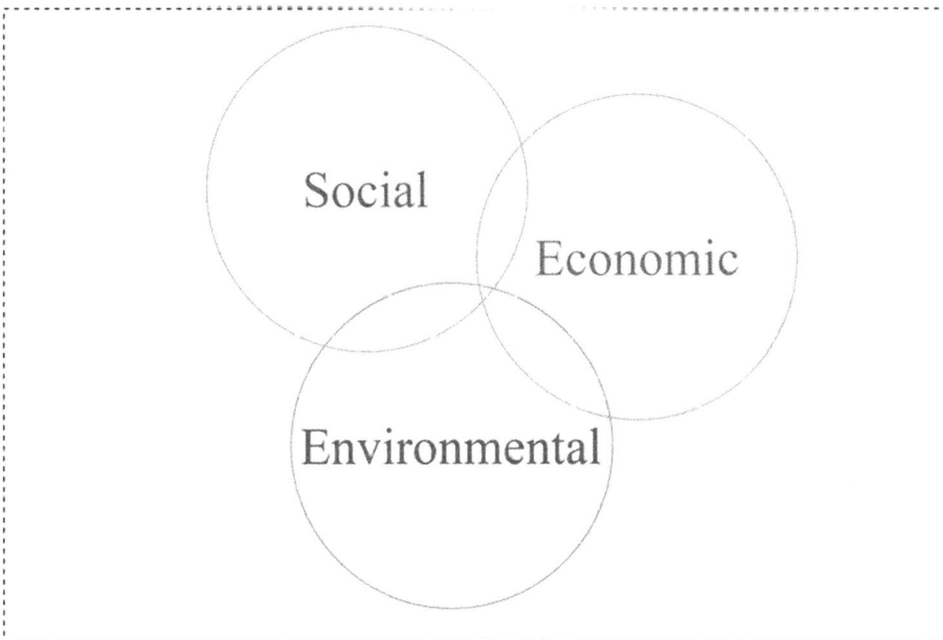

Relationship Between Sustainable Development and Entrepreneurship

Establishing the connections between the three facets of sustainable development and entrepreneurship is the second step. This is accomplished by looking at how entrepreneurship can help achieve sustainable development objectives while also balancing economic, environmental, and social sustainability. Different roles can be played by entrepreneurship in accomplishing sustainable development objectives. By creating new businesses, products, or services with considerations for economic, environmental, and social sustainability, entrepreneurship can make significant contributions towards sustainable development goals. The contribution of entrepreneurship to the three components of sustainable development is elaborated as follows.

Economic Sustainability

Entrepreneurs are involved in various activities to contribute to economic sustainability. They can establish new businesses to meet customer demands and promote sustainable economic growth. Additionally, entrepreneurs can develop new employment prospects within their organizations to support a sustainable economy. Entrepreneurs must, however, take care to compromise social or environmental sustainability in the pursuit of economic sustainability.

Environmental Sustainability

Entrepreneurs need to evaluate the potential impacts of their new businesses or products on the environment while working on environmental sustainability. Entrepreneurs need to adopt practices that favour the minimizing the of environmental footprint of their ventures. These practices can include utilizing sustainable materials, reducing waste generation, and effectively managing hazardous emissions. Furthermore, to establish environmentally sustainable businesses, entrepreneurs need to inspire businesses such as those specializing in eco-tourism, sustainable agriculture, the green economy, or renewable energy. By incorporating these strategies, entrepreneurs can make meaningful contributions to a more sustainable future.

Social Sustainability

Entrepreneurs play a substantial role in promoting social sustainability when establishing new businesses. They can contribute to social sustainability by prioritizing and enhancing vital social services such as healthcare and education.

Table 2. Sustainable entrepreneurship

Sustainability Components	Entrepreneurship Contribution
Economic Sustainability	• Establishing new businesses • Creating new jobs • Addressing consumer demands for products and services. • Supporting enduring, financially efficacious, and sustainable businesses.
Environmental Sustainability	• Establishing businesses that have minimal impact on the environment. • Establishing businesses that support environmental sustainability.
Social Sustainability	• Establishing businesses that support social services. • Creating new jobs for marginalized groups.

Additionally, entrepreneurs can support social sustainability by providing new job opportunities, particularly for marginalized groups. By actively addressing social needs and fostering inclusivity, entrepreneurs can make substantial contributions to social sustainability.

The possible contributions of entrepreneurship toward three pillars of sustainable development to achieve wide-ranging and stable sustainable entrepreneurship are summarized in Table 2.

However, entrepreneurs must consider the delicate balance between economic, environmental, and social sustainability when striving to achieve sustainable development. Therefore, entrepreneurs need to establish businesses and launch new products that are not only financially viable but also environmentally responsible and socially inclusive.

Role of Spatial Data in Sustainable Entrepreneurship

The third step entails establishing the roles of spatial data in sustainable entrepreneurship. This is accomplished by examining how spatial data contributes to sustainable entrepreneurship and supports the achievement of sustainable development goals. Spatial data plays a crucial role in fostering economic, environmental, and social sustainability within the realm of entrepreneurial activities. The potential contribution of spatial data to sustainable entrepreneurship in terms of economic, environmental, and social sustainability is elaborated as follows:

Economic Sustainability

Spatial data possesses numerous capabilities that support economic sustainability within the realm of sustainable entrepreneurship. For instance, it aids in gaining insights into identifying new market opportunities and determining specific geographic regions with high consumer demand for particular products or services. Entrepreneurs can leverage these insights to expedite business opportunities and foster economic growth. The potential contributions of spatial data to sustainable entrepreneurship, specifically in terms of economic sustainability, are elaborated as follows:

Market Analysis: Spatial data can be effectively utilized for market analysis and serve as a lens to magnify and identify specific geographic regions with a high demand for targeted products and services. Entrepreneurs can leverage spatial data to make informed decisions regarding the establishment or expansion of businesses within specific geographic boundaries, offering products or services that cater to the needs of the targeted market in those areas (Wieland, 2017).

Supply Chain Management: Effective supply chain management relies on identifying optimal transportation routes and logistics tracks to minimize costs and enhance efficiency. Spatial data plays a crucial role in achieving this by enabling network analysis of transportation networks. By analyzing spatial data, businesses can identify the most effective routes, transportation modes, and locations of warehouses (Yusianto et al., 2020). This not only reduces transportation and logistics costs but also improves the overall sustainability of the supply chain by reducing carbon footprint.

Job Creation: Spatial data holds the unique ability to pinpoint geographic areas with high unemployment rates. This information is valuable for entrepreneurs and businesses looking to create job opportunities and contribute to economic growth in those areas. By analyzing spatial data on job availability and unemployment rates, entrepreneurs can strategically establish or expand businesses in regions that would benefit from increased employment opportunities. This, in turn, helps to boost local economies, improves the living standards of people, and promotes sustainable development in the area (Pinto et al., 2023).

Agricultural Efficiency: The application of spatial data can greatly enhance the efficiency of the agricultural sector, leading to economic growth (Singh et al., 2022). For instance, businesses can leverage spatial data to monitor soil quality, identify areas for improvement, and implement precision agriculture techniques. By analyzing spatial data on soil composition, moisture levels, and other factors, farmers can optimize irrigation patterns, lessen water usage, and improve crop yields.

Real Estate Development: The real estate industry serves as a significant driver of economic activity in specific areas. By conducting feasibility studies using spatial data, real estate developers can enhance their understanding of the market. Spatial data enables developers to identify underutilized urban areas with untapped potential (Rabiei-Dastjerdi et al., 2021). By combining spatial data with demographic and infrastructure information, entrepreneurs can devise effective strategies to improve affordable housing or commercial spaces in the targeted area, unlocking the economic growth of the local communities.

Small Business Support: Smaller businesses often face challenges in accessing the necessary resources compared to larger enterprises. Leveraging spatial data allows for the identification of regions with low or high concentrations of small businesses. This data-driven approach empowers authorities to plan and allocate resources effectively to support smaller businesses in need. By leveraging spatial data, policymakers and organizations can foster an environment that promotes economic growth and sustains entrepreneurship among small businesses (Iskandar et al., 2023).

Infrastructure Planning: By combining spatial data with other factors such as population density, traffic patterns, and transportation networks, governments can gain valuable insights into areas with limited infrastructure development (Brandt et al., 2022). This information assists in planning the placement of new transportation infrastructure. For instance, identifying areas with high traffic congestion allows the government to prioritize infrastructure improvements, stimulating economic activities in those regions.

Tourism: Tourism serves as a vital economic driver in almost every part of the world. Entrepreneurs involved in travel and tourism businesses can utilize spatial data to develop new, sustainable tourism plans. Spatial data enables the analysis of specific geographic areas to identify locations that offer eco-friendly experiences, cultural immersion, or other unique attractions (Liu & Gao, 2022). By leveraging spatial data, entrepreneurs not only generate revenue for their businesses but also contribute to the economic growth and preservation of cultural heritage in local communities.

Disaster Response: During natural disasters, spatial data plays a crucial role in identifying affected areas and planning rescue and rehabilitation services. Real-time spatial data from satellites, weather stations, and other sources can be integrated to explore the particular locations and magnitude of disasters such as floods, wildfires, or earthquakes (Hu et al., 2019). This empowers prompt response, immediate relief efforts, minimal economic losses, and swift recovery.

Energy Efficiency: Entrepreneurs engaged in energy management services can leverage spatial data to identify geographical regions with low or high energy consumption (Aldhshan et al., 2021). By combining spatial data with factors such as energy usage, building densities, and infrastructure planning, they can develop strategies to meet energy demands efficiently. The implementation of energy-efficient practices not only benefits individual businesses but also contributes to broader environmental sustainability and economic efficiency (Aldhshan et al., 2021).

Environmental Sustainability

Spatial data can be utilized to achieve environmental sustainability within the context of sustainable entrepreneurship. For example, it can assist in the management and monitoring of various natural resources, including water, forests, minerals, mines, wildlife, and more. Entrepreneurs can harness this valuable spatial information to establish and enhance their businesses focused on sustainable resource management. The potential contributions of spatial data to sustainable entrepreneurship, particularly in terms of environmental sustainability, are elaborated as follows:

Sustainable Resource Management: Incorporating spatial data into the management and monitoring of natural resources can enhance effectiveness (Chaula, 2019). For example, spatial data can be leveraged for the analysis of the availability of water resources, forest cover, and wildlife habitats to enable data-driven decisions for the conservation of these valuable resources. By implementing, these data-driven strategies, entrepreneurs can establish economically and environmentally sustainable businesses.

Renewable Energy: Spatial data offers valuable insights into areas with high potential for renewable energy sources, such as solar or wind power. Leveraging the power of spatial data, entrepreneurs can identify optimal locations to establish and promote renewable energy projects (Aldhshan et al., 2021). For example, spatial analysis can be employed to assess solar radiation, wind patterns, and land suitability, enabling informed decisions regarding the installation and operation of renewable energy infrastructure for environmentally friendly solutions.

Green Infrastructure: Spatial data analysis plays a crucial role in identifying suitable areas for the implementation of green infrastructure, thereby promoting environmental sustainability (Firehock & Walker, 2019). By analyzing spatial data on land use and water resources, entrepreneurs can locate regions ideal for green infrastructure initiatives, such as green roofs, rain gardens, or permeable pavements. Implementing these practices allows for the reduction of stormwater runoff, improvement of air quality, and mitigation of urban heat, thereby contributing to the enhancement of environmental conditions in urban areas.

Reducing Carbon Footprint: Entrepreneurs can contribute to reducing their carbon footprint by utilizing spatial data to optimize logistic networks, thereby reducing fuel consumption and greenhouse gas emissions. For example, spatial data can assist in optimizing traffic patterns, distances, and delivery destinations, resulting in increased overall operational efficiency and a reduced carbon footprint.

Sustainable Agriculture: Business ventures involved in sustainable agriculture can employ spatial data, along with other factors such as soil health, topography, and climate data, to improve farming methods, leading to better crop yields, reduced water usage, and minimized reliance on toxic chemicals (Singh et al., 2022). This effective resource allocation contributes to a sustainable economy.

Wildlife Conservation: Spatial patterns can be developed to track and record wildlife behavior and migration trends, aiding in the protection of endangered species (Y. Hu et al., 2021). By utilizing data-driven analytics, entrepreneurs and governments can formulate conservation strategies to preserve and restore vital wildlife ecosystems.

Waste Management: Spatial data is used to identify garbage dumping sites, track waste disposal trends, minimize waste generation, establish garbage collection sites, and set up recycling facilities. These measures improve waste management policies and promote sustainable waste reduction and management methods (Singh, 2019).

Ocean Conservation: Spatial data can be leveraged to monitor ocean tides, water quality, and marine biodiversity for conservation purposes. Analyzing satellite imagery along with other datasets, such as sensor data, helps address challenges of ocean conservation (Sarker et al., 2019) related to pollution levels, harmful algal blooms, and fishing activities, leading to better solutions for preserving ocean resources for future generations.

Environmental Education: Spatial data can enhance capacity by facilitating the creation of interactive educational materials that promote environmental awareness and conservation methods. It improves educational opportunities, cultivates a sense of environmental stewardship, and motivates future generations to take action for a sustainable future.

Social Sustainability

Spatial data plays a significant role in achieving social sustainability within the realm of sustainable entrepreneurship. It can be effectively utilized to locate and identify geographical areas where there is a need for the establishment of essential social services, such as healthcare and education, among others. Entrepreneurs can harness this valuable spatial information to establish or enhance businesses related to social services. The potential contributions of spatial data to sustainable entrepreneurship, specifically in terms of social sustainability, are elaborated as follows:

Healthcare Services: Identifying regions with limited access to healthcare services is crucial for social sustainability. Entrepreneurs and governments can utilize spatial data insights to establish and improve healthcare facilities for underserved communities. Spatial data contributes to social sustainability by ensuring that quality healthcare is accessible to all (Diggle, 2005).

Educational Services: Spatial data analysis empowers entrepreneurs to identify geographical areas with limited access to quality educational services and address educational gaps (Muzaffar et al., 2017). Entrepreneurs can leverage this information to create and improve businesses that offer educational services to underserved communities, contributing to social sustainability by empowering communities through education.

Social Inclusion: Spatial data analysis enables entrepreneurs and governments to identify areas with high poverty concentrations. By utilizing spatial data, entrepreneurs can promote social inclusion by providing employment opportunities and developing products or services tailored to the needs of underprivileged communities. Spatial data analysis reduces inequalities and encourages diversity, contributing to social sustainability (Akinyemi, 2007).

Elder Care: Businesses involved in elder care can leverage spatial data to identify areas with a high concentration of senior citizens. By analyzing population statistics, healthcare facilities, and accessibility, entrepreneurs can establish eldercare facilities or provide specialized care and support services to improve social sustainability (Norouzi et al., 2019).

Community Development: Spatial data can be used to identify areas with inadequate housing or infrastructure facilities. By analyzing housing conditions, infrastructure availability, and community resources, better strategies and practices can be devised for socially sustainable infrastructure development and community empowerment.

Disaster Relief: By integrating spatial data analysis with natural disaster recovery programs and emergency response plans, the social sustainability of communities can be enhanced. The spatial analysis provides insights into historical disaster patterns, population density, and critical infrastructure, helping minimize the impact of disasters and save lives (Jitt-Aer et al., 2022).

Mental Health Services: Spatial analysis can be utilized to recognize areas with limited access to mental health services. By analyzing spatial data on mental health facility locations, population demographics, and connectivity, entrepreneurs can improve mental health services in specific geographical locations (Diggle, 2005).

Cultural Preservation: Entrepreneurs can employ spatial data to locate areas with rich cultural heritage and develop initiatives to preserve and promote the cultural value of those localities. By analyzing cultural sites, historical landmarks, and tourism patterns, better strategies for heritage conservation and tourism can be devised (Zabulis et al., 2020).

Gender Equality: Spatial data analysis can help to classify areas with diverse gender inequalities. By conducting spatial analysis of gender ratios, employment rates, and access to resources, entrepreneurs can develop and improve businesses that provide employment opportunities, skill development programs, and support services to promote gender equality (Jo et al., 2020).

The diverse roles of spatial data in supporting sustainable entrepreneurship in terms of social, economic, and environmental sustainability are presented. Figure 2 provides a visual representation of these roles of spatial data in sustainable development, offering a comprehensive overview.

Figure 2 illustrates how spatial data can support entrepreneurship in achieving sustainability across three key aspects: economic, social, and environmental. The detailed insights presented in Figure 2 highlight the potential roles of spatial data

Figure 2. Role of spatial data for sustainable entrepreneurship

in entrepreneurship, offering valuable suggestions for establishing, improving, and sustaining businesses. By leveraging spatial data, entrepreneurs can make informed decisions in business development while simultaneously addressing economic viability, environmental concerns, and social inclusiveness. Spatial data provides opportunities to balance these crucial factors and promote sustainable entrepreneurial practices.

Developing the Model

The final step entails the development of a model of spatial sustainable entrepreneurship by integrating the findings derived from the preceding three steps. This is achieved by connecting the three components of sustainable development with the roles of entrepreneurship and establishing a link between spatial data and sustainable entrepreneurship. Figure 3 elaborates on the model of spatial sustainable entrepreneurship, illustrating spatial data as a catalyst that drives entrepreneurial growth toward the attainment of sustainable development goals.

Figure 3 provides a concise integration of the key concepts developed in the four steps of model development and depicts the interrelationship between spatial data, entrepreneurship, and sustainable development. The significance of spatial data for sustainable entrepreneurship becomes evident when examining the presented model. The development of this model can be regarded as an initial step toward further exploration of the potential of spatial data in facilitating sustainable entrepreneurship development. The model serves to inform policymakers about the knowledge in the field of sustainable entrepreneurship and the achievement of sustainable development goals through the utilization of spatial data inputs.

Figure 3. Model of spatial sustainable entrepreneurship

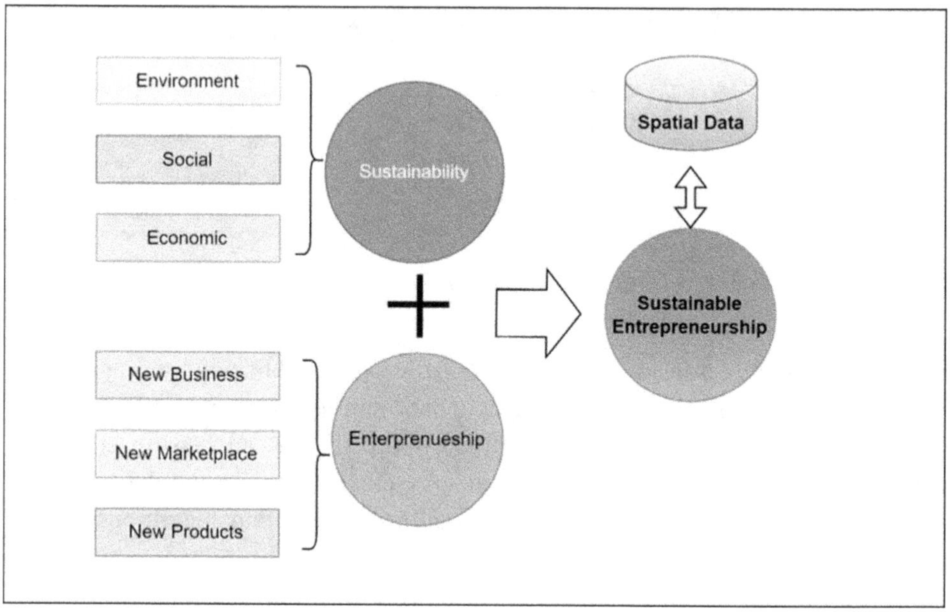

RECOMMENDATIONS

Spatial data holds immense potential for enhancing sustainable entrepreneurship. Building upon the results of the previous section, this section provides detailed recommendations for policymakers, entrepreneurs, and researchers. These recommendations aim to leverage the power of spatial data to drive sustainable entrepreneurial practices forward.

Policymakers' Recommendations

The potential of spatial data for policymakers to contribute to sustainable development and entrepreneurial growth is undeniable. Policymakers can play an instrumental role in harnessing the power of spatial data to address various policy issues. They can devise policies that facilitate the effective collection, management, sharing, and utilization of spatial data. To this end, the following key recommendations are proposed for policymakers:

Using Spatial Data in Business Planning

Policymakers can facilitate the utilization of spatial data in business planning by implementing effective mechanisms. By endorsing the provision of training programs and allocating resources to entrepreneurs, policymakers can support them in understanding the significance of business planning and harnessing spatial data for sustainable entrepreneurship. These efforts can help entrepreneurs magnify environmentally and socially responsible business opportunities, thereby contributing to both economic growth and sustainable development.

Developing Open Data Policies

Open data policies are great promoters of spatial data sharing among different stakeholders, including entrepreneurs, researchers, academia, and the general public. Policymakers can advocate for the development of open data policies focused on sharing unclassified spatial data. These policies can encourage transparency, collaboration, and innovation through the use of readily accessible spatial data.

Establishing Data Interoperability Standards

Spatial data often faces interoperability challenges that can limit its use among various users. Policymakers can promote the development of data interoperability standards to facilitate the sharing and integration of spatial data with different formats across different platforms. In this way, data interoperability standards can help reduce data duplication practices, improve data quality, and enhance decision-making.

Promoting Data Privacy and Security

Spatial data often represents classified information; thus, it needs to be protected to avoid misuse. Policymakers can endorse the regular use of data privacy and data security regulations to control the collection, use, and sharing of classified spatial data.

Encouraging Public-Private Partnerships

Public-private partnerships have the potential for the effective use of spatial data in entrepreneurial growth and sustainable development. Policymakers can boost the formation of such partnerships by advocating for governments to provide incentives, such as funding opportunities.

Supporting Spatial Data Education and Training

Specialized technical skills and knowledge are prerequisites for conducting spatial data analysis. Policymakers can encourage training programs to prepare entrepreneurs, researchers, and policymakers to collect, manage, analyze, share, and visualize spatial data in their work or tasks. This can help build a skilled workforce capable of effectively utilizing spatial data for entrepreneurship and sustainable development.

Investing in Spatial Data Infrastructure

Spatial data infrastructure serves as a medium to make spatial data available to a wider audience through information and communication technologies. Policymakers can advocate for investing in the development of spatial data infrastructure to support data collection, storage, and processing facilities in the context of entrepreneurship and sustainable development.

Hence, there is a significant role for policymakers in maximizing the utilization of spatial data for sustainable development and entrepreneurial growth. The aforementioned recommendations serve as guidelines to assist policymakers in formulating policies that fully harness the potential of spatial data. By implementing these recommendations, policymakers can promote innovation, foster economic growth, and drive social progress through the effective utilization of spatial data.

Entrepreneurs' Recommendations

Entrepreneurs have the opportunity to apply innovative spatial business models and strategies to drive sustainable development. To assist entrepreneurs in fully harnessing the potential of spatial data in their entrepreneurial endeavors, the following recommendations have been formulated:

Acquire the Necessary Skills and Expertise

To effectively harness the benefits of spatial data, entrepreneurs need to equip themselves with skills and expertise in spatial data analysis, visualization, and interpretation. This involves gaining knowledge about spatial data collection, analysis, visualization, sharing, and interpretation.

Foster Collaborations and Partnerships

Entrepreneurs should collaborate with both public and private sectors to share expertise and resources, maximizing the benefits of spatial data. They are also encouraged to partner with academia and research institutions to develop skills in cutting-edge spatial data research and development.

Advocate for Open Data Policies

Entrepreneurs should utilize open data policies to facilitate the availability and accessibility of spatial data in research, development, and innovation practices. This includes advocating for the use of agreed-upon standards in spatial data collection and storage to promote interoperability and sharing of spatial data.

Address Ethical and Legal Considerations

Entrepreneurs should be mindful of ethical and legal considerations related to the use and sharing of spatial data. They should encourage compliance with relevant spatial data laws, rules, and regulations.

Embrace Emerging Technologies

Entrepreneurs need to equip themselves with the latest emerging technologies and understand their potential applications for spatial data. They should be prepared to capitalize on new emerging technologies to facilitate effective spatial data collection, analysis, and interpretation.

Utilize Spatial Data for Market Intelligence

Entrepreneurs should leverage spatial data to gain insights into market trends, customer behavior, and competitive landscapes to make data-driven decisions and develop effective business strategies.

Therefore, entrepreneurs can play a significant role in maximizing the utilization of spatial data for sustainable entrepreneurial growth. The aforementioned recommendations can assist entrepreneurs in exploring emerging technologies, addressing ethical considerations, developing efficient business strategies, and many more. By implementing these recommendations, entrepreneurs can unlock the full potential of spatial data and drive sustainable entrepreneurial success.

Researchers' Recommendations

The potential of spatial data for researchers to contribute to sustainable development and entrepreneurial growth is undeniable. Researchers play a crucial role in proposing innovative solutions to maximize the utilization of spatial data in entrepreneurial growth and sustainable development. To guide researchers in their endeavors, the following key recommendations are put forth:

Develop Standardized Data Formats and Models

Compatibility and interoperability between various spatial datasets are crucial for the effective use of spatial data. Researchers need to create standardized data formats and models to facilitate data interoperability.

Conduct Research on Emerging Technologies

Researchers should undertake the development of innovative applications using cutting-edge technologies such as blockchain, virtual reality, and artificial intelligence to effectively utilize spatial data for sustainable development.

Develop Data Analysis Tools

Handling vast and complex spatial datasets presents a challenge. Researchers should develop advanced geographic dataset processing applications to make the most of various formats of spatial data. This will enable entrepreneurs and decision-makers to make informed decisions even when dealing with intricate spatial datasets.

Encourage Open Data Policies

Researchers should advocate the development of open data policies. Such policies will facilitate all stakeholders to use, share, and interpret spatial data in various decision-making tasks.

Foster Interdisciplinary Collaborations

Researchers should promote interdisciplinary interactions between various domains such as geography, economics, and computer science to develop innovative spatial applications.

Hence, researchers hold a significant position in shaping the utilization of spatial data for sustainable entrepreneurial growth. The aforementioned recommendations can assist researchers in encouraging open data policies, fostering interdisciplinary collaborations, developing data analysis tools, and much more. By implementing these recommendations, researchers can expose the full value of spatial data and drive sustainable entrepreneurship.

The presented recommendations serve as a guide for policymakers, entrepreneurs, and researchers to facilitate the transformation of entrepreneurial growth and sustainable development. It is crucial for these stakeholders to collaborate and collectively address the challenges while leveraging the opportunities presented by spatial data. By implementing the right policies, embracing appropriate technologies, and fostering collaborations, spatial data can become a powerful catalyst for driving entrepreneurial growth and sustainable development.

CONCLUSION

Spatial data is indispensable for decision-making and market intelligence, as it offers information on the location and features of various markets and potential clients in the context of entrepreneurship development. It is a vital tool for entrepreneurs to make informed decisions when launching new ventures or expanding their current business setups. For example, spatial data can assist entrepreneurs in determining the most lucrative sites for their operations, comprehending the traits of their target audiences, and monitor the activities of their competitors. Moreover, spatial data can aid entrepreneurs in optimizing their supply chains, reducing operating expenses, and enhancing the overall effectiveness of their business operations.

A model of spatial sustainable entrepreneurship has been developed by integrating the three dimensions of sustainable development with entrepreneurship and spatial data. This model provides valuable insights into the interplay between these dimensions. Spatial data is instrumental in promoting economic sustainability within sustainable entrepreneurship. It enables entrepreneurs to identify market opportunities, optimize supply chains, create jobs, enhance agricultural efficiency, drive real estate development, support small businesses, plan infrastructure, boost tourism, facilitate disaster response, and promote energy efficiency. By leveraging spatial data, entrepreneurs can make informed decisions, drive economic growth, and contribute to sustainable development.

Spatial data also plays a crucial role in achieving environmental sustainability within sustainable entrepreneurship. Entrepreneurs can utilize spatial data for sustainable resource management, renewable energy projects, green infrastructure initiatives, reducing carbon footprint, sustainable agriculture, wildlife conservation, waste management, ocean conservation, and environmental education. By leveraging spatial data, entrepreneurs can make informed decisions, implement environmentally friendly practices, and contribute to a sustainable future.

Moreover, spatial data is inevitable in achieving social sustainability in sustainable entrepreneurship. Entrepreneurs can utilize spatial data to identify areas in need of essential social services like healthcare and education. By leveraging spatial data, entrepreneurs can establish businesses that address these social needs, contributing to social sustainability. Spatial data enables the improvement of healthcare and educational services, promotes social inclusion, supports elder care, facilitates community development, aids in disaster relief efforts, enhances mental health services, preserves cultural heritage, and promotes gender equality. Through the use of spatial data, entrepreneurs can foster social equity and create businesses that positively impact communities.

To effectively exploit spatial data in business planning and foster entrepreneurship and sustainable development, policymakers should devise mechanisms such as endorsing training programs, developing open data policies, establishing interoperability standards, and encouraging public-private partnerships. These efforts enable informed decision-making, collaboration, and innovation, leading to economic growth and sustainable development. Entrepreneurs need to acquire skills in spatial data analysis, visualization, and interpretation to effectively utilize its benefits. Collaborations with public and private sectors, academia, and research institutions can maximize spatial data's advantages. Advocating for open data policies ensures accessibility while addressing ethical and legal considerations. Embracing emerging technologies enhances spatial data collection and analysis. Leveraging spatial data for market intelligence empowers entrepreneurs to make data-driven decisions and develop successful business strategies. Researchers should develop standardized data formats and models to enhance compatibility among spatial datasets. Exploring emerging technologies like blockchain, virtual reality, and artificial intelligence can lead to inventive applications utilizing spatial data. Advanced geographic dataset processing tools enable informed decision-making. Open data policies encourage spatial data utilization and collaboration among stakeholders. Interdisciplinary collaborations between geography, economics, and computer science drive innovative spatial applications.

Future work in the field of spatial sustainable entrepreneurship should focus on integrating advanced spatial data analytics techniques, addressing privacy and security concerns, fostering stakeholder collaboration and engagement, developing comprehensive impact assessment frameworks, advancing spatial data infrastructure, encouraging cross-disciplinary research, and establishing policy and governance frameworks. By exploring these areas, the full potential of spatial data in fostering entrepreneurship and sustainable development, driving economic growth, environmental sustainability, and social equity, can be unlocked.

REFERENCES

Akinyemi, F. O. (2007). Spatial data needs for poverty management. *Research and Theory in Advancing Spatial Data Infrastructure Concepts, 5.*

Aldhshan, S. R. S., Abdul Maulud, K. N., Wan Mohd Jaafar, W. S., Karim, O. A., & Pradhan, B. (2021). Energy consumption and spatial assessment of renewable energy penetration and building energy efficiency in Malaysia: A review. *Sustainability (Basel), 13*(16), 9244. doi:10.3390u13169244

Ali, A., Imran, M., Jabeen, M., Ali, Z., & Mahmood, S. A. (2021). Factors influencing integrated information management: Spatial data infrastructure in Pakistan. *Information Development*. doi:10.1177/02666669211048483

Argiolas, M. (2014). Surveying housing market supply affordability using a spatial data mining approach. *WIT Transactions on Ecology and the Environment, 191*, 125–137. doi:10.2495/SC140111

Armenakis, C., & Nirupama, N. (2013). Prioritization of disaster risk in a community using GIS. *Natural Hazards, 66*(1), 15–29. Advance online publication. doi:10.100711069-012-0167-8

Ayoub, N., & Musharavati, F. (2014). A Future Prospect for Domestic Waste Management in Qatar. *International Conference on Earth, Environment and Life Sciences (EELS-2014)*. 10.15242/IICBE.C1214080

Basiri, A., Haklay, M., Foody, G., & Mooney, P. (2019). Crowdsourced geospatial data quality: challenges and future directions. In International Journal of Geographical Information Science, 33(8). doi:10.1080/13658816.2019.1593422

Birkin, M. (2019). Spatial data analytics of mobility with consumer data. *Journal of Transport Geography, 76*, 245–253. doi:10.1016/j.jtrangeo.2018.04.012

Bordogna, G., Kliment, T., Frigerio, L., Brivio, P. A., Crema, A., Stroppiana, D., Boschetti, M., & Sterlacchini, S. (2016). A spatial data infrastructure integrating multisource heterogeneous geospatial data and time series: A study case in agriculture. *ISPRS International Journal of Geo-Information, 5*(5), 73. Advance online publication. doi:10.3390/ijgi5050073

Brandt, D., Alnyme, O., & Heldt, T. (2022). Building a spatial decision support system for tourism and infrastructure planning: Technical solution and data integration challenges. *European Journal of Geography, 13*(1), 94–108. doi:10.48088/ejg.d.bra.13.1.094.108

Brown, G. (2005). Mapping spatial attributes in survey research for natural resource management: Methods and applications. In Society and Natural Resources, 18(1). doi:10.1080/08941920590881853

Bunce, R. G. H., Metzger, M. J., Jongman, R. H. G., Brandt, J., De Blust, G., Elena-Rossello, R., Groom, G. B., Halada, L., Hofer, G., Howard, D. C., Kovář, P., Mücher, C. A., Padoa-Schioppa, E., Paelinx, D., Palo, A., Perez-Soba, M., Ramos, I. L., Roche, P., Skånes, H., & Wrbka, T. (2008). A standardized procedure for surveillance and monitoring European habitats and provision of spatial data. *Landscape Ecology*, *23*(1), 11–25. doi:10.100710980-007-9173-8

Chaula, J. (2019). Opportunities and Challenges for Integrating Statistical and Spatial Data in East African Countries. *Technology, and Sciences (ASRJETS) American Scientific Research Journal for Engineering, 59*(1).

Cheng, J., Turkstra, J., Peng, M., Du, N., & Ho, P. (2006). Urban land administration and planning in China: Opportunities and constraints of spatial data models. *Land Use Policy*, *23*(4), 604–616. doi:10.1016/j.landusepol.2005.05.010

Ciampi, F., Marzi, G., Demi, S., & Faraoni, M. (2020). The big data-business strategy interconnection: a grand challenge for knowledge management. A review and future perspectives. In Journal of Knowledge Management, 24(5). doi:10.1108/JKM-02-2020-0156

Cliquet, G. (2020). From Geomarketing to Spatial Marketing. In *Spatial Economics* (Vol. II). Applications., doi:10.1007/978-3-030-40094-1_10

Diggle, P. (2005). Applied Spatial Statistics for Public Health Data. *Journal of the American Statistical Association*, *100*(470), 702–703. doi:10.1198/jasa.2005.s15

Dubman, R. (2019). The digital governance of data-driven smart cities: Sustainable urban development, big data management, and the cognitive internet of things. *Geopolitics, History, and International Relations*, *11*(2), 34. doi:10.22381/GHIR11220195

Firehock, K., & Walker, R. A. (2019). *Green infrastructure: map and plan the natural world with GIS*. Esri Press Redlands.

Franch-Pardo, I., Napoletano, B. M., Rosete-Verges, F., & Billa, L. (2020). Spatial analysis and GIS in the study of COVID-19. A review. *The Science of the Total Environment*, *739*, 140033. doi:10.1016/j.scitotenv.2020.140033 PMID:32534320

Goodchild, M., Haining, R., & Wise, S. (1992). Integrating GIS and spatial data analysis: Problems and possibilities]. *International Journal of Geographical Information Systems*, *6*(5), 407–423. doi:10.1080/02693799208901923

Grimshaw, D. J. (2001). Harnessing the power of geographical knowledge: The potential for data integration in an SME. *International Journal of Information Management*, *21*(3), 183–191. . doi:10.1016/S0268-4012(01)00004-4

Hu, X., Gong, J., Renard, E. G., & Parashar, M. (2019). Two-Stage Framework for Big Spatial Data Analytics to Support Disaster Response. *Proceedings - 2019 IEEE International Conference on Big Data, Big Data 2019*. IEEE. 10.1109/BigData47090.2019.9005613

Hu, Y., Fan, H., Chen, Y., Chang, J., Zhan, X., Wu, H., Zhang, B., Wang, M., Zhang, W., Yang, L., Hou, X., Shen, X., Pan, T., Wu, W., Li, J., Hu, H., & Wei, F. (2021). Spatial patterns and conservation of genetic and phylogenetic diversity of wildlife in China. *Science Advances*, *7*(4), eabd5725. doi:10.1126ciadv.abd5725 PMID:33523945

Iskandar, D. Suliyanto, Dinanti, A., Naufalin, L. R., Krisnaresanti, A., & Chasanah, N. (2023). SME's market development through digital mapping to stimulate business growth. AIP Conference Proceedings, 2482(1), 100003.

Jitt-Aer, K., Wall, G., Jones, D., & Teeuw, R. (2022). Use of GIS and dasymetric mapping for estimating tsunami-affected population to facilitate humanitarian relief logistics: A case study from Phuket, Thailand. *Natural Hazards*, *113*(1), 185–211. doi:10.100711069-022-05295-x

Jo, A., Lee, S. K., & Kim, J. (2020). Gender gaps in the use of urban space in seoul: Analyzing spatial patterns of temporary populations using mobile phone data. *Sustainability (Basel)*, *12*(16), 6481. doi:10.3390u12166481

Karan, E. P., & Irizarry, J. (2014). Developing a spatial data framework for facility management supply chains. *Construction Research Congress 2014: Construction in a Global Network - Proceedings of the 2014 Construction Research Congress*. IEEE. 10.1061/9780784413517.239

Krohn, J., Davis-Manigaulte, J., Fulcher, C., & Tiffany, J. S. (2022). Visualizing Diversity: Spatial Data as a Resource Enabling Extension to Better Engage Communities. *Journal of Human Sciences and Extension*. doi:10.54718/AEKV5040 le Roux, A., Khuluse-Makhanya, S., Arnold, K., Engelbrecht, F., Paige-Green, P., & Verhaeghe, B. (2019). A framework for assessing the risks and impacts of rural access roads to a changing climate. *International Journal of Disaster Risk Reduction*, *38*, 101175. doi:10.1016/j.ijdrr.2019.101175

Li, D., Wang, S., & Li, D. (2016). Spatial data mining: Theory and application. In *Spatial Data Mining*. Theory and Application. doi:10.1007/978-3-662-48538-5

Li, H., Tang, B., Lu, H., Cheema, M. A., & Jensen, C. S. (2022). Spatial Data Quality in the IoT Era: Management and Exploitation. *Proceedings of the ACM SIGMOD International Conference on Management of Data*. ACM. 10.1145/3514221.3522568

Li, X., & Kao, C. (2022). Spatial Analysis and Modeling of the Housing Value Changes in the U.S. during the COVID-19 Pandemic. *Journal of Risk and Financial Management*, *15*(3), 139. doi:10.3390/jrfm15030139

Liu, S., & Gao, B. (2022). Econometric Analysis of Leisure Agriculture and Rural Tourism Scenery Based on Spatial Data Analysis. *Computational Intelligence and Neuroscience*, *2022*, 1–8. doi:10.1155/2022/2286803 PMID:35898780

Muzaffar, H. M., Tahir, A., Ali, A., Ahmad, M., & McArdle, G. (2017). Quality assessment of volunteered geographic information for educational planning. In *Volunteered Geographic Information and the Future of Geospatial Data*. IGI Global. doi:10.4018/978-1-5225-2446-5.ch005

Nizeyimana, E. (2020). Remote Sensing and GIS Integration. In Managing Human and Social Systems. doi:10.1201/9781003053514-17

Norouzi, N., Jarrott, S., & Chaudhury, H. (2019). Designing intergenerational space through a human-development lens. *Journal of Architectural and Planning Research*, *36*(1).

Omali, T. U., & Umoru, K. (2019). A Survey of Remote Sensing and GIS Integration Method. *International Journal of Latest Technology in Engineering*, *VIII*(I).

Pinto, D. G. L., Loureiro, C. F. G., de Matos Sousa, F. F. L., & Motte-Baumvol, B. (2023). The effects of informality on socio-spatial inequalities in accessibility to job opportunities: Evidence from Fortaleza, Brazil. *Journal of Transport Geography*, *108*, 103577. doi:10.1016/j.jtrangeo.2023.103577

Prasetyo, P. E. (2019). Role of entrepreneurial culture as the driver of economic growth. *International Journal of Economics and Financial Issues*, *9*(3), 237–243. doi:10.32479/ijefi.7882

Rabiei-Dastjerdi, H., McArdle, G., Matthews, S. A., & Keenan, P. (2021). Gap analysis in decision support systems for real-estate in the era of the digital earth. *International Journal of Digital Earth, 14*(1), 121–138. doi:10.1080/17538947.2020.1808719

Rajabifard, A., & Williamson, I. P. (2001). Spatial data infrastructures: concept, SDI hierarchy and future directions. *Proceedings of Geomatics*, 10. https://doi.org/10.1.1.9.1919

Ramachandra, T. V., Bharath, A. H., & Sowmyashree, M. V. (2015). Monitoring urbanization and its implications in a mega city from space: Spatiotemporal patterns and its indicators. *Journal of Environmental Management, 148*, 67–81. doi:10.1016/j.jenvman.2014.02.015 PMID:24768450

Ranjbari, M., Morales-Alonso, G., Esfandabadi, Z. S., & Carrasco-Gallego, R. (2019). Sustainability and the sharing economy: Modelling the interconnections. *Direccion y Organizacion, 68*, 33–40. doi:10.37610/dyo.v0i68.549

Report, F. (2010). Rapid Urbanization and Mega Cities: The Need for Spatial Information Management. In The International Federation of Surveyors (FIG) Kalvebod Brygge 31–33, DK-1780 Copenhagen V DENMARK (48).

Rikalovic, A., Soares, G. A., & Ignjatic, J. (2018). Spatial analysis of logistics center location: A comprehensive approach. *Decision Making: Applications in Management and Engineering, 1*(1), 38–50. doi:10.31181/dmame180138r

Sarker, S., Rahman, M. M., Yadav, A. K., & Islam, M. M. (2019). Zoning of marine protected areas for biodiversity conservation in Bangladesh through socio-spatial data. *Ocean and Coastal Management, 173*, 114–122. doi:10.1016/j.ocecoaman.2019.03.002

Shabanpour, N., Razavi-Termeh, S. V., Sadeghi-Niaraki, A., Choi, S. M., & Abuhmed, T. (2022). Integration of machine learning algorithms and GIS-based approaches to cutaneous leishmaniasis prevalence risk mapping. *International Journal of Applied Earth Observation and Geoinformation, 112*, 102854. doi:10.1016/j.jag.2022.102854

Shang, S., Du, S., Du, S., & Zhu, S. (2021). Estimating building-scale population using multi-source spatial data. *Cities (London, England), 111*, 103002. doi:10.1016/j.cities.2020.103002

Singh, A. (2019). Remote sensing and GIS applications for municipal waste management. *Journal of Environmental Management, 243*, 22–29. doi:10.1016/j.jenvman.2019.05.017 PMID:31077867

Singh, A., Mehrotra, R., Rajput, V. D., Dmitriev, P., Singh, A. K., Kumar, P., Tomar, R. S., Singh, O., & Singh, A. K. (2022). Geoinformatics, artificial intelligence, sensor technology, big data: emerging modern tools for sustainable agriculture. *Sustainable Agriculture Systems and Technologies*, 295–313.

Stewart Fotheringham, A., & Rogerson, P. A. (1993). GIS and spatial analytical problems. *International Journal of Geographical Information Systems*, *7*(1), 3–19. doi:10.1080/02693799308901936

Wan, S., Chen, Y., Xiao, Y., Zhao, Q., Li, M., & Wu, S. (2021). Spatial analysis and evaluation of medical resource allocation in China based on geographic big data. *BMC Health Services Research*, *21*(1), 1084. doi:10.118612913-021-07119-3 PMID:34641850

Widaningrum, D. L., Surjandari, I., & Sudiana, D. (2020). Discovering spatial patterns of fast-food restaurants in Jakarta, Indonesia. *Journal of Industrial and Production Engineering*, *37*(8), 403–421. doi:10.1080/21681015.2020.1823495

Wieland, T. (2017). Market area analysis for retail and service locations with MCI. *The R Journal*, *9*(1), 298. doi:10.32614/RJ-2017-020

Yusianto, R., Marimin, Suprihatin, & Hardjomidjojo, H. (2020). Intelligent spatial decision support system concept in the potato agro-industry supply chain. *2020 International Conference on Computer Science and Its Application in Agriculture, ICOSICA 2020*. IEEE. 10.1109/ICOSICA49951.2020.9243233

Zabulis, X., Meghini, C., Partarakis, N., Beisswenger, C., Dubois, A., Fasoula, M., Nitti, V., Ntoa, S., Adami, I., Chatziantoniou, A., Bartalesi, V., Metilli, D., Stivaktakis, N., Patsiouras, N., Doulgeraki, P., Karuzaki, E., Stefanidi, E., Qammaz, A., Kaplanidi, D., & Galanakis, G. (2020). Representation and preservation of heritage crafts. *Sustainability (Basel)*, *12*(4), 1461. doi:10.3390u12041461

Chapter 5
The Derived Modification of the Exponentiated Weighted Weibull Distribution

Donne Komla Muddey
https://orcid.org/0000-0002-9667-045X
Ho Technical University, Ghana

Solomon Sarpong
University of Environment and Sustainable Development, Somanya, Ghana

Salifu Katara
University for Development Studies, Tamale, Ghana

Vincent Tulasi
Ho Technical University, Ghana

ABSTRACT

Statistically, to retain the originality of data, appropriate probability distribution must be employed rather than transforming the existing dataset. Classical Distributions (CD) lack the ability to model and describe some important real-life events. In this study, weighted Weibull distribution (WWD) is modified using Exponentiated Weighted Weibull distribution (EWWD). The shapes of the probability density function and hazard rate functions are investigated together with some structural statistical properties of the distribution. The performances of the maximum likelihood estimation of the unknown parameters of the distribution were done by implementing the Monte Carlo simulation method. We provide applications to two datasets to prove the flexibility of the model. The processes were performed using R-software. The EWWD is considered to be more tractable and capable than the WWD. The Exponentiated weighted Weibull distribution assumes a superior performance among the compared distributions as evidenced by the Kolmogorov-Smirnov statistics and Akaike Information Criteria values.

DOI: 10.4018/978-1-6684-9843-9.ch005

INTRODUCTION

In modeling real life situation for practices concerning probability theory, two particular characteristics are required. Thus, either the probability distribution is flexible or the distribution is tractable (Oguntunde et al. 2016; Ofosu et al. 2016). It is very important to develop and improve WD, because; first, it enhances the theoretical interpretation of the data and secondly, it improves judgment on appropriate model fit by introducing a model whose empirical fit better suits a particular dataset. The theory of weighted distribution provides a technique for fitting models to the unknown weight functions (Saghir et al. 2017).

In modeling monotone hazard rates, the Weibull distribution has proven to be powerful probability distribution that received appreciable usage in reliability analysis (Nasiru, 2018; Amalkia & Yuan, 2012; Zhang & Xie, 2007).

The standard distributions such as Weibull, exponentiated and gamma distributions have monotonic hazard rate functions in solving reliability problems, but they do not provide parametric fit to practical applications. Statistically, it is necessary to retain the originality of the dataset by employing appropriate probability distribution for a good fit rather than to transform the existing data. Consequently, previous literature had shown that there is the need to ensure that, existing classical distributions are extended and modified (Merovci, 2013a). Modified weighted Weibull distributions (MWWD) often fit the first and middle phases of the hazard functions. However, MWWD lack the credibility to fit the last phase of the bathtub and the modified unimodal shapes. Hence, the general objective of this research is to study and develop some modification of the weighted Weibull distribution. Recently, various weighted versions of Weibull and exponential distributions have been proposed in the literature. For example, Azzalini (1985) first proposed a method of obtaining weighted and the method has been used extensively for several symmetric and non-symmetric distributions (Mudholkar & Srivastava, 1993; Abbas et al., 2019).

The generalized Exponentiated weighted Weibull (EWW) is the modified form of the weighted Weibull (WW) distribution proposed by Nasiru (2015). GEWW provides a better flexibility and tractability of the mathematical concepts of the weighted Weibull distribution.

The Exponentiated Generalized family of distribution

Gupta et al. (1998) proposed an exponentiated generalized family of distribution technique to enhance flexibility of the existing model for modeling life-time dataset.

Definition: Suppose a random variable $R_{it} = \ln\left(P_{it}\right) - \ln\left(P_{i(t-1)}\right)$ has an arbitrary baseline distribution G(x), the cumulative distribution function (CDF) and probability distribution function (PDF) of the distributions are

$$C_{ij} = \frac{n(\sum r_i r_j) - (\sum r_i)(\sum r_j)}{\sqrt{\left[n\sum r_i^2 - (\sum r_i)^2\right]\left[n\sum r_j^2 - (\sum r_j)^2\right]}} \tag{1}$$

and

$$-1 \le C_{ij} \le 1 \tag{2}$$

where $d_{ij} = \sqrt{2\left(1 - C_{ij}\right)}$ is the shape parameter, G(x) and g(x) are the CDF and PDF of the baseline distribution.

Conceptualization of the Weighted Weibull Distributions

Weibull distribution commonly applied lifetime distributions in different areas or fields with many applications. The PDF and CDF are traceable and closed forms which result into simple expressions of survival and hazard functions.

The density (pdf) and distribution function (cdf) of the Weibull distribution are given by (Weibull, 1939)

$$0 \le d_{ij} \le 2 \tag{3}$$

$$F\left(x\right) = 1 - \exp\left(\alpha x^{\theta}\right), \quad x > 0 \tag{4}$$

where $\alpha > 0$ and $0 > 0$ are the scales and shape parameter respectively. The corresponding survival function $S(x)$ is given by the equation

$$\bar{F} = \exp\left(\alpha x^{\theta}\right), \; x > 0 \tag{5}$$

and its hazard function $h\left(x\right)$ is defined as the ratio of $f\left(x\right)$ to $S\left(x\right)$

$$h(x) = \frac{f(x)}{S(x)}$$

Thus, the hazard function is

$$h(x) = \alpha\theta x^{\theta-1}, \quad x > 0 \tag{6}$$

Hazard rate function can be increasing, decreasing or constant depending on θ.

MATERIALS AND METHODS

Developing Exponentiated Weighted Weibull Distribution

Suppose the random X follows the exponentiated weighted Weibull (EWW) distribution. Then the CDF is:

$$F(x) = \left(1 - e^{-\alpha x^{\gamma}(1+\lambda^{\gamma})}\right)^{\beta}, \alpha > 0, \lambda > 0, \gamma > 0, x > 0 \tag{7}$$

The corresponding probability density function (PDF) is obtained by differentiating equation (7):

$$f(x) = \alpha\beta\gamma(1+\lambda^{\gamma})x^{\gamma-1}e^{-\alpha x^{\gamma}(1+\lambda^{\gamma})}\left(1 - e^{-\alpha x^{\gamma}(1+\lambda^{\gamma})}\right)^{\beta-1}, x > 0 \tag{8}$$

Figure 1 displays the pdf plots of the EWW that can be monotonically decreasing, increasing, initially decreasing then increasing – decreasing. It can be seen that the density of the EWW distribution exhibits unimodal shapes.

The survival and hazard rate function of the EWW distribution are respectively:

$$S(x) = 1 - \left(1 - e^{-\alpha x^{\gamma}(1+\lambda^{\gamma})}\right)^{\beta}, x > 0 \tag{9}$$

And

Figure 1.

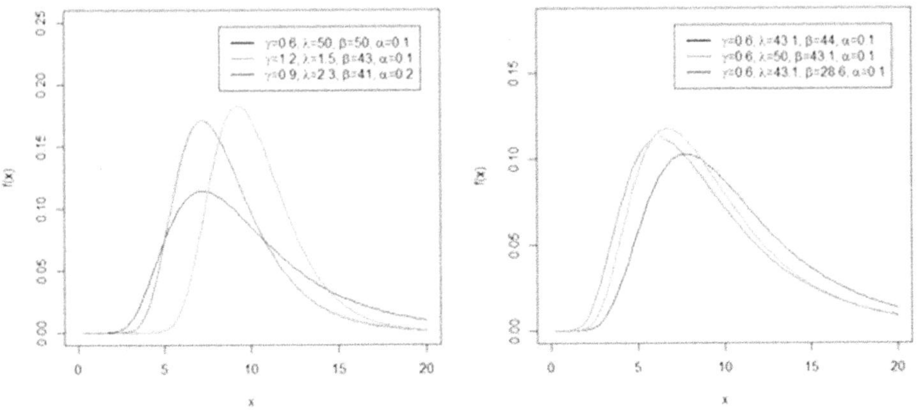

Figure 2.

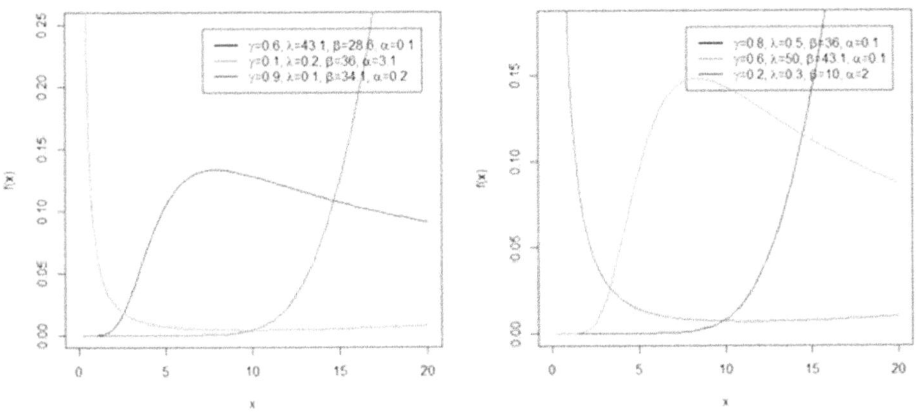

$$h(x)=\frac{\alpha\beta\gamma(1+\lambda^{\gamma})x^{\gamma-1}e^{-\alpha x^{\gamma}(1+\lambda^{\gamma})}\left(1-e^{-\alpha x^{\gamma}(1+\lambda^{\gamma})}\right)^{\beta-1}}{1-\left(1-e^{-\alpha x^{\gamma}(1+\lambda^{\gamma})}\right)^{\beta}},x>0 \tag{10}$$

The plots of the hazard function display various attractive shapes such as monotonically decreasing, monotonically increasing, upside-down bathtub, bathtub shapes for different combinations of the parameter values. The patterns of the different shapes of the EWW make it more suitable for modeling monotonic and non-monotonic failure rates application for real-life situations.

Mixture Representation

Preposition 1: The density of EWW distribution can be expressed as

$$f(x)=\alpha\beta\gamma(1+\lambda^{\gamma})x^{\gamma-1}\sum_{i=0}^{\infty}(-1)^{i}\binom{\beta-1}{i}e^{-\alpha(1+\lambda^{\gamma})(1+i)x^{\gamma}} \tag{11}$$

Proof. By definition, using equation 4

$$f(x)=\alpha\beta\gamma(1+\lambda^{\gamma})x^{\gamma-1}e^{-\alpha x^{\gamma}(1+\lambda^{\gamma})}\left(1-e^{-\alpha x^{\gamma}(1+\lambda^{\gamma})}\right)^{\beta-1},x>0.$$

Using the power series expansion

let $z=e^{-\alpha x^{\gamma}(1+\lambda^{\gamma})}$

$$(1-z)^{\beta-1}=\sum_{i=0}^{\infty}(-1)^{i}\binom{\beta-1}{i}z^{i}, \qquad |z|<1$$

By substitution, we obtain

$$f(x)=\alpha\beta\gamma(1+\lambda^{\gamma})x^{\gamma-1}e^{-\alpha x^{\gamma}(1+\lambda^{\gamma})}\sum_{i=0}^{\infty}(-1)^{i}\binom{\beta-1}{i}e^{-i\alpha x^{\gamma}(1+\lambda^{\gamma})}$$

Hence,

$$f(x)=\alpha\beta\gamma(1+\lambda^{\gamma})x^{\gamma-1}\sum_{i=0}^{\infty}(-1)^{i}\binom{\beta-1}{i}e^{-\alpha(1+\lambda^{\gamma})(1+i)x^{\gamma}}$$

STATISTICAL PROPERTIES

The researchers discussed some important statistical properties of the proposed distribution such as quantile, moments, moment generating function, incomplete moment, and Rényi entropy.

Quantile

Quantile is the inverse cumulative distribution function used to generate random samples for the probability distribution. It can also serve as an alternative to the probability density function.

Proposition 2: The quantile function of the EWW distribution can be expressed as:

$$x = \left[-\frac{\log(1 - u^{\frac{1}{\beta}})}{\alpha(1+\lambda^{\gamma})} \right]^{\frac{1}{\gamma}}, \qquad u \in (0,1) \tag{12}$$

Proof. The quantile function is obtained by solving the equation

$$(1 - e^{-\alpha x_u^{\gamma}(1+\lambda^{\gamma}})^{\beta} = u$$

Hence, $x = \left[-\frac{\log(1 - u^{\frac{1}{\beta}})}{\alpha(1+\lambda^{\gamma})} \right]^{\frac{1}{\gamma}}$

This completes the proof.

Moments

Moments are used to compute various measures of central tendency, variation, kurtosis, and skewness among others.

Proposition 3: The moment of a random variable X of an exponentiated weighted Weibull distribution is given by

$$\mu_r' = \beta\sum_{i=0}^{\infty}(-1)^i\binom{\beta-1}{i}\frac{\Gamma\left(\frac{r}{\gamma}+1\right)}{[\alpha(1+\lambda^{\gamma})]^{\frac{r}{\gamma}}(1+i)^{\frac{r}{\gamma}+1}} \tag{13}$$

Proof. By definition

$$\mu_r' = \int_0^{\infty} x^r f(x)dx$$

where f(x) represents the PDF in equation 5.
This implies that

$$\mu_r' = \alpha\beta\gamma(1+\lambda^{\gamma})\sum_{i=0}^{\infty}(-1)^i\binom{\beta-1}{i}\int_0^{\infty} x^r.x^{\gamma-1}e^{-\alpha(1+\lambda^{\gamma})(1+i)x^{\gamma}}dx$$

let $y = \alpha(1+\lambda^{\gamma})(1+i)x^{\gamma}, \quad x\to 0, y\to 0, x\to\infty, y\to\infty$

$$x = \left[\frac{y}{\alpha(1+\lambda^{\gamma})(1+i)}\right]^{\frac{1}{\gamma}}$$

$$\frac{dy}{dx} = \alpha\gamma(1+\lambda^{\gamma})(1+i)x^{\gamma-1}$$

$$dx = \frac{dy}{\alpha\gamma(1+\lambda^{\gamma})(1+i)x^{\gamma-1}}$$

By substituting x and dx, we obtain,

$$\mu_r' = \alpha\beta\gamma(1+\lambda^{\gamma})\sum_{i=0}^{\infty}(-1)^i\binom{\beta-1}{i}\int_0^{\infty} x^{r+\gamma-1}e^{-\alpha(1+\lambda^{\gamma})(1+i)x^{\gamma}}\frac{dy}{\alpha\gamma(1+\lambda^{\gamma})(1+i)x^{\gamma-1}}$$

Simplifying, we have

$$\mu_r' = \beta\sum_{i=0}^{\infty}(-1)^i\binom{\beta-1}{i}\frac{1}{[\alpha(1+\lambda^{\gamma})]^{\frac{r}{\gamma}}(1+i)^{\frac{r}{\gamma}+1}}\int_0^{\infty} y^{\frac{r}{\gamma}+1-1}e^{-y}dy$$

Recall,

$$\Gamma(a)=\int_0^\infty t^{a-1}e^{-t}dt$$

$$\mu'_r=\beta\sum_{i=0}^{\infty}(-1)^i\binom{\beta-1}{i}\frac{\Gamma\left(\frac{r}{\gamma}+1\right)}{[\alpha(1+\lambda^\gamma)]^{\frac{r}{\gamma}}(1+i)^{\frac{r}{\gamma}+1}]}$$

This completes the proof.

Incomplete Moments

The incomplete moments are also known as the conditional moments which usefully describe the shape of many distributions. It plays an important role in measuring inequality, for example; income quantiles, Lorenz, and Bonferroni curves.

Proposition 4: The incomplete moment of Y can be expressed as

$$\rho_r(x)=\sum_{i=0}^{\infty}(-1)^i\binom{\beta-1}{i}\frac{1}{\left[\alpha(i+1)(1+\lambda^\gamma)\right]^{\frac{\gamma}{\gamma_1}}}\int_0^{\alpha(i+1)(1+\lambda^\gamma)x^\gamma}\left[z\right]^{\frac{\gamma}{\gamma_1}}e^{-z}\frac{dz}{(i+1)} \tag{14}$$

Proof. By definition

$$\rho_r(x)=\int_0^x y^r f(y)dy$$

$$\rho_r(x)=\int_0^x y^r\alpha\beta\gamma(1+\lambda^\gamma)y^{\gamma-1}e^{-\alpha y^\gamma(1+\lambda^\gamma)}(1-e^{-\alpha y^\gamma(1+\lambda^\gamma)})^{\beta-1}dy$$

$$\rho_r(x)=\int_0^x y^r\alpha\beta\gamma(1+\lambda^\gamma)y^{\gamma-1}\sum_{i=0}^{\infty}(-1)^i\binom{\beta-1}{i}e^{-\alpha(i+1)(1+\lambda^\gamma)y^\gamma})^{\beta-1}dy$$

$$\rho_r(x) = \sum_{i=0}^{\infty} (-1)^i \binom{\beta-1}{i} \int_0^{\alpha(i+1)(1+\lambda^\gamma)x^\gamma} \left[\frac{z}{\alpha(i+1)(1+\lambda^\gamma)}\right]^{\frac{\gamma}{\gamma_1}} e^{-z} \frac{dz}{(i+1)}$$

$$\rho_r(x) = \sum_{i=0}^{\infty} (-1)^i \binom{\beta-1}{i} \frac{1}{\left[\alpha(i+1)(1+\lambda^\gamma)\right]^{\frac{\gamma}{\gamma_1}}} \int_0^{\alpha(i+1)(1+\lambda^\gamma)x^\gamma} [z]^{\frac{\gamma}{\gamma_1}} e^{-z} \frac{dz}{(i+1)}$$

This completes the proof.

Rényi Entropy

The entropy is used to measure the randomness of the systems. To examine the uncertainty of the random variable Rényi entropy can be employed. It is derived as follows:

Proposition 5: Rényi entropy is defined by

$$I_R = \frac{1}{1-\delta} \log\left[(1-\theta)^\delta (1+\lambda^\beta)^\delta (\alpha\beta)^\delta \sum_{i=0}^{\infty} \binom{2\delta+i-1}{i} \theta^i \frac{\Gamma\left(\frac{\delta(\beta-1)-\beta+1}{\beta}\right)}{\left[\alpha(i+1)(1+\lambda^\beta)\right]^{\left(\frac{\delta(\beta-1)+1}{\beta}\right)}}\right], \tag{15}$$

Proof: Rényi entropy is defined by

$$I_R = \frac{1}{1-\delta} \log\left[\int_0^\infty f_{\mathbb{R}} f^\delta(\mathrm{y})\,\mathrm{dy}\right], \quad \delta > 0,\ \delta \neq 1.$$

PARAMETER ESTIMATION

In this section MLE method was employed to estimate unknown parameters of the GWW four parameters distribution.

Maximum Likelihood Estimation (MLE)

The maximum likelihood estimation method is the most widely used classical approach for estimating the parameters of a probability distribution model and is based on a likelihood function. It enjoys many desirable properties including consistency, asymptotic efficiency, invariance property, and asymptotic normality (Nassar et al, 2018).

Definition: Let $x_1 ..., x_n$ be a random sample of size n from the EWW $(\gamma,\alpha,\beta,\lambda)$ distribution. The log-likelihood function for the vector of parameters $\xi = (\gamma,\alpha,\beta,\lambda)^T$ becomes

$$L = \prod_{i=1}^{n} \alpha\beta\gamma(1+\lambda^{\gamma})x^{\gamma-1}e^{-\alpha x^{\gamma}(1+\lambda^{\gamma})}\left(1-e^{-\alpha x^{\gamma}(1+\lambda^{\gamma})}\right)^{\beta-1} \quad (16)$$

$$L = \alpha^n\beta^n\gamma^n(1+\lambda^{\gamma})^n\left(\prod_{i=1}^{n} x^{\gamma-1}\right)\left(\prod_{i-1}^{n} e^{-\alpha x^{\gamma}(1+\lambda^{\gamma})}\right)\left(\prod_{i-1}^{n}\left[1-e^{-\alpha x^{\gamma}(1+\lambda^{\gamma})}\right]\right)^{\beta-1}$$

Taking logarithm of both sides

$$L - n\ln\alpha + n\ln\beta + n\ln\gamma + n\ln(1+\lambda^{\gamma}) + (\gamma-1)\sum_{i=1}^{n}\ln x_i$$
$$-\alpha(1+\lambda^{\gamma})\sum_{i-1}^{n} x_i^{\gamma} + (\beta-1)\sum_{i-1}^{n}\ln\left(1-e^{-\alpha x_i^{\gamma}(1+\lambda^{\gamma})}\right) \quad (17)$$

Components of the score vector $U(\xi)$ are obtained by partially differentiating the log-likelihood function. The numerical result is derived directly with the application of lifetime data set using statistical software R. A good set of initial values is essential.

DISCUSSION OF FINDING AND RESULTS (NUMERICAL ILLUSTRATION)

In this section, random numbers are generated and two datasets are considered.

Simulation Study

In this section, we study the performance of the MLE estimators of the unknown parameters of the EWW distribution by implementing the Monte Carlo simulation method. A random sample of sizes $n = 50,100,150$ and 200 is generated from the quantile function given in equation (8) of the distribution. The simulation is made for different values of the parameters α, γ, λ and β. Comparisons of different methods are made using the estimated values of parameters, the average Bias, and the mean square error MSE for 1000 replicates. The processes were performed using the R Code Software.

The purpose of generating random numbers is to know if:

- There exists any difference between the average estimates and the true values, whether it would be small or not.
- The MLE would converge to the true value in all cases when the sample size increases.
- The standard errors of the MLEs would decrease as the sample size increases.

Table 1 reveals the average BIAS and Root Mean Square Error (in the parentheses) for the maximum likelihood estimators $(\gamma, \alpha, \beta, \lambda) = (7.5, 0.3, 0.2, 7.1)$ for $n = 50,100,150,200$. ABIAS decreases for estimators of some parameters as the sample size increases while for others it fluctuates. The RMSE for the estimators on average decreases for all parameters as the sample size increases. The process is repeated, in this case, the parameter values are fixed at $(\gamma, \alpha, \beta, \lambda) = (7.6, 0.3, 0.2, 7.1)$ for $n = 50,100,150,200$. The ABIAS for some estimators of the parameters also decreases as the sample size increases while it fluctuates others.

The RMSE for the estimators of all the parameters showed a decreasing pattern.

Applications

In order to investigate the advantage of the EWW proposed model, we apply two lifetime datasets; twenty electronic components and twenty weight of the diamond stones in carat. We consider criteria like the Kolmogorov-Smirnov test statistics (K-S), Log-likelihood Ratio Test (lnL), Akaike Information Criteria (AIC), Bayesian Information Criteria (BIC), Cramér-Von Mises (W*), and Anderson Darling (A^*). The better distribution corresponds to the smallest (W^*), (K-S), (lnL), (AIC), (BIC), and (A^*) values. The computations were done using R Software.

Table 1. MLEs, ABIAS and RMSE (in parentheses) for EWW simulation study for different parameters values

Parameter Values				*N*	ABIAS/RMSE			
α	γ	β	λ		$\hat{\alpha}$	$\hat{\beta}$	$\hat{\lambda}$	$\hat{\gamma}$
0.3	**7.5**	**0.2**	**7.1**	*50*	0.1775 (0.7908)	0.1860 (0.4882)	-1.6118 (2.3212)	0.3770 (0.8755)
				100	0.07704 (0.1765)	0.1132 (0.1548)	-1.3071 (1.7839)	0.2462 (0.5015)
				150	0.05215 (0.1426)	0.0948 (0.1236)	-1.1773 (1.5848)	0.2015 (0.4073)
				200	0.0313 (0.1223)	0.0802 (0.10267)	-1.0019 (1.3933)	0.1704 (0.3540)
α	γ	β	λ	*N*	$\hat{\alpha}$	$\hat{\beta}$	$\hat{\lambda}$	$\hat{\gamma}$
0.3	**7.6**	**0.2**	**7.1**	*50*	0.1442 (0.3274)	0.1959 (0.3584)	-1.555 (2.3174)	0.4233 (1.2247)
				100	0.0557 (0.1729)	0.1174 (0.1682)	-1.2376 (1.8304)	0.2706 (0.6123)
				150	0.0227 (0.1235)	0.0928 (0.1196)	-1.0549 (1.4962)	0.1994 (0.4173)
				200	0.0069 (0.0959)	0.0809 (0.1005)	-0.9762 (1.3248)	0.1788 (0.3661)

Table 2. Lifetimes of 20 electronic components

0.03	0.22	0.73	1.25	1.52	1.8	2.38	2.87	3.14	4.72
0.12	0.35	0.79	1.41	1.79	1.94	2.4	2.99	3.17	5.09

Table 3. Descriptive statistics of electronic components life data

Minimum	Mean	Maximum	Variance	Skewness	Kurtosis
0.03	1.94	5.09	2.06	0.60	2.72

Table 4. Maximum likelihood estimates and the standard errors for electronic components

Distribution	Estimate	Standard error	*z-value*	*p-value*
EWW	$\hat{\alpha} = 0.0082$	0.07476	0.1093	0.9130
	$\hat{\lambda} = 0.5743$	5.2515049	0.1094	0.9129
	$\hat{\gamma} = 3.4149$	3.8971	0.8763	0.3809
	$\hat{\beta} = 0.2365$	0.3281	0.7207	0.4711
WW	$\hat{\alpha} = 0.1221$	4.9928	0.0245	0.9805
	$\hat{\lambda} = 2.1456$	102.766	0.0209	0.9833
	$\hat{\theta} = 1.1961$	0.2248	5.3197	$1.04\times10^{-7*}$
EGWW	$\hat{\alpha} = 0.0037$	9.9347e-03	0.3700	0.7114
	$\hat{\lambda} = 0.0371$	6.2953e-04	58.9293	$<2\times10^{-16*}$
	$\hat{\theta} = 2.5948$	1.5907e+0	1.6312	0.1028
	$\hat{a} = 10.1457$	1.9592e-03	5178.4089	$<2\times10^{-16*}$
	$\hat{b} = 0.3294$	2.5999e-01	1.2669	0.2052

*: means significant at 5% significance level

Application One: Electronic Components Failure Rate Data

Table 3 shows the average electronic components failure rate at 1.94 and having a coefficient of skewness of 0.60 and kurtosis of the value of 2.72 of the data which have positive skewness and kurtosis.

The maximum likelihood estimates for the parameters of the fitted models with their corresponding standard errors and z-value are given in Table 4. The accuracy of the approximation of the standard error of the MLEs was verified by using the standard error test which states that for a parameter to be significant at the 5% significance level the standard error should be less than half the parameter values.

Table 5. Goodness - of - fit statistics

Model	Log-likelihood	AIC	BIC	CAIC	A*	K-S	p-value	W*
EWW	-31.47	70.9	74.2	73.6	0.214	0.1039	0.9667	0.03226
WW	-32.79	71.6	74.6	73.1	0.419	0.1271	0.8638	0.0712
EGWW	-31.57	73.1	78.1	77.4	0.219	0.1052	0.963	0.0344

Figure 3.

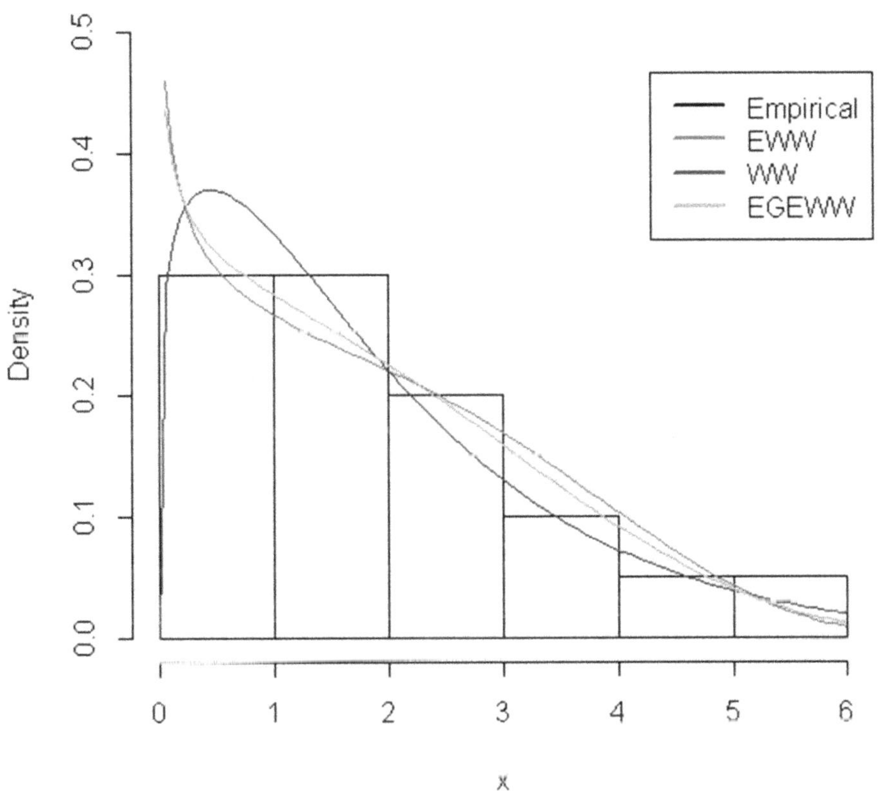

Table 6. Weight of the diamond stones in carat

1.4575	0.3092	0.3642	0.0119	0.0664	2.6125	0.6027	0.1693	0.5894	0.1558
0.7701	0.0626	0.5350	0.1352	0.4024	0.2872	1.2177	2.6257	0.3954	0.4107

This output shows that MLEs for the parameters for EWW was not statistically significant. For the WW model only θ was significant and for the EGWW, the parameters λ and ạ were significant.

Table 5 reveals that EWW distribution was the best model for the dataset since it has the highest log-likelihood value of -31.47 and the smallest values for AIC, BIC and CAIC. To check the appropriateness of the model, Kolmogorov Smirnov statistics was considered at the value of 0.1309.

The Anderson-Darling (AD) statistics at 0.214 also demonstrated that the exponentiated weighted Weibull (EWW) fits the data set better than the Weighted Weibull (WW) as quoted in Nasiru (2015) and other sub-models.

In order to gain more insight into the EWW distribution, a visual comparison of the histogram of the data with the fitted density functions was developed. Similarly, it is clear in Figure 3 that the EWW fits the left and right peaks of the histogram better.

Application Two: Weight of the Diamond Stones in Carat

In this section, we illustrate the performance of the proposed model with data obtained from Singfat (1996) which represents the weight of the diamond stones in carat.

To show the applicability and the assessment of the merit of the proposed model, we use the data set indicated in Table 6.

Table 7, reveals that the two-information criterion employed for the fitted model to the data, the EWW is more flexible than EGWW and AddW but unfavorable to Weighted Weibull distribution. According to Oguntunde and Adepoju (2015), the lowest information criteria are considered to be the best fit, more flexible, and capable than the other models.

This result is in accordance with what is obtainable in other notable researches with respect to the probability characteristic of the distribution. For example, according to Oguntunde et al. (2016), the characteristics of the model could either be flexible or tractable. Hence, The EWW distribution is considered to be more tractable and capable than the Weighted Weibull distribution defined by Nasiru (2015) and more flexible as compared to EGWW and AddW to the Weight of the Diamond Stones in the Carat dataset.

Table 7. MLEs, Standard errors (in parentheses) and measures AIC, and log-likelihood test on the weight of the diamond stones in carat dataset

Model	$\hat{\alpha}$	$\hat{\lambda}$	$\hat{\beta}$	$\hat{\gamma}$	$\hat{\theta}$	$\hat{a}$	$\hat{b}$	AIC	lnL
EWW	0.76	2.84	2.02	0.65	-	-	-	30.7	-11.35
	(-28.8)	(254.19)	(2.47)	(0.39)	-	-	-		
WW	0.96	0.55	-	-	0.92	-	-	29.1	-11.54
	(101.17)	(172.29)	-	-	(0.16)	-	-		
EGWW	1.12	1.34	-	-	0.64	0.91	2.02	32.7	-11.36
	(39.45)	(109.33)	-	-	(0.39)	(32.01)	-2.48		
AddW	1.21	-	0.30	0.92	0.92	-	-	31.1	-11.55
	(0.17)	-	(0.17)	(0.71)	(0.23)	-	-		

CONCLUSION

A new modification of the exponentiated weighted Weibull (EWW) distribution is derived and its probability characteristics were obtained. The shapes of its probability density function and hazard rate functions are investigated. Statistical properties of EWW are derived. The method of maximum likelihood estimation is used to estimate the unknown parameters. The performance of the MLE estimators of the unknown parameters of the distribution by implementing the Monte Carlo simulation method. We provide applications to two datasets to prove the flexibility of the model. The EWW distribution is considered to be more tractable and capable than the weighted Weibull distribution.

ACKNOWLEDGEMENT

The authors wish to express their profound gratitude to Prof. Suleman Nasiru for his enormous contributions to the success of this work. He provided supports and insight to the weighted Weibull distributions.

REFERENCES

Abbas, S., Ozal, G., Shahbaz, S. H., & Shahbaz, M. Q. (2019). A new generalized weighted Weibull distribution. *Pakistan Journal of Statistics and Operation Research, 15*(1), 161-178.

Akaike, A. (1974). A new look at the statistical model identification. *IEEE Transactions on Automatic Control, 19*(6), 716–723. doi:10.1109/TAC.1974.1100705

Almalki, S. J., & Yuan, J. (2013). A new modified Weibull distribution. *Reliability Engineering & System Safety, 111*, 164–170. doi:10.1016/j.ress.2012.10.018

Azzalini, A. (1985). A class of distributions which includes the normal ones. *Scandinavian Journal of Statistics, 12*(2), 171–178.

Fisher, R. A. (1922). On the mathematical foundations of theoretical statistics. *Philosophical Transactions of the Royal Society of London. Series A, Mathematical and Physical Sciences*, 309–368.

Gupta, R. C., Gupta, P. I., & Gupta, R. D. (1998). Modelling failure time data by Lehmann alternatives. *Communications in Statistics. Theory and Methods, 27*(4), 887–904. doi:10.1080/03610929808832134

Gupta, R. D., & Kundu, D. (2009). A new class of weighted exponential distributions. *Statistics, 43*(6), 621–634. doi:10.1080/02331880802605346

Merovci, F., & Elbatal, I. (2015). Weibull-Rayleigh distribution: Theory and applications. *Applied Mathematics & Information Sciences, 9*(5), 1–11.

Mudholkar, G. S. & Srivastava, D. K. (1993). Exponentiated weibull family for analyzing bathtub failure-rate data. *Reliability, IEEE Transactions on, 42*(2), 299-302.

Nadarajah, S., & Kotz, S. (2006b). The exponentiated type distributions. *Acta Applicandae Mathematicae, 92*(2), 97–111. doi:10.100710440-006-9055-0

Nasiru, S. (2015). Another weighted weibull distribution from azzalinis family. *European Scientific Journal, 11*(9)

Nasiru, S., Mwita, P. N., & Ngesa, O. (2018). Discussion on generalized modified inverse Rayleigh distribution. *Applied Mathematics & Information Sciences, 12*(1), 113–124. doi:10.18576/amis/120110

Ofosu, J. B., Otchere, F., & Hesse, C. A. (2016). *Intermediate Statistical Methods*. Excellent Publication and Printing.

Oguntunde, P.E., Adejumo, A.O. Okagbue, H.I. & Rastogi, M.K. (2016). Statistical properties and application of a new Lindley exponential distribution. *Gazi University journal of Sciences, 29*(4), 831-838.

R'enyi, A. (1961). On measures of entropy and information. In *Proceedings of the Fourth Berkeley Symposium on Mathematical Statistics and Probability*, (pp. 547-561). University of California Press, Berkeley, CA.

Ramachandran, K. M., & Tsokos, C. P. (2015). *Mathematical Statistics with Applications in R* (2nd ed.). Elsevier.

Saghir, A. M., Tazeem, S., & Ahad, I. (2017). The weighted exponentiated inverted Weibull distribution: Properties and application. *Journal of Information and Mathematical Sciences*, *9*(1), 137–151.

Weibull, W. A. (1951). A statistical distribution function of wide applicability. *Journal of Applied Mechanics*, *18*(3), 293–297. doi:10.1115/1.4010337

Zhang, T., & Xie, M. (2007). Failure data analysis with extended Weibull distribution. *Communications in Statistics. Simulation and Computation*, *36*(3), 579–592. doi:10.1080/03610910701236081

Chapter 6
Antibiotics in the Ghanaian Environment:
Occurrence, Uptake, Risk Assessment, and Removal

James Darmey
Kumasi Technical University, Ghana

David Azanu
Kwame Nkrumah University of Science and Technology, Ghana

Nene Kwabla Amoatey
Kumasi Technical University, Ghana

Abena Owusuaa Adu
Kumasi Technical University, Ghana

Nasser Dia
Kumasi Technical University, Ghana

ABSTRACT

The use of antibiotics is widespread throughout the world, and Ghana is not an exception. They are one of the most frequently prescribed medications. Antibiotic residue concentrations are rising in the environment and are found in practically all water bodies, including lakes and rivers, because of inadequate management of residential and medical wastes. Contrarily, there is a substantial chance of medication toxicity and resistance. Therefore, removal of these pollutants from water bodies is essential, yet conventional wastewater or water treatment methods frequently fall short of efficiently removing these chemicals. Antibiotics will be used as a case study in this chapter to discuss the concentration of pharmaceutical pollutants in the environment, specifically in Ghanaian waterbodies, and risk evaluations of these pollutants. Additionally, this chapter will look into the possible removal technologies report and establish the present investigation on removing these pollutants with microplastics.

DOI: 10.4018/978-1-6684-9843-9.ch006

1. INTRODUCTION

Pharmaceuticals are chemicals whose utility is based on specific biological activity with the aim of promoting health and high living standard (Kümmerer, 2010). According to the Organization for Economic Cooperation and Development (2018), the word pharmaceutical which is used interchangeably with medicine or drug can be defined as any substance used for medicinal purposes like delaying the onset of and curing diseases, relieving symptoms, and preventing complications. Other uses or functions of pharmaceuticals specified in Encyclopedia Britannica are diagnosis, treatment, and prevention of diseases; as well as regenerating, correcting, and modifying organic functions. The effect of pharmaceuticals depends on their rightful use.

Bacterial infections have long been a significant cause of morbidity and mortality worldwide. The development of antibiotics has played a huge success in improving the outcomes of bacterial infections as well as improvements in nutrition, sanitation and vaccination (Azanu et al., 2018). Between 2000 and 2018, consumption of antibiotic saw a 46% increase from 9.8 defined daily doses per 1000 population per day in the year 2000 to 14.3 defined daily doses per 1000 population per day in 2018 (Jia et al., 2022). The widespread, improper use and easy accessibility of antibiotics often lead to the resistant microorganism strains emerging as well as the potential for drug toxicity and increased morbidity and mortality. The positive impacts of various antibiotics on health are threatened by resistance against these drugs.

In Ghana, antibiotics have been extensively used for treatment in both humans and animals. Afriyie et al. (2015) reported a high use of antibiotics with penicillin having the highest frequency of prescription in Ghana. The availability of these drugs over the counter without prescription or the appropriate indication has contributed widely to their use and misuse. In veterinary practice, antibiotics reserved for human infections are used as prophylaxis, growth promoters, and treatment of livestock, which further adds to their increased presence in the environment (Azanu et al., 2018).

Field and laboratory studies by Schoenfuss et al. (2016) revealed that pharmaceutical mixtures can reduce the population of fishes through varied effect of such mixtures on sexes. Thus, pharmaceutical mixtures have different impacts on males and females, which may not be coherent.

Removal of pharmaceuticals from water and the environment is a critical environmental concern due to the potential presence of various contaminants, including active pharmaceutical ingredients (APIs) and their metabolites. The conventional wastewater treatment processes are often insufficient to remove these compounds effectively. Therefore, advanced treatment technologies have been developed to address this issue. Several methods have been investigated and applied for the treatment of pharmaceutical wastewater, including biological, physical, and chemical processes.

One commonly employed biological treatment method is the activated sludge process, which utilizes microorganisms to degrade organic pollutants. This process can be enhanced by incorporating additional treatment units, such as membrane bioreactors (MBRs), to improve the removal efficiency of pharmaceutical compounds (Kümmerer, 2010). MBRs combine conventional activated sludge treatment with membrane filtration, resulting in a more efficient removal of contaminants, including pharmaceutical residues (Fatta-Kassinos et al., 2011).

Advanced oxidation processes (AOPs) have also shown promising results in pharmaceutical wastewater treatment. AOPs involve the generation of highly reactive hydroxyl radicals to oxidize and degrade organic pollutants. Methods such as ozonation, photocatalysis, and Fenton reaction have been investigated for their effectiveness in removing pharmaceutical compounds from wastewater (Boix et al., 2012; Carra et al., 2013; Ağar et al., 2022). These processes have the advantage of being able to mineralize organic compounds to less harmful byproducts.

In addition to biological and oxidative treatments, physical processes such as adsorption and membrane filtration are commonly used for pharmaceutical wastewater treatment. Adsorption methods, such as activated carbon adsorption, have been found effective in removing pharmaceutical compounds from wastewater (Teixeira et al., 2012). Membrane filtration, including reverse osmosis (RO) and nanofiltration (NF), can selectively remove pharmaceuticals based on their size and charge (Radjenović et al., 2013). These physical processes can be combined with biological or oxidative treatments to achieve higher removal efficiencies. This chapter will look to address the occurrence, in terms of concentration, of pharmaceutical pollutants, using antibiotics as case study, in the environment specifically in waterbodies found in Ghana and risk assessments of these pollutants in the environment. Also, the possible removal of these pollutants and the current study of removing these pollutants with microplastics will be established in this chapter.

2. ANTIBIOTICS

According to the U.S. Department of Health and Human Service (2020), antibiotics are medicines used for treating and preventing bacterial infections by either killing bacteria or inhibiting their growth. Antibiotics can be put into groups depending on their source, chemical structure, target, function and range of activity; and these are penicillin (beta-lactams e.g. amoxicillin, flucloxacillin); tetracycline (e.g. doxycycline); cephalosporin (e.g. cefuroxime); quinolones (e.g. ciprofloxacin); lincomycins (e.g. clindamycin); macrolides (e.g. erythromycin, azithromycin);

Table 1. Classification of antibiotics

Basis	Classes	References
Chemical structure	Beta-lactams, e.g., carbapenems, cephalosporins and penicillins	Britannica Encyclopedia (2023)
Function	Narrow-spectrum (e.g., penicillins) Broad-spectrum (e.g., tetracyclines) Extended-spectrum (e.g.,extendedspectrum penicillins, beta-lactamase, etc.)	Willacy (2020)
Source	Natural (aminoglycosides) Semi-synthetic (carbapenems, cephalosporins and penicillins) Artificial (sulfonamides and quinolones)	Willacy (2020)
Target	Enzymatic function (sulfonamides and quinolones), Growth (cephalosporins and penicillins), Protein synthesis (tetracyclines and macrolides)	Willacy (2020)

sulfonamides (e.g. co-trimoxazole and trimethoprim); glycopeptides (e.g. vancomycin); aminoglycosides (e.g. gentamycin); carbapenems (e.g. imipenem and cilastatin); nitrofurantoin and metronidazole (Willacy, 2020). Based on the sources, there are natural (aminoglycosides), semi-synthetic (carbapenems, cephalosporins and penicillins) and artificial (sulfonamides and quinolones) ones (Table 2.1). Based on their structure, carbapenems, cephalosporins and penicillins are beta-lactams: and based on their function, there are narrow-spectrum (penicillins), broad-spectrum (tetracyclines) and extended-spectrum (Britannica Encyclopedia, 2023). Antibiotics work by targeting bacterial growth process or function, thus cephalosporins and penicillins target growth; sulfonamides and quinolones target enzymatic function; and tetracyclines and macrolides inhibit protein synthesis.

Antibiotics are effective for the treatment of bacterial infections, but their use can lead to the death of beneficial bacteria of the human body and other health problems like diarrhea, allergies, nausea, and rashes (Forbes et al., 2018). The Centers for Disease Control and Prevention (2021) has reported that the abuse of antibiotics can lead to antibiotic resistance; a situation where an antibiotic becomes less effective against the target bacteria. The inappropriate use of antibiotics has contributed to the increase in the number of resistant bacterial infections (Klein et al., 2018) and Kwak et al. (2015) reports of an increasing antibiotic resistance. According to Kostich et al. (2014), antibiotics inhibit the growth of some beneficial microorganisms and can contribute to antibiotic resistance (Iweriebor et al., 2015; Morris et al., 2008). Antibiotic resistance is a grave worldwide public health problem.

Some researchers have called for interventions to curb the disposal of pharmaceuticals in the environment because of the presence of high levels of antibiotics and its associated resistant microbes (Borquaye et al., 2019; Nawrat,

2018; Urevig, 2019). A report by the BIO Intelligence Service (2013) identifies antibiotics as among the notable pharmaceuticals with ecotoxicological effect. When antibiotics occur in soils, they can delay germination, lower biomass allocation, and alter composition of plants species (Minden et al., 2017). Their absorption by plants can lead to an interference of physiological processes like photosynthesis through chloroplast gene expression and cell proliferation; and mitochondrial functioning through oxidative stress response (Wang et al., 2015).

3. OCCURRENCE OF PHARMACEUTICALS IN THE ENVIRONMENTS

3.1 Occurrence and Fate of Pharmaceuticals in Hospital Effluents

Effluent simply means wastewater and is defined by United States Environmental Protection Agency (US EPA) as treated or untreated wastewater often discharged into aquatic environment; hence hospital effluent connotes wastewater from hospital. Hospital effluent or wastewater is the residual collection from medical and non-medical activities or services like diagnosis, first-aid, surgery, laboratory examinations, laundry, and drug dispensing (Orias, 2015), which involve the use of water in a hospital. It is composed of micro-pollutants (heavy metals, microorganisms, radioactive elements, and toxic chemicals); and macro-pollutants (Biochemical Oxygen Demand, Chemical Oxygen Demand, Total Nitrogen, Total Organic Concentration, carbonates, and phosphorus) thereby presenting biological, chemical, and physical risks or danger (Carraro et al., 2016).

Pharmaceuticals are micro-pollutants, as they are considered as chemicals detected at concentrations of ng/L to μg/L (Azuma et al., 2016). Comparatively, hospital effluents have higher concentrations of pharmaceutically active compounds (PhACs) and their metabolites (Balakrishna et al., 2017; Chonova et al., 2016; Prabhasnkar et al., 2016). As shown in Figure 1, pharmaceuticals get into hospital effluents through excretion (urine and faeces) after consumption and disposal of unused or expired medications (Owens, 2015). From the principles of pharmacokinetics, antibiotics are not completely absorbed or metabolized after intake, as such, about 30 to 90% of all ingested antibiotics are excreted with urine and faeces and are expelled into the environment (Gu et al., 2021). Hospital services also contribute to the pharmaceutical contaminants of hospital effluents (Orias & Perrodin, 2014) as a study by Akin (2016) revealed that wastewater generated from laboratory service increases the pharmaceutical contamination of hospital effluent. The quantities of hospital wastewater are rising because of increase and advancement in medical services (Amoaei et al, 2015); as such the constituent pollutants which include antibiotics, and their residues are increasing. This has deepened concerns about the presence of antibiotics and other pharmaceuticals in aquatic environments as they serve as sinks or reservoirs for hospital wastewaters. Concentrations of pharmaceuticals in hospital effluents depend on the population and size of the hospital (Al Aukidy et al., 2017); the type and the use of drugs prescribed, range of activities, season of year, and time of day (Fedaku et al., 2019). Bigger hospitals with several specialized units like maternity, psychiatric, pediatric wards produce more and wider range of pharmaceutical residues confirming report by Oliveira et al. (2017) that the occurrence of classes of pharmaceuticals depend on the healthcare services available at the hospital.

Different classes of pharmaceuticals have been detected in hospital effluents. Mendoza et al. (2015) identified the presence of acetaminophen, diclofenac, ibuprofen, and naproxen (analgesics); furosemide (diuretic); clarithromycin, ofloxacin and trimethoprim (antibiotics); propranolol (β-blocker) and prophyphenazone (anti-inflammatory) in hospital wastewater from a medium-size hospital in Spain. In Greece, Papageorgiou et al. (2019) reported on the occurrence of diclofenac, ibuprofen, paracetamol, and salicylic acid (analgesics); amoxicillin, ampicillin, azithromycin, and clarithromycin (antibiotics); irbesartan, telmisartan and valsartan (anti-hypertensives); atenolol, metoprolol and pindolol (beta blockers); furosemide (diuretic); carbamazepine, citalopram, and venlafaxine (psychiatric drugs); and caffeine (stimulant) in hospital effluent. A comparison of the two reports shows that the occurrence of pharmaceutical residues in hospital effluent is increasing rendering hospital effluents more toxic. Kosma et al. (2020) has declared antibiotics as the most identified class of pharmaceuticals in hospital effluents. Consequently, in their review, Majumder et al. (2021) declare analgesics and antibiotics as the most prevalent pharmaceuticals in hospital effluents.

The occurrence of pharmaceuticals in hospital effluents have been identified in a number of developed and developing countries (Amoaei et al, 2015; aus der Beek et al., 2016); making the issue a global concern. In a recent review, Fekadu et al. (2019) identified diclofenac, ibuprofen, paracetamol (analgesics); naproxen (anti-inflammatory); venlafaxine, clarithromycin (antibiotic); sulfamethoxazole, carbamazepine, trimothoprim, ketoprofen as the ten most frequently detected pharmaceuticals in Africa and Europe.

Azanu et al. (2018) and Gyesi et al., (2022) have reported the concentrations of pharmaceuticals in hospital effluents in Ghana. Ciprofloxacin (antibiotic) was detected in Kumasi South Hospital (KSH) and KATH (Komfo Anokye Teaching Hospital) effluent samples with a highest detection concentration of 0.232 mg/L and this can be attributed to its broad spectrum nature and use for treatment of multi-drug resistant bacterial infections (Johnson & Sherrell, 2022). There has been report on its detection in some hospitals effluents by Azanu et al. (2018) and Serna-Galvis et al. (2022) an indication of its widespread nature. Metronidazole (antibiotic) was the pharmaceutical with the least detected concentration of 0.042 mg/L found in KSH effluent sample. Metronidazole might have a high tendency to be adsorbed on to sediment or soil particles as a study by Matongo et al. (2015) showed its detection in sediments and bio-soils. Thus, some pharmaceuticals adsorb to solids suspended in wastewater while the soluble ones remain in solution. This agrees with a report by Gyesi et al. (2022) that most pharmaceuticals are soluble in water with few being insoluble while others get attached to suspended particles. The presence of antibiotics in hospital effluents makes them more likely to be found in other compartments. This helps antibiotic-resistant bacteria (ARB) and antibiotic-resistant genes (ARG) grow in these compartments (Ahmed et al.,2022; Borquaye et al., 2019; Gyesi et al., 2022). Correlations between the physicochemical parameters and pharmaceutical concentrations were low. The high risk of ciprofloxacin to algae is an indication of its potential hazards in aquatic environment, hence the need for immediate attention as suggested by Gyesi et al. (2022). Though the risks of other pharmaceuticals considered were low, there is the possibility of their exhibition of medium to high risk since they are becoming rampant and can accumulate as indicated by O'Flynn et al. (2021). Both ciprofloxacin and metronidazole have exhibited resistance risk with the former being higher than the latter. Hospital effluents therefore serve as major sources of pharmaceutical contamination of the environment (Krakkó et al., 2019), water bodies and have been reported over the last three years (2012–2022).

3.2 Pharmaceuticals in Wastewater, Surface Water, and Groundwater

The occurrence of pharmaceuticals in naturally occurring water bodies has been the subject of several studies in Africa. Chaib et al. (2019) determined the occurrence of seven antibiotics in surface waters in the city of Fez, Morocco, in which amoxicillin was present at the highest concentrations along with sulfamethoxazole and erythromycin. Also, Kermia et al. (2016) analyzed the presence of anti-inflammatory in rivers in Algeria (ibuprofen, naproxen, ketoprofen, and diclofenac) and discovered that naproxen was present at the highest concentrations. Findings of a study confirmed the presence of amoxicillin, trimethoprim, diazepam, ibuprofen, cefuroxime, and acetaminophen in the surface water for vegetable irrigation in Sunyani Ghana (Azanu et al., 2021). Out of 31 research documents published on the environmental monitoring studies conducted in Africa, 61% of the articles report pharmaceuticals present in wastewater while 48% are present in river water (Madikizela et al., 2020). The occurrence of pharmaceuticals in dam and ground water has been reported in South Africa (Rimayi et al., 2018; Wanda et al., 2017), Zimbabwe (Teta et al.,

2018) and Cameroon (Branchet et al., 2019) and in drinking water sources such as tap water and well-water in Nigeria (Olaitan et al., 2017).

3.3 Pharmaceuticals in Sediments

Pharmaceuticals are known to sorb onto solids and to partition between river waters and sediments (Fountouli and Chrysikopoulos, 2018; Xiang et al., 2021). The maximum concentration of the pharmaceuticals ranged from 0.01 to 1.17×103 ng/g, with an average concentration of 52 ng/g. Pharmaceuticals in the river or lake adsorb onto the soil/sediments, dilute, and undergo biological and/or photochemical transformations (Balakrishna et al. 2016; Onesios et al., 2008).

4. RISK ASSESSMENT

4.1 Environmental Risk Assessment

According to the European Medicines Agency (2015), environmental risk assessment (ERA) of pharmaceuticals is the study of the potential effects of pharmaceuticals on the environment and the precautionary measures to be taken. It involves an estimation of the exposure (based on bioaccumulation and persistence) to and analysis of the fate and effect of pharmaceuticals; and depicts a function of exposure and hazard (Oelkers, 2020; Vestel et al., 2016). Contemporary ERA procedures have led to the acquisition of more insight in environmental issues and aid in protecting environmental health (Lee & Choi, 2019); thus, managing the environment and making decisions with respect to pharmaceuticals depends on their ERA (He et al., 2020).

Environmental risk assessment is a legal requirement before a drug is allowed for marketing; to restrict its potential ill effect on the environment (European Medicines Agency, 2015). Though ERA, which includes the ranking of drugs for subsequent monitoring and control measures in the environment is needed for the management of pharmaceuticals (Oelkers & Floeter, 2019); information on the potential risk of most pharmaceuticals for humans is non-existent (Gunnarsson et al., 2019). Consequently, Zhu et al. (2019) in their study revealed that the environmental risks of most Active Pharmaceutical Ingredients (APIs) are not known though Lee & Choi (2019) suggest that risk assessment of active ingredients of drugs is a better option. Han & Lee (2017) suggest that the metabolites of pharmaceuticals should be considered in ERA of Active Pharmaceutical Ingredients (APIs) as they bear equal or higher risk. Such data, according to Rand-Weaver et al. (2013) will help ascertain the authenticity of generally accepted read-across environmental risk hypothesis.

Again, ERA of pharmaceuticals should be carried out periodically at intervals of five years as indicated by Pereira et al. (2017).

4.2 Guidelines and Frameworks for Risk Assessment of Pharmaceuticals

There are various guidelines or frameworks for assessing the risk of pharmaceuticals; the commonly used ones being European Medicinal Agency and U.S. Food and Drug Administration guidelines (Jose et al., 2020; Lee & Choi, 2019; Walter & Mitkidis, 2018). In their review, Berkner et al. (2016) considered the risk assessment of nano pharmaceuticals using the guidelines of European Union which is based on active pharmaceutical ingredient though Ågerstrand et al. (2015) had proposed certain recommendations for improvement of the guidelines. While Pereira et al. (2017) suggest of the inclusion of consumption and excretion data for improvement, Lee & Choi (2019) point out that the focus should be on active ingredient instead of the product. In comparing regulatory conceptual structure of ERA for human pharmaceuticals, Lee & Choi (2019) revealed that while Canada focuses on substances, European Union (EU) and United States of America (USA) consider products. The estimation of environmental risk of pharmaceuticals is based on aquatic concentrations that transcend the Expected Environmental Concentration (EEC) of 0.1 µg/L in the USA or Predicted Environmental Concentration (PEC) of 0.01 µg/L as well as the Persistence, Bioaccumulation and Toxicity (PBT) of the pharmaceutical in the EU (Vestel et al., 2016).

4.3 Procedure for Environmental Risk Assessment

Assessing the environmental risk of pharmaceutical residues involves identification of hazard; assessment of exposure which gives the PEC; characterization of hazard which gives the Predicted No Effect Concentration (PNEC); and characterization of risk which is the Risk Quotient (RQ) value given by the ratio of PEC to PNEC, i.e. PEC/PNEC for water, sediment and biota (Amiard & Amiard-Triquet, 2015). In estimating RQ value, the PEC in the equation can be replaced with Measured Environmental Concentration (MEC) in situations where data on environmental concentration is available (Peake et al., 2016), from field studies or modeled values (Belden, 2020).The RQ value becomes a ratio of MEC to PNEC, i.e., MEC/PNEC (Palma et al., 2014) which according to Zhou et al. (2019) is a proposed estimation by the European Commission in 2003 and is commonly used.

Predicted No Effect Concentration (PNEC) is defined as the concentration below which, there will be no adverse effect on an organism (European Commission, 2003) and can be derived using two approaches: that is either from Assessment Factors

Table 2. Assessment factor scheme for PNEC estimation

Toxicity Data	Assessment Factor
One short-term EC_{50} from each trophic level	1000
Long-term NOEC from one trophic level	100
Long-term NOEC from two trophic levels	50
Long-term NOEC from three trophic levels	10
Field data or modeled ecosystem	Based on scenario

Source: European Commission (2003) in Smit et al. (2005)

(AFs) or Species Sensitivity Distribution (SSD) (Figure 4) (Smit et al., 2005). Using the AFs, PNEC is estimated as a ratio of Median Lethal Concentration value (LC_{50}), Median Effective Concentration (EC_{50}), No-Observed Effect Concentration (NOEC) or the Lowest-Observed Effect Concentration (LOEC) to AFs (European Chemicals Agency, 2008). Thus, Belden (2020) confirms that LC_{50}, EC_{50}, NOEC or LOEC of a pharmaceutical are the toxicological dose descriptors used for estimating PNEC. While LC_{50} and EC_{50} are obtained from short term studies and therefore provide acute toxicity data, NOEC and LOEC are obtained from long term studies and provide chronic toxicity data (Zhou et al., 2019).

The value of the AFs depend on the environmental compartment and its associated trophic levels under consideration as well as duration (long or short term), as shown in Table 4.1; and is meant to deal with uncertainties regarding natural environmental conditions as compared to laboratory or test or controlled conditions, extrapolating from short to long term toxicity and variations among intra and intra species (Edinburgh Centre for Toxicology, 2015; European Chemicals Agency, 2009; European Commission, 2003). Thus, Bebianno & Gonzalez-Rey (2015) confirm that calculation of PNEC is based on acute and chronic toxicity data. Algae, daphnia, and fish are the trophic levels considered for aquatic environment (European Chemicals Agency, 2009). The AFs cater for the assumptions that the sensitivity of ecosystem is dependent on the most sensitive species and the level of protection of ecosystem structure is adequate for the protection of functioning of community.

The SSD is a model that is used to deduce PNEC statistically and it depicts the different sensitivities of multiple species to a toxicant or stressor (Kitsis, 2015). Species Sensitivity Distribution (SSD) in Figure 4.1 is a description of the changes in hazard of a substance to organism and shown as frequency distribution or cumulative curve using NOEC values (Smit et al., 2005). It is a dose-effect curve in which data for toxicity are changed into log and matched to a distribution function. There are several distribution functions, but the choice depends on results of best fit. In a normal cumulative distribution, NOEC values for species are matched to a

Figure 1. Species sensitivity distribution
(Smit et al., 2005)

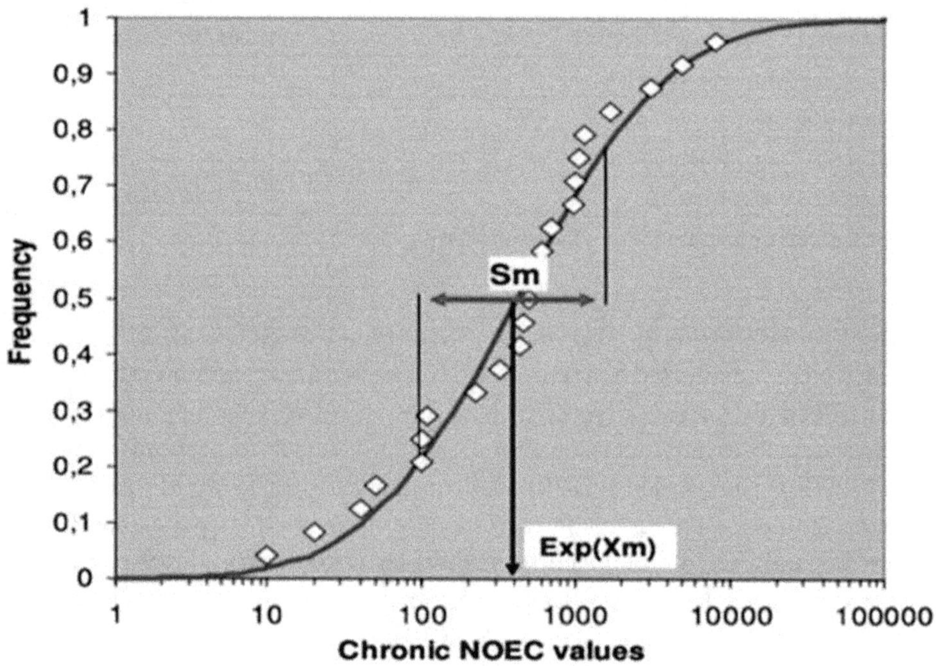

logarithmic scale (Smit et al., 2005). The PNEC value can be statistically obtained from SSD using a percentile. The concentration that corresponds to the point on the SSD curve, which 5% of species occur (Figure 4), is considered as intermediate value to determine PNEC (Smit et al., 2005). Xiaowei et al. (2011) in their study revealed that using AFs in determining PNEC gives a secure value than the use of SSD model.

4.4 Risk Assessment Values

According to Zhou et al. (2019), environmental risk is low when RQ < 0.1, moderate when 0.1 < RQ < 1 and high when 1 < RQ < 10 in Table 4.2. Thus, some pharmaceuticals have low environmental risk whiles others have high environmental risk (Chen et al., 2016). Paracetamol, ibuprofen, ciprofloxacin, clarithromycin, omeprazole and triclosan are pharmaceuticals that exhibit risk in aquatic environment and wastewater treatment plants (Ortiz de García et al., 2014), and Chen et al. (2016) identified ibuprofen as a drug with high environmental risk (Table 4.2). As shown in Table 5, Ashfaq et al. (2017) have shown that diclofenac, ibuprofen and ofloxacin pose high risk. According to Carvalho et al. (2014), caution should be taken when

Table 3. Description of risk quotient values

Value	Risk	Pharmaceuticals
RQ < 0.1	Low	Anti-inflammatories, ß-blockers, diuretics, paracetamol (Chen et al.,2016)
0.1 < RQ < 1	Moderate	Clarithromycin, erythromycin, caffeine, ciprofloxacin, roxithromycin, salicylic acid and sulfamethoxazole, triclosan (Zhou et al., 2019)
1 < RQ < 10	High	Diclofenac, ethinylestradiol, paraxanthine, ibuprofen, amoxicillin, atorvastatin, carbamazepine, estriol, ofloxacin, ranitidine, spiramycin, venlafaxine and zidovudine (Ashfaq et al., 2017; Chen et al., 2016; Zhou et al., 2019)

assessing the risk of several chemicals occurring at a time due to cocktail effect and synergistic interaction, though the latter is regarded as less important (Cedergreen, 2014). However, Claessens et al. (2013) have reported on the non-existence of risk posed by pharmaceutical mixtures.

According to study done by Azanu et al, 2016, applying the risk ranking scale (Hernando et al., 2006), on RQ of all antibiotics calculated for algae, showed ciprofloxacin with RQ of 1.3 x 10^{-1} could cause median risk to algae in aquatic environment. The rest of the antibiotics would cause minimal risk to the algae, daphinds and fish in the aquatic environment. Comparatively a study focused on the occurrence of pharmaceuticals in water from Pego-Oliva Marsh, Spain, found ciprofloxacin RQ to be 6.9 in algae (Vazquez-Roig et al., 2012). Other studies have reported high RQs in surface water due to the presence of antibiotics in high concentrations, sulfamethoxazole (García-Galán et al., 2011), amoxicillin and oxytetracycline (Jones et al., 2002). These data all indicate that the potential toxicological effects posed by antibiotics for the aquatic environment could be reduced when the hydraulic characteristics of the receiving water bodies are taken into consideration in their management, as supported by Gros et al., (2010). This risk evaluation has its limitations, such as the lack of long-term toxicological studies and the unfeasibility to carry out chronic studies during the lifespan of the organisms (especially in fishes), but on the other hand since mixture of compounds with the same pharmacological mechanism is present in waters, synergistic effects could be expected (Gros et al., 2010), making the real hazard greater than that calculated for individual antibiotics. For example, assuming similar effects and concentration addition for the 4 tetracyclines investigated, the RQs for this group would be 1.4 x 10^{-7} for algae, 1.3 x 10^{-7} for daphnids, 3.6 x 10^{-8} for fish and 9.4 x 10^{-2} for antibiotic resistance development. Nevertheless, the presence of antibiotics in the environment is not limited only to an ecological problem but most importantly to antibiotics resistance.

Antibiotics resistance due to bacteria exposure to sub-MIC could occur in the study area, because Gullberg et al. (2011) found minimum selective concentration (MSC) for antibiotic resistant mutant to be 100 ng/L for ciprofloxacin and 15 μg/L for tetracycline suggesting that there is medium to high risk of antibiotics resistance development in the environment. The fact that antibiotic levels several hundred-fold below the MIC of the susceptible strains can select resistant bacteria means that the sub-MIC selective window is much larger than the traditional selective window (Andersson & Hughes, 2014). Further work would be required to determine the extent of exposure and the risk associated. Since these point sources of antibiotics in Ghana are discharged into streams and nearby rivers used for farming, there is a need to investigate the possibility of contamination of foods. Use of these surface water studied for vegetable production could be taken up by plants (Azanu et al., 2016b), eventually, contaminating the food quality and again exposing the vegetable consumers to antibiotics.

Antibiotic presence in the environment can significantly contribute to the emergence and spread of resistant strains of bacteria increasing the threat to public health ((Azanu et al., 2018)).

5. POSSIBLE REMOVAL TECHNOLOGIES

5.1 Electrochemical Processes for the Treatment of Pharmaceutical Wastewater

Electrochemical processes play a crucial role in the treatment of pharmaceutical wastewater due to their ability to efficiently remove contaminants and pollutants. Several mechanisms are involved in these processes, including electrocoagulation, electrooxidation, and electrochemical advanced oxidation processes (EAOPs). The mechanisms involved in the electrochemical processes are;

5.1.1 Electrocoagulation (EC)

Electrocoagulation involves the use of an electric current to destabilize and aggregate colloidal particles and dissolved substances in wastewater. This process occurs through the following steps: (i) generation of coagulant species (metal ions or hydroxides) at the anode, (ii) transport of these species towards the cathode, (iii) neutralization of charged species, and (iv) formation of flocs that can be easily separated from the wastewater. EC is effective in removing suspended solids, heavy metals, and certain organic pollutants (Bazrafshan et al., 2015)

5.1.2 Electrooxidation

Electrooxidation involves the direct or indirect oxidation of organic pollutants present in wastewater using an electric current. Direct electrooxidation occurs at the anode, where organic compounds are oxidized to simpler and less toxic species. Indirect electrooxidation involves the generation of reactive oxidizing species (e.g., hydroxyl radicals) at the anode, which then react with organic pollutants in the bulk solution. Electrooxidation is effective in removing various organic contaminants, including pharmaceutical residues (Mousset et al., 2016)

5.1.3 Electrochemical Advanced Oxidation Processes (EAOPs)

EAOPs combine electrochemical and advanced oxidation techniques to enhance the degradation of recalcitrant organic pollutants. These processes typically involve the electro generation of highly reactive oxidizing species, such as hydroxyl radicals, through indirect or direct means. EAOPs can efficiently degrade a wide range of persistent organic pollutants, including pharmaceuticals, by promoting oxidation reactions. Examples of EAOPs include electro-Fenton, electro-peroxone, and electrochemical ozonation (Sirés et al., 2014)

Below are the advantages and disadvantages of electrochemical processes in the treatment of pharmaceutical wastewater,

5.1.4 Advantages of Electrochemical Processes

i. Efficient removal of pollutants: Electrochemical processes offer high removal efficiency for various pharmaceutical compounds present in wastewater, including antibiotics, hormones, and organic pollutants (Nabgan et al., 2022).
ii. Wide applicability: Electrochemical processes can be applied to treat a wide range of pharmaceutical wastewater compositions, including low and high concentration wastewater and complex mixtures of pharmaceutical compounds (Gogate, & Pandit, 2004).
iii. Versatile process conditions: Electrochemical processes can be operated under different process conditions, allowing flexibility in treatment. Factors such as current density, pH, temperature, and electrode materials can be adjusted to optimize the removal efficiency for specific pharmaceutical compounds (Panizza & Cerisola, 2005).
iv. Selective degradation: Electrochemical processes can be tailored to selectively target specific pharmaceutical compounds, which is beneficial in treating wastewater with multiple contaminants (Raj et al., 2021)

5.1.5 Disadvantages of Electrochemical Processes

i. High energy consumption: Electrochemical processes typically require significant energy input due to the need for electricity to drive the electrochemical reactions. The energy consumption can be high, especially when treating large volumes of wastewater or dealing with highly recalcitrant pharmaceutical compounds (Ganiyu et al., 2015).
ii. Electrode fouling: During the treatment of pharmaceutical wastewater, electrode fouling can occur, leading to reduced treatment efficiency and increased maintenance requirements. Fouling can be caused by the deposition of reaction by-products or the formation of biofilms on the electrode surface (Burgos-Castilo et al., 2018)
iii. Generation of disinfection by-products: Some electrochemical processes, particularly those involving chloride-containing wastewater, can lead to the formation of disinfection by-products (DBPs) with environmental and health concerns, necessitating further treatment steps (Huang et al., 2021).
iv. Limited scale-up experience: Large-scale application and commercialization of electrochemical processes for pharmaceutical wastewater treatment are still limited, requiring further research and development for optimization (Zhang et al., 2019).

5.2 Biofiltration Removal Method

Biofiltration is an effective biological treatment process used in the removal of organic and inorganic pollutants from wastewater, including pharmaceutical wastewater. It involves the use of microbial communities to degrade and transform contaminants into less harmful substances. The mechanism of biofiltration typically consists of several steps, including adsorption, microbial degradation, and final product formation.

i. Adsorption: When pharmaceutical wastewater flows through a biofilter, contaminants present in the water, such as pharmaceutical compounds, are initially adsorbed onto the surface of the filter media. The filter media typically consists of a porous material, such as activated carbon or compost, which provides a large surface area for adsorption to occur (Calheiros et al., 2009).
ii. Microbial degradation: Microorganisms play a crucial role in the biofiltration process by degrading the adsorbed contaminants. Bacteria and fungi present in the biofilm on the filter media utilize the organic compounds as a source of energy and carbon. The microbial community consists of various species with different metabolic capabilities, allowing for the degradation of a wide range of pharmaceutical compounds (Kalyuzhnyi & Gladchenko 2014).

a. Aerobic degradation: In the presence of oxygen, aerobic microorganisms utilize the pharmaceutical compounds as a carbon source and break them down into simpler compounds through enzymatic reactions. These reactions involve processes such as hydroxylation, oxidation, deamination, and decarboxylation. The end products of aerobic degradation are typically carbon dioxide (CO2), water (H2O), and inorganic compounds (Shah et al., 2015).

b. Anaerobic degradation: In the absence of oxygen, anaerobic microorganisms can also contribute to the degradation of pharmaceutical compounds. These microorganisms employ various anaerobic metabolic pathways, such as fermentation and methanogenesis, to convert complex organic compounds into simpler products. The end products of anaerobic degradation may include methane (CH4), carbon dioxide (CO2), and other volatile organic compounds (Mrozik & Piotrowska 2015).

iii. Final product formation: Throughout the biofiltration process, the degradation of pharmaceutical compounds results in the formation of less harmful substances. The transformation products formed depend on the specific compounds present in the wastewater and the metabolic capabilities of the microbial community. In some cases, the final products may still have some degree of toxicity, but they are generally less harmful and more amenable to further treatment or removal (Matamoros et al., 2019).

5.2.1 Advantages of Biofiltration in Pharmaceutical Wastewater Treatment

The advantages of biofiltration in pharmaceutical wastewater treatment include:

i. High removal efficiency
ii. Cost-effectiveness
iii. Sustainability
iv. Versatility
v. Scalability

5.2.2 Disadvantages of Biofiltration Processes

While biofiltration is an effective method for treating pharmaceutical wastewater, it also has some limitations and disadvantages that need to be considered. These disadvantages include:

i. Long start-up period

ii. Sensitivity to operational conditions
iii. Limited removal of recalcitrant compounds
iv. Potential formation of metabolites
v. Space requirements

5.3 Membrane Separation Processes

A membrane is a barrier that separates two phases from each other by restricting the transportation of constituents through it in a selective manner (Ezugbe & Rathilal, 2020). Membranes, which have been in existence from the 18th century has undergone various forms of modification and transformations to increase their suitability for the gamut of applications in the treatment of wastewater with recent interests in emerging contaminants such as pharmaceutical active ingredients (Takht Ravanchi et al., 2009; Fane et al., 2011).

The commonly used membranes can be categorized into: conventional membranes and commercial composite thin film (TFC)membranes (Cevallos-Mendoza et al., 2022). A thin active layer of polyamide (PA) (<200 nm), which can be obtained by interfacial polymerization, is deposited on a porous layer of polyethersulfone (PES) or polysulfone (PSU) with a pore size of about 50 microns (Hosseini et al., 2016).

Membrane filtration systems differ in terms of pore size and surface properties. In descending order of pore sizes, they comprise microfiltration, ultrafiltration, nanofiltration, reverse osmosis and forward osmosis. (Rezakazemi et al., 2018). The selection of the type of membrane filtration depends on the properties of the contaminants to be separated. Properties such as particle size, charge, and hydrophobicity (Alfonso-Muniozguren et al., 2021).

Membrane filtration of liquid effluent feed will generate two different streams namely retentate and permeate. The retentate is a concentrated stream containing components that are rejected by the membrane and the permeate stream is the water that passes through the membrane (Rezakazemi et al., 2018). Due to the size and molecular weight of the emerging pharmaceutical contaminants, it has been observed that microfiltration and ultrafiltration systems are inefficient (Ganiyu et al., 2015). However the efficiency of the diminution of active pharmaceutical ingredients from water is significantly higher when Nanofiltration, reverse osmosis or forward osmosis systems are used.(Alfonso-Muniozguren et al., 2021). Antibiotic removal rates of between 90% and 99% have been attained using this technique (Sahar et al., 2011; Dolar et al., 2009).

5.4 Nanofiltration

Nanofiltration membranes came to the forefront of membrane separation processes in the late 1980 with properties between Reverse Osmosis and Ultra filtration. Pressure is applied to transport the molecular mixture to the surface of the membrane. The solvent and some solutes with low molecular weights penetrate through the membrane while the other components are retained (Bera et al., 2022)

5.5 Artificial Wetlands

It is an artificial ecosystem designed to treat wastewater before it enters wetlands. To do so, soil, plants, and microorganisms work in concert. Filtration, adsorption, co-precipitation, and ion exchange are used to purify wastewater microbial decomposition, plant adsorption, and exchange (Vymazal, 2005). According to the direction of water flow, this treatment can be divided into three categories: Free Water Surface flow Constructed Wetlands, Horizontal Subsurface Flow Constructed Wetlands, and Vertical Subsurface Flow Constructed Wetlands. (Huang et al., 2021)

5.6 Adsorption

Among tertiary treatments, adsorption offers benefits including lower energy usage and easier operating conditions, making it stand out as a potential treatment option. In the field of pollution control and waste management, pharmaceutical removal from water and effluents has been regarded promising since adsorption enables straightforward design and operation and does not introduce unwanted byproducts to the system (Grassi et al., 2012). Despite receiving substantial research as a pharmaceutical adsorbent, commercial activated carbon's large-scale applicability is constrained by its inflated cost. As a result, numerous research for pharmaceuticals absorption from water and wastewater have focused on adsorbents based on clays, charcoal, chitosan, agricultural and industrial wastes, and metalorganic frameworks. These non-conventional low-cost materials have also been examined in more detail (Andrade et al., 2018). Molecular imprinted polymers (MIPs), thermoresponsive gel, and magnetic nanoparticles are some other more recent materials that have been reported; but, in comparison, they have not yet been developed or have case studies to support them (Mansouri et al., 2021). Studies have been done using the biochar (Sun et al. 2018; Li et al., 2018; Han et al., 2013) from different precursors, Granular Activated Carbon (GAC) (Liu et al., 2017; Kovalova et al., 2013) and Powdered Activated Carbon (PAC) (Yang et al., 2011) adsorbents to remove antibiotics from wastewater and hospital wastewater. Using cellulose membranes is another adsorption alternative. In a study using ciprofloxacin and this approach, 27% eradication was attained (Sulbaran et al., 2019). Nanoparticles (Chao et al., 2014; Rathod et al., 2015) have been used to removal gatifloxaxin, amoxicillin cephalexin, tetracycline, and penicillin G. Also granulated plastics (Dia et al., 2023) have been used in the adsorption of ciprofloxacin. Zeolites have also been employed in tests to adsorb ciprofloxacin. Antibiotic removal rates of between 90% and 97% have been attained using this technique (de Ilurdoz et al., 2022). Therefore, these have been demonstrated to be effective in the removal of antibiotic residues and other pharmaceuticals with similar structural features in pharmaceutical company effluent and water bodies.

5.7 Ongoing Study on the Removal of Antibiotics from Water Using Microplastics

Microplastics can behave as adsorbents to remove the antibiotics pollutant in water bodies. This study investigated the feasibility of removing amoxicillin (AMX), and metronidazole (MET) antibiotics from aqueous media using microplastics polyethylene terephthalate (PET bottles) and high-density polyethylene (HDPE, water sachets). Scanning Electron microscopy, XRD and FT-IR spectroscopy were used to characterize the microplastic. Antibiotics sorption studies were carried out, and the data fitted with adsorption models of Linear, Langmuir, Freundlich, Timken, Elovich, Jovanovic, and Hurking-Jura isotherm. The result of the experiment showed that the equilibrium time for sorption was 30 min. The adsorption isotherm models that best fit the data was Freundlich isotherm models with R^2 =0.940 for AMX and R^2 =0.968 for MET. The sorption study revealed that microplastics sorption decreases with pH increase. This could be due to electrostatics attraction between them and it found that the shape of adsorption is type 2 sinusoidal. In conclusion using plastics as adsorbent can be a suitable way to remove antibiotics pollutant from water bodies and a recycling method to reduce the amount of plastic waste released into the environment.

6. CONCLUSION REMARKS

The outcomes of this chapter suggest there could be indirect exposure of humans to antibiotics through vegetable consumption and drinking water in Ghana. Although the levels found in the environment could not cause toxic effects to human, further research needs to be investigated since low levels of antibiotics in food and low-quality water could contribute to development of bacterial resistance. Also, it is evident that a variety of methods accomplish great eradication rates of pharmaceutical residues. So, whether to utilize one method or the other would rely on the amenities currently present at the site or country. Therefore, the commercialization of these method needs to be studied to identify the economical feasible removal (treatment) methods for Ghana.

REFERENCES

Afriyie, E. O., Edusei, A. K., Mensah, K. A., Nkoom, B., Okyere, P., Adjei, R. O., & Appiah-Brempong, E. (2015). Antibiotics Availability and Usage in Health Facilities: A Case of the Offinso-South Municipality of Ghana. *Journal of Biology*, *5*(2), 132–138.

Ågerstrand, M., Berg, C., Björlenius, B., Breitholtz, M., Brunström, B., Fick, J., Gunnarsson, L., Larsson, D. G. J., Sumpter, J. P., Tysklind, M., & Rudén, C. (2015). Improving environmental risk assessment of human pharmaceuticals. *Environmental Science & Technology*, *49*(9), 5336–5345. doi:10.1021/acs.est.5b00302 PMID:25844810

Akin, B. S. (2016). Contaminant properties of hospital clinical laboratory wastewater: A physiochemical and microbiological assessment. *Journal of Environmental Protection*, *7*(5), 635–642. doi:10.4236/jep.2016.75057

Al Aukidy, M., Al Chalabi, S., & Verlicchi, P. (2017). Hospital wastewater treatments adopted in Asia, Africa, and Australia. In P. Verlicchi (Ed.), *Hospital Wastewaters. The Handbook of Environmental Chemistry* (Vol. 60). Springer., doi:10.1007/698_2017_5

Alfonso-Muniozguren, P., Serna-Galvis, E. A., Bussemaker, M., Torres-Palma, R. A., & Lee, J. (2021). A review on pharmaceuticals removal from waters by single and combined biological, membrane filtration and ultrasound systems. *Ultrasonics Sonochemistry*, *76*, 105656. doi:10.1016/j.ultsonch.2021.105656 PMID:34274706

Amiard, J., & Amiard-Triquet, C. (2015). Conventional risk assessment of environmental contaminants. In: Advancing tools for dealing with emerging risks. Aquatic Ecotoxicology, 2, 25-49. doi:10.1016/B978-0-12-800949-9.00002-4

Amoaei, A., Asgharnia, H., Fallah, H., Faraji, H., Barari, R., & Naghipour, D. (2015). Characteristics of effluent wastewater in hospitals of Babol University of Medical Sciences, Babol, Iran. *Health Scope*, *4*(2), e23222. doi:10.17795/jhealthscope-23222

Anderson, L. A. (2019). Antibiotics guide. *Drugs.com*. https://www.drugs.com/article/antibiotics.html. Accessed on 25/12/2021.

Andrade, S. N., Veloso, C. M., Fontan, R. C. I., Bonomo, R. C. F., Santos, L. S., Brito, M. J. P., & Diniz, G. A. (2018). Chemical-activated carbon from coconut (Cocos nucifera) endocarp waste and its application in the adsorption of ß-lactoglobulin protein. *Revista Mexicana de Ingeniería Química*, *17*(2), 463–475. doi:10.24275/10.24275/uam/izt/dcbi/revmexingquim/2018v17n2/Andrade

aus der Beek, T., Weber, F. A., Bergmann, A., Hickmann, S., Ebert, I., Hein, A., & Küster, A. (2016). Pharmaceuticals in the environment—Global occurrences and perspectives. *Environmental Toxicology and Chemistry*, *35*(4), 823–835. doi:10.1002/etc.3339 PMID:26666847

Azanu, D., Styrishave, B., Darko, G., Weisser, J. J., & Abaidoo, R. C. (2018). Occurrence and risk assessment of antibiotics in water and lettuce in Ghana. *The Science of the Total Environment*, *622*, 293–305. doi:10.1016/j.scitotenv.2017.11.287 PMID:29216470

Azuma, T., Arima, N., Tsukada, A., Hirami, S., Matsuoka, R., Moriwake, R., Ishiuchi, H., Inoyama, T., Teranishi, Y., Yamoako, M., Mino, Y., Hayasshi, T., Fujita, Y., & Masada, M. (2016). Detection of pharmaceuticals and phytochemicals together with their metabolites in hospital effluents in Japan, and their contribution to sewage treatment plant influents. *The Science of the Total Environment*, *548-549*, 189–197. doi:10.1016/j.scitotenv.2015.12.157 PMID:26802347

Balakrishna, K., Rath, A., Praveenkumarreddy, Y., Guruge, K. S., & Subedi, B. (2017). A review of the occurrence of pharmaceuticals and personal care products in Indian water bodies. *Ecotoxicology and Environmental Safety*, *137*, 113–120. doi:10.1016/j.ecoenv.2016.11.014 PMID:27915141

Bazrafshan, E., Mahvi, A. H., & Naseri, S. (2015). Evaluation of Electrocoagulation Process Efficiency for Direct and Indirect Dyes Removal from Synthetic Wastewater. *Journal of Environmental Health Science & Engineering*, *13*(1), 43. doi:10.118640201-015-0191-6

Bebianno, M. J., & Gonzalez-Rey, M. (2015). Ecotoxicological risk of personal care products and pharmaceuticals. In *Advancing Tools for Dealing with Emerging Risks 16* (pp. 383–416). Aquatic Ecotoxicology., doi:10.1016/B978-0-12-800949-9.00016-4

Belden, J. (2020). Introduction to ecotoxicology. In: An Introduction to Interdisciplinary Toxicology: From Molecules to Man, 5, 381-393. doi:10.1016/B978-0-12-813602-7.00028-4

Bera, S. P., Godhaniya, M., & Kothari, C. (2022). Emerging and advanced membrane technology for wastewater treatment: A review. *Journal of Basic Microbiology*, *62*(3-4), 245–259. doi:10.1002/jobm.202100259 PMID:34496068

Berkner, S., Schwirn, K., & Voelker, D. (2016). Nanopharmaceuticals: Tiny challenges for the environmental risk assessment of pharmaceuticals. *Environmental Toxicology and Chemistry*, *35*(4), 780–787. doi:10.1002/etc.3039 PMID:25931425

BIO Intelligence Service. (2013). *Study on the environmental risks of medicinal products, Final Report.* Executive Agency for Health and Consumers. https:// ec.europa.eu>files > files >study_environment.

Borquaye, L. S., Ekuadzi, E., Darko, G., Ahor, H. S., Nsiah, S. T., Lartey, J. A., Mutala, A. H., Boamah, V. E., & Woode, E. (2019). Occurrence of antibiotics and antibiotic-resistant bacteria in landfill sites in Kumasi, Ghana. *Journal of Chemistry*, *2019*, 2019. doi:10.1155/2019/6934507

Editors of Encyclopaedia. (2023, May 21). antibiotic. *Encyclopedia Britannica.* https://www.britannica.com/science/antibiotic

Burgos-Castillo, R. C., Sirés, I., Sillanpää, M., & Brillas, E. (2018). Application of electrochemical advanced oxidation to bisphenol A degradation in water. Effect of sulfate and chloride ions. *Chemosphere*, *194*, 812–820. doi:10.1016/j. chemosphere.2017.12.014 PMID:29268102

Calheiros, C. S. C., Rangel, A. O. S. S., & Castro, P. M. L. (2009). The effect of water and wastewater composition on the adsorption of pharmaceuticals onto activated carbons. *The Science of the Total Environment*, *407*(12), 3793–3802. PMID:19345984

Carraro, E., Bonetta, S., Bertine, C., Lorenzi, E., Bonetta, S., & Gilli, G. (2016). Hospital effluents management: Chemical, physical, microbiological risks and legislation in different countries. *Journal of Environmental Management*, *168*, 185–199. doi:10.1016/j.jenvman.2015.11.021 PMID:26708649

Carvalho, R. N., Arukwe, A., Ait-Aissa, S., Bado-Nilles, A., Balzamo, S., Baun, A., Belkin, S., Blaha, L., Brion, F., Conti, D., Creusot, N., Essig, Y., Ferrero, V. E. V., Flander-Putrle, V., Fürhacker, M., Grillari-Voglauer, R., Hogstrand, C., Jonáš, A., Kharlyngdoh, J. B., Loos, R., Lundebye, A., Modig, C., Olsson, P., Pillai, S., Polak, N., Potalivo, M., Sanchez, W., Schifferli, A., Schirmer, K., Sforzini, S., Stürzenbaum, S. R., Søfteland, L., Turk, V., Viarengo, A., Werner, I., Yagur-Kroll, S., Zounková, R. & Lettieri, T. (2014). Mixtures of chemical pollutants at European Legislation Safety Concentrations: How safe are they? *Toxicological Sciences*, *141*(1), 218–233, https://doi.org/10.1093/toxsci/kfu118 doi:.

Carra, I., Ortega-Clemente, L. A., Perales, J. F., & Giménez, J. (2013). Optimization and validation of a photoreactor for the degradation of diclofenac in wastewater treatment plant effluents. *Chemical Engineering Journal*, *230*, 79–88.

Cedergreen, N. (2014). Quantifying synergy: A systematic review of mixture toxicity studies within environmental toxicology. *PLoS One*, *9*(5), e96580. doi:10.1371/journal.pone.0096580 PMID:24794244

Centers for Disease Control and Prevention. (2021). *Antibiotic resistance questions and answers*. CDC. https://www.cdc.gov/antibiotic-use/antibiotic-resistance.html

Cevallos-Mendoza, J., Amorim, C. G., Rodríguez-Díaz, J. M., & Montenegro, M. D. C. B. (2022). Removal of contaminants from water by membrane filtration: A review. *Membranes*, *12*(6), 570. doi:10.3390/membranes12060570 PMID:35736277

Chao, Y., Zhu, W., Wu, X., Hou, F., Xun, S., Wu, P., Ji, H., Xu, H., & Li, H. (2014). Application of graphene-like layered molybdenum disulfide and its excellent adsorption behavior for doxycycline antibiotic. *Chemical Engineering Journal*, *243*, 60–67. doi:10.1016/j.cej.2013.12.048

Chen, Y., Vymazal, J., Březinová, T., Koželuh, M., Kule, L., Huang, J., & Chen, Z. (2016). Occurrence, removal and environmental risk assessment of pharmaceuticals and personal care products in rural wastewater and treatment wetlands. *The Science of the Total Environment*, *566-567*, 1660–1669. doi:10.1016/j.scitotenv.2016.06.069 PMID:27342641

Chonova, T., Keck, F., Labanowski, J., Montuelle, B., Rimet, F., & Bouchez, A. (2016). Separate treatment of hospital and urban wastewaters: A real scale comparison of effluents and their effect on microbial communities. *The Science of the Total Environment*, *542*, 965–975. doi:10.1016/j.scitotenv.2015.10.161 PMID:26562343

Claessens, M., Vanhaecke, L., Wille, K., & Janssen, C. R. (2013). Emerging contaminants in Belgian marine waters: Single toxicant and mixture risks of pharmaceuticals. *Marine Pollution Bulletin*, *71*(1-2), 41–50. doi:10.1016/j.marpolbul.2013.03.039 PMID:23643408

de Ilurdoz, M. S., Sadhwani, J. J., & Reboso, J. V. (2022). Antibiotic removal processes from water & wastewater for the protection of the aquatic environment-a review. *Journal of Water Process Engineering*, *45*, 102474. doi:10.1016/j.jwpe.2021.102474

Dia, N., Azanu, D., Darmey, J., & Sanogo, A. (2023). Adsorption of ciprofloxacin using composite microplastics. *Research, Education and Emerging Trends for Technological and Sustainable Development, 7th International Conference on Applied Sciences and Technology 2023-KsTU*.

Dolar, D., Košutić, K., Pavlović, D. M., & Kunst, B. (2009). Removal of emerging contaminants of industrial origin by NF/RO-A pilot scale study. *Desalination and Water Treatment*, *6*(1-3), 197–203. doi:10.5004/dwt.2009.636

Edinburgh Centre for Toxicology. (2015). Environmental Risk Assessment. *UNEP/IPCS Training Module No. 3. SectionB*. UNEP. http://www.chem.unep.ch/irptc/Publications/riskasse/B2text.pdf.

European Chemicals Agency. (2008). Guidance on information requirements and chemical safetyassessment. *Chapter R.10: Characterisation of dose [concentration]-response for environment*. Europea. https://echa.europa.eu/documents/10162/13632/information_requirements_r10_en.pdf.

European Medicines Agency. (2015). Environmental risk-assessment of medicines. *Science Health Medicines*. www.ema.europa.eu.

Ezugbe, E. O., & Rathilal, S. (2020). Membrane technologies in wastewater treatment: A review. *Membranes, 10*(5), 89. doi:10.3390/membranes10050089 PMID:32365810

Fane, A. G. T., Wang, R., & Jia, Y. (2011). *Membrane and Desalination Technologies*. Membrane and Desalination Technologies., doi:10.1007/978-1-59745-278-6

Fedaku, S., Alemayehu, E., Dewil, R., & Van der Bruggen, B. (2019). Pharmaceuticals in freshwater aquatic environments. In: A comparison of the African and European challenge. *The Science of the Total Environment, 654*, 324–337. doi:10.1016/j.scitotenv.2018.11.072 PMID:30448654

Ganiyu, S. O., Van Hullebusch, E. D., Cretin, M., Esposito, G., & Oturan, M. A. (2015). Coupling of membrane filtration and advanced oxidation processes for removal of pharmaceutical residues: A critical review. *Separation and Purification Technology, 156*, 891–914. doi:10.1016/j.seppur.2015.09.059

García, J., García-Galán, M. J., Day, J. W., Boopathy, R., White, J., Wallace, S., & Hunter, R. G. (2020). A review of emerging organic contaminants (EOCs), antibiotic resistant bacteria (ARB) and antibiotic resistance genes (ARGs) in the environment: Increasing removal with wetlands and reducing environmental impacts. *Bioresource Technology, 307*, 123228. doi:10.1016/j.biortech.2020.123228 PMID:32247686

Gogate, A., & Pandit, A. S. (2004). A review of imperative technologies for wastewater treatment II: Hybrid methods. *Advances in Environmental Research, 8*(3-4), 553–597. doi:10.1016/S1093-0191(03)00031-5

Grassi, M., Kaykioglu, G., Belgiorno, V., & Lofrano, G. (2012). Removal of Emerging Contaminants from Water and Wastewater by Adsorption Process. In G. Lofrano (Ed.), *Emerging Compounds Removal from Wastewater. SpringerBriefs in Molecular Science(). Springer*. doi:10.1007/978-94-007-3916-1_2

Gu, J., Chen, C., Huang, X., Mo, J., Xie, Q., & Zeng, Q. (2021). Occurrence and risk assessment of tetracycline antibiotics in soils and vegetables from vegetable fields in Pearl River Delta, South China. *The Science of the Total Environment, 776*, 145959. doi:10.1016/j.scitotenv.2021.145959

Gunnarsson, L., Snape, J. R., Verbruggen, B., Owen, S. F., Kristiansson, E., Margiotta-Casaluci, L., Österlund, T., Hutchinson, K., Leverett, D., Marks, B., & Tyler, C. (2019). Pharmacology beyond the patient – the environmental risks of human drugs. *Environment International, 129*, 320–332. doi:10.1016/j.envint.2019.04.075 PMID:31150974

Han, E. J., & Lee, D. S. (2017). Significance of metabolites in the environmental risk assessment of pharmaceuticals consumed by human. *The Science of the Total Environment, 592*, 600–607. doi:10.1016/j.scitotenv.2017.03.044 PMID:28318699

Han, X., Liang, C. F., Li, T. Q., Wang, K., Huang, H. G., & Yang, X. E. (2013). Simultaneous removal of cadmium and sulfamethoxazole from aqueous solution by rice straw biochar. *Journal of Zhejiang University. Science. B., 14*(7), 640–649. doi:10.1631/jzus.B1200353 PMID:23825150

Hosseini, S. S., Bringas, E., Tan, N. R., Ortiz, I., Ghahramani, M., & Shahmirzadi, M. A. A. (2016). Recent progress in development of high-performance polymeric membranes and materials for metal plating wastewater treatment: A review. *Journal of Water Process Engineering, 9*, 78–110. doi:10.1016/j.jwpe.2015.11.005

Huang, A., Yan, M., Lin, J., Xu, L., Gong, H., & Gong, H. (2021). A review of processes for removing antibiotics from breeding wastewater. *International Journal of Environmental Research and Public Health, 18*(9), 4909. doi:10.3390/ijerph18094909 PMID:34062980

Iweriebor, B. C., Iwu, C. J., Obi, L. C., Nwodo, U. U., & Okoh, A. I. (2015). Multiple antibiotic resistances among Shiga toxin producing Escherichia coli O157 in feces of dairy cattle farms in Eastern Cape of South Africa. *BMC Microbiology, 15*(1), 1–9. doi:10.118612866-015-0553-y PMID:26475706

Jia, W. L., Song, C., He, L. Y., Wang, B., Gao, F. Z., Zhang, M., & Ying, G. G. (2022). Antibiotics in soil and water: Occurrence, fate, and risk. *Current Opinion in Environmental Science & Health*, 100437.

Jin, K. R., & Ji, Z.-G. (2015). An integrated environment model for a constructed wetland – Hydrodynamics and transport processes. *Ecological Engineering, 84*, 416–426. doi:10.1016/j.ecoleng.2015.09.022

Jose, J., Pinto, J. S., Kotian, B., Thomas, A. M., & Charyulu, R. N. (2020). Comparison of the regulatory outline of ecopharmacovigilance of pharmaceuticals in Europe, USA, Japan and Australia. *The Science of the Total Environment*, *709*, 134815. doi:10.1016/j.scitotenv.2019.134815 PMID:31887508

Kalyuzhnyi, S., & Gladchenko, M. (2014). Aerobic biodegradation of pharmaceutical compounds. In *Comprehensive Biotechnology* (2nd ed., Vol. 5, pp. 525–538). Academic Press.

Kitsis, G. (2015). Trophic Response to Multiple Stressors Using Species Sensitivity (SSD) Distribution Models. *Scholars Week, 24*. https://cedar.wwu.edu/scholwk/2015/Day_one/24.

Klein, E. Y., Van Boeckel, T. P., Martinez, E. M., Pant, S., Gandra, S., Levin, S. A., Goossens, H., & Laxminarayan, R. (2018). Global increase and geographic convergence in antibiotic consumption between 2000 and 2015. *Proceedings of the National Academy of Sciences of the United States of America*, *115*(15), E3463–E3470. doi:10.1073/pnas.1717295115 PMID:29581252

Kosma, C. I., Kapsi, M. G., Konstas, P. G., Trantopoulos, E. P., Boti, V. I., Konstantinou, I. K., & Albanis, T. A. (2020). Assessment of multiclass pharmaceutical active compounds (PhACs) in hospital WWTP influent and effluent samples by UHPLC-Orbitrap MS: Temporal variation, removals and environmental risk assessment. *Environmental Research*, *191*, 110152. doi:10.1016/j.envres.2020.110152 PMID:32877707

Kostich, M. S., Batt, A. L., & Lazorchak, J. M. (2014). Concentrations of prioritized pharmaceuticals in effluents from 50 large wastewater treatment plants in the US and implications for risk estimation. *Environmental Pollution*, *184*, 354–359. doi:10.1016/j.envpol.2013.09.013 PMID:24095705

Kovalova, L., Siegrist, H., Von Gunten, U., Eugster, J., Hagenbuch, M., Wittmer, A., Moser, R., & McArdell, C. S. (2013). Elimination of micropollutants during post-treatment of hospital wastewater with powdered activated carbon, ozone, and UV. *Environmental Science & Technology*, *47*(14), 7899–7908. doi:10.1021/es400708w PMID:23758546

Krakkó, D., Licul-Kucera, V., Záray, G., & Mihucz, V. G. (2019). Single-run ultra-high performance liquid chromatography for quantitative determination of ten popular active pharmaceutical ingredients by quadrupole time-of-flight mass spectrometry after offline preconcentration by solid phase extraction from drinking and river as well as treated wastewater. *Microchemical Journal*, *148*, 108–119. doi:10.1016/j.microc.2019.04.047

Kümmerer, K. (2010). Pharmaceuticals in the environment. *Annual Review of Environment and Resources*, *35*(1), 57–75. doi:10.1146/annurev-environ-052809-161223

Kwak, Y. K., Colque, P., Byfors, S., Giske, C. G., Möllby, R., & Kühn, I. (2015). Surveillance of antimicrobial resistance among Escherichia coli in wastewater in Stockholm during 1 year: Does it reflect the resistance trends in the society? *International Journal of Antimicrobial Agents*, *45*(1), 25–32. doi:10.1016/j.ijantimicag.2014.09.016 PMID:25465520

Lee, D., & Choi, K. (2019). Comparison of regulatory frameworks of environmental risk assessment s of human pharmaceuticals in EU, USA and Canada. *The Science of the Total Environment*, *671*, 1026–1035. doi:10.1016/j.scitotenv.2019.03.372

Li, R., Wang, J. J., Gaston, L. A., Zhou, B., Li, M., Xiao, R., Wang, Q., Zhang, Z., Huang, H., Liang, W., Huang, H., & Zhang, X. (2018). An overview of carbothermal synthesis of metal–biochar composites for the removal of oxyanion contaminants from aqueous solution. *Carbon*, *129*, 674–687. doi:10.1016/j.carbon.2017.12.070

Li, S., Shi, W., You, M., Zhang, R., Kuang, Y., Dang, C., Sun, W., Zhou, Y., Wang, W., & Ni, J. (2019). Antibiotics in water and sediments of Danjiangkou Reservoir, China: Spatiotemporal distribution and indicator screening. *Environmental Pollution*, *246*, 435–442. doi:10.1016/j.envpol.2018.12.038 PMID:30579212

Liu, Y., Liu, X., Dong, W., Zhang, L., Kong, Q., & Wang, W. (2017). Efficient adsorption of sulfamethazine onto modified activated carbon: A plausible adsorption mechanism. *Scientific Reports*, *7*(1), 12437. doi:10.103841598-017-12805-6 PMID:28963547

Majumder, A., Gupta, A. K., Ghosai, P. S., & Varma, M. (2021). A review on hospital wastewater treatment: A special emphasis on occurrence and removal of pharmaceutically active compounds, resistant microorganisms and SARS-CoV-2. *Journal of Environmental Chemical Engineering*, *9*(2), 104812. doi:10.1016/j.jece.2020.104812 PMID:33251108

Mansouri, F., Chouchene, K., Roche, N., & Ksibi, M. (2021). Removal of pharmaceuticals from water by adsorption and advanced oxidation processes: State of the art and trends. *Applied Sciences (Basel, Switzerland)*, *11*(14), 6659. doi:10.3390/app11146659

Matamoros, V., Arias, C. A., & Brix, H. (2019). *Removal of pharmaceutical compounds in constructed wetlands: Mechanisms, performance influencing factors, and implications for treatment design.* Research Gate.

Mendoza, A., Aceña, J., Pérez, S., López De Alda, M., Barceló, D., Gil, A., & Valcárcel, Y. (2015). Pharmaceuticals and iodated contrast medium in a hospital wastewater: A case study to analyse their presence and characterize their environmental risk and hazard. *Environmental Research*, *140*, 225–241. doi:10.1016/j.envres.2015.04.003 PMID:25880605

Minden, V., Deloy, A., Volkert, A. M., Leonhardt, S. D., & Pugal, G. (2017). Antibiotics impact plants traits even at small concentrations. *AoB Plants*, *9*(2), plx010. doi:10.1093/aobpla/plx010 PMID:28439396

Morris, D., Harris, S., Morris, C., Commins, E., & Cormican, M. (2008). Hospital effluent: impact on the microbial environment and risk to human health. *Environment protection agency. Ireland*. https://www.epa.ie/publications/research/environment--health /EPA-162-final-web.pdf. Accessed on 01/30/2023.

Mousset, E., Oturan, N., & Oturan, M. A. (2016). Electrochemical advanced oxidation processes: Today and tomorrow. A review. *Environmental Science and Pollution Research International*, *23*(14), 13626–13651. doi:10.100711356-016-6447-y PMID:27300165

Mrozik, A., & Piotrowska-Cyplik, A. (2015). Anaerobic microbial degradation of pharmaceuticals—A review. *Environment International*, *79*, 105–118.

Nawrat, A. (2018). Pharma and the environment: pollution continues despite public pressure. *Pharmaceutical Technology*. https:// www.pharmaceutical-technology.com/features/pharma-an d-the-environment-pollution-trend/

Oelkers, K. (2020). The accessibility of data on environmental risk assessment of pharmaceuticals – are environmental risk assessments information on emissions with respect to international and European environmental information law? *Regulatory Toxicology and Pharmacology*, *111*, 104571. doi:10.1016/j.yrtph.2019.104571 PMID:31893528

Oelkers, K., & Floeter, C. (2019). The accessibility of data on environmental risk assessment of pharmaceuticals: Is the marketing authorization procedure in conflict with the international right of access to environmental information? *Environmental Sciences Europe*, *31*(1), 58. doi:10.118612302-019-0256-3

Organization for Economic Cooperation and Development. (2018). *Pharmaceutical innovation and access to medicines*. OECD. www.oecd.org health systems.

Orias, F. (2015). Contribution to the evaluation of the ecotoxicological risks of hospital effluents: bioconcentration, bioaccumulation and biomagnification of pharmaceutical residues. *Ecotoxicologie,* École Nationalle des Travaux Publics de l'État [ENTPE], Vaulx-en-Velin France. https://tel.archives-ouvertes.fr/tel-01778521

Orias, F., & Perrodin, Y. (2014). Pharmaceuticals in hospital wastewater: Their ecotoxicity and contribution to the environmental hazard of the effluent. *Chemosphere, 115*, 31–39. doi:10.1016/j.chemosphere.2014.01.016 PMID:24502927

Ortiz de García, S. A., Pinto Pinto, G., García-Encina, P. A., & Irusta-Mata, R. (2014). Ecotoxicity and environmental risk assessment of pharmaceuticals and personal care products in aquatic environments and wastewater treatment plants. *Ecotoxicology (London, England), 23*(8), 1517–1533. doi:10.100710646-014-1293-8 PMID:25064485

Owens, B. (2015). Pharmaceuticals in the environment: A growing problem. *The Pharmaceutical Journal, 294*, 7850. doi:10.1211/PJ.2015.20067898

Palma, P., Köck-Schulmeyer, M., Alvarenga, P., Ledo, L., Barbosa, I. R., de Alda, M. L., & Barceló, D. (2014). Risk assessment of pesticides detected in surface water of the Alqueva reservoir (Guadiana basin, southern of Portugal). *The Science of the Total Environment, 488–489*, 208–219. doi:10.1016/j.scitotenv.2014.04.088 PMID:24836129

Panizza, J., & Cerisola, G. (2005). Direct and mediated anodic oxidation of organic pollutants. *Chemical Reviews, 109*(12), 6541–6569. doi:10.1021/cr9001319 PMID:19658401

Papageorgiou, M., Zioris, I., Danis, T., Bikiaris, D., & Lambropoulou, D. (2019). Comprehensive investigation of wide range of pharmaceuticals and personal care products in urban and hospital wastewaters in Greece. *The Science of the Total Environment, 694*, 133565. doi:10.1016/j.scitotenv.2019.07.371 PMID:31401503

Peake, B. M., Braund, R., Tong, A. Y. C., & Tremblay, L. A. (2016). Impact of pharmaceuticals on the environment. In book: The Life-Cycle of Pharmaceuticals in the Environmentchapter, 5, 109-152. doi:10.1016/B978-1-907568-25-1.00005-0

Pereira, A. M. P. T., Silva, L. J. G., Lino, C. M., Meisel, L. M., & Pena, A. (2017). A critical evaluation of different parameters for estimating pharmaceutical exposure seeking an improved environmental risk assessment. *The Science of the Total Environment, 603-604*, 226–236. doi:10.1016/j.scitotenv.2017.06.022 PMID:28628814

Prabhasnkar, V. P., Joshua, D. I., Balakrishna, K., Siddiqui, I. F., Taniyasu, S., Yamashita, N., Kannan, K., Akiba, M., Praveenkumarreddy, Y., & Guruge, K. S. (2016). Removal rates of antibiotics in four sewage treatment plants in South India. *Environmental Science and Pollution Research International, 23*(9), 8679–8685. doi:10.100711356-015-5968-3 PMID:26797959

Raj, R., Tripathi, A., Das, S., & Ghangrekar, M. M. (2021). Removal of caffeine from wastewater using electrochemical advanced oxidation process: A mini review. *Case Studies in Chemical and Environmental Engineering, 4*, 100129. doi:10.1016/j.cscee.2021.100129

Rand-Weaver, M., Margiotta-Casaluci, L., Patel, A., Panter, G. H., Owen, S. F., & Sumpter, J. P. (2013). The read-across hypothesis and environmental risk assessment of pharmaceuticals. *Environmental Science & Technology, 47*(20), 11384–11395. doi:10.1021/es402065a PMID:24006913

Rathod, M., Haldar, S., & Basha, S. (2015). Nanocrystalline cellulose for removal of tetracycline hydrochloride from water via biosorption: Equilibrium, kinetic and thermodynamic studies. *Ecological Engineering, 84*, 240–249. doi:10.1016/j.ecoleng.2015.09.031

Rezakazemi, M., Khajeh, A., & Mesbah, M. (2018). Membrane filtration of wastewater from gas and oil production. *Environmental Chemistry Letters, 16*(2), 367–388. doi:10.100710311-017-0693-4

Sahar, E., David, I., Gelman, Y., Chikurel, H., Aharoni, A., Messalem, R., & Brenner, A. (2011). The use of RO to remove emerging micropollutants following CAS/UF or MBR treatment of municipal wastewater. *Desalination, 273*(1), 142–147. doi:10.1016/j.desal.2010.11.004

Santos, L. H. M. L. M., Gros, M., Rodriguez-Mozaz, S., Delerue-Matos, C., Pena, A., Barceló, D., & Montenegro, M. C. B. S. M. (2013). Contribution of hospital effluents to the load of pharmaceuticals in urban wastewaters: Identification of ecologically relevant pharmaceuticals. *The Science of the Total Environment, 461–462*, 302–316. doi:10.1016/j.scitotenv.2013.04.077 PMID:23732224

Schoenfuss, H. L., Furlong, E. T., Phillips, P. J., Scott, T. M., Kolpin, D. W., Cetkovic-Cvrlje, M., Lesteberg, K. E., & Rearick, D. C. (2016). Complex mixtures, complex responses: Assessing pharmaceutical mixtures using field and laboratory approaches. *Environmental Toxicology and Chemistry, 35*(4), 953–965. doi:10.1002/etc.3147 PMID:26561986

Shah, M. P., Yao, K. M., & Seshadri, R. (2015). Aerobic microbial degradation of organic pollutants: Revisiting the biochemistry behind the pathway. *Applied Microbiology and Biotechnology*, *99*(10), 4155–4173.

Sirés, I., Brillas, E., & Oturan, M. A. (2014). Electrochemical advanced oxidation processes: today and tomorrow. In *Electrochemical Water Treatment Methods* (pp. 261–304). Royal Society of Chemistry. doi:10.1039/9781782620113-00261

Smith, M. W. (2015). The 10 most-prescribed and top-selling medications. *Drug & Medications*. https://www.webmd.com/drug-medication/news/20150508.

Sulbarán Rangel, B. C., Madrigal Olveira, A. E., Romero Arellano, V. H., & Guzmán González, C. A. (2019). Cellulosic nanomaterials for adsorption of emerging contaminants. *Tecnura*, *23*(62), 13–20. doi:10.14483/22487638.15451

Sun, P., Li, Y., Meng, T., Zhang, R., Song, M., & Ren, J. (2018). Removal of sulfonamide antibiotics and human metabolite by biochar and biochar/H2O2 in synthetic urine. *Water Research*, *147*, 91–100. doi:10.1016/j.watres.2018.09.051 PMID:30300785

Takht Ravanchi, M., Kaghazchi, T., & Kargari, A. (2009). Application of membrane separation processes in petrochemical industry: A review. *Desalination*, *235*(1–3), 199–244. doi:10.1016/j.desal.2007.10.042

Teixeira, S., Delerue-Matos, C., & Santos, L. (2012). Removal of sulfamethoxazole from solution by raw and chemically treated walnut shells. *Environmental Science and Pollution Research International*, *19*(8), 3096–3106. doi:10.100711356-012-0853-9 PMID:22744161

Urevig, A. (2019). 5 Things we can do now to reduce the threat of drugs in our environment. *Ensia*.. https://ensia.com/notable/pharmaceuticals-drugs-medications-antiobiotics-endocrine-water-pollution/

Vestel, J., Cadwell, D. J., Constantine, L., D'Aco, V. J., Davidson, T., Dolan, D. G., Millard, S. P., Murray-Smith, R., Parke, N. J., Ryan, J. J., Straub, J. O., & Wilson, P. (2016). Use of acute and chronic ecotoxicity data in risk assessment of pharmaceuticals. *Environmental Toxicology and Chemistry*, *35*(5), 1201–1212. doi:10.1002/etc.3260 PMID:26403382

Vymazal, J. (2005). Constructed wetlands for wastewater treatment. *Ecological Engineering*, *25*(5), 475–477. doi:10.1016/j.ecoleng.2005.07.002

Walter, S., & Mitkidis, K. (2018). The risk assessment of pharmaceuticals in environment: EU and US Regulatory Approach. *European Journal of Risk Regulation*, *9*(3), 527–547. doi:10.1017/err.2018.33

Wang, X., Rye, D., Houtkooper, R. H., & Auwerx, J. (2015). Antibiotic use and abuse: A threat to mitochondria and chloroplasts with impact on research, health, and environment. *BioEssays*, *37*(10), 1045–1053. doi:10.1002/bies.201500071 PMID:26347282

WillacyH. (2020). Antibiotics. In *Patient*. https://patient.info/infections/antibiotics-leaflet

World Health Organisation. (2018). *Resource mobilisation for antimicrobial resistance (AMR): Getting AMR into plans and budgets of government and development partners*. WHO. www.who.int. Accessed on 03/23/2023.

Xiaowei, J., Zha, J., Xu, Y., Giesy, J. P., Richardson, K. L., & Wang, Z. (2011). Derivation of predicted no effect concentrations (PNEC) for 2,4,6-throchlorophenol base on Chinese resident species. *Chemosphere*, *86*, 17–23. https://www.usask.ca/toxicology/jgiesy/pdf/publications/JA-6 24 PMID:21955353

Yang, X., Flowers, R. C., Weinberg, H. S., & Singer, P. C. (2011). Occurrence and removal of pharmaceuticals and personal care products (PPCPs) in an advanced wastewater reclamation plant. *Water Research*, *45*(16), 5218–5228. doi:10.1016/j.watres.2011.07.026 PMID:21864879

Zhang, Y. N., Niu, Q., Gu, X., Yang, N., & Zhao, G. (2019). Recent progress on carbon nanomaterials for the electrochemical detection and removal of environmental pollutants. *Nanoscale*, *11*(25), 11992–12014. doi:10.1039/C9NR02935D PMID:31140537

Zhou, S., Paolo, C. D., Wu, X., Shao, Y., Seiler, T. B., & Hollert, H. (2019). Optimization of screening-level risk assessment and priority selection of emerging pollutants – The case of pharmaceuticals in European surface waters. *Environment International*, *128*, 1–10. doi:10.1016/j.envint.2019.04.034 PMID:31029973

Chapter 7
Antioxidant and Anticancer Potential of Natural Cocoa Extract

Jacob Apibilla Ayembilla
Accra Technical University, Ghana

Phyllis Naa Yarley Otu
Accra Technical University, Ghana

Eunice A. Dotse
Noguchi Memorial Institute for Medical Research, Ghana

Abigail Aning
https://orcid.org/0000-0001-9465-2488
Noguchi Memorial Institute for Medical Research, Ghana

Emmanuel Kweku Ofori
https://orcid.org/0000-0001-7375-8830
University of Ghana, Ghana

Regina Appiah-Opong
Noguchi Memorial Institute for Medical Research, Ghana

Henry Asare-Annane
University of Ghana, Ghana

ABSTRACT

The aim of this study was to investigate antioxidant and anticancer potentials of natural cocoa extracts. Antioxidant properties of the extracts were evaluated by DPPH assay, total phenolic content by Folin-Ciocalteau and reduced glutathione content by O-phthalaldehyde assays. Antiproliferative activity was also evaluated on leukaemia and prostate cancer cell lines using MTT assay. The hydroethanolic extracts of cocoa roots and beans and aqueous cocoa leaves showed high total antioxidant activities with the root extract having the strongest scavenging effect on DPPH and the highest total phenolics content. The hydroethanolic cocoa bean crude extract showed the highest anticancer activity against LNCaP cells, whiles the ethyl acetate fraction of hydroethanolic cocoa leaf exhibited the highest anticancer

DOI: 10.4018/978-1-6684-9843-9.ch007

against PC3. Jurkat and PC3 cells had similar susceptibilities to the natural cocoa extracts. Thus, cocoa has antioxidant and anticancer activity, and the specific bioactive compounds responsible for these activities need to be elucidated in future research.

INTRODUCTION

Cancer is a disease state that stems from molecular events that lead to the alteration of normal cell properties and function. Cancer cells acquire hallmarks including uncontrollable proliferative potential, self-sustaining in growth signaling, resistance to apoptotic and anti-proliferatives cues because of series of mutagenic processes (Luo et al., 2009). The world today is facing a huge increase in cancers including breast, prostate, cervical cancer, leukemia, etc which are impacting negatively on the global economy. World Health Organization (WHO) data shows that cancer is the second leading cause of morbidity and mortality claiming about 9.6 million lives each year (Janardhanan et al., 2020). The GLOBOCAN 2020 cancer incidence and mortality estimated 19.3million new cancer cases and 10.0million death occurred in 2020 (Chhikara & Parang, 2022; Wild et al., 2020).

In Ghana, 24,009 new cases of cancer were recorded with 15,802 deaths in 2020 (Wild et al., 2020). Also, Laryea et al., (2014) reported increasing rate of cancers in Ghana, 21.1% liver cancer and 13.2% prostate cancer among men; 33.9% breast cancer and 29.4% cervical cancer among women.

Couple with these statistics, are challenges associated with efficacy and safety profile of cytotoxic agents. Despite improved treatment modalities over the years, cancer mortality is still on the increase especially in developing countries and the treatments are not affordable (Jemal et al., 2011). Conventional cancer treatments including chemotherapy and radiotherapy are associated with undesirable side effects due to poor selectivity towards cancer cells.

Ghana is the second highest producer of the world cocoa. Cocoa is a potent nutritional and medicinal plant and therefore the attention of many researchers has been drawn to investigating the possible bioactive compounds in the plant responsible for these nutritional and medicinal properties and their mechanism of action. The use of cocoa as medicinal product dates back to ancient times where it was used to treat snakebite, heart pain and burns (Abrokwah et al., 2009).

It is worth noting also that some effective chemotherapeutic drugs used in modern day treatment, vinblastine and vincristine, taxol, and camptothecin were all derived from natural plants *Catharanthus roseus, Taxus brevifolia* and *Camptheca acuminate* respectively (Colegate & Molyneux, 2008). Screening natural products including cocoa for their anticancer and antioxidant potentials brings us close to discovering alternative sources of chemotherapeutic agents. This will possibly be less expensive, affordable and have mild side effects since they are readily available in our backyard and are edible.

Inasmuch as the pharmacological activity of cocoa have been studied elsewhere, the anticancer activity of cocoa has remained a subject of controversy. However, there is scarcity of data on the anticancer activity of cocoa from Sub-Saharan Africa. Besides, that, due to the variation of climatic, soil conditions and composition, we anticipate that the biological constitutes in Ghanaian cocoa may vary from other regions of the globe. This thought is buttressed by (Twilley et al., 2020), who reported that *R. hirsuta* and *C. contracta* had antitumour activity when tested on in vivo tumour model in rats however, when a second preparation of the plants' extracts were done at a different season, no activity was observed in the same tumour models. This signifies that climatic variation affects the chemical compositions of plants.

The motivation for this study stem from the erratic increase in cancers in Ghana and the lack of scientific and empirical data on the potency of cocoa and its products in the treatment and management of cancer in Ghana. The objective for this study was to investigate the antioxidant and anticancer potential of natural cocoa extract.

MATERIALS AND METHODS

Materials

Rose Park Memorial Institute (RPMI)-1640 and Fetal Bovine Serum (FBS), Folin-Ciocalteau reagent, Curcumin and 2, 2- diphenyl-1-picryl hydrazyl free radical reagent (DPPH) were purchased from Sigma-Aldrich Company (Missouri, USA). The 3-(4, 5-dimethylthiazol-2-yl)-2, 5-diphenyltetrazolium bromide (MTT) and Trypan Blue were obtained from (Wako Pure Chem. Ind., Tokyo, Japan), whereas dimethyl sulphoxide (DMSO), butylated hydroxyltoluene (BHT) and gallic acid were obtained from Sigma-Aldrich (Hamburg, Germany), Human leukemia-immortalized T lymphocytes (Jurkat), human prostate carcinoma (PC3), human androgen-sensitive prostate carcinoma (LNCaP) and Chang liver cells (Hela cells) were obtained from RIKEN BioResources Centre Cell Bank (Japan). All other reagents used for the study were obtained from standard suppliers.

Methods

Sample Collection

Fresh natural cocoa leaves, roots, and sun-dried natural cocoa bean samples were collected from the Cocoa Research Institute, Akim-Tafo in the Eastern Region of

Ghana (voucher specimen number UG/CHS/HM1/2016/C01). The leaves and roots samples were rinsed with distilled water and the roots were chopped into small pieces and dried at room temperature for 2 weeks. Commercially available roasted and defatted natural cocoa powder was also purchased from Kakawa Food Enterprise Ltd, Ghana with Lot Number KK1701A. The samples were coded as follows: CB_{ET} = Cocoa Bean Hydroethanolic, CB_{AQ} = Cocoa Bean Aqueous, CL_{ET} = Cocoa Leaf Hydroethanolic, CL_{AQ}= Cocoa Leaf Aqueous, CR_{ET}= Cocoa Root Hydroethanolic, CR_{AQ} = Cocoa Root Aqueous, CPI = Cocoa Powder from Industry

Preparation of Extracts

Distilled water and 70% ethanol were used as solvents for the extraction of bioactive compounds in the natural cocoa parts. Sun-dried natural cocoa beans were cleaned and dehusked. The unroasted nibs were milled into powder. Air-dried natural cocoa leaves and roots were also pulverized. The pulverized samples (25g) were extracted by cold maceration with 70% ethanol (v/v) for 48 h. After extraction the mixtures were filtered with No.1 Whatman filter paper and the filtrate was evaporated at 40°C using a Büchi B-490 rotary evaporator (Switzerland). To obtain aqueous extracts, the pulverized samples were heated and freeze-dried as described by (Appiah-Opong et al., 2016).

Determination of Total Phenolic Content

Total phenolic content of natural cocoa extracts was determined as previously described by (Acheampong et al., 2015). The concentration of the crude extract tested was 5 mg/ml. To 10 µL of each sample aliquoted into 12-well microplates, 790 µL of distilled water was added. This was followed with addition of 50 µL of Folin-Ciocaltaeu. The plates were shaken and incubated at room temperature (26°C) in the dark for 8 min. Subsequently, 150 µL of 7% Na_2CO_3 solution was added to each well and incubation was continued for 2 h in the dark at room temperature. Absorbance was read at 750 nm wavelength using Tecan Infinite M200 (Austria) microplate reader. Gallic acid was used as standard phenolic compound. The total phenolic content of each crude natural cocoa extract was calculated from the regression equation of the gallic acid calibration curve ($y = 0.013x - 0.0003$, $R^2 = 0.9985$) and expressed as gallic acid equivalents (GAE). All experiments were carried out in triplicates.

In Vitro Total Antioxidant Activity

DPPH Free Radical Scavenging Activity

The effect of crude extracts of natural cocoa on the stable DPPH free radical was determined using a modified protocol by (Brand-Williams & Cuvelier, 1995). Aliquots of 100 µL of each extract were pipetted into 96-well microplates in triplicates and 100 µL of 0.5 mM DPPH was added to each well and shaken gently. The plates were incubated in the dark at room temperature for 20 min. Subsequently, the absorbance was measured at a wavelength of 517 nm using the microplate reader. The percentage of DPPH free radical scavenging (Antioxidant) activity was calculated using the formula:

$$\text{Antioxidant activity (\%)} = \left(\frac{\text{Absorbance}_{\text{control}} - \text{Absorbance}_{\text{sample}}}{\text{Absorbance}_{\text{control}}}\right) \text{X } 100$$

Butylated hydroxytoluene (BHT) was used as a positive control. The EC_{50} value, which is the effective concentration at which 50% free radical scavenging activity was obtained was determined from the graph of percentage antioxidant activity versus concentration of test sample. All experiments were carried out in triplicates.

Determination of Reduced Glutathione Levels

The O-phthalaldehyde (OPT) fluorometric procedure described by (Acheampong et al., 2015) was adopted for the determination of the amount of reduced glutathione (GSH) in each of the crude natural cocoa extracts, with slight modification. To aliquots of 180 µL 0.1 mM GSH buffer (pH 8.0) in 96-well microtiter plates, 10 µL of 10 mg/ml of the extracts were added to each well. Aliquots of 10 µL of 1 mg/ml of OPT were added to each well and the reaction mixture was incubated at room temperature for 15 min. Fluorescence emission was measured at excitation wavelength of 340 nm and emission wavelength of 420 nm using the microplate reader. Reduced GSH was used as the standard. The GSH concentration in the extracts were determined using the GSH standard curve ($y = 1x - 1E^{-0.7}$, $R^2 = 0.9983$).

Bioactivity Guided Fractionation

Bioactive hydroethanolic cocoa extracts were selected for bioassay guided fractionation. The fractions were coded as follows: CB_{PET}= Cocoa Bean Petroleum Ether, CB_{C}= Cocoa Bean Chloroform, CB_{ETOAC}= Cocoa Bean Ethyl acetate, CL_{PET}=

Cocoa Leaf Petroleum Ether, CL_C= Cocoa Leaf Chloroform, CL_{ETOAC}= Cocoa Leaf Ethyl acetate, CR_{PET}= Cocoa Root Petroleum Ether, CR_C= Cocoa Root Chloroform, CR_{ETOAC}= Cocoa Root Ethyl acetate. Fractionation of the hydroethanolic extracts of natural cocoa parts was carried out in separating funnels using solvents of increasing polarity; petroleum ether, chloroform, and ethyl acetate. A volume of 100 ml, each of the extracts was partitioned with equal volumes of petroleum ether, followed by chloroform and finally ethyl acetate. The remaining portion (70% ethanol fractions) were concentrated under reduced pressure using a rotary evaporator (Buchi Rotavapor R-205, Switzerland).

Cell Culture

All cell lines were maintained in RPMI-1640 supplemented with 10% FBS and 1% penicillin streptomycin L-glutamine. Cells were incubated in a humidified incubator with 5% CO_2 at 37°C and sub-cultured at 80-90% confluence (Appiah-Opong et al., 2016).

In Vitro Cell Viability Assay

The cytotoxic effect of natural cocoa extracts and fractions on cancer and Chang liver cell lines were determined using the tetrazolium-based colorimetric assay (MTT). Chang liver cells were used instead of normal human cells since it was available and growth rates of both cells are similar. The MTT assay as described by (Appiah-Oppong et al., 2016) was used. Cells were seeded in 96-well plates at a concentration of 1×10^5 cells/ml. The cells were then treated with different concentrations of the extracts (62.5- 1000 µg/ml) and fractions (6.25-100 µg/ml). The plates were incubated as described above for 72 h and then the cells were treated with 20 µL of 2.5 mg/ml MTT solution and incubated again for 4 h. A volume of 150 µL of acidified isopropanol was added to each well and the plates were incubated in the dark overnight at room temperature to dissolve any formazan crystals formed from MTT. The absorbance was read using the microplate reader at a wavelength of 570 nm. Curcumin was used as a positive control. The negative control consisted of 1% of DMSO which was the dissolution solvent. All experiments were performed in triplicate.

The percentage cell viability was evaluated using the formula:

$$\%\ \text{Cell viability} = \left(\frac{\text{Absorbance}_{\text{sample}} - \text{Absorbance}_{\text{blank}}}{\text{Absorbance}_{\text{control}} - \text{Absorbance}_{\text{blank}}}\right) \text{X } 100$$

The IC_{50} values, which were concentrations at which 50% inhibition of cell growth occurred were used to determine the selectivity indices (SI) of the extracts and fractions using the formula:

$$\mathbf{SI} = \frac{IC_{50}\text{of normal cell line}}{IC_{50}\text{of cancer cell line}}$$

SI values ≥ 2 were considered as significant (Badisa et al., 2009).

Statistical Analysis

The data were analyzed using Microsoft Office Excel 2013 and GraphPad Prism 6.01 (GraphPad Software Inc, San Diego, CA) software. The data were presented as mean ± Standard Error of the Mean, (SEM). *P*-values < 0.05 were considered statistically significant.

RESULTS

Cocoa Extract Yield

Solvent extraction and freeze-drying 25 g each of air-dried pulverized cocoa leaves, roots and bean samples yielded CB_{ET} = 2.1561g, CB_{AQ} = 1.7415g, CL_{ET} = 2.1820g, CL_{AQ} = 1.9065g, CR_{ET} = 0.8924g, CR_{AQ} = 0.4909g, and CPI = 0.9799g of powdered extracts, respectively, representing a percentage yield of 8.62%, 6.97%, 8.73%, 7.63%, 3.57%, 1.96%, and 3.92% (w/w), respectively. The hydroethanolic extracts recorded higher yields than their aqueous extracts with CB_{ET} being the highest and CR_{AQ} the least.

Total Phenolic Content

Figure 1 shows that all the cocoa extracts contained varied amounts of total phenolics. The hydroethanolic cocoa root extract recorded the highest total phenolic content (8854.55 ± 990.83 mg GAE /100 g) out of the seven extracts tested, followed by hydroethanolic cocoa bean extract (7803.12 ± 860.37 mg GAE/100 g). The aqueous cocoa root extract recorded the least total phenolic content (1221.24 ± 482.13 mg GAE/100 g).

Figure 1. Total phenolic content (TPC) of cocoa extracts. Results are represented as mean ± SEM of triplicate experiments

**p < 0.01, **p < 0.001 represent significant difference of hydroethanolic extracts compared to their aqueous extracts. +p < 0.01 represent significant difference of CB_{ET}, CR_{ET} compared to CL_{ET}. Brackets represent significant difference of aqueous extracts to CPI (*P < 0.05, **P < 0.001).*

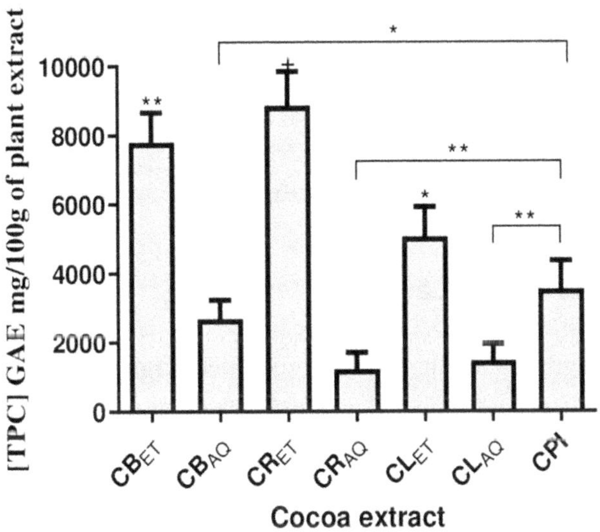

Total Antioxidant Activity

The results of the natural cocoa extracts on the DPPH scavenging activity are shown in Table 1. Generally, the hydroethanolic extracts showed stronger free radical scavenging (antioxidant) activity than the aqueous extracts. However, the aqueous cocoa leaf extract recorded stronger activity than the hydroethanolic extract. The hydroethanolic cocoa root extract showed the strongest antioxidant activity with the least effective concentration ($EC_{50} = 0.40 \pm 0.30$mg/ml), however, this was about 6-fold weaker than the standard, BHT. This strength in activity is followed by hydroethanolic cocoa bean extract, aqueous cocoa leaf extract, and the industrially processed cocoa powder. Concentration-dependent increases in antioxidant activity of the extracts was observed. This trend was similar for all the extracts. The hydroethanolic cocoa roots extract had stronger antioxidant activity than the aqueous extracts. The aqueous cocoa leaf extract had antioxidant activity which was stronger than the hydroethanolic extract with $EC_{50} = 0.83 \pm 0.02$mg/ml. The hydroethanolic cocoa bean extract was effective free radical scavenger than the aqueous with $EC_{50} = 0.41 \pm 0.03$mg/ml. The EC_{50} value for industrially processed cocoa was 1.04 ± 0.002 mg/ml.

Table 1. Antioxidant activity of cocoa extract determined by DPPH method

EC_{50} Values (mg/ml) *			
Sample	**Aqueous**	**Hydroethanolic**	**BHT**
Cocoa Bean	>1000	0.41 ± 0.03	0.07 ± 0.00
Cocoa Leaf	0.83 ± 0.02	>1000	
Cocoa Root	>1000	0.40 ± 0.30	
CPI	1.04 ± 0.00		
p-Value	>0.05	>0.05	>0.05

*EC_{50} Values are mean ± standard error of mean, n = 3. EC_{50} of hydroethanolic extracts not significant different compared to aqueous extracts and BHT ($P > 0.05$)

Reduced Glutathione Level

Generally, the hydroethanolic extracts recorded higher GSH levels than the corresponding aqueous extracts at the same concentration tested except for cocoa aqueous and hydroethanolic leaf extracts (Figure 2). The hydroethanolic cocoa bean extract recorded the highest GSH content of 4835.7 mg/100 g of cocoa extract. Industrially processed cocoa powder recorded lower level than cocoa bean extracts. The other extracts recorded minimal level of GSH.

Figure 2. Glutathione content of cocoa extracts. Results are represented as mean ± SEM of triplicate experiments

**p < 0.01 represent significant difference of hydroethanolic extracts compared to their aqueous extracts. Brackets represent significant difference of aqueous extracts to CPI (*P < 0.001).*

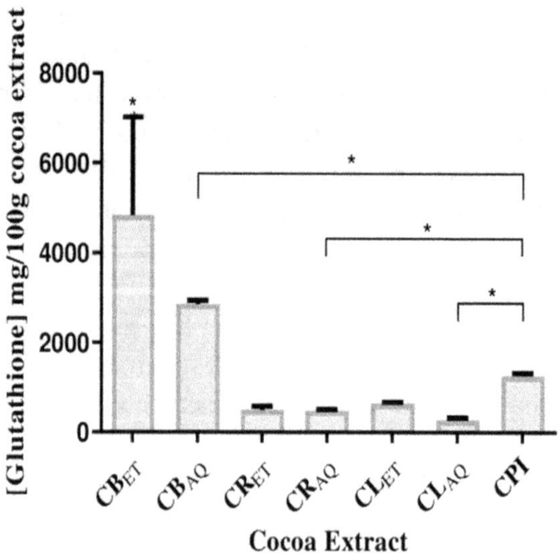

In Vitro Cell Viability Assay

The MTT assay was used to determine the cytotoxic effect of cocoa extracts on Chang liver (representing normal cell line due to slow growth), leukemia (Jurkat) and prostate (LNCaP and PC3) cancer cell lines. All the hydroethanolic extracts showed a significant cytotoxic effect on all the three cancer cell lines except hydroethanolic root which did not show any significant activity on LNCaP cells (Table 2).

Hydroethanolic cocoa leaves extract was the most cytotoxic to all the cancer cell lines tested. Its highest and lowest inhibition activities were observed on PC3 and Jurkat with IC_{50} values of 39.5 ± 3.5 μg/ml and 201.6 ± 7.9 μg/ml, respectively. On the other hand, the aqueous extract of the cocoa leaves was not cytotoxic towards the cells. The hydroethanolic cocoa root extract showed cytotoxic effect on only two cell lines, Jurkat and PC3 with IC_{50} value of 105.0 ± 6.1 μg/ml and 131.1 ± 4.5 μg/ml respectively. The aqueous cocoa root extract, however, had no cytotoxic effect on all the cell lines. Both the aqueous and the hydroethanolic extracts of cocoa bean had cytotoxic effect on all cell lines including the normal cells (Chang liver). The hydroethanolic cocoa bean extract exhibited a strong cytotoxic effect on LNCaP cell lines with an IC_{50} value of 37.3 ± 0.8 μg/ml and lowest cytotoxic effect on PC3 cell lines with an IC_{50} value of 361.5 ± 6.3 μg/ml. The aqueous bean extract showed stronger activity on PC3 cells with an IC_{50} value of 53.1 ± 1.1 μg/ml and lowest activity on Jurkat cells with an IC_{50} value of 838.1 ± 8.0 μg/ml.

Aqueous cocoa bean, hydroethanolic cocoa bean, and leaves extracts showed weak cytotoxic activities toward the Chang liver cells with IC_{50} values of 597.0 ± 58.1 μg/ml, 227.6 ± 15.9 μg/ml and 171.2 ± 7.5 μg/ml, respectively. The industrially processed cocoa powder had no cytotoxic effect on all the cell lines including the normal cells within the concentration range tested. There was an inverse relation between the concentration of the curcumin and the percent cell viability. It recorded an IC_{50} = 6.6 ± 0.5 μg/ml.

The selectivity indices of the natural cocoa extracts were determined as the ability of extracts to specifically target and inhibit the proliferation of only cancer cells without affecting normal cells. It was determined by comparing the IC_{50} values of each extract against the cancer cells (Jurkat, LNCaP, and PC3) with that of the normal cell line (Chang liver). The extracts that had selectivity index (SI) greater than or equal to two (2) were considered to be preferred drug candidates to the particular cancer cell line. Aqueous cocoa bean showed the highest selectivity against LNCaP and PC3 cell lines with SI values of 16.0 and 11.3, respectively. It however, showed weak selectivity against Jurkat cells with SI value of 0.7 (Table 2).

Table 2. Cytotoxic activity of natural cocoa extract on normal and cancer cell lines

IC_{50} Values (μg/ml) *				
Sample	**Jurkat**	**PC3**	**LNCaP**	**Chang liver**
Curcumin	5.0 ± 0.7 (1.3)	2.7 ± 0.2 (2.4)	11.8 ± 0.5 (0.6)	6.6 ± 0.5
CB_{ET}	203.8 ± 5.7 (1.1)	361.5 ± 6.3 (0.6)	37.3 ± 0.8 (0.4)	227.6 ± 15.9[a]
CB_{AQ}	838.1 ± 8.0 (0.7)	53.1 ± 1.1 (11.3)	626.0 ± 11.7 (16.0)	597.0 ± 58.1[a]
CL_{ET}	201.6 ± 7.7 (0.9)	39.5 ± 2.0 (4.3)	77.6 ± 7.7 (2.2)[b]	171.2 ± 7.5
CL_{AQ}	>1000	>1000	>1000	>1000
CR_{ET}	105.0 ± 3.5 (9.6)	131.1 ± 8.4 (7.6)[a]	>1000	>1000
CR_{AQ}	>1000	162.5 ± 19.4 (6.2)[b]	>1000	>1000
CPI	>1000	>1000	>1000	>1000
p-value	$p > 0.05$	$P < 0.001^a, 0.01^b$	$P < 0.01^b$	$p < 0.001^a$

*Values are mean ± standard error of mean, n = 3. SI values are in parenthesis, ($p < 0.001^a$, $p < 0.01^b$) represent significant differences in cytotoxicity of the different extracts compared to curcumin on all cell lines. Active ≤ 20μg/ml, moderately active > 20-100μg/ml, weakly active > 100-1000μg/ml, inactive > 1000μg/ml.

MTT Assay of Hydroethanolic Cocoa Fractions

The MTT assay was used to determine the cytotoxic effects of the fractions of the cocoa extracts. All the fractions had no significant cytotoxic effect on the Chang liver cells. The hydroethanolic cocoa leaf ethyl acetate fraction recorded the least cytotoxic activity on the Jurkat cells with an IC_{50} value of 75.9 ± 4.4 μg/ml and the highest activity was observed in LNCaP cell with an IC_{50} value of 35.3 ± 5.0 μg/ml. It had no significant cytotoxic effect on PC3 and the normal cell lines. The hydroethanolic cocoa root petroleum ether fraction on the other hand, was least cytotoxic to Jurkat cells with an IC_{50} value of 74.9 ± 5.2 μg/ml and highest on PC3 with an IC_{50} value of 37.4 ± 4.1 μg/ml but had no significant effect on LNCaP and Chang liver cells with IC_{50} values >1000 μg/ml. The ethyl acetate fraction of the hydroethanolic cocoa root extract, however, showed a strong cytotoxic activity on Jurkat cells with an IC_{50} value of 66.9 ± 1.3 μg/ml and the least activity on PC3 cell lines with an IC_{50} value of 69.7 ± 0.0 μg/ml. No significant cytotoxic effect was observed with the rest of the fractions on all the cell lines tested. All fractions that recorded cytotoxic activity had good selectivity indices (SI > 2) against the normal cell lines (Chang liver). Their cytotoxicity was statistically significant compared with the standard, curcumin ($p < 0.001^a, 0.05^b$).

Table 3. Cytotoxic activity of natural cocoa fractions on normal and cancer cell lines

Cell line	Extract	IC_{50} Values (mg/ml) *			Curcumin
		Fractions			
		Petroleum ether	Chloroform	Ethyl acetate	
Jurkat	CB_{ET}	>1000	>1000	>1000	5.5 ± 0.2 (3.5)
	CL_{ET}	>1000	>1000	75.9 ± 4.4	
	CR_{ET}	74.9 ± 5.2 (13.4)[a]	>1000	66.9 ± 1.3 (14.9)	
PC3	CB_{ET}	>1000	>1000	>1000	10.6 ± 0.4 (1.8)
	CL_{ET}	>1000	>1000	>1000	
	CR_{ET}	37.4 ± 4.1 (26.7)[b]	>1000	69.7 ± 0.0 (14.3)	
LNCaP	CB_{ET}	>1000	>1000	>1000	20.8 ± 0.7 (0.9)
	CL_{ET}	>1000	>1000	35.3 ± 5.0 (14.3)[b]	
	CR_{ET}	>1000	>1000	>1000	
Chang liver	CB_{ET}	>1000	>1000	>1000	19.4 ± 0.6
	CL_{ET}	>1000	>1000	>1000	
	CR_{ET}	>1000	>1000	>1000	

*IC_{50} Values are mean ± standard error of mean, n = 3. SI values are in parenthesis, ($p < 0.001^a$, $p < 0.05^b$) represent significant differences in cytotoxicity of the different fractions compared to curcumin on all cell lines

DISCUSSION

For several centuries medicinal plants have been used to manage cancer because of their phytochemical contents. These phytochemicals may have antioxidant properties; thus, acting as chemo-protectors or radioprotectors by mopping up free radicals (Agrawal et al., 2017).

The results from phytochemical analysis of the cocoa extracts used in this study revealed that the hydroethanolic extracts had higher total phenolic contents than their corresponding aqueous extracts except aqueous cocoa leaves that showed higher total phenolic content than the hydroethanolic extract. Hydroethanolic cocoa root recorded the highest total phenolic content (8854.6 ± 990.4 mg GAE /100g extract) and aqueous cocoa beans recorded the least (1221.2 ± 482.1 mg GAE /100g extract). This result is consistent with (Baharum et al., 2014), who reported that methanolic cocoa root extract recorded the highest total phenol content (22.0 ± 1.1 g GAE/100g extract) than other cocoa parts. The differences in the amount of total phenolic content determined from the cocoa root between our results and (Baharum et al., 2014) could be as a result of the different solvents used, genetic diversity and the different geographical locations of the cocoa. A study by (Redovniković et al., 2009), on cocoa liquor showed the following: Madagascar > Mexico > Ecuador >

Venezuela > Sao Tome > Ghana in decreasing order of total phenolic content of cocoa by both HPLC and Folin Cioceltaeus methods. Same order was observed for DPPH activity. A similar study by (Othman et al., 2010) reported that Sulawesi > Malaysia ≈ Ghana ≈ Cote d'Ivoire in decreasing order of total phenolic content of cocoa. CB_{ET} showed the second highest antioxidant activity which contradict with findings by Jayanthi & Singh (2015), where antioxidant activity was in the order of pods > leaves >Seeds. It appears that the phytochemicals in the cocoa constituents are more soluble in the organic solvents used, therefore, the hydroethanol could extract more of the phytochemicals which have antioxidant activity.

The hydroethanolic extracts recorded EC_{50} of $CB_{ET} = 0.41 \pm 0.03$ mg/ml and $CR_{ET} = 0.40 \pm 0.30$ mg/ml and the aqueous, $CL_{AQ} = 0.83 \pm 0.02$ mg/ml and CPI = 1.04 ± 0.00 mg/ml. However, no EC_{50} values were recorded for CB_{AQ}, CL_{ET} and CR_{AQ}, as very low to no antioxidant activity was detected in these extracts. The cocoa root hydroethanolic extract thus had the highest antioxidant activity. The result of this study agreed with findings by (Baharum et al., 2014), who report that methanolic cocoa root extract showed the highest antioxidant activity and total phenolic content than other parts of cocoa with $EC_{50} = 358.00 \pm 7.00$ μg/ml. However, results of the present study contradict the outcome of a study which reported that aqueous cocoa bean extract had higher level of antioxidant activity compared to the ethanolic extract (Othman et al., 2010). Comparing the DPPH free radical activity of the extracts to BHT (positive control), BHT had the highest antioxidant activity with $EC_{50} = 0.07.00 \pm 0.00$ mg/ml because it is a pure antioxidant compound. The result for CPI was interesting because well fermented cocoa beans have lower phenolic content and processing of cocoa beans alters its composition and chemistry (Wollgast & Anklam, 2000). Astringency of cocoa is related to low phenol content and the purple colour of unfermented cocoa beans to anthocyanins which are both undesirable in cocoa products thus cocoa is processed to reduce the content of phenols and anthocyanins (Bauer et al., 2016). However, CPI had higher phenolic content than all aqueous extracts, thus, this study contradicts that of (Bauer et al., 2016). There was a significant difference ($p < 0.01$) between the quantity of total phenolic content in hydroethanolic extracts and the aqueous extracts (Fig. 1). However, there was no significant difference (($p > 0.05$) between the antioxidant activity of the hydroethanolic extract and the aqueous extracts (Table 1). In this study, aqueous cocoa root, aqueous cocoa bean, and hydroethanolic cocoa leaf extracts had weak antioxidant activities. The strong antioxidant activity shown by the hydroethanolic root extract suggest that 70% ethanol is a better extractor of compounds with antioxidant activity in the cocoa parts studied than aqueous medium.

Glutathione is an important antioxidant, and the body's major scavenger, responsible for maintaining other antioxidants in their reduced state and neutralizing the effect of oxidative stressors. According to (Traverso et al., 2013), glutathione is important for the maintenance of the protein-SH groups. In this study, CB_{ET}, CB_{AQ} and CPI extracts possessed the highest GSH content in decreasing order whiles the other extracts did not record significant amounts. Using the same protocol, (Acheampong et al., 2015) did not detect any GSH in *A. conyzoides* extracts. However, (Martin et al., 2013) reported protective effect of cocoa extract on cellular GSH content of HepG2 cells where pretreatment of cells with cocoa extract significantly prevented oxidative stress by evoking GSH recovery. There was a significant difference ($p < 0.01$) between the quantity of glutathione in hydroethanolic extract and the aqueous extract (Fig. 2). It supports the observation in this study that CL_{AQ} which had minimal glutathione level had $EC_{50} = 0.83 \pm 0.02$ mg/ml lower than CPI.

Some secondary metabolites which include flavonoids, phenols, alkaloids, and tannins have been shown to possess anticancer properties (Eid et al., 2015). Hazirah et al. (2013) reported that *T. cacao* bean is a promising anticancer agent and (Baharum et al., 2014) validated the *in vitro* anticancer activity of methanolic extract of cocoa parts which revealed that the methanolic extract of the cocoa leaf recorded the highest antiproliferative potential. In this study, the IC_{50} were categorized as follows: Active ≤ 20μg/ml, moderately active > 20-100μg/ml, weakly active > 100-1000μg/ml, inactive > 1000μg/ml (Atjanasuppat et al., 2009). Thus, the MTT screening of the cocoa extracts showed that none of the extract had active cytotoxic activity. However, the following extracts were moderately active against the cancer cell lines $IC_{50} <$ 100μg/ml: CB_{AQ} (53.1 ± 1.1 μg/ml against PC3 cell lines), CL_{ET} (39.5 ± 2.0 μg/ml against PC3 cell lines), CB_{ET} (37.3 ± 0.8 μg/ml against LNCaP cell lines), and CL_{ET} (77.6 ± 7.7 μg/ml against LNCaP cell lines). According to the United States National Cancer Institute plant screening program, crude extract is considered to have cytotoxic activity if its IC_{50} is < 40 μg/ml. Thus, only CB_{ET} (37.3 ± 0.8 μg/ml against LNCaP cell lines) and CL_{ET} (39.5 ± 2.0 μg/ml against PC3 cell lines) meet this requirement. CB_{ET} was the most potent among all the extracts with poor selectivity (SI = 0.4) against Chang liver cells. Thus, CL_{ET} with good selectivity (SI = 4.4) is a better drug candidate than CB_{ET}. This agrees with (Baharum et al., 2014) who found cocoa leaf to be the most potent with good selectivity against MCF-7 (41.4 ± 3.3 μg/ml). However, Jayanthi & Singh (2015), reported that cocoa pod was most cytotoxic among seed and leaf extracts against MG63 osteosarcoma cell lines. CPI was not cytotoxic against any of the cell lines with IC_{50} > 1000. This is consistent with (Bauer et al., 2016) study where roasted cocoa bean extract had very minimal cytotoxicity against A549 cell lines. Results from this study demonstrated that the hydroethanolic extracts of cocoa have a stronger cytotoxic effect on the cell lines tested than aqueous extracts except CB_{AQ}. Suggesting that polar organic solvents have higher extraction yield of anticancer compounds from cocoa (Baharum et al.,

2016). It is worth nothing that high yields may not necessarily translate to high activity (Anwar et al., 2013). The cytotoxic activity was statistically significant ($p < 0.01$) between the natural cocoa extracts and the curcumin for all the cell lines except Jurkat which was not significant ($p > 0.05$) (Table 2).

The bioassay guided fractionation strategy was adopted to fractionate the hydroethanolic extracts to improve their anticancer activity. Ethyl acetate fraction of CL_{ET} was the most cytotoxic among all the fractions with $IC_{50} = 35.3 \pm 5.0$ μg/ml against LNCaP cell line and good selectivity (SI = 14.3) against Chang liver cell lines. The CR_{ET} petroleum ether and ethyl acetate fractions had moderately active cytotoxicity $IC_{50} < 100$ mg/ml and good selectivity (SI > 10) against Jurkat and PC3 cell lines. None of the fractions was cytotoxic to the Chang liver cell lines. The cytotoxicity of the active fractions was, however, lower than the standard, curcumin. This study demonstrated that CR_{ET} recorded the highest total phenolic and antioxidant activity. It is thus not surprising that two fractions of CR_{ET} from petroleum ether and ethyl acetate each recorded moderately active cytotoxicity $IC_{50} < 100$ μg/ml against these cell lines than the other fractions. These cytotoxic activities are however lower that the standard, curcumin. From this study, it could be deduced that crude hydroethanolic extracts of the various cocoa parts used had higher cytotoxic effect on the cell lines used than their fractions. Crude extracts are reported to be pharmacologically active than their isolated active principles (Karagöz et al., 2007). This could be due to the principle of synergistic effect. In the crude extracts, there are a lot of bioactive compounds which produce their own effect, and these effects complement each other and so the overall effect of the extracts is higher than the effect of the individual bioactive compounds. The study showed that five ethyl acetate fractions had activity whiles two petroleum fractions had activity. The chloroform fraction had no activity against any of the cell lines. This results in agreement with (Krishnamurthy et al., 2015), where ethyl acetate fractions recorded the highest activity and chloroform fractions the poorest among organic solvents. Sandanasamy et al., (2015), also reported that ethyl acetate fractions are most potent cytotoxic fractions among other organic solvents extracts from cocoa. A similar study (Yahya et al., 2022) by revealed that ethyl acetate fraction from methanolic T. cacao pod husk extract had moderate anticancer activity against MCF-7 cells and weak activity against Hela cells. These suggest that ethyl acetate is the best solvent for extraction of compounds from cocoa with anticancer properties.

This study has shown that various cocoa components possess anticancer activity, and this property resides mostly in the hydroethanolic root, leaf, and bean extracts.

The limitation of this study was that we were unable to identify the specific compounds responsible for the biological activities observed.

FURTHER RESEARCH DIRECTIONS

It is recommended that further studies be conducted to isolate, characterize, and elucidate the chemical structures of the bioactive compounds in cocoa responsible for these activities and investigate the anticancer mechanism(s) of action of these compounds.

CONCLUSION

Findings from this study indicate that natural cocoa extracts have antioxidant activities, and the activities of cocoa roots and beans are higher than other parts of cocoa. The study revealed that the antioxidant activity of the hydroethanolic extracts were stronger than the aqueous extracts. Also, GSH concentration was higher in the hydroethanolic and aqueous cocoa extracts than the other extracts tested. In this study, aqueous cocoa bean, hydroethanolic cocoa root and hydroethanolic cocoa leaf extracts had good selectivity towards PC3/LNCaP, Jurkat/PC3 and, PC3/LNCaP cell lines respectively. Only petroleum ether and ethyl acetate fractions of hydroethanolic cocoa root and ethyl acetate fraction of cocoa leaf had good selectivity towards Jurkat and PC3, and LNCaP respectively.

Funding

This research received no specific grant from any funding agency in the public, commercial, or not-for-profit sectors.

ACKNOWLEDGEMENT

The authors wish to thank Cocoa Research Institute of Ghana, Akim-Tafo for providing cocoa parts free of charge to Jacob Apibilla Ayembilla to conduct this study as part of his MPhil research work. The authors also thank staff of the Department of Clinical Pathology, Noguchi Memorial Institute for Medical Research (NMIMR) and staff of Department of Chemical Pathology, School of Biomedical and Allied Health Sciences for their technical support during the study.

REFERENCES

Abrokwah, F. K., Asamoah, K. A., & Esubonteng, P. K. A. (2009). Effects of the intake of natural cocoa powder on some biochemical and haematological indices in the rat. *Ghana Medical Journal, 43*(4), 164–168. PMID:21326996

Acheampong, F., Larbie, C., Arthur, F. K. N., Appiah-Opong, R., & Tuffour, I. (2015). Antioxidant and anticnacer study of ageratum conyzoides aqueous extracts. *Journal of Global Biosciences, 4*(1), 1804–1815.

Agrawal, N., Pandey, P., Mishra, D., & Maheshwaris, K. (2017). *Anticarcinogenic and Chemopreventive effect of AndrographisPaniculata extract in Swiss albino mice. 7*(6).

Anwar, F., Kalsoom, U., Sultana, B., Mushtaq, M., Mehmood, T., & Arshad, H. A. (2013). Effect of drying method and extraction solvent on the total phenolics and antioxidant activity of cauliflower (Brassica oleracea L.) Extracts. *International Food Research Journal, 20*(2), 653–659.

Appiah-Opong, R., Asante, I. K., Safo, D. O., Tuffour, I., Ofori-Attah, E., Uto, T., & Nyarko, A. K. (2016). Cytotoxic effects of Albizia zygia (DC) J. F. Macbr, a Ghanaian medicinal plant, against human T-lymphoblast-like leukemia, prostate and breast cancer cell lines. *International Journal of Pharmacy and Pharmaceutical Sciences, 8*(5), 392–396.

Appiah-Opong, R., Tuffour, I., Annor, K. G., Blankson-Darku, A. D., Cramer, P., Kissi-Twum, A. A., & Uto, T. O. A. (2016). Antiproliferative, antioxidant activities and apoptosis induction by Morinda lucida and Taraxacum officiale in Human HL-60 leukemia cells. *J. Global Biosci, 5*(7), 4281–4291.

Atjanasuppat, K., Wongkham, W., Meepowpan, P., Kittakoop, P., Sobhon, P., Bartlett, A., & Whitfield, P. J. (2009). In vitro screening for anthelmintic and antitumour activity of ethnomedicinal plants from Thailand. *Journal of Ethnopharmacology, 123*(3), 475–482. doi:10.1016/j.jep.2009.03.010 PMID:19473794

Badisa, R. B., Darling-Reed, S. F., Joseph, P., Cooperwood, J. S., Latinwo, L. M., & Goodman, C. B. (2009). Selective cytotoxic activities of two novel synthetic drugs on human breast carcinoma MCF-7 cells. *Anticancer Research, 29*(8), 2993–2996. PMID:19661306

Baharum, Z., Akim, A. M., Taufiq-Yap, Y. H., Hamid, R. A., & Kasran, R. (2014). In vitro antioxidant and antiproliferative activities of methanolic plant part extracts of Theobroma cacao. *Molecules (Basel, Switzerland), 19*(11), 18317–18331. doi:10.3390/molecules191118317 PMID:25389662

Baharum, Z., Md Akim, A., Hin, T. Y. Y., Hamid, R. A., & Kasran, R. (2016). Theobroma cacao: Review of the extraction, isolation, and bioassay of its potential anti-cancer compounds. *Tropical Life Sciences Research*, *27*(1), 21–42. PMID:27019680

Bauer, D., De Abreu, J. P., Oliveira, H. S. S., Goes-Neto, A., Koblitz, M. G. B., & Teodoro, A. J. (2016). Antioxidant Activity and Cytotoxicity Effect of Cocoa Beans Subjected to Different Processing Conditions in Human Lung Carcinoma Cells. *Oxidative Medicine and Cellular Longevity*, *2016*, 1–11. Advance online publication. doi:10.1155/2016/7428515 PMID:27034742

Brand-Williams, W., Cuvelier, E., & Berset, C. (1995). Use of a free radical method to evaluate antioxidant activity. *Lebensmittel-Wissenschaft + Technologie*, *28*(1), 25–30. doi:10.1016/S0023-6438(95)80008-5

Chhikara, B. S., & Parang, K. (2022). Global Cancer Statistics 2022: The trends projection analysis. In *Chemical Biology Letters Chem.* []. https://pubs.thesciencein.org/cbl]. *Biol. Lett*, *2023*(1).

Eid, S. Y., El-Readi, M. Z., Fatani, S. H., Mohamed Nour Eldin, E. E., & Wink, M. (2015). Natural Products Modulate the Multifactorial Multidrug Resistance of Cancer. *Pharmacology &amp. Pharmacology & Pharmacy*, *06*(03), 146–176. doi:10.4236/pp.2015.63017

Hazirah, A. R., Zainal, B., & Abdah, M. A. (2013). Total phenolic content, antioxidant and cytotoxic activity of cocoa (Theobroma cacao L.) polyphenols extracts on cancer cell lines. *Malaysian Journal of Nutrition*, *19*(2), 223–232.

Janardhanan, K. K., Ravikumar, K. S., & Karuppayil, S. M. (2020). Medicinal mushroom bioactives: Potential sources for anti-cancer drug development. *International Journal of Applied Pharmaceutics*, *12*(Special Issue 4), 40–45. doi:10.22159/ijap.2020.v12s4.40103

Jayanthi, A., Singh, N., Datta, S., Dey, A., & Chowdhury, A. R. (2015). Antimicrobial activity and cytotoxicity of theobroma cacao extracts. *Der Pharmacia Lettre*, *7*(7), 287–294.

Jemal, A., Bray, F., & Ferlay, J. (2011). Global Cancer Statistics: 2011. *CA: a Cancer Journal for Clinicians*, *49*(2), 1, 33–64. doi:10.3322/caac.20107.Available

Karagöz, A., Turgut-Kara, N., Çakır, Ö., Demirgan, R., & Arı, Ş. (2007). Cytotoxic activity of crude extracts from astragalus chrysochlorus (leguminosae). *Biotechnology, Biotechnological Equipment*, *21*(2), 220–222. doi:10.1080/13102818.2007.10817449

Krishnamurthy, P. T., Vardarajalu, A., Wadhwani, A., & Patel, V. (2015). Identification and characterization of a potent anticancer fraction from the leaf extracts of moringa oleifera L. *Indian Journal of Experimental Biology*, *53*(2), 98–103. PMID:25757240

Laryea, D. O., Awuah, B., Amoako, Y. A., Osei-Bonsu, E., Dogbe, J., Larsen-Reindorf, R., Ansong, D., Yeboah-Awudzi, K., Oppong, J. K., Konney, T. O., Boadu, K. O., Nguah, S. B., Titiloye, N. A., Frimpong, N. O., Awittor, F. K., & Martin, I. K. (2014). Cancer incidence in Ghana, 2012: Evidence from a population-based cancer registry. *BMC Cancer*, *14*(1), 1–8. doi:10.1186/1471-2407-14-362 PMID:24884730

Luo, J., Solimini, N. L., & Elledge, S. J. (2009). Principles of cancer therapy: Oncogene and non-oncogene addiction, the current state of cancer research. *Cell*, *136*(5), 823–837. doi:10.1016/j.cell.2009.02.024 PMID:19269363

Martin, M. A., Goya, L., & Ramos, S. (2013). Potential for preventive effects of cocoa and cocoa polyphenols in cancer. *Food and Chemical Toxicology*, *56*, 336–351. doi:10.1016/j.fct.2013.02.020 PMID:23439478

Othman, A., Jalil, A. M. M., Weng, K. K., Ismail, A., Ghani, N. A., & Adenan, I. (2010). Epicatechin content and antioxidant capacity of cocoa beans from four different countries. *African Journal of Biotechnology*, *9*(7), 1052–1059. doi:10.5897/AJB09.1219

Redovniković, I. R., Delonga, K., Mazor, S., Dragović-Uzelac, V., Carić, M., & Vorkapić-Furač, J. (2009). Polyphenolic content and composition and antioxidative activity of different Cocoa liquors. *Czech Journal of Food Sciences*, *27*(5), 330–337. doi:10.17221/119/2008-CJFS

Sandanasamy, J., Hamid Nour, A., Nizam Tajuddin, S., Hamid Nour, A., Abd Hamid, H., & Ziad Sulaiman, A. (2015). Bioassay guided isolation and identification of a cytotoxic compound form azadirachta indica leaves. *Journal of the Chinese Chemical Society (Taipei)*, *62*(8), 680–688. doi:10.1002/jccs.201400461

Traverso, N., Ricciarelli, R., Nitti, M., Marengo, B., Furfaro, A. L., Pronzato, M. A., Marinari, U. M., & Domenicotti, C. (2013). Role of glutathione in cancer progression and chemoresistance. *Oxidative Medicine and Cellular Longevity*, *2013*, 1–10. doi:10.1155/2013/972913 PMID:23766865

Twilley, D., Rademan, S., & Lall, N. (2020). A review on traditionally used South African medicinal plants, their secondary metabolites and their potential development into anticancer agents. *Journal of Ethnopharmacology*, *261*, 113101. doi:10.1016/j.jep.2020.113101 PMID:32562876

Wild, P. C., Wilederpass, E., & Sterwart, W. B. (2020). *World Cancer Report : Cancer research for cancer prevention*. https://doi.org/Licence: CC BY-NC-ND 3.0 IGO

Wollgast, J., & Anklam, E. (2000). Review on polyphenols in theobroma cacao: Changes in composition during the manufacture of chocolate and methodology for identification and quantification. *Food Research International*, *33*(6), 423–447. doi:10.1016/S0963-9969(00)00068-5

Yahya, M., Ginting, B., & Saidi, N. (2022). Anticancer and antiretroviral activities of methanolic extract from theobroma cacao L pod husk: Focusing on the ethyl acetate partition. *F1000 Research*, *11*, 1395. doi:10.12688/f1000research.128048.1

Chapter 8
Polysaccharides of Moringa oleifera Lam. Leaves Disrupted With Ultrasound:
Optimization, Characterization and Bioactivity

Phyllis Naa Yarley Otu
Accra Technical University, Ghana

Jacob Apibilla Ayembilla
Accra Technical University, Ghana

Doreen Dedo Adi
Accra Technical University, Ghana

ABSTRACT

The aim of this study was to apply ultrasound technology to extract bioactive compounds from Moringa oleifera Lam. leaves. Bringing modern trends of extraction into perspective, ultrasound-assisted extraction (UAE) of polysaccharides of Moringa leaves (MLP) was established engaging response surface methodology (RSM). The experimental yield of 9.24% of crude polysaccharide was close to the predicted 9.57%. Three graded ethanol precipitation of the UAE polysaccharides with corresponding molecular weights (55,250, 60,330 and 34,260 g/mol) was obtained. Despite similar Fourier transform infrared spectroscopy among fractions there was different molar ratios of monosaccharides composition and different surface morphology under scanning electron microscopy. All fractions were not of helical structures as demonstrated by the Congo red assay. MLP precipitated at 80% was more viscous (G" loss modulus) and had the highest radical scavenging ability, thus can prevent damage of biomolecules. The polysaccharides extract obtained may be incorporated in pharmaceutical and nutraceutical products.

DOI: 10.4018/978-1-6684-9843-9.ch008

INTRODUCTION

The species *Moringa oleifera* Lam. belongs to the monogeneric family Moringaceae, discovered to be widely spread and naturalized in many geographical locations in the world including the tropics and subtropics of Asia and Africa (Adisakwattana & Chanathong, 2011). It has all its parts useful in the treatment of ailment in traditional medicine: root, leaf, flowers, seed, gum, bark, and seed oil. The leaves has been used traditionally in the treatment of sore throat, piles, fevers, wiped on the temples for headaches and dressing to sores (Fuglie, 2001). Aside its traditional medicinal purpose, it has been established that some varieties are also consumed as nutritious vegetables in some countries. It possess considerable levels of carotene, iron and ascorbic acid (Chawla, Saxena, & Seshadri, 1988), 30.30% crude protein level in dried leaves (Moyo et al., 2011) and 43.88% carbohydrate (Owusu, Ellis, & Oduro, 2008).

Current extraction trends of essential biomolecules from such medicinal plants are devoid of organic solvents, one of such applications is ultrasound. Ultrasonication is mainly concerned with cell disruption, the reason for short time and easy permeate of solvent into cellular materials producing a high mass transfer. Ultrasonic waves causes cavitation disruption of cell walls and assist the discharge of matrix components. Thus ultrasound ensures permeabilization of cell membranes to ions (Mummery, 1978) rendering cell membranes less selective. Most currently, polysaccharide has been extracted from the dry ripped fruit, *Trischosanthes fructus* using ultrasound-assisted enzymatic extraction (Chen et al., 2017). A total of 6.58% of crude polysaccharide was extracted at a temperature of 45 °C. The temperature under which extraction was made was a limitation since denaturation and structural changes occurs in proteins with the application of temperatures above 65 °C (Oakes, 1976). The level of extraction yield may therefore be higher if the extraction operating temperature of the ultrasound is increased in the absence of an enzyme. Furthermore, (Chen et al., 2017), also used microwave-assisted technique to optimally extract polysaccharide from *M. oleifera* and obtained 2.98 ± 0.11% of crude polysaccharide, which was notably low. Therefore this project seeks to establish a higher extraction yield of crude polysaccharide from *M. oleifera* Lam. leaves using solely ultrasound at a higher ultrasound operational temperature.

Monosaccharide polymers possess very much needed properties for the food and pharmaceutical industry which includes gel formation, high stability, non-toxicity, biodegradability and hydrophilicity (Coviello et al., 2001). Furthermore, natural plant source complex carbohydrate have recorded numerous biological activities such as stimulation of the immune system, wound healing, impacts on the hematopoietic system and enhancement of the reticuloendothelial system (Talmadge et al., 2004). Interestingly, extraction methods directly affects the structure and bioactivity of polysaccharides from natural sources (Coviello et al., 2012) (Ying, Han, & Li, 2011) Thus, this work also sort to investigate the possible impact on structure and antioxidant activity of the ultrasound-assisted extracted polysaccharides.

Knowledge on the most productive process parameters that churns out the targeted product is referred to as optimization. Response surface methodology (RSM) is an extremely efficient mathematical and statistical tool for evaluation of a number of experimental parameters interacting simultaneously (Hammi et al., 2015; Vuong et al., 2015) (Vuong et al., 2015). Thus, the objectives of this study were as follows: (I) Investigate the individual and interactive effects of ultrasound-assisted process variables including ultrasound power density, temperature, frequency and time on the yield of *M. oleifera* leaf polysaccharides (II) Optimize the operational parameters to obtain the maximum yield of polysaccharides using Box-Behnken Designs (BBD) (III) Characterize the yielded polysaccharides of the different isolated fractions (IV) Determine scavenging ability of the different isolated fractions of polysaccharides.

MATERIALS AND METHODS

Materials

Moringa oleifera Lam., commercial blender, 100-mesh screen, desiccator, soxhlet System, hot air oven, ultrasound water bath, rotary evaporator, freeze dryer, whatman # 1, UV-vis spectrophotometer, gas chromatography, ampoules, filters (0.22 μm), fourier transform-infrered spectroscopy, scanning electron microscopy, high performance size-exclusion chromatography, DHR-1 rheometer, distilled water, phenol, sulfuric acid, ethanol, coomassie blue G-250, albumin (standard), phosphoric acid, trifluoroacetic acid, methanol, hydroxylamine hydrochloride, pyridine, acetic anhydride, D-glucose, D-xylose, D-mannose, D-galactose, D-galacturonic acid, L-arabinose (Standards), potassium bromide, sodium hydroxide pellets, congo red, gold palladium, Iron sulfate, salicylic acid, hydrogen peroxide, ABTS, and potassium peroxydisulfate.

Methods

Preparation of Sample Powder

M. oleifera leaves was adopted from Ghana. The leaves was air-dried, pulverized into powder and sieved through a 100-mesh screen. The powdered sample was stored in a desiccator for experimental use. Majority of monosaccharides, pigments and fats were removed from the powdered *M. oleifera* leaves using ethanol (90% v/v) in a

Soxhlet system at 70 °C for 4 h. The filtered residue was dried at 40 °C in a hot air oven to obtain the pretreated dry sample powder.

Polysaccharide Extraction

Pretreated sample was extracted by water at a sample: water ratio (1:20) using designed operational parameters: power, frequency, temperature and time of a water bath ultrasound under pulsating mode (5 seconds on, 2 seconds off). Extracted sample in water was separated using centrifugation at 4000 rpm for 15 min. The supernatant was filtered to remove any particles. The polysaccharide content was measured using the phenol-sulfuric method. The polysaccharide content was calculated using the formula:

$$\text{Yield } (\%) = C*V/W*100 \quad (1)$$

Where C (g/ml) is the concentration of polysaccharide solution, V (ml) is the volume of polysaccharide solution and W (g) is the weight of dried sample.

Single-Factor Experimental Design

As a preliminary tool, single-factor design was established with extraction factors X_1 (extraction temperature: 10, 30, 50, 70, 90 °C), X_2 (extraction power: 60, 120, 180, 240, 300 W) for 1 L of water used in the water bath, X_3 (extraction frequency: 20, 40, 60 kHz) and X_4 (extraction time: 10, 30, 50, 70, 90 mins), each with a control group. Each experimental factor was varied as the other experimental factors were kept constant. All experiments were conducted in triplicate and the dependent variable used was the extraction yield of crude polysaccharide.

Table 1. Variables and the corresponding three levels employed in Box Behnken design

Variable	Units	Symbol	Coded Levels		
			-1	0	1
Temperature	°C	X_1	10	50	90
Power density	W/L	X_2	60	150	240
Frequency	KHz	X_3	20	40	60
Time	Mins	X_4	10	50	90

Table 2. Experimental design of response surface analysis and its experimental values

Run	X_1 Temperature (°C)	X_2 Power Density (W/L)	X_3 Frequency (KHz)	X_4 Extraction Time (mins)	Y Extracted Polysaccharide (%)
1	50	60	40	90	6.15
2	50	150	20	90	4.14
3	10	60	40	50	5.70
4	50	150	40	50	5.32
5	10	240	40	50	5.53
6	90	150	40	10	7.28
7	10	150	60	50	6.45
8	90	60	40	50	6.74
9	50	150	20	10	4.01
10	50	150	40	50	5.97
11	10	150	20	50	5.13
12	10	150	40	90	6.97
13	50	240	40	90	7.47
14	50	60	60	50	6.55
15	50	60	20	50	3.17
16	50	240	40	10	5.68
17	50	240	20	50	4.53
18	50	60	40	10	6.51
19	90	240	40	50	8.15
20	90	150	20	50	5.56
21	50	150	40	50	6.47
22	90	150	60	50	9.09
23	50	150	40	50	7.22
24	50	150	40	50	6.05
25	50	150	60	90	7.36
26	50	150	60	10	6.38
27	10	150	40	10	5.95
28	50	240	60	50	6.53
29	90	150	40	90	7.97

Optimization of Experimental Design

Four variables were assigned codes (X_1, X_2, X_3, X_4) and examined in three levels (Table1) based on the experimental results from the single-factor experiment. A response surface methodology was then conducted using the Box-Behnken Design (BBD). It comprised of 29 experimental runs, conducted in triplicates and its experimental results has been presented (Table 2).

Purification of Polysaccharide (Ethanol Graded Precipitation Method)

The pretreated pulverized *M. oleifera* leaves were extracted using optimum extraction conditions obtained previously. It was then deproteinized using the TCA method (repeated three times) and the supernatant was sequentially fractionated by ethanol graded precipitation at different ethanol concentrations of 60%, 70% and 80% (v/v) which was assigned MLP 60, MLP 70, and MLP 80. Water was added to the precipitates to aid in the removal of the ethanol by the use of the rotary-evaporator at 37 °C and then lypholized.

Chemical Composition

Total Carbohydrate, Protein, Polyphenol Content

The total carbohydrate and protein contents of graded fractions MLP 60, MLP 70 and MLP 80 were determined according to phenol-sulfuric method (Dubois, Gilles, Hamilton, Rebers, & Smith, 1956) and Bradford Method (Bradford, 1976). The phenolic contents were measured as described by (Kumar, Ganesan, & Rao, 2008). All experiments were carried out in triplicate.

Monosaccharide Composition

Monosaccharide composition was conducted using Gas Chromatography method described by (Wang, Lian, Yu, Wei, & Kang, 2017). Briefly, samples (10 mg) were hydrolyzed in ampoules using 2 M trifluoroacetic acid (2 ml) for 3 h at 110 °C. It was then evaporated and methanol was added to remove TFA. Hydrolyzates were mixed thoroughly with (10 mg) hydroxylamine hydrochloride and (0.5 ml) pyridine and incubated at 90 °C for 30 min. Acetic anhydride (0.5 ml) was added and incubated at 90 °C for 30 min. The mixtures were cooled to room temperature, and filtered with (0.22 μm) filters. The resulting alditol acetates were then analyzed by GC, which was performed on a Thermo TRACE1300 instrument fitted with

FID (280 °C) and equipped with HP-5 column (30 m × 0.25 mm × 0.25 μm). The column temperature was maintained at 110 °C for 5 min, and increased to 190 °C for 4 min at a rate of 5 °C/min, then increased to 210 °C for 10 min at a rate of 3 °C/min. Standard monosaccharide used included (D-glucose, D-xylose, D-mannose, D-galactose, D-galacturonic acid and L-arabinose). These were also prepared and subjected to GC analysis separately in the same way.

Spectral Analysis

Fourier Transform-Infrared (FT-IR) Spectroscopy

The infrared spectra of the samples were measured as described by (Zhou et al., 2016) using FT-IR spectrophotometer (Rayleigh, WQF-510A). Briefly, samples were prepared by grinding polysaccharide with KBr (1:100 mg) and then pressed into a translucent pellet and then scanned with a resolution of 4 cm^{-1} in a region of 4000-400 cm^{-1}.

Ultraviolet (UV) Spectroscopy

The ultraviolet spectra of the three fraction solutions (1.0 mg/ml) were obtained with a UV-vis spectrophotometer (UV- 1601) in the 200-400 nm region in 1.00 cm quartz cell against distilled water as blank.

Structural Analysis

Tertiary Structure (Congo Red Assay)

The conformational structures of MLP 60, MLP 70 and MLP 80 was determined as described by (Qiu et al., 2013) with little modification. Briefly, different sets of solutions were prepared containing polysaccharides (1 mg/ml, 2 ml) in (0.0 - 0.5 M) NaOH (increasing stepwise by 0.05 M increments using (4 M) NaOH solution) and (100μM, 2 ml) of Congo red. These were analyzed in the range of 400-700 nm with a UV-Vis spectrophotometer (UV- 1601) and the maximum absorption wavelengths were recorded as a function of NaOH concentration. Congo red in NaOH served as the negative control.

Scanning Electron Microscopy (SEM)

The morphological characteristics of the graded polysaccharides was observed as described by (Zhou et al., 2016). Briefly, samples were coated with a conductive layer of gold-palladium and observed with a scanning electron microscope (JSM-7001F, JEOL, Tokyo, Japan).

Molecular Weight (HPSEC)

The molecular weights (Mw) of MLP 60, MLP 70 and MLP 80 were determined by High Performance Size-Exclusion Chromatography (HPSEC) using 3 columns: OHpak SB-806 HQ column (8.0mm ID x 300mm), OHpak SB-805 HQ column (8.0mm ID x 300mm) and OHpak SB-806 HQ column (8.0mm ID x 300mm), Shodex®, Japan. The performance conditions were: column temperature (25 °C), injection volume (15 µl, 2 mg/ml w/v) and flow rate (0.500 ml/min) of the flow solution (1 mol/l NaCl, 2% NaN_3) and Detector (RI). Data from the light scattering instrument (HELEOS) was collected and analyzed using ASTRA 6.1.7.15 (Wyatt Technology), calculating the molecular weight moments.

Rheological Properties

The steady shear and oscillatory tests of MLP 60, MLP 70, MLP 80 at a concentration of (5 mg/ml) were conducted with a DHR-1 rheometer (TA Instruments, Surrey, UK) using a cone-and-plate geometry. Conditions set included a cone angle of (1°), a diameter of (50 mm), and a gap of (1.0mm). The steady share rate measurement was performed at (25 °C) between (1 and $100s^{-1}$). Oscillatory (dynamic) tests were performed using small deformation oscillatory rheometry at (25 °C) between (10) and (0.01 Hz). The viscoelastic properties, storage modulus (G′), loss modulus (G″) and tan δ were recorded versus frequency. All measurements were done in triplicates.

Antioxidant Activity In-Vitro

Hydroxyl Radical Scavenging Assay

The hydroxyl radical assay was measured as described by (Jen, Leu, & Yang, 1998) with little modification. Briefly, Samples were dissolved in distilled water (5 – 30 mg/ml). The sample solution (1 ml) was mixed with $FeSO_4$ (9 mM, 1 ml) and 9 mM salicylic acid solution (1 ml, 50% ethanol). Then 8.8 mM H_2O_2 (1 ml) was added to start the reaction. The mixture was kept in water bath at 37 °C for 1 h. The background was mixed as described above except 50% ethanol (1 ml) was used in place of the salicylic acid solution. For the control distilled water (1 ml) was substituted for the polysaccharide solution. After warming in a water bath, the absorbance of the mixture was measured at 510 nm. The hydroxyl radical scavenging rate was calculated using the following formula:

$$\text{Scavenging Rate } (\%)=[1-(A_1 - A_2)/A_0]\times 100\% \quad (2)$$

Where A_0 is the absorbance of the control group (without polysaccharides), A_1 is the absorbance of the test group, A_2 is the absorbance of the background group.

Radical Scavenging Assay

The ABTS radical scavenging assay was measured using (Zhou et al., 2011) described methodology. Briefly, ABTS (50 ml, 7 mM) was mixed with 140 mM potassium peroxydisulfate (890 μl), then kept in dark at room temperature for 12–16 h before use. The samples were prepared in a variety of concentrations (5 –30 mg/ml). The polysaccharides solution (0.2 ml) was added to the ABTS·+ solution (5 ml). The absorbance of the mixture was measured at 734 nm namely after holding at room temperature for 6 min. The ABTS radical scavenging rate was calculated using the following formula:

$$\text{Scavenging Rate } (\%)=[1-(A_1 - A_2)/A_0]\times 100\% \quad (3)$$

Where A_0 is the absorbance of the control group (without polysaccharides), A_1 is the absorbance of the test group, A_2 is the absorbance of the background group.

RESULTS AND DISCUSSION

Single-Factor Experiments

Effect of Ultrasound Temperature

The study of *M. oleifera* crude polysaccharides yield based on the effect of ultrasound extraction temperature under fixed conditions (ultrasound power: 240 W, ultrasound frequency: 60 KHz, ultrasound time: 90 mins) has been presented in (Fig 1A). Increase in polysaccharide yield displayed consistent significant difference ($p < 0.05$) from 10 °C to 90 °C. The experimental temperatures was not increase beyond 90 °C because research has proved that temperatures near 100 °C influences the physiological activity of polysaccharide and further causes degradation (Tan et al., 2011), hence an optimal temperature of 90 °C was selected.

Effect of Ultrasound Power

The results of *M. oleifera* crude polysaccharides yield based on the effect of ultrasound extraction power under fixed conditions (ultrasound temperature: 50 °C, ultrasound frequency: 40 KHz, ultrasound time: 50 mins) has also been displayed in (Fig 1B). Significant difference ($p < 0.05$) was observed in the increase of extraction yield from a power of 60 W to 240 W. (Ratoarinoro et al., 1992) has reported that an increase in power directly ensures an increase in reaction yield and then after cavitation power threshold, it remains nearly constant. Thus, under ultrasound power from 240 W to 300 W there was no significant difference ($p > 0.05$) in polysaccharide, establishing an optimal power density of 240 W.

Effect of Ultrasound Frequency

M. oleifera crude polysaccharides extraction yield based on the effect of extraction frequency under fixed conditions (ultrasound temperature: 50 °C, ultrasound power: 240 W, ultrasound time: 50 mins) has been depicted in (Figure 1C). Constant significant increase ($p < 0.05$) in extraction yield was also observed from the lowest frequency (20 KHz) to the highest frequency (60 KHz). At a lower frequency (KHz), cavitation zones are more violent whiles at a much higher frequency (MHz), acoustic streaming are observed in place of cavitation (Patist & Bates, 2008). The best liquid cavitation was observed at 60 KHz, ensuring fragmentation of polysaccharide macromolecules. The highest operational frequency of the water bath ultrasound used in this work, 60 KHz, was thus chosen as the optimal frequency.

Effect of Ultrasound Time

The yield of *M. oleifera* crude polysaccharides based on the effect of ultrasound extraction time under fixed conditions (ultrasound temperature: 50 °C, ultrasound power: 240 W, ultrasound frequency: 40 KHz) has been presented in (Fig 1D). Significant increase ($p < 0.05$) in extraction yield was observed from 10 mins to 90 mins of ultrasound irradiation exposure but, at 110 mins a significant ($p < 0.05$) reduction of extraction yield was observed. This may be due to degradation of polysaccharide after the time 90 mins. Hence, 90 mins was selected as the optimal extraction time.

Figure 1. Effect of A) temperature, B) power, C) frequency, and D) time on the polysaccharide extraction yield of Moringa leaves powder

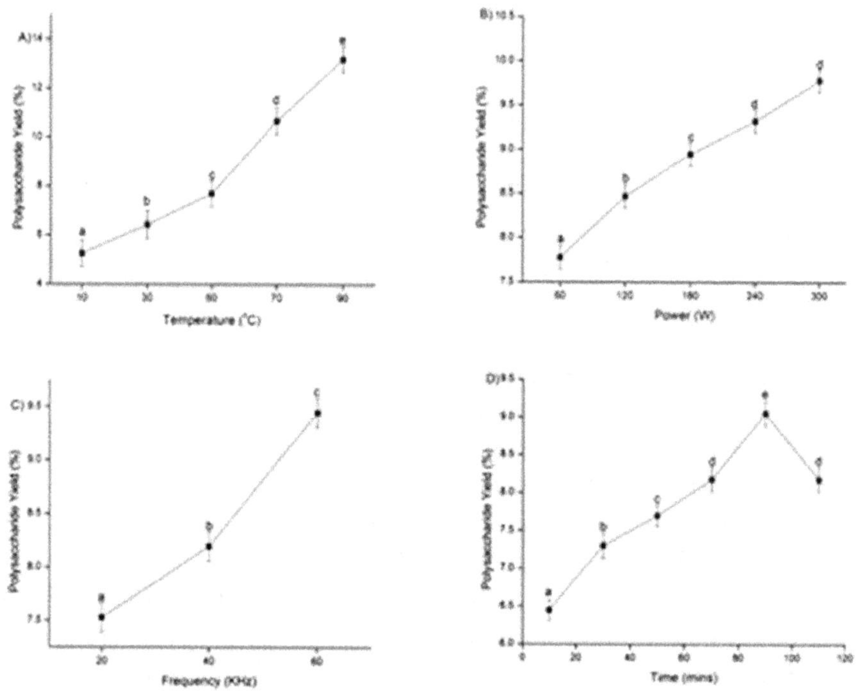

Optimization using Box-Behnken Design

Fitting the Model

Analysis of variance (ANOVA) was performed to evaluate the quality of the fitted model (Table 3). In this model, the second-order polynomial model for the extraction of crude polysaccharide was statistically significant with a small model ($p < 0.0001$) and satisfactory coefficient of determination ($R^2 = 0.9241$). The linear parameters (X_1, X_3) and (X_4) were significant at the level of ($p < 0.0001$) and ($p < 0.03$) respectively. Quadratic parameter (X_1^2, X_3^2) were also significant at the level of ($p < 0.001$; $p < 0.003$) respectively, whiles the interaction parameter (X_1X_3) were significant at the level of ($p < 0.05$). The "Lack of Fit-Value" of the model was not significant with a p-value of 0.9348. The significant regression and non-significant lack of fit indicated that the regression equation is adequate to represent the actual relationship between the response values (Y) and four independent variables. The quadratic regression equation was obtained as follows; Equation (4):

Table 3. ANOVA for response surface quadratic model

Source	Sum of Squares	DF	Mean Square	F Value	p-value Prob > F
Model	42.79	14	3.06	12.17	< 0.0001
X_1	6.87	1	6.87	27.36	0.0001
X_2	0.79	1	0.79	3.14	0.0982
X_3	20.86	1	20.86	83.08	< 0.0001
X_4	1.51	1	1.51	6.01	0.0279
X_1X_2	0.63	1	0.63	2.50	0.1365
X_1X_3	1.23	1	1.23	4.89	0.0442
X_1X_4	0.027	1	0.027	0.11	0.7462
X_2X_3	0.48	1	0.48	1.91	0.1889
X_2X_4	1.15	1	1.15	4.60	0.0500
X_3X_4	0.18	1	0.18	0.73	0.4084
X_1^2	3.68	1	3.68	14.66	0.0018
X_2^2	0.32	1	0.32	1.28	0.2771
X_3^2	3.15	1	3.15	12.55	0.0032
X_4^2	0.20	1	0.20	0.78	0.3921
Residual	3.52	14	0.25		
Lack of Fit	1.56	10	0.16	0.32	0.9348
Pure Error	1.95	4	0.49		
Cor Total	46.31	28			
R-Squared	0.9241				
Adj R-Squared	0.8482				

$$Y = 6.20 + 0.76X_1 + 0.26X_2 + 1.32X_3 + 0.35X_4 + 0.40X_1X_2 + 0.55X_1X_3 + 0.083X_1X_4 - 0.35X_2X_3 + 0.54X_2X_4 + 0.21X_3X_4 + 0.75X_1^2 - 0.22X_2^2 + 0.70X_3^2 + 0.17X_4^2 \quad (4)$$

Analysis of Response Surfaces

Response surface plots are displayed in (Figure 2). The interaction between various features can be seen directly from the response surface plots. A temperature of 50 °C interaction with power of 150 W or with frequency of 40 KHz initiated a steady rise in yield of polysaccharide extraction. Progressively, as temperature increased to 90 °C and then also power and frequency moved close to 240 W and 60 KHz respectively, polysaccharide was then observed to exhibit the highest and then continued with a constant yield (Figure 2A & 2B). Time at 50 mins and temperature at 50 °C demonstrated a rise in polysaccharide yield and continued steadily to 90 min and 90 °C respectively (Figure 2C). Meanwhile, extraction yield of polysaccharide increased from 50 mins of time and power of 150 W but, after a power of 195 W the polysaccharide yield showed a constant yield (Figure 2D). The combination

Figure 2. Response surface model plot showing the effects of independent variables on Moringa Leaves Polysaccharide yield: (A) temperature and power, (B) temperature and frequency, (C) temperature and time, (D) power and frequency

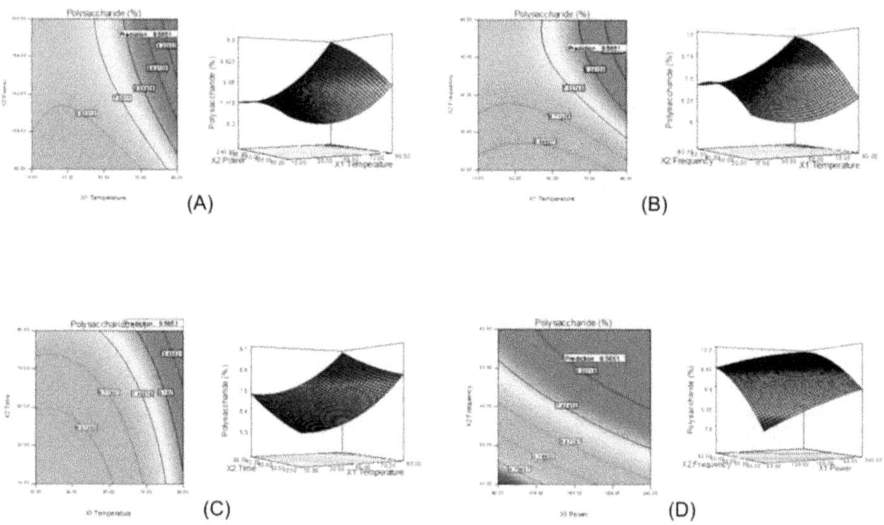

of the analysis of variance (ANOVA) (Table 3) and response surfaces (Figure 2) however, indicted that the interaction effect between ultrasound irradiation frequency and temperature were statistically significant. Thus, it was concluded that effect of ultrasound irradiation temperature and frequency were significant than the ultrasound irradiation power and time on the extraction yield of crude polysaccharide.

Verification of Predicted Value of the Models

The optimal conditions for simultaneous ultrasonic extraction were predicted at 90 °C temperature, 210.15 W power, 49.93 KHz frequency and 88.54 mins time of ultrasonic treatment with results of 9.57% of crude polysaccharide. The UEA process was repeated at near optimum conditions by modifying the extraction power density from 210.15 W to 240 W, extraction frequency from 49.93 KHz to 60 KHz and extraction time from 88.54 mins to 90 mins. Table 4, summarizes the amounts of crude polysaccharide extracted under the optimal predicted conditions and actual experimental conditions. The experimental and predicted values were very close. Hence, the model can be used to optimize the process of extracting crude polysaccharide from *M. oleifera*.

Table 4. Predicted and actual experimental values of Moringa oleifera leaves polysaccharide (%) extracted under modified optimal extraction conditions

Type	Extraction Variables				Polysaccharide (%)
	X_1 (°C)	X_2 (W/L)	X_3 (KHz)	X_4 (mins)	
Optimum (Predicted)	90	210.15	49.93	88.54	9.57
Modified Optimum (Experimental)	90	240	60	90	9.24 ± 0.06

Chemical Composition of Graded Fractions

Total Carbohydrate, Protein, Polyphenol Content

The chemical composition of MLP 60, MLP 70 and MLP 80 in terms of the total carbohydrate, proteins and polyphenol content has been presented on (Table 5). The results indicated that increase in ethanol concentration used in the isolation resulted in increase of chemical component.

Monosaccharide Composition

The polysaccharide fractions also shown on (Table 5) contained little L-arabinose and D-mannose. MLP 60 contained mainly D-xylose (31.82%), D-galactose (49.07%) and D-glucose (13.18%). MLP 70 and MLP 80 on the other hand, were mainly D-xylose (70.31% and 40.97%) and D-galactose (22.15% and 36.27%) and a little D-glucose (4.60% and 5.00%) respectively. D-galacturonic content for MLP 60, MLP 70 and MLP 80 were (3.56%, 1.59% and 9.48%) respectively, similar to report of higher levels of D-galacturonic acid detected in the highest ethanol concentration precipitated polysaccharide from *M. oleifera* leaves extracted by microwave-assisted extraction (C. Chen et al., 2017).

Spectral Analysis

FT-IR Spectra

The functional groups of extracted polysaccharide was studied using the FT-IR Spectroscopy (Fig. 3). It revealed that the precipitated polysaccharide from *Moringa* leaves using ethanol at different concentration had similar characteristics. Easily recognized and usually associated with polysaccharides was the hydroxyl stretching vibration (Geddes & Lakowicz, 2007) by the bands at 3350 $cm^{-1}$and the stretching

Table 5. Preliminary characterization of MLP 60, MLP 70, and MLP 80 fractions

Composition	MLP 60	MLP 70	MLP 80
Carbohydrate (%)	24.18 ± 0.07	74.64 ± 1.75	85.39 ± 0.58
Protein (%)	1.33 ± 0.019	2.24 ± 0.03	2.58 ± 0.04
Polyphenol (%)	8.96 ± 0.02	9.97 ± 0.04	22.23 ± 0.02
Monosaccharide composition (molar ratio%)[a]			
D-Xylose	31.82	70.31	40.97
D-Mannose	1.88	1.00	2.22
D-Glucose	13.18	4.60	5.00
D-Galactose	49.07	22.15	36.27
L-Arabinose	0.49	0.35	6.06
Uronic acid content (molar ratio %)[a]			
Galacturonic acid	3.56	1.59	9.48

[a]Samples were not analyzed in duplicate.

Figure 3. The FT-IR spectra of MLP 60, MLP 70, and MLP 80 polysaccharides

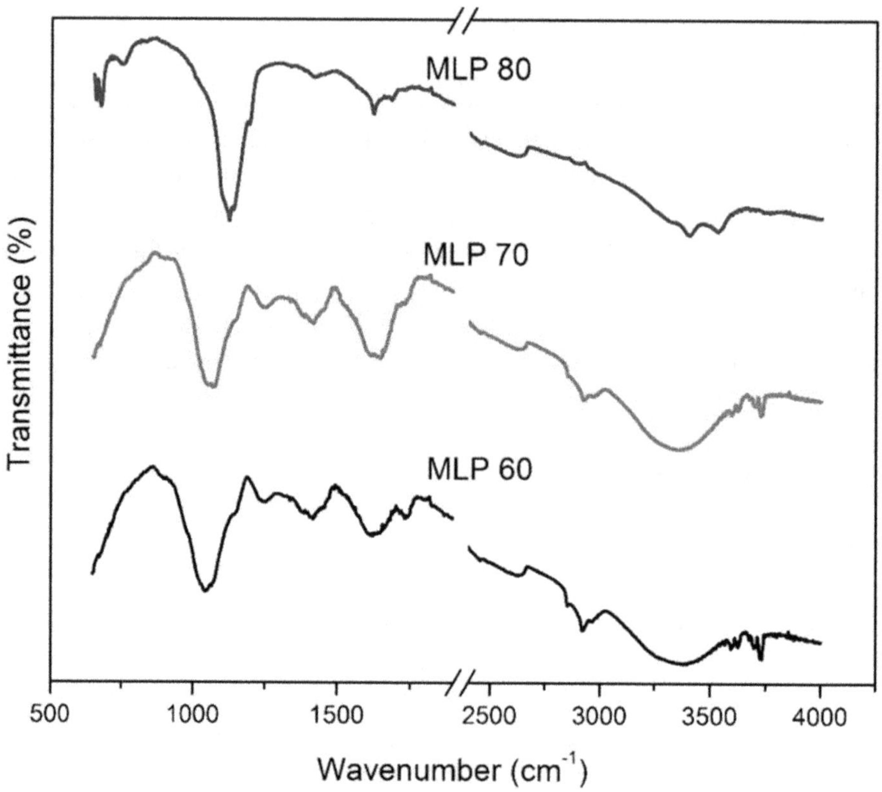

asymmetric vibration of C-H (Geddes & Lakowicz, 2007) and then also the stretching symmetric vibration of C-H (Dopico & Tigyi, 2007) at the bands ranging 2900 cm^{-1} - 2920 cm^{-1} . Furthermore, carbonyl group was revealed at 1730 cm^{-1} and a small peak at 1625 cm^{-1} which originates from associated water (Sun et al., 2008). This was followed by the bending stretch of the C-H at the band 1400 cm^{-1} (Geddes & Lakowicz, 2007). The bands at 1230 cm^{-1} and 1050 cm^{-1} proposed peak related to C-C stretching vibration and C-O stretching vibration respectively (Zou et al., 2013).

UV Spectra

The UV spectra of MLP 60, MLP 70 and MLP 80 as shown in (Figure 4) displayed virtually, no absorption peak at 260 and 280 nm, depicting small amount of nucleic acid and protein (Cai et al., 2013). The results were in accordance of the protein content results (Table 5).

Figure 4. UV-vis spectra of the three M. Oleifera Leaves Polysaccharide fractions

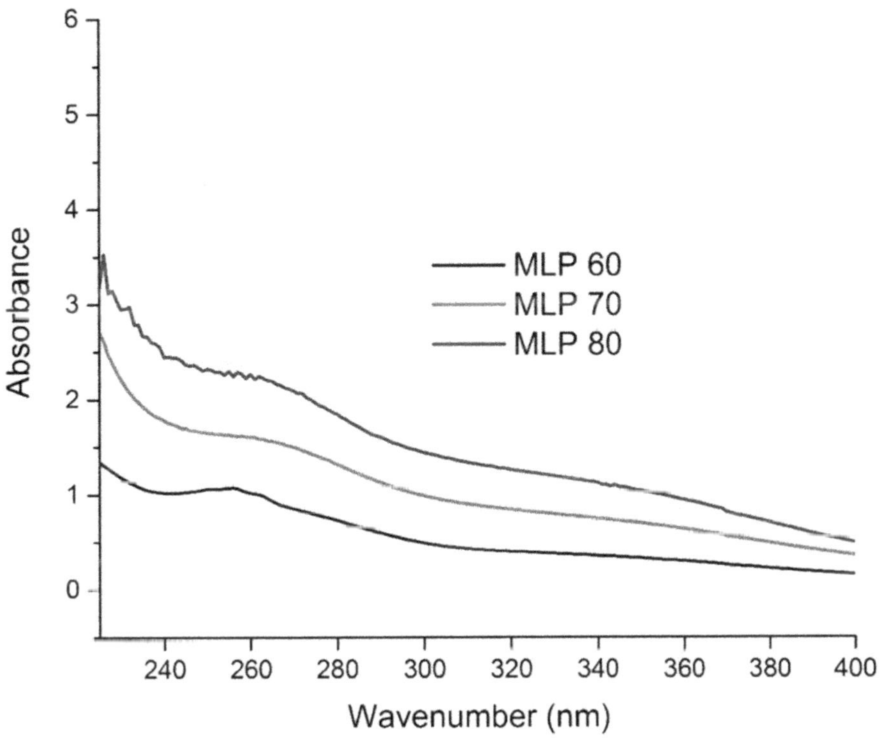

Structural Analysis

Tertiary Structure

Congo red is known to react with glucans with polysaccharide triple helix structure. The complex is well noted for a shift of the absorption maximum of the Congo red solutions to longer wavelengths (Villares, Mateo-Vivaracho, & Guillamon, 2012) – Bathochromic shift or decrease to shorter wavelength (Semedo, Karmali, & Fonseca, 2015) – hypsochromic shift. All samples MLP 60, MLP 70, MLP 80 displayed neither hypsochromic nor bathochromic shift (Fig. 5), thus the absence of triple-helical structure.

Figure 5. Helix-coil transition analysis of MLP 60, MLP 70, and MLP 80

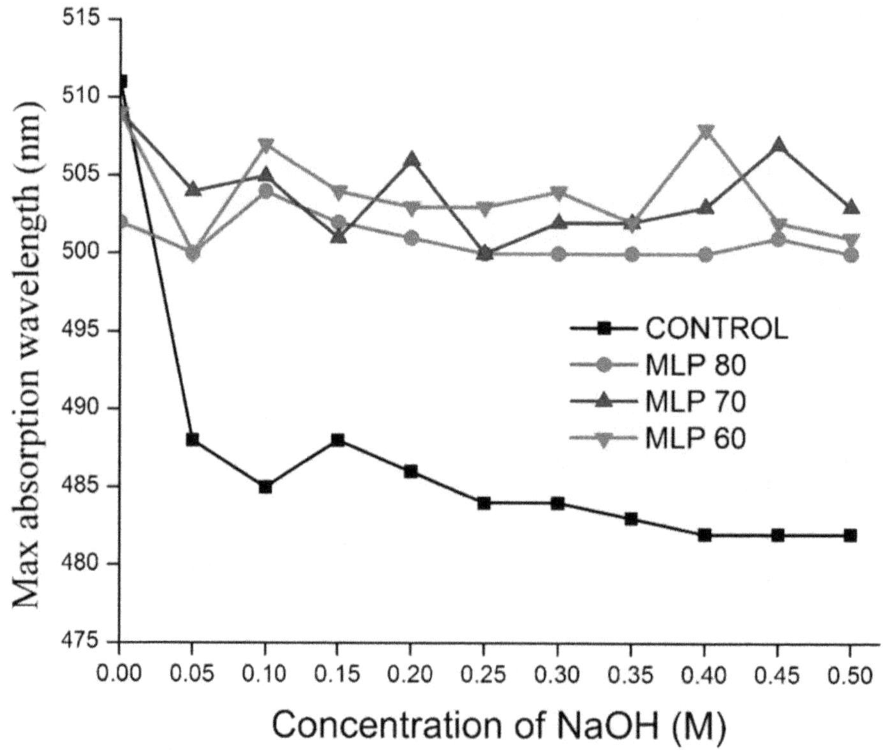

SEM Analysis

The fraction, MLP 60 displayed a relatively compact thick flat slice morphology (Figure 6). The second extracted fraction, MLP 70 showed a flat slitting thin surface with numerous round shaped structures beneath whiles MLP 80 showed compact round shaped morphological structures. The observed morphological structures was very much related to described shapes reported for extracted polysaccharides from *M. oleifera* Lam. leaves using microwave-assistance (Chen et al., 2017). Polysaccharide with higher mass percentage ratio of protein content had round group-like shapes whiles sample with less protein mass percentage ratio presented flat thin-slice shapes.

Figure 6. The SEM of MLP 60, MLP 70, and MLP 80

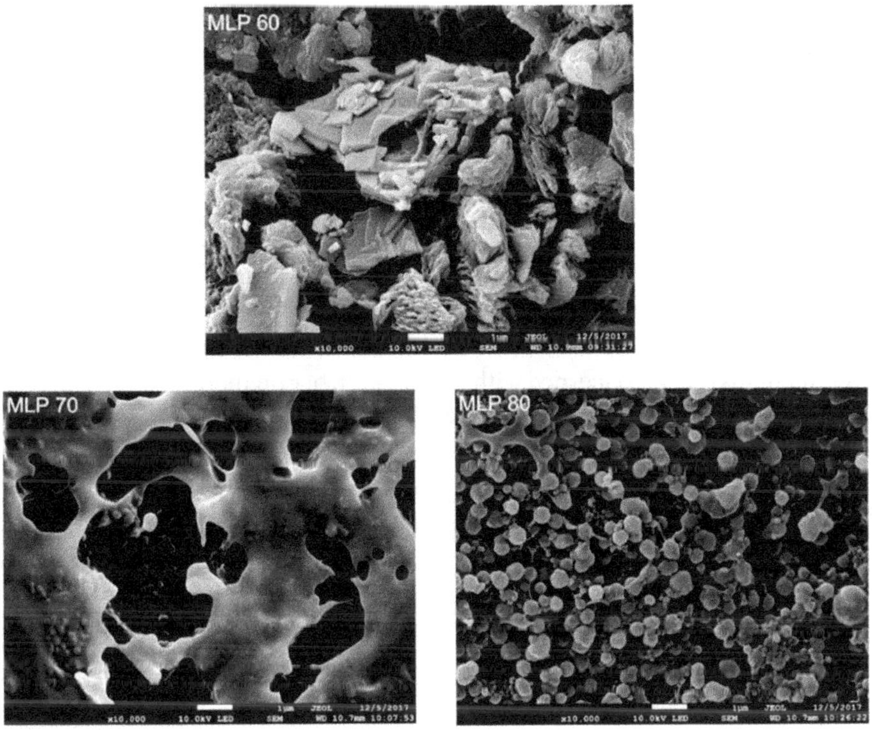

Molecular Weight

Flat shaped polymers have been reported to display high molecular weight with broader distribution of polymers whiles round shaped polymers display low molecular weight with narrow distribution. Furthermore, bimodal and multimodal distributions (high polydispersity) have also been observed during splitting of polymers from flat to round (Kenawy et al., 2002). This report is in agreement with observed morphological shapes, polydispersity and molecular weight in this work (Figure 6, Table 6). Isolated fractions 60% (MLP 60), 70% (MLP 70) and 80% (MLP 80) (v/v) of ethanol solution displayed corresponding molecular weights (55,250, 60,330 and 34,260 g/mol) and MLP 70 recorded a polydispersity of (6.586).

Table 6. Molecular weight of polysaccharides from leaves of Moringa oleifera

Samples	Molecular weight (g/mol)				
	Mn	*Mp*	*Mw*	*Mz*	*Mw/Mn*
MLP 60	41,640	2,6170	55,250	73,740	1.327
MLP 70	9,160	779	60,330	11,7700	6.586
MLP 80	33,780	34,510	34,260	34,960	1.014

Rheological Properties

Storage modulus (G′) is concerned with deformation of polymers caused by an exterior force (Oechsle et al. 2014). As can be observed from (Figure 7a), all polysaccharides extracted at different ethanol concentrations recorded increase in storage modulus with increase in frequency (1 to 10 Hz), suggesting strong elasticity. The loss modulus (G″) on the other hand, is concerned with energy dissipation, a typical characteristic of viscous properties (Friess & Schlapp, 2001). As shown in (Figure 7b), MLP 60 and MLP 70 (with highest Mw), displayed more elasticity showing lower G″ values. MLP 80 (with lowest Mw) however, displayed the most viscous property showing the highest G″ values (Fig. 7b). Figure (7c), shows change in tan δ as a function of frequency. The tan δ values of MLP 60, MLP 70 and MLP 80 decreased rapidly under frequency from 0.1 Hz to 0.2 Hz. At a frequency less than 1 Hz, a tan δ $<$ 1 was recorded. Samples with tan δ $>$ 1 usually exhibit more damping effect because the loss modulus is greater than the storage modulus (Korhonen et al., 2001), which means it dissipates more energy. The results thus indicate that all three fractions had the elastic component of the complex modulus to be more dominant. The viscosities of MLP 60, MLP 70 and MLP 80 dispersions also presented decrease speedily with

Figure 7. The storage modulus G' (a), loss modulus G" (b) and loss tangent delta (tan δ) (c) in dynamic frequency sweep test of MLP 60, MLP 70, and MLP 80

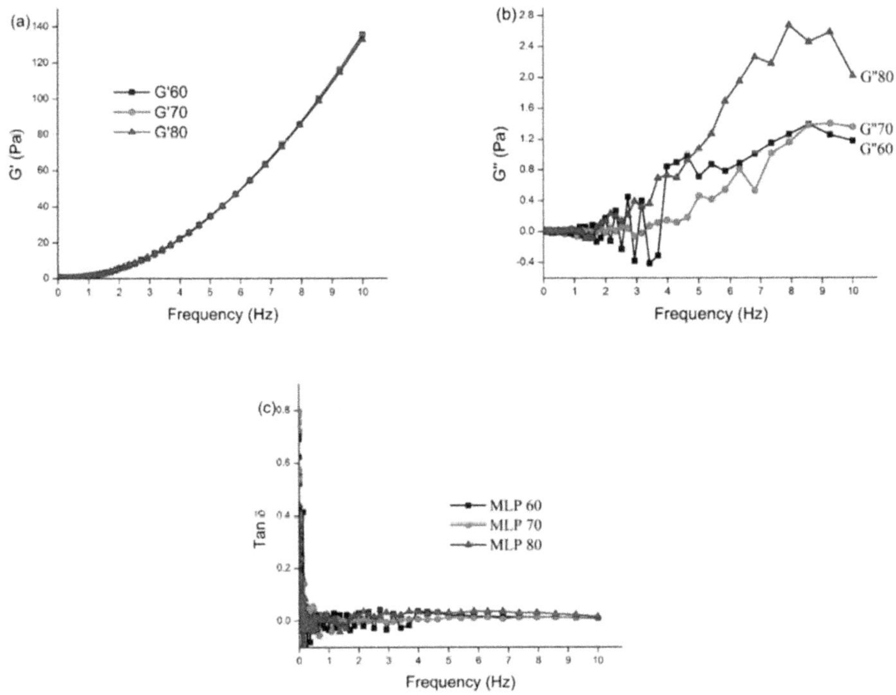

Figure 8. Steady shear flow curves of MLP 60, MLP 70, and MLP 80

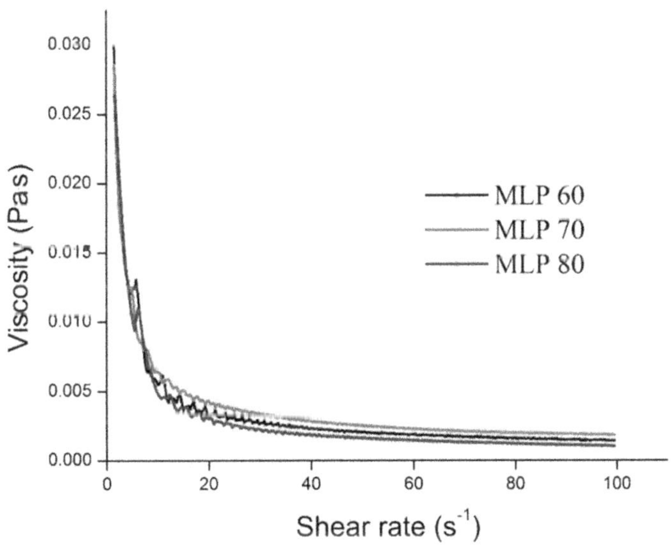

increase in shear rate (Figure 8), a characteristic of pseudoplastic fluids displaying shear thinning flow behavior.

Antioxidant Activity

Scavenging Effect on Hydroxyl Radical

Figure 9, displayed the hydroxyl radical scavenging activity of Ascorbic acid, MLP 60, MLP 70 and MLP 80 at different concentrations (5 – 30 mg/ml). At a concentration of 10 mg/ml, the scavenging rate of MLP 80 and ascorbic acid was similar (79%). At a concentration of 25 mg/ml MLP 70, MLP 80 and ascorbic acid also demonstrated similar (79%) scavenging rate. However, IC_{50} of MLP 60, MLP 70, MLP 80 and ascorbic acid on hydroxyl radical were 30, 21.5, 5.11 and 5 mg/ml respectively. This clearly shows that purified polysaccharide MLP 80 possess a much higher hydroxyl scavenging potential among the three fractions.

Figure 9. Scavenging effects of MLP 60, MLP 70, and MLP 80 polysaccharide on hydroxyl radicals

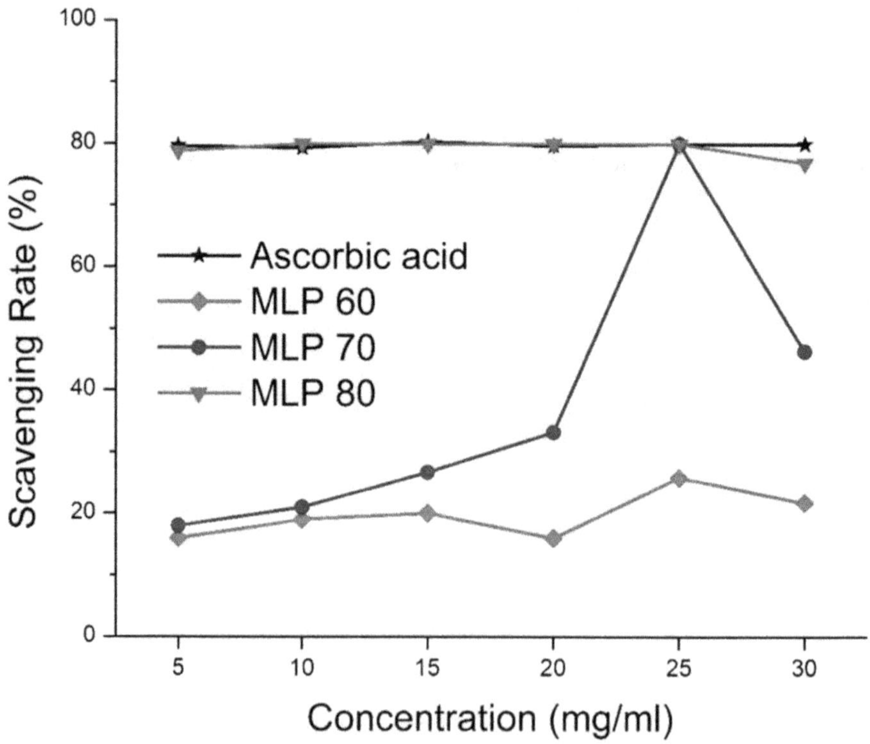

Scavenging Effect on ABTS Radical

Similarly, Figure (10), displayed the ABTS radical scavenging activity of Ascorbic acid, MLP 60, MLP 70 and MLP 80 at different concentrations (5 – 30 mg/ml).The sample, MLP 80, was again noted to have similar scavenging rate (99%) as ascorbic acid at concentration of 10 mg/ml. Calculated IC_{50} of MLP 60, MLP 70, MLP 80 and ascorbic acid on ABTS radical were 27, 21.75, 5.11 and 5 mg/ml respectively. This indicated that MLP 80 had very close ABTS radicals scavenging power as ascorbic acid.

Figure 10. Scavenging effects of MLP 60, MLP 70, and MLP 80 polysaccharide on ABTS radicals

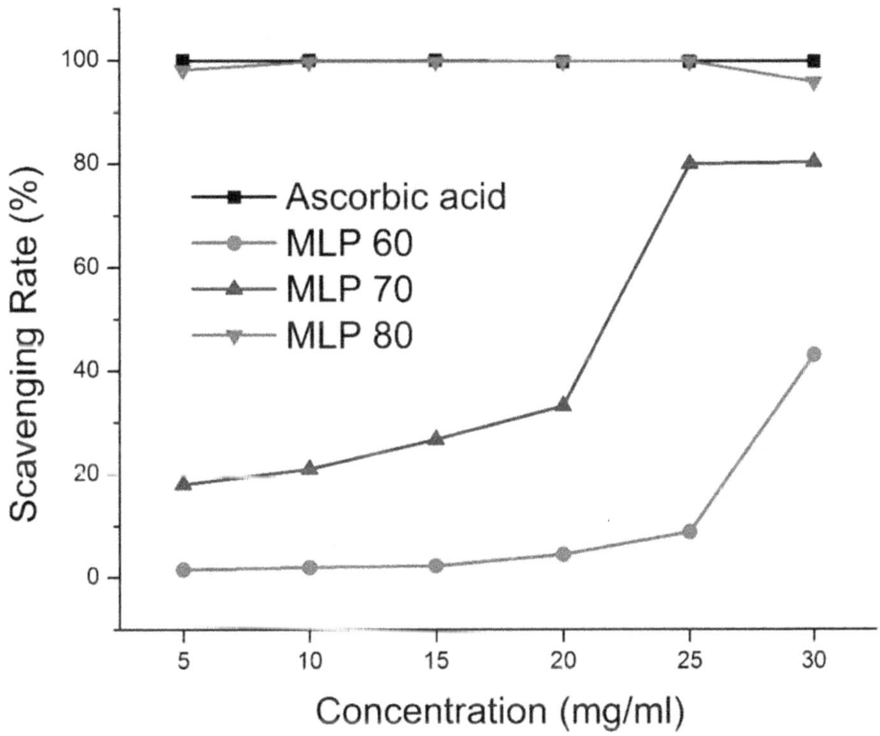

FUTURE RESEARCH DIRECTIONS

In the near future, the polysaccharide extract obtained in this study will be incorporated in pharmaceutical products such medicinal topical creams and soaps. Formulation of nutraceutical beverages will be another research direction to consider. It will serve as an excellent vehicle to distribute its bioactive compounds. Details of these compounds will also be established.

CONCLUSION

Crude polysaccharide extracted under ultrasound application using a higher operational temperature resulted in an appreciable amount (9.24%). The molecular bonds within all three isolated polysaccharides were similar according to spectral analysis. Total carbohydrate, protein and polyphenol of precipitated polysaccharides increased with increase in ethanol concentration used in precipitation. Tertiary structure and rheological properties was also found to be similar among the three isolated fractions except MLP 80 was more viscous. Conversely, in place of MLP 60 extracted with the lowest ethanol concentration, MLP 70 was observed morphologically to be under splitting process (from flat to round shape) and therefore had the highest molecular mass with the highest polydispersity (bimodal or multimodal distribution) and thus more elastic. Moreover, polysaccharide extracted at 80% (v/v) ethanol was most effective in scavenging hydroxyl and ABTS radicals.

REFERENCES

Adisakwattana, S., & Chanathong, B. (2011). Alpha-glucosidase inhibitory activity and lipid-lowering mechanisms of Moringa oleifera leaf extract. *European Review for Medical and Pharmacological Sciences*, *15*(7), 803–808. PMID:21780550

Bradford, M. M. (1976). A rapid and sensitive method for the quantitation of microgram quantities of protein utilizing the principle of protein-dye binding. *Analytical Biochemistry*, *72*(1-2), 248–254. doi:10.1016/0003-2697(76)90527-3 PMID:942051

Cai, W., Xie, L., Chen, Y., & Zhang, H. (2013). Purification, characterization and anticoagulant activity of the polysaccharides from green tea. *Carbohydrate Polymers*, *92*(2), 1086–1090. doi:10.1016/j.carbpol.2012.10.057 PMID:23399132

Chawla, S., Saxena, A., & Seshadri, S. (1988). In-vitro availability of iron in various green leafy vegetables. *Journal of the Science of Food and Agriculture*, *46*(1), 125–127. doi:10.1002/jsfa.2740460112

Chen, C., Zhang, B., Huang, Q., Fu, X., & Liu, R. H. (2017). Microwave-assisted extraction of polysaccharides from Moringa oleifera Lam. leaves: Characterization and hypoglycemic activity. *Industrial Crops and Products*, *100*, 1–11. doi:10.1016/j.indcrop.2017.01.042

Chen, F., Li, D., Shen, H., Wang, C., Li, E., Xing, H., & Nguyen, H. (2017). Polysaccharides from Trichosanthes Fructus via Ultrasound-Assisted Enzymatic Extraction Using Response Surface Methodology. *BioMed Research International*. PMID:29147656

Chen, R., Li, S., Liu, C., Yang, S., & Li, X. (2012). Ultrasound complex enzymes assisted extraction and biochemical activities of polysaccharides from Epimedium leaves. *Process Biochemistry (Barking, London, England)*, *47*(12), 2040–2050. doi:10.1016/j.procbio.2012.07.022

Coviello, T., Matricardi, P., Marianecci, C., & Alhaique, F. (2007). Polysaccharide hydrogels for modified release formulations. *Journal of Controlled Release*, *119*(1), 5–24. doi:10.1016/j.jconrel.2007.01.004 PMID:17382422

Dopico, A. M., & Tigyi, G. J. (2007). *A glance at the structural and functional diversity of membrane lipids Methods in Membrane Lipids*. Springer.

Dubois, M., Gilles, K. A., Hamilton, J. K., Rebers, P., & Smith, F. (1956). Colorimetric method for determination of sugars and related substances. *Analytical Chemistry*, *28*(3), 350–356. doi:10.1021/ac60111a017

Friess, W., & Schlapp, M. (2001). Effects of processing conditions on the rheological behavior of collagen dispersions. *European Journal of Pharmaceutics and Biopharmaceutics*, *51*(3), 259–265. doi:10.1016/S0939-6411(01)00136-9 PMID:11343891

Fuglie, L. J. (2001). *The miracle tree: the multiple attributes of Moringa*.

Geddes, C. D., & Lakowicz, J. R. (2007). *Glucose sensing* (Vol. 11). Springer Science & Business Media.

Hammi, K. M., Jdey, A., Abdelly, C., Majdoub, H., & Ksouri, R. (2015). Optimization of ultrasound-assisted extraction of antioxidant compounds from Tunisian Zizyphus lotus fruits using response surface methodology. *Food Chemistry*, *184*, 80–89. doi:10.1016/j.foodchem.2015.03.047 PMID:25872429

Jen, J.-F., Leu, M.-F., & Yang, T. C. (1998). Determination of hydroxyl radicals in an advanced oxidation process with salicylic acid trapping and liquid chromatography. *Journal of Chromatography. A*, *796*(2), 283–288. doi:10.1016/S0021-9673(97)01019-4 PMID:9474786

Kenawy, E.-R., Bowlin, G. L., Mansfield, K., Layman, J., Simpson, D. G., Sanders, E. H., & Wnek, G. E. (2002). Release of tetracycline hydrochloride from electrospun poly (ethylene-co-vinylacetate), poly (lactic acid), and a blend. *Journal of Controlled Release*, *81*(1-2), 57–64. doi:10.1016/S0168-3659(02)00041-X PMID:11992678

Korhonen, M., Hellen, L., Hirvonen, J., & Yliruusi, J. (2001). Rheological properties of creams with four different surfactant combinations-effect of storage time and conditions. *International Journal of Pharmaceutics*, *221*(1-2), 187–196. doi:10.1016/S0378-5173(01)00675-5 PMID:11397580

Kumar, K. S., Ganesan, K., & Rao, P. S. (2008). Antioxidant potential of solvent extracts of Kappaphycus alvarezii (Doty) Doty–An edible seaweed. *Food Chemistry*, *107*(1), 289–295. doi:10.1016/j.foodchem.2007.08.016

Moyo, B., Masika, P. J., Hugo, A., & Muchenje, V. (2011). Nutritional characterization of Moringa (Moringa oleifera Lam.) leaves. *African Journal of Biotechnology*, *10*(60), 12925–12933. doi:10.5897/AJB10.1599

Mummery, C. L. (1978). The effect of ultrasound on fibroblasts' In Vitro'. King's College London (University of London).

Oakes, J. (1976). Thermally denatured proteins. Nuclear magnetic resonance, binding isotherm and chemical modification studies of thermally denatured bovine serum albumin. *Journal of the Chemical Society, Faraday Transactions 1. Physical Chemistry in Condensed Phases*, *72*, 228–237.

Oechsle, A. M., Wittmann, X., Gibis, M., Kohlus, R., & Weiss, J. (2014). Collagen entanglement influenced by the addition of acids. *European Polymer Journal*, *58*, 144–156. doi:10.1016/j.eurpolymj.2014.06.015

Owusu, D., Ellis, W. O., & Oduro, I. (2008). *Nutritional potential of two leafy vegetables: Moringa oleifera and Ipomoea batatas leaves.*

Patist, A., & Bates, D. (2008). Ultrasonic innovations in the food industry: From the laboratory to commercial production. *Innovative Food Science & Emerging Technologies*, *9*(2), 147–154. doi:10.1016/j.ifset.2007.07.004

Qiu, T., Ma, X., Ye, M., Yuan, R., & Wu, Y. (2013). Purification, structure, lipid lowering and liver protecting effects of polysaccharide from Lachnum YM281. *Carbohydrate Polymers*, *98*(1), 922–930. doi:10.1016/j.carbpol.2013.07.014 PMID:23987429

Ratoarinoro, N., Wilhelm, A., Berlan, J., & Delmas, H. (1992). Effects of ultrasound emitter type and power on a heterogeneous reaction. *The Chemical Engineering Journal*, *50*(1), 27–31. doi:10.1016/0300-9467(92)80003-S

Semedo, M. C., Karmali, A., & Fonseca, L. (2015). A high throughput colorimetric assay of β-1, 3-d-glucans by Congo red dye. *Journal of Microbiological Methods*, *109*, 140–148. doi:10.1016/j.mimet.2014.12.020 PMID:25555819

Sinha, V., & Kumria, R. (2001). Polysaccharides in colon-specific drug delivery. *International Journal of Pharmaceutics*, *224*(1-2), 19–38. doi:10.1016/S0378-5173(01)00720-7 PMID:11472812

Sun, Y., Wang, S., Li, T., Li, X., Jiao, L., & Zhang, L. (2008). Purification, structure and immunobiological activity of a new water-soluble polysaccharide from the mycelium of Polyporus albicans (Imaz.) Teng. *Bioresource Technology*, *99*(4), 900–904. doi:10.1016/j.biortech.2007.01.029 PMID:17368889

Talmadge, J., Chavez, J., Jacobs, L., Munger, C., Chinnah, T., Chow, J. T., Williamson, D., & Yates, K. (2004). Fractionation of Aloe vera L. inner gel, purification and molecular profiling of activity. *International Immunopharmacology*, *4*(14), 1757–1773. doi:10.1016/j.intimp.2004.07.013 PMID:15531292

Tan, S., Xu, Q., Luo, Z., Liu, Z., Yang, H., & Yang, L. (2011). Inquiry of water-soluble polysaccharide extraction conditions from grapefruit skin. *Engineering (London)*, *3*(11), 1090.

Villares, A., Mateo-Vivaracho, L., & Guillamón, E. (2012). Structural Features and Healthy Properties of Polysaccharides Occurring in Mushrooms. *Revista de Agricultura (Piracicaba)*, *2*(4), 452–471.

Vuong, Q. V., Thanh, D. T., Bhuyan, D. J., Goldsmith, C. D., Sadeqzadeh, E., Scarlett, C. J., & Bowyer, M. C. (2015). Optimization of ultrasound-assisted extraction conditions for cuphol from the medicinal plant, Euphorbia tirucalli, using response surface methodology. *Industrial Crops and Products*, *63*, 197–202. doi:10.1016/j.indcrop.2014.09.057

Wang, J., Lian, P., Yu, Q., Wei, J., & Kang, W.-y. (2017). Purification, characterization and procoagulant activity of polysaccharides from Angelica dahurice roots. *Chemistry Central Journal*, *11*(1), 17. doi:10.118613065-017-0243-y PMID:28246546

Ying, Z., Han, X., & Li, J. (2011). Ultrasound-assisted extraction of polysaccharides from mulberry leaves. *Food Chemistry, 127*(3), 1273–1279. doi:10.1016/j.foodchem.2011.01.083 PMID:25214126

Zhou, C., Li, Y., Yu, X., Yang, H., Ma, H., Yagoub, A. E. A., Cheng, Y., Hu, J., & Otu, P. N. Y. (2016). Extraction and characterization of chicken feet soluble collagen. *Lebensmittel-Wissenschaft + Technologie, 74*, 145–153. doi:10.1016/j.lwt.2016.07.024

Zhou, H.-C., Lin, Y.-M., Li, Y.-Y., Li, M., Wei, S.-D., Chai, W.-M., & Tam, N. F.-y. (2011). Antioxidant properties of polymeric proanthocyanidins from fruit stones and pericarps of Litchi chinensis Sonn. *Food Research International, 44*(2), 613–620. doi:10.1016/j.foodres.2010.12.016

Zou, P., Yang, X., Huang, W.-W., Zhao, H.-T., Wang, J., Xu, R.-B., Hu, X.-L., Shen, S.-Y., & Qin, D. (2013). Characterization and bioactivity of polysaccharides obtained from pine cones of Pinus koraiensis by graded ethanol precipitation. *Molecules (Basel, Switzerland), 18*(8), 9933–9948. doi:10.3390/molecules18089933 PMID:23966080

Chapter 9
Commonly Used Public Facility Devices as Potential Sources of Infections

Maxwell Hubert Antwi
Koforidua Technical University, Ghana

Dorcas Serwaa
University of Melbourne, Australia

Francisca Serwaa Adomako
Garden City University College, Ghana

Aaron Siaw Kwakye
Koforidua Technical University, Ghana

Samuel Adomako Ocrah
Garden City University College, Ghana

Isaac Darban
https://orcid.org/0009-0005-2007-9822
Koforidua Technical University, Ghana

ABSTRACT

Surfaces and devices in public institutions are likely to be contaminated with various microorganisms as people congregate there for various reasons. Swab samples from devices like ATMs, seats, teller counters, door handles, pens, writing desks, toilet flush handles, and tap heads were obtained from banks and churches. 60 samples in all were cultured and isolates were identified using Gram stain reaction, morphological, and biochemical characteristics. Results indicated a 100% microbial contamination on all surfaces with Staphylococcus aureus having the highest frequency, 35(31.5%). Bacilli spp had the next highest frequency, 23(20.7%). Klebsiella spp 13 (11.7%), Salmonella spp 13(11.7%), Enterobacter spp 13(11.7%), Serratia spp 6(5.4%), Citrobacter spp 4(3.6%), Proteus 3(2.8%) and Streptococcus 1(0.90%) followed in that order. Commonly used public facilities' devices could serve as potential sources of infections due to their microbial contamination. It is highly recommended amidst this pandemic to have frequent proper hand hygiene to avoid unknowing contamination.

DOI: 10.4018/978-1-6684-9843-9.ch009

INTRODUCTION

Public facilities play a crucial role in communities as they serve as important gathering places, providing a wide range of services and amenities. In recent times, there has been a growing awareness of the escalating risks posed by infectious diseases on a global scale (Baker et al., 2022; Bloom & Cadarette, 2019). The emergence and rapid transmission of various pathogens have underscored the significance of comprehending and effectively addressing potential sources of infection. Despite extensive research on healthcare-associated infections, the role of commonly utilized public utility devices as potential sources of infection remains a crucial area that demands additional exploration (Bloom & Cadarette, 2019; Bonadonna et al., 2017). Public utility devices, including ATMs, pens, writing desks, chairs, doorknobs, touch screens, and others, serve as frequent points of contact for individuals from various backgrounds, potentially facilitating the transmission of pathogens (Bloom & Cadarette, 2019; Bonadonna et al., 2017).

Bacteria are ubiquitous microorganisms that establish colonies in a diverse array of environments, including commonly touched utility devices found in public spaces (Bonadonna et al., 2017; Cave et al., 2021; Russotto et al., 2015). Utility devices frequently harbor *Staphylococcus aureus*, which stands out as one of the most commonly detected bacteria, with contamination rates varying from 10% to 60% (Cave et al., 2021; Jabłońska-Trypuć et al., 2022; Suleyman et al., 2018). *Staphylococcus aureus* can lead to various infections, encompassing skin and soft tissue infections (SSTI), respiratory infections, bloodstream infections and the fatal Toxic Shock Syndrome (TSS) in pregnant women (Kwiecinski & Horswill, 2020; Linz et al., 2023). *Escherichia coli*, a common inhabitant of the intestinal tract has been detected on utility devices, with contamination rates ranging from 5% to 30% (Dawodu & Akanbi, 2021; Dsani et al., 2020). Whiles most of the strains are harmless, certain pathogenic strains can cause gastrointestinal illnesses. *Pseudomonas aeruginosa*, an opportunistic pathogen which is frequently found in moist environment has also been detected on utility device with contamination rates ranging from 10% to 50%. *Enterococcus faecalis*, a common member of the human gastrointestinal microbiota has been found on various utility devices with contamination rates ranging from 10% to 40% (Dawodu & Akanbi, 2021; Dsani et al., 2020).

Contaminated devices can serve as sources of bacterial infections leading to community-wide outbreaks (Cudjoe et al., 2022). The handling of food with contaminated hands has been identified as a contributing factor to the widespread occurrence of foodborne diseases in Ghana, resulting in a significant loss of lives (Cudjoe et al., 2022).

The virulence of microorganisms isolates from fomites which are surfaces or objects that can potentially harbour and transmit pathogens can vary depending on the specific microorganism and its characteristics (Abid et al., 2023; Wißmann et al., 2021). Several factors contribute to the virulence of these microorganism including their ability to survive on fomites, their pathogenicity and their resistance to environmental conditions (Abid et al., 2023; Wißmann et al., 2021). *Staphylococcus aureus* and *Escherichia coli* can persist on surfaces for hours or even days, remaining viable and infectious (Harris et al., 2021; Katzenberger et al., 2021; Wißmann et al., 2021).

Factors that increase the chances of public devices in spreading infections include but not limited to the following; high touch surfaces of public devices, lack of hand hygiene, lack of regular cleaning and ineffective disinfection of these devices, frequently shared public device accessories, lack of clear instructions and reminders regarding hygiene practices and preventive measures and limited awareness and compliance (Ssekitoleko et al., 2020).

Technological advancements in public places have revolutionized the way we interact, access services and navigate through various environments (Ssekitoleko et al., 2020). They offer significant benefits but also introduce challenges related to infection transmission. Maintaining a delicate equilibrium between harnessing technological advancements and prioritizing public health and safety is paramount to the creation of inclusive and infection-resistant environments (Ssekitoleko et al., 2020). Understanding the prevalence, distribution and transmission dynamics of bacteria on these devices is essential for implementing effective preventive measures.

A total of sixty (60) samples were collected. This was made up of swabs from 5 Automated Teller Machines (ATM), 5 pens, 5 bank door handles, 5 writing desks, 10 seats, 5 teller counters, 10 church seats, 5 toilet flush handles, 5 tap heads and 5-bathroom door knobs. The obtained results showed that all swabbed surfaces (100%) had microbial contamination as shown in Table 1 below.

Staphylococcus aureus was found to be a common contaminant at all swabbed areas representing a corresponding high proportion (31.5%) of the bacteria isolated. In contrast, *Streptococcus* spp. was isolated only on writing desk. Seats was found to be the most bacteria contaminated public utility (3.8±3.55) followed by flush handle (1.5±1.33). It was observed that pens were the least contaminated with *Staphylococcus aureus* and *Bacillus* spp. as shown in Table 2 below.

Table 3 shows the morphological characteristics, Gram stain reaction and biochemical test results of the nine different bacteria isolates obtained from the various surfaces. Some of the isolates are Gram Positive whilst others are Gram Negative. The biochemical characteristics of the isolates following Urease, Citrate, Coagulase, Catalase, Indole and Oxidase tests are shown in (Table 3) below.

Table 1. Shows Microbial contaminations on sampled objects

INSTITUTIONS	OBJECT	TOTAL NUMBER OF COLLECTED SWABS	MICROBIAL CONTAMINATION (%)
BANKS	ATM	5	100
	Pens	5	100
	Door handles	5	100
	Writing desks	5	100
	Seats	10	100
	Teller counters	5	100
CHURCHES	Seats	10	100
	Toilet Flush handles	5	100
	Toilet Tap heads	5	100
	Bathroom door knobs	5	100
	Total	60	100

Table 2. Shows the Distribution of bacterial isolates on utility devices/surfaces

Organism	Frequency (%)	Shared devices/utilities								
		ATM	Pen	Door handle	Writing desk	Seat	Teller counter	Flush handle	Tap head	Bathroom door knob
Staphylococcus aureus	35 (31.5)	2	5	4	3	8	6	3	4	0
Klebsiella pneumoniae	13 (11.7)	2	0	1	1	4	0	3	1	1
Streptococcus spp.	1 (0.90)	0	0	0	1	0	0	0	0	0
Serratia spp.	6 (5.4)	0	0	0	1	1	2	0	1	1
Salmonella spp.	13 (11.7)	3	0	3	1	3	1	1	1	0
Baccillus spp.	23 (20.7)	0	4	2	3	4	4	3	1	2
Enterobacter spp.	13(11.7)	2	0	1	1	3	0	2	2	2
Citrobacter spp.	4(3.6)	0	0	0	0	1	0	2	1	0
Proteus spp.	3(2.8)	0	0	0	0	1	0	1	1	0
Total	111(100)	9	9	11	11	25	13	15	12	6
Mean+SD		1 ± 1.22	1 ± 2.00	1.2 ± 1.48	1.2 ± 1.09	3.8 ± 3.55	1.5 ± 2.12	1.5 ± 1.33	1.3 ± 1.11	0.77 ± 0.83

Table 3. Shows morphological characteristics of isolated organisms

								T.S.I				
MORPHOLOGICAL CHARACTERISTICS	**GRAM REACTION**	**CATALASE**	**COAGULASE**	**OXIDASE**	**UREASE**	**CITRATE**	**INDOLE**	**SLOPE**	**BUTT**	**H_2S**	**GAS**	**ISOLATED ORGANISM**
Small, round, convex, golden/ yellow colonies with smooth appearance and entire edges, beta-hemolytic on Blood agar.	+	+	+	-								*Staphylococcus aureus*
Small, round, pink convex, translucent, smooth and mucoid colonies with entire edges on MacConkey agar	-	+	-	-	+	+	+	Y	Y	-	+	*Klebsiella pneumoniae*
Circular, raised, mucoid and shiny colonies with depressed center, entire edges and alpha haemolytic on blood agar	+	-	-	-								*Streptococcus spp*
Small, round, smooth, convex colonies with crenated edges on MacConkey agar. Non-lactose fermenters but produces a red pigment.	-			-	+	+	-	Y	Y	-	D	*Serratia spp*
Small, round, colourless, transparent, mucoid colonies with entire edges on MacConkey agar. Non lactose fermenters.	-			-	-	+	-	R	Y	+	-	*Salmonella spp*
Small, rod-shape, no pigment, endospore-forming colonies, beta-hemolytic on Blood agar.	+	+	-	-	+	+	-	R	Y	-	-	*Baccillus* spp.
Rod-shape, non-spore-forming, colonies either red or colourless on MacConkay agar, Lactose fermenter	-	+	-	-	-	+	-	Y	Y	-	+	*Enterobacter* spp.
Small, circular, convex dark pink colonies on MacConkay agar	-	+	-	-	-	+	-	Y	Y	+	+	*Citrobacter* spp.
Strong fishy odor, rod-shape swarming growth when grown on blood agar plate, on MacConkay agar, smooth, pale or colourless colonies	-	-	-	-	+	D	-	R	Y	+	D	*Proteus* spp.

Problem Statement

It is a common mistake that people judge cleanliness of objects and surfaces by their physical appearance. Due to this biased view, it is a common occurrence that people overlook hand washing and disinfection of hands after handling these so called 'clean' objects and surfaces. Objects and surfaces handled by many people can harbour harmful pathogens from infected people who previously handled them, and is one major way through which diseases, mostly food borne, spread through communities (Al-Ghamdi et al., 2011; Jones, 2017; Okoro et al., 2012; Todd et al., 2010).

Research Questions

This study seeks to answer the following questions;

Are microbes present on the surfaces of shared devices?
If yes, which types of microorganisms are present?
What is the pathogenic potential of these organisms? (Are they normal environmental flora or disease-causing organisms?)
What can be done to avoid spread of these potential pathogenic organisms?

Justification

This study assessed the pathogenic risks associated with these shared devices through isolation and identification of microorganisms found on the sample sites and also suggested measures that can be put in place by both the institutions and individuals using the facilities to reduce the spread of certain diseases.

LITERATURE REVIEW AND RELATED WORK

Although most microorganisms are beneficial in producing oxygen, decomposing organic materials, providing nutrients for plants and maintaining human health by aiding digestion among other things, there are still some that are pathogenic and cause diseases in plants and animals (Asrat et al., 2014; Mitchell & Isberg, 2017). The study focused on bacteria causing disease and limited its work to such pathogens in human transmissions.

Pathogenic and Non-Pathogenic Bacteria

The human body is host to many microbes, many of which are beneficial and do not cause any disease, however, a few microbes do cause infections and diseases (Isberg, 2014). Pathogenic bacteria are usually distinct from normal flora in that, normal flora can only invade and cause diseases when a host is immunocompromised or is able to gain access to a sterile part of the body. Pathogens on the other hand have developed highly specialized mechanisms for crossing cellular and biochemical barriers, irrespective of the status of the individual's immune system and eliciting specific responses from the host organism that contributes to the survival and multiplication of the pathogen (Alberts et al., 2002; Isberg, 2014).

Biofilm Formation and Ineffective Decontamination

Biofilms are multicellular communities of microorganisms held together by a self-produced extracellular matrix. The ability of microbes to form biofilm is a universal, ubiquitous, and dynamic process. This dynamic process of biofilms establishes an important strategy to withstand and survive harsh environmental conditions and antimicrobial agents (Chen et al., 2013; Su et al., 2022). Its presence may protect microorganism from disinfectant action and contribute to failure of decontamination (Chen et al., 2013; Su et al., 2022). Bacteria residing within biofilms are many times more resistance to chemical inactivation than in suspension (da Costa Luciano et al., 2016; Pajkos et al., 2004). Due to the continuous rise in biofilm-related infections, biofilms seriously threaten human health (Chen et al., 2013; Su et al., 2022). The formation of biofilms makes conventional antibiotics ineffective and dampens immune clearance. Therefore, it is important to understand the mechanisms of biofilm formation and develop novel strategies to treat biofilms more effectively (Chen et al., 2013; Su et al., 2022).

Factors Influencing Bacterial Cross Contamination Between Surfaces

Bacteria are easily transferred from one surface to another and it had been shown that a number of factors come together to facilitate this phenomenon. Some bacterial species have high affinity for certain surfaces as reported by Mackintosh et al, 1984 that *Staphylococcus saprophyticus* may have an affinity for the human skin that is not shared by other bacterial species (Mackintosh & Hoffman, 1984; Rusin et al., 2002; Sattar et al., 2001); (Rutala & Weber, 2019b). *Escherichia coli* and *Streptococcus pyogenes* did not transfer well over all. Although *E. coli* was picked up well from contaminated fabrics, it does not survive well on skin. It has been established that, moisture levels of the surfaces play a role as a greater number bacteria are transferred when the surfaces in question are moist (Gill & Jones, 2002). Dry surfaces usually have lower numbers of colonising organisms with a corresponding lower number of transfers. Moist surfaces also support bacteria so they survive longer on moist surfaces (Rutala & Weber, 2019b).

Sattar et al., (2001) also discussed that, higher levels of transfer of *Staphylococcus aureus* were observed with moist fabrics than dry ones (Sattar et al., 2001). They also discussed friction as another factor that influences bacterial cross contamination as there was a corresponding increase in transfer of organisms when friction was applied during contact.

Fomites, inanimate objects capable of transmitting disease, are recognized reservoirs of fecal pathogens. Elevated levels of fecal contamination in the environment are attributed to inadequate sanitation, a leading cause of child morbidity and mortality from diarrheal diseases, malnutrition, and stunting (Pickering et al., 2012; Rutala & Weber, 2019b).

Time post inoculation is also another factor. Scott et al, 1990 reported that, survival and transfer of certain bacteria was affected by time (Scott & Bloomfield, 1990). They noticed in their study that, on soiled surfaces, Gram positive and some Gram-negative species survived for up to 4 hours and in some cases, up to 24 hours after coming into contact with the inanimate surface (Anderson et al., 2020; Criscuolo et al., 2021). These organisms may survive longer if the contact is with human skin. Other factors include, the type of surface in question and pressure (Chen et al., 2001; Gill & Jones, 2002; Isberg, 2014; Montville & Schaffner, 2003).

Decontamination Effect on Microorganism

Antiseptics and Disinfectants

Antiseptics and disinfectants inactivate or inhibit the proliferation of microorganisms; they are applied to the epidermis, mucous membranes, and inanimate surfaces (fomites), respectively (Castaño et al., 2021; Lompo et al., 2023; Rutala & Weber, 2013b). Alcohols, sodium hypochlorite (chlorine), hydrogen peroxide, phenolics, etc. are examples of disinfectants (Rutala & Weber, 2019b).

Research on disinfection methods and disinfectants utilized in public facilities has demonstrated their effectiveness in deactivating pathogens on various surfaces, thus effectively halting the spread of harmful microorganisms. (Anderson et al., 2020; Criscuolo et al., 2021). The most effective preventive measures against the spread of pathogens through hand contact are hand washing with detergents and water, along with the use of alcohol-based hand sanitizers. An alcohol-based sanitizer containing 75-80% ethanol concentration proves to be active and effective in combating pathogens (Singh et al., 2020).

A 0.1% v/v concentration of sodium hypochlorite solution has been proven highly effective in swiftly deactivating pathogens on surfaces (Singh et al., 2020). Studies demonstrate that this solution can efficiently neutralize a wide range of harmful microorganisms, including bacteria, viruses, and fungi, within just one minute of application (Fadaei, 2021). Moreover, the 0.1% v/v concentration strikes a balance between efficacy and safety, ensuring that it efficiently eradicates pathogens while remaining non-toxic to humans and most surfaces (Fadaei, 2021). Hydrogen peroxide, a powerful oxidizing agent, exhibits potent antimicrobial properties, making it an excellent choice for surface disinfection. Research has demonstrated that a 0.5% concentration of hydrogen peroxide is remarkably effective in rapidly clearing a wide range of pathogens on various surfaces within just one minute of application (Kampf et al., 2020). Mileto and colleagues found that the combination of hydrogen peroxide with acetic and citric acids creates a synergistic effect, leading to a more potent antimicrobial action. The stabilized acidic composition of the mixture contributes to improved pathogen inactivation, making it a valuable approach for surface disinfection and infection control measures (Mileto et al., 2021).

Ultra-Violet (UV) or Radiation

As research continues to improve our understanding of UV radiation and its applications, it is evident that when used correctly, it offers a potent and environmentally friendly approach to combating pathogens and maintaining hygienic conditions in various settings (Gidari et al., 2021; Heilingloh et al., 2020). According to Gidari et al. (2021), the effectiveness of UV radiation in disinfecting various materials varies, polymers and stainless steel hold effective pathogens longer on its surface than glass does, and also inferred that, chemical and physical cleaning approaches are efficient in neutralizing pathogens on surfaces (Gidari et al., 2021).

The major way to prevent contact transfer of bacteria is through frequent hand washing and disinfection. A study conducted by Mackintosh (1984) revealed that, hand washing with soap and water was marginally superior to rubbing the hands with ethanol or isopropanol. Soap and water, according to his study produced on average, a 10,000-fold reduction in transfer, while ethanol and isopropanol produced 3,200-fold and 1,400-fold reduction respectively (King et al., 2020; Mackintosh & Hoffman, 1984; Rusin et al., 2002; Sattar et al., 2001). It was also noted that, the effectiveness of the rub with isopropanol was increased by increasing the volume used. When increased from 0.3ml to 0.5ml, it then has the same efficacy in transfer interruption as soap and water. Analysis suggested that, much of the variations were due to differences in susceptibilities of the different organisms. *Staphylococcus saprophyticus* appeared to be generally more resistant to the disinfection procedure than gram negatives (King et al., 2020; Mackintosh & Hoffman, 1984; Rusin et al., 2002; Sattar et al., 2001)(Burton et al., 2011).

Epidemiological Significance of Pathogen on Surfaces

It is a well-known phenomenon that pathogens can spread throughout the population by a number of mechanisms. Air, water, food, and soil are all common media; it is also possible for people to pass pathogens directly or indirectly through physical contact and contaminated surfaces respectively (Mathai et al., 2010; Randall et al., 2021). A common mistake is the habit of linking microbes only to medical equipment and different surfaces in healthcare facilities (Reynolds et al., 2005; Van Seventer & Hochberg, 2017). Other normal looking surfaces can harbour pathogenic microorganisms and therefore, the need to make people aware of the dangers of not practising hygienic hand washing. Seeing as infections transmitted by the hands and through food is easily spread through the community, frequent hand washing practices are necessary to prevent the spread of such infections (Mathai et al., 2010; Randall et al., 2021).

A study conducted to investigate hand washing among female college students examined the bacterial contamination in one of the female rest rooms. The study reported that, the faucet handles of the sink contained the greatest number of bacteria, (167 colonies) followed by the toilet seats (108 colonies), whereas the least number of bacteria was found on the door handle (12 colonies). The study also revealed that 63% of the users of the washroom frequently washed their hands with either water alone or soap and water, while the remaining 37% did not wash their hands at all, and this depended on circumstances such as the availability of water and soap, whether the sink was in use by another person or how much of a hurry the individual is in (Drankiewicz & Dundes, 2003). Sreshtaa & Geetha, (2021), in their study of the role of handwashing in preventing infection also reported and associated ineffective hand sanitation to increased microbial contamination and infection.

Another study, investigating the biogeographically pattern of bacteria across different surfaces in public restrooms such as door handles, faucet handles, toilet seats, soap dispensers etc. revealed that human skin associated bacteria dominated the microbial colonies of all samples collected (Flores et al., 2011). Other studies have investigated the presence of microbes on surfaces other than restrooms and hospital surfaces. It has been shown that, large numbers of bacteria can colonise objects such as ATMs, shopping carts, restaurant menus, elevator buttons, hand rails, mobile phones, computer keys and mice (Elsergany et al., 2015; Giovinazzo et al., 2017). In view of this, a study was conducted on bacterial contamination of different surfaces in shopping malls, the results indicated that, out of 224 samples collected, 192, representing 86% were positive for contamination (Elsergany et al., 2015).

A number of other studies have also considered the type of microorganisms encountered on these surfaces. In one, swabs collected from shopping cart handles and elevator buttons were positive for bacteria in 93% and 96% of samples respectively. Coagulase negative *Staphylococcus* dominated 84-87% of the samples gram positive *Bacilli* was found on 55-60% of samples; *Staphylococcus aureus*, (11-14%), *Psuedomonas spp*. (10%) and Gram-negative *Bacilli*, (8%) (Al-Ghamdi et al., 2011). A further study included 10 restaurants and found that, 70% of table surfaces were contaminated with coliforms, while 20% were with *E. coli* (Yepiz-

Gomez et al., 2006). Some researchers have reported some pathogenic bacteria on food handling surfaces from different areas which corroborated previous studies (Akonor & Akonor, 2013; Gutiérrez et al., 2012; Paul et al., 2021; Todd et al., 2009).

In 2004, another study concentrated on surface contamination of different toys in a paediatric hospital, it included 70 children's toys made of different materials. Cultures obtained from them showed a 100% microbial contamination with organisms such as, coagulase negative *Staphylococcus,* (78%), *Bacillus* spp. (37%), *Staphylococcus aureus* (18%), alpha hemolytic *Streptococcus* (11%), and *Pseudomonas* spp (9%) (Avila-Aguero et al., 2004). Microbial contamination on fomites, its possibility as a vehicle for spread of bacterial infection and their implications on human health has also been reported by researchers in different studies (Boone & Gerba, 2007; Cunningham et al., 2022; Olise & Simon-Oke, 2018; Simmonds et al., 2020; Stephens et al., 2019).

Current Issues and Advance Approaches

Studies involved in ''no touch'' decontamination technologies comprising Ultra-violet and Hydrogen Peroxide (HP) have been reported (Rutala & Weber, 2019a). UV irradiation has been used for the control of pathogenic microorganisms in a variety of applications, such as control of Legionellosis, as well as disinfection of air, surfaces, and instruments. At certain wavelengths, UV light will break the molecular bonds in DNA, thereby destroying the organism (Rutala & Weber, 2019a, 2019b).

Several systems that produce HP (eg, HP vapor, aerosolized dry mist HP) have been studied for their ability to decontaminate environmental surfaces, objects in hospital rooms, and unused medical supplies. These investigators found that HP systems are a highly effective method for eradicating various pathogens (eg, MRSA, Mycobacterium tuberculosis, Serratia, Clostridium botulinum spores) from inoculated carriers, rooms, furniture and equipment (Rutala & Weber, 2013a, 2018, 2019b). Considering that routine cleaning of room surfaces by environmentally safe solution is frequently inadequate, continuous room decontamination methods are being evaluated.

The intent of this technology is to make surfaces hygienically clean (not sterile), that is free of pathogens in sufficient numbers to prevent human disease (Rutala & Weber, 2019b). These include: visible light disinfection (high intensity narrow-spectrum light); low concentration HP; persistent disinfectants; and self-disinfection surfaces (eg, copper) (Rutala & Weber, 2018). Creating policies and procedures for environmental cleaning and disinfection are an integral part of preventing transmission of pathogens (Rutala & Weber, 2019b).

MATERIALS AND METHODS / RESEARCH METHODOLOGY

Study Design

The study was an observational one, by making concrete scientific findings on the various samples taken from the surfaces. Documentation and analysis followed the pattern of results obtained after the laboratory findings.

Study Area

This study was conducted in Adum and Bantama in the Kumasi metropolis in Ashanti region of Ghana from February to April 2019.

Study Population

A total of 60 different swab samples were collected from various surfaces of 10 different institutions in Bantama and Adum in Kumasi metropolis. 35 were collected from 5 randomly selected banks and the remaining 25 from 5 randomly selected churches.

Inclusion and Exclusion Criteria

The study biasly selected bank facilities with the presumption of low patronage of these facilities for possibly lower contagion. There are a lot of facilities especially public ones in Kumasi metropolis and so it excluded mostly patronized public bank facilities whose names were not mentioned even though they are located in the study area

Sampling Collection and Processing

Sample were taken from door handles, tap heads, pens, counters, writing desk surfaces etc. using normal saline moistened cotton swabs. The swabs were moistened to ensure proper adherence of the microorganisms to the cotton swab sticks. Each swab was quickly transferred into a labelled tube of Stuart's transport media. Each label indicated date of collection, the facility sample was collected from and the kind of surface that was sampled. The samples were then transported to the Garden City University College microbiology laboratory for analysis.

Bacterial Isolation

Each swab sample was directly plated on blood agar first, then MacConkey agar, using the streak plate method. The inoculated plates were incubated at 37^0C for 24- 48 hours before identification was done on them.

Bacterial Identification

Morphological characteristics of the isolates were noted. The first step in the identification process was gram staining. Smears were prepared from the isolates by picking single colonies and emulsifying in a drop of normal saline. The smears were allowed to air dry and then heat fixed by passing it a few times through the Bunsen burner flame. Gram staining was done on the smears and observed under the light microscope using ×100 oil immersion.

Gram positive organisms, as well as gram negatives were noted to enable further tests to be done on them. Catalase test was done on all Gram-positive isolates to help differentiate *Staphylococcus* from *Streptococcus*. Care was taken when picking colonies from blood agar to ensure that the red cells in the agar did not interfere with the test to produce false positives as red cells also have catalase activities.

Coagulase test was done on all catalase positive organisms to again differentiate *Staphylococcus aureus* from other *Staphylococcus spp*. The slide method was used for this study.

Gram negative isolates on the other hand were each inoculated in citrate, Tripple Sugar Indole (TSI) and urea using the stab/stab and streak method. Each was observed for their respective colour changes and organisms were identified and reported using their biochemical properties as per the chart.

Statistical Analysis

Categorical variables were presented and reported as simple frequencies with their corresponding percentages. Means with their standard deviation and median (ranges) were used to summarize continuous variables. The morphological characteristics, gram stain reaction and biochemical characteristics of the nine different bacteria isolates were used to identify the bacteria from the various surfaces.

Ethical Approval and Consent to Participate

Approval for this study was obtained from the Faculty of Health Sciences, the Research Committee of Garden City University College and from the management of the banks and churches that were involved in this study.

DISCUSSION AND CONCLUSION OF RESULT FINDINGS

Microorganisms are living things which are found everywhere including the environment and the human body. In these environments they live either freely or as parasites (Baker et al., 2022; Bloom & Cadarette, 2019). Human body harbor a number of microbes including several species of bacteria, viruses, fungi and protozoa. The sites where bacteria are found include; skin (*Staphylococci* and *Bacteroides*), Oropharynx (*Streptococci,* anaerobes), large intestine (Enteric *Bacilli*) and vagina (*Lactobacilli*) (Beaugerie & Petit, 2004; Simmonds et al., 2020). Normal microbial flora usually does not cause infection in body but can cause infection due to individual susceptibility and environmental factors (Ducel et al., 2002; Randall et al., 2021). Human hands aid in the transmission of microorganisms from place to place and from person to person. Human hands usually constitute microorganisms both as part of the body normal flora as well as transient microbes contracted from the environment (Mathai et al., 2010; Burton et al., 2011; Randall et al., 2021). Hence, this study aimed at assessing the pathogenic risks associated with shared devices isolated on the sample sites.

The study randomly took in mind of bank facilities whose patronage is low with the expectation of very minimal possible contaminations on the shared surfaces of these facilities. To our surprise, these very low patronage facilities had a hundred percentage contamination on their shared devices surfaces. This is an indication that the most patronized public banks are likely to harbor more possible infectious agents and this could make generalization of findings true. In this study, shared devices in both banks and churches were found to be hundred percentage contaminated with bacterial isolates. The hundred percent microbial contamination from the study could be due to these possible reasons; inadequate decontamination processes and procedures put in place by the facility management, infrequent hand hygiene on the part of visitors that use the facility, the less stringent and flexible measures ensuring that visitors comply with decontamination protocols as they use the facility and unhygienic environmental conditions surrounding the facility that could possibly contribute to high contamination.

The findings from this study are in agreement with research carried out by Bright et al., (2010) and Badger-Emeka et al., (2015) who have also reported high percentage of bacteria contaminant on similarly shared devices (Bright et al., 2010; Lorina et al., 2015; Cunningham et al., 2022). The microorganisms isolated in the study included *Staphylococcus aureus*, *Klebsiella pneumoniae*, *Streptococcus* spp, *Serratia* spp, *Salmonella* spp, *Bacilli* spp, *Enterobacter* spp, *Citrobacter* spp, and *Proteus* spp which are all potential pathogens depending on several factors that could lead to their pathogenicity (Baker et al., 2022; Bloom & Cadarette, 2019).

The most prevalent bacterial contaminant found in this study was *Staphylococcus* species (31.5%). This may be due to the fact that it is a major component of the normal flora of the skin and nostrils and it can be easily discharged by several human activities (Chikere et al., 2008; Lorina et al., 2015; Cunningham et al., 2022). Besides, these shared devices are routinely touched with hands. This observation is again in consistent with the findings of other researchers (Dibua et al., 2014; Ducel et al., 2002; Nworie et al., 2012).

A high percentage of *Bacillus* species (20.7%) were also observed in this study and could be explained by the fact that *Bacillus* species are ubiquitous in nature and they are spore forming bacteria. These spores are able to resist environmental changes, withstand dry heat and certain chemical disinfectants for prolonged periods. The findings again are in agreement with the research carried out by Dibua et al.,

(2014) and Onwubiko and Chinyeaka (2015) who reported *Bacillus* species to be the predominant organism isolated from door handles (Dibua et al., 2014; Onwubiko & Chinyeaka, 2015).

The current study recorded more Gram-positive organisms than Gram-negative ones (Table 2). The prevalence of more Gram-positive organisms compared to Gram negative ones in this study was in accordance with studies done by (Chikere et al., 2008; Lorina et al., 2015; Cunningham et al., 2022). The study was again in consistent with studies done by Inweregbu et al. (2005) where Gram-positive bacteria were the predominant bacteria isolated from fomites (Inweregbu et al., 2005). The reason for this occurrence is probably because Gram positive organisms are the members of the body flora of both asymptomatic carriers and sick persons (Chikere et al., 2008; Lorina et al., 2015; Cunningham et al., 2022). These organisms can be transmitted by the hand, expelled from the respiratory tract or spread by animate or inanimate objects (Chikere et al., 2008; Lorina et al., 2015; Cunningham et al., 2022). Their main sources of colonization on the fomites might likely be nasal carriage (Graham III et al., 2006; Baker et al., 2022; Bloom & Cadarette, 2019) which is likely to be transmitted by hand-to-mouth or hand-to-nose contact while using these fomites, or by poor hand-washing habits (Microbiology, 2005; Baker et al., 2022; Bloom & Cadarette, 2019) .

Limitation

The study did not do susceptibility test for the various bacteria isolates to be able to determine the kind of drugs that would work effectively on them. Notwithstanding this test not being done, the findings however depict the presence of possible infectious agents of significant concern and how hand and personal hygiene should be of great importance when these public facilities are being visited. Again, the study was done before the COVID-19 pandemic and as such, hand decontamination and sanitizing awareness and consciousness were very low as the pandemic brought these awakenings to place stringent and strict personal hygiene measures on us when going to crowd places.

CONCLUSION

The results from this study show that microbial contamination on shared devices in these areas could serve as sources of infections if they are transmitted from one individual to another through hands. ATMs, teller counters, writing desks and seats harbor bacterial contaminants. This can be attributed to the exposure of these devices to the environment and frequent use. Contaminated and improperly washed hands contaminate door handles, flush handles, bathroom door knobs and tap heads. It is therefore highly recommended amidst this pandemic to have a frequent proper hand washing to avoid unknowing contamination before, during and after visiting these facilities. On the basis of the above findings, it is highly recommended that good hygienic practices, including effective cleaning of the washrooms, proper hand washing, cleaning and disinfecting contaminated surfaces, must be taught and strictly adhered to. This can be done through occasional organization of seminars to educate employees and the people patronising the various institutions on the dangers of improper hygiene. Though it is difficult to completely eliminate bacteria from surfaces exposed to air, strict adherence to hand washing and disinfection after using shared devices will reduce the spread of infectious agents. New simple but effective technological ideas like placing of Veronica buckets at vantage public places for hand washing, frequent hand sanitizers applications in reducing microbial contamination need to be still in place as there are possible contagious microbes that could be transmitted when hand or personal hygiene is compromised. Any advance techniques that hardly use contact thus ''no touch'' system done elsewhere in the western countries be employed and existing ones be improved upon in these public facilities as these can minimize high microbial contamination.

Competing Interest

The authors of this publication declare there are no competing interests.

Funding

This research received no specific grant from any funding agency in the public, commercial, or not-for-profit sectors. Funding for this research was covered by the author(s) of the article.

ACKNOWLEDGMENTS

We acknowledge the Department of Medical Laboratory Technology, Faculty of Health Sciences, Garden City University College, the management of the various banks and churches in Kumasi -Ghana for allowing us to use their facilities to conduct this study.

REFERENCES

Abid, S. A., Aziz, S. N., Saeed, N. A.-H. A. A., Mizil, S. N., Al-Kadmy, I. M. S., Hussein, N. H., Al-Saryi, N., Ibrahim, S. A., & Hussein, J. D. (2023). Investigation of virulence factors in microbial organisms that are associated with public health risk isolates from different regions. *Al-Mustansiriyah Journal of Science*, *33*(5), 1–7. doi:10.23851/mjs.v33i5.1303

Akonor, P. T., & Akonor, M. A. (2013). *Food Safety Knowledge: The case of domestic food handlers in Accra*.

Al-Ghamdi, A. K., Abdelmalek, S. M. A., Ashshi, A. M., Faidah, H., Shukri, H., & Jiman-Fatani, A. A. (2011). Bacterial contamination of computer keyboards and mice, elevator buttons and shopping carts. *African Journal of Microbiological Research*, *5*(23), 3998–4003. doi:10.5897/AJMR11.770

Alberts, B., Johnson, A., Lewis, J., Raff, M., Roberts, K., & Walter, P. (2002). The plant cell wall. In *Molecular Biology of the Cell* (4th ed.). Garland Science.

Anderson, E., Hughes, G., & Patterson, E. (2020). *Inactivation of SARS-CoV-2 on surfaces and in solution with Virusend (TX-10), a novel disinfectant*.

Asrat, S., de Jesús, D. A., Hempstead, A. D., Ramabhadran, V., & Isberg, R. R. (2014). Bacterial pathogen manipulation of host membrane trafficking. *Annual Review of Cell and Developmental Biology*, *30*(1), 79–109. doi:10.1146/annurev-cellbio-100913-013439 PMID:25103867

Avila-Aguero, M. L., German, G., Paris, M. M., & Herrera, J. F., (2004). Toys in a pediatric hospital: Are they a bacterial source? *American Journal of Infection Control*, *32*(5), 287–290. doi:10.1016/j.ajic.2003.10.018 PMID:15292894

Baker, R. E., Mahmud, A. S., Miller, I. F., Rajeev, M., Rasambainarivo, F., Rice, B. L., Takahashi, S., Tatem, A. J., Wagner, C. E., Wang, L. F., Wesolowski, A., & Metcalf, C. J. E. (2022). Infectious disease in an era of global change. *Nature Reviews. Microbiology*, *20*(4), 193–205. doi:10.103841579-021-00639-z PMID:34646006

Beaugerie, L., & Petit, J.-C. (2004). Antibiotic-associated diarrhoea. *Best Practice & Research. Clinical Gastroenterology*, *18*(2), 337–352. doi:10.1016/j.bpg.2003.10.002 PMID:15123074

Bloom, D. E., & Cadarette, D. (2019). Infectious disease threats in the twenty-first century: Strengthening the global response. *Frontiers in Immunology*, *10*(MAR), 1–12. doi:10.3389/fimmu.2019.00549 PMID:30984169

Bonadonna, L., Briancesco, R., & Coccia, A. M. (2017). Analysis of microorganisms in hospital environments and potential risks. *SpringerBriefs in Public Health*, *9783319491592*, 53–62. doi:10.1007/978-3-319-49160-8_5

Boone, S. A., & Gerba, C. P. (2007). Significance of fomites in the spread of respiratory and enteric viral disease. *Applied and Environmental Microbiology*, *73*(6), 1687–1696. doi:10.1128/AEM.02051-06 PMID:17220247

Bright, K. R., Boone, S. A., & Gerba, C. P. (2010). Occurrence of bacteria and viruses on elementary classroom surfaces and the potential role of classroom hygiene in the spread of infectious diseases. *The Journal of School Nursing: the Official Publication of the National Association of School Nurses*, *26*(1), 33–41. doi:10.1177/1059840509354383 PMID:19903773

Burton, M., Cobb, E., Donachie, P., Judah, G., Curtis, V., & Schmidt, W.-P. (2011). The effect of handwashing with water or soap on bacterial contamination of hands. *International Journal of Environmental Research and Public Health*, *8*(1), 97–104. doi:10.3390/ijerph8010097 PMID:21318017

Castaño, N., Cordts, S. C., Kurosu Jalil, M., Zhang, K. S., Koppaka, S., Bick, A. D., Paul, R., & Tang, S. K. Y. (2021). Fomite transmission, physicochemical origin of virus–surface interactions, and disinfection strategies for enveloped viruses with applications to SARS-CoV-2. *ACS Omega*, *6*(10), 6509–6527. doi:10.1021/acsomega.0c06335 PMID:33748563

Cave, R., Cole, J., & Mkrtchyan, H. V. (2021). Surveillance and prevalence of antimicrobial resistant bacteria from public settings within urban built environments: Challenges and opportunities for hygiene and infection control. *Environment International*, *157*, 106836. doi:10.1016/j.envint.2021.106836 PMID:34479136

Chen, M., Yu, Q., & Sun, H. (2013). Novel strategies for the prevention and treatment of biofilm related infections. *International Journal of Molecular Sciences*, *14*(9), 18488–18501. doi:10.3390/ijms140918488 PMID:24018891

Chen, Y., Jackson, K. M., Chea, F. P., & Schaffner, D. W. (2001). Quantification and variability analysis of bacterial cross-contamination rates in common food service tasks. *Journal of Food Protection, 64*(1), 72–80. doi:10.4315/0362-028X-64.1.72 PMID:11198444

Chikere, C. B., Omoni, V. T., & Chikere, B. O. (2008). Distribution of potential nosocomial pathogens in a hospital environment. *African Journal of Biotechnology, 7*(20).

Cudjoe, D., Balali, G. I., Titus, O. O., Osafo, R., & Taufiq, M. (2022). Food Safety in Sub-Sahara Africa, An insight into Ghana and Nigeria. *Environmental Health Insights, 16*. doi:10.1177/11786302221142484 PMID:36530486

Criscuolo, E., Signorelli, C., Mancini, N., & Clementi, M. (2021). *Fast inactivation of SARS-CoV-2 by UV-C and ozone exposure on di ff erent materials. 10*, 2–5.

Cunningham, B., O'Neill, V., Devereux, M., McGann, D., & O'Hora, J. (2022). Use of a door handle disinfection system to reduce the risks associated with microbial loads on fomites in a healthcare setting. *The Journal of Hospital Infection, 130*, 104–107. doi:10.1016/j.jhin.2022.09.003 PMID:36113634

da Costa Luciano, C., Olson, N., Tipple, A. F. V., & Alfa, M. (2016). Evaluation of the ability of different detergents and disinfectants to remove and kill organisms in traditional biofilm. *American Journal of Infection Control, 44*(11), e243–e249. doi:10.1016/j.ajic.2016.03.040 PMID:27234012

Dawodu, O. G., & Akanbi, R. B. (2021). Isolation and identification of microorganisms associated with automated teller machines on Federal Polytechnic Ede campus. *PLoS ONE, 16*(8), 1–15. doi:10.1371/journal.pone.0254658

Dibua, U. M. E., Onyemerela, I. S., & Nweze, E. I. (2014). Frequency, urinalysis and susceptibility profile of pathogens causing urinary tract infections in Enugu State, southeast Nigeria. *Revista do Instituto de Medicina Tropical de São Paulo, 56*(1), 55–59. doi:10.1590/S0036-46652014000100008 PMID:24553609

Drankiewicz, D., & Dundes, L. (2003). Handwashing among female college students. *American Journal of Infection Control, 31*(2), 67–71. doi:10.1067/mic.2003.6 PMID:12665738

Dsani, E., Afari, E. A., Danso-Appiah, A., Kenu, E., Kaburi, B. B., & Egyir, B. (2020). Antimicrobial resistance and molecular detection of extended spectrum β-lactamase producing Escherichia coli isolates from raw meat in Greater Accra region, Ghana. *BMC Microbiology, 20*(1), 1–8. doi:10.118612866-020-01935-z PMID:32795260

Ducel, G., Fabry, J., & Nicolle, L. (2002). Prevention of hospital acquired infections: a practical guide. Prevention of Hospital Acquired Infections: A Practical Guide., Ed. 2.

Elsergany, M., Ben Moussa, M., Ahsan, A., Khalfan, A., & Eissa, A. (2015). Exploratory study of bacterial contamination of different surfaces in four shopping malls in Sharjah, UAE. *Journal of Environmental and Occupational Science*, *4*(2), 101–105. doi:10.5455/jeos.20150611030039

Fadaei, A. (2021). *Viral Inactivation with Emphasis on SARS-CoV-2 Using Physical. 2021.*

Flores, G. E., Bates, S. T., Knights, D., Lauber, C. L., Stombaugh, J., Knight, R., & Fierer, N. (2011). Microbial biogeography of public restroom surfaces. *PLoS One*, *6*(11), e28132. doi:10.1371/journal.pone.0028132 PMID:22132229

Gidari, A., Sabbatini, S., Bastianelli, S., Pierucci, S., Busti, C., Stabile, A. M., Monari, C., Galli, F., Rende, M., Cruciani, G., & Francisci, D. (2021). *SARS-CoV-2 Survival on Surfaces and the Effect of UV-C Light*. 2–9.

Gill, C. O., & Jones, T. (2002). Effects of wearing knitted or rubber gloves on the transfer of Escherichia coli between hands and meat. *Journal of Food Protection*, *65*(6), 1045–1048. doi:10.4315/0362-028X-65.6.1045 PMID:12092720

Giovinazzo, R., Caradonna, L., Giaquinta, G., Guerrera, E., Mameli, M., Mansi, A., Marena, G., Mastromartino, T., Sarto, D., & Tomao, P. (2017). guidance values for microbiological monitoring on surfaces: A literature overview. *Biomedicine & Prevention*, *4*(135), 174–180.

Graham, P. L. III, Lin, S. X., & Larson, E. L. (2006). A US population-based survey of Staphylococcus aureus colonization. *Annals of Internal Medicine*, *144*(5), 318–325. doi:10.7326/0003-4819-144-5-200603070-00006 PMID:16520472

Gutiérrez, D., Delgado, S., Vázquez-Sánchez, D., Martínez, B., Cabo, M. L., Rodríguez, A., Herrera, J. J., & García, P. (2012). Incidence of Staphylococcus aureus and analysis of associated bacterial communities on food industry surfaces. *Applied and Environmental Microbiology*, *78*(24), 8547–8554. doi:10.1128/AEM.02045-12 PMID:23023749

Harris, D., Taylor, K. P., Napierkowski, K., & Zechmann, B. (2021). Indoor finish material influence on contamination, transmission, and eradication of methicillin-resistant staphylococcus (MRSA). *HERD*, *14*(1), 118–129. doi:10.1177/1937586720952892 PMID:32867539

Heilingloh, C. S., Aufderhorst, U. W., Schipper, L., Dittmer, U., Witzke, O., Yang, D., Zheng, X., Sutter, K., Trilling, M., Alt, M., Steinmann, E., & Krawczyk, A. (2020). *Susceptibility of SARS-CoV-2 to UV irradiation. January.*

Inweregbu, K., Dave, J., & Pittard, A. (2005). Nosocomial infections. *Continuing Education in Anaesthesia, Critical Care & Pain*, *5*(1), 14–17. doi:10.1093/bjaceaccp/mki006

Isberg, R. (2014). *Pathogenic Bacteria: What Distinguishes a Pathogen from a Non-Pathogen?* All Talks in Microbiology. https://www.ibiology.org/microbiology/pathogenic-bacteria-distinguishes-pathogen-non-pathogen/

Jabłońska-Trypuć, A., Makuła, M., Włodarczyk-Makuła, M., Wołejko, E., Wydro, U., Serra-Majem, L., & Wiater, J. (2022). Inanimate surfaces as a source of hospital infections caused by fungi, bacteria and viruses with particular emphasis on SARS-CoV-2. *International Journal of Environmental Research and Public Health*, *19*(13), 8121. doi:10.3390/ijerph19138121 PMID:35805776

Jones, J. B. (2017). *Shopping Carts and Restaurant Menus as Community Fomites: A Pilot Study in Henrico County*. AT Still University of Health Sciences.

Kampf, G., Todt, D., Pfaender, S., & Steinmann, E. (2020). *Persistence of coronaviruses on inanimate surfaces and their inactivation with biocidal agents.*

Katzenberger, R. H., Rösel, A., & Vonberg, R. P. (2021). Bacterial survival on inanimate surfaces: A field study. *BMC Research Notes*, *14*(1), 1–10. doi:10.1186/13104-021-05492-0 PMID:33722269

King, M., López-García, M., Atedoghu, K. P., Zhang, N., Wilson, A. M., Weterings, M., Hiwar, W., Dancer, S. J., Noakes, C. J., & Fletcher, L. A. (2020). Bacterial transfer to fingertips during sequential surface contacts with and without gloves. *Indoor Air*, *30*(5), 993–1004. doi:10.1111/ina.12682 PMID:32329918

Kwiecinski, J. M., & Horswill, A. R. (2020). Staphylococcus aureus bloodstream infections: Pathogenesis and regulatory mechanisms. *Current Opinion in Microbiology*, *53*, 51–60. doi:10.1016/j.mib.2020.02.005 PMID:32172183

Linz, M. S., Mattappallil, A., Finkel, D., & Parker, D. (2023). Clinical impact of staphylococcus aureus skin and soft tissue infections. *Antibiotics (Basel, Switzerland)*, *12*(3), 1–27. doi:10.3390/antibiotics12030557 PMID:36978425

Lompo, P., Heroes, A.-S., Agbobli, E., Kühne, V., Tinto, H., Affolabi, D., & Jacobs, J. (2023). Bacterial contamination of antiseptics, disinfectants and hand hygiene products in healthcare facilities in high-income countries: A scoping review. *Hygiene*, *3*(2), 136–175. doi:10.3390/hygiene3020012

Lorina, I. B. E., Abdulrahman, A. A. S., Hajer, S. A. D., Nouf, K. A. H., Fatimah, A. A. N., & Hani, M. A. F. (2015). Potential pathogenic bacterial contaminants of shared utility devices in a university setting at Al-Hofuf, Saudi Arabia. *African Journal of Microbiological Research*, *9*(41), 2139–2144. doi:10.5897/AJMR2015.7577

Mackintosh, C. A., & Hoffman, P. N. (1984). An extended model for transfer of micro-organisms via the hands: Differences between organisms and the effect of alcohol disinfection. *Epidemiology and Infection*, *92*(3), 345–355. PMID:6429238

Mathai, E., Allegranzi, B., Kilpatrick, C., & Pittet, D. (2010). Prevention and control of health care-associated infections through improved hand hygiene. *Indian Journal of Medical Microbiology*, *28*(2), 100–106. doi:10.4103/0255-0857.62483 PMID:20404452

Microbiology, A. S. for. (2005). *Women Better At Hand Hygiene Habits, Hands Down*. ScienceDaily. https://www.sciencedaily.com/releases/2005/09/050922020856.htm

Mileto, D., Mancon, A., Staurenghi, F., Rizzo, A., Econdi, S., Gismondo, M. R., & Guidotti, M. (2021). Inactivation of SARS-CoV-2 in the Liquid phase: Are aqueous hydrogen peroxide and sodium percarbonate efficient decontamination agents? *Journal of Chemical Health and Safety*, *28*(4), 260–267. doi:10.1021/acs.chas.0c00095 PMID:37556234

Mitchell, G., & Isberg, R. R. (2017). Innate immunity to intracellular pathogens: Balancing microbial elimination and inflammation. *Cell Host & Microbe*, *22*(2), 166–175. doi:10.1016/j.chom.2017.07.005 PMID:28799902

Montville, R., & Schaffner, D. W. (2003). Inoculum size influences bacterial cross contamination between surfaces. *Applied and Environmental Microbiology*, *69*(12), 7188–7193. doi:10.1128/AEM.69.12.7188-7193.2003 PMID:14660365

Nworie, A., Ayeni, J. A., Eze, U. A., & Azi, S. O. (2012). Bacterial contamination of door handles/knobs in selected public conveniences in Abuja metropolis, Nigeria: A public health threat. *Continental Journal of Medical Research*, *6*(1), 7–11.

Okoro, C. K., Kingsley, R. A., Connor, T. R., Harris, S. R., Parry, C. M., Al-Mashhadani, M. N., Kariuki, S., Msefula, C. L., Gordon, M. A., De Pinna, E., Wain, J., Heyderman, R. S., Obaro, S., Alonso, P. L., Mandomando, I., MacLennan, C. A., Tapia, M. D., Levine, M. M., Tennant, S. M., & Dougan, G. (2012). Intracontinental spread of human invasive Salmonella Typhimurium pathovariants in sub-Saharan Africa. *Nature Genetics*, *44*(11), 1215–1221. doi:10.1038/ng.2423 PMID:23023330

Olise, C. C., & Simon-Oke, I. (2018). Fomites: Possible vehicle of nosocomial infections. *Journal of Public Health and Nutrition*, *1*(1).

Onwubiko, N. E., & Chinyeaka, A. H. (2015). Isolation and identification of bacterial contaminants from door handles in a tertiary institution in Umuahia, Abia state, Nigeria. *Nigerian Journal of Microbiology*, *29*, 3139–3147.

Pajkos, A., Vickery, K., & Cossart, Y. (2004). Is biofilm accumulation on endoscope tubing a contributor to the failure of cleaning and decontamination? *The Journal of Hospital Infection*, *58*(3), 224–229. doi:10.1016/j.jhin.2004.06.023 PMID:15501338

Paul, A., Rahma, M. M., & Ahmed, T. (2021). Identification of pathogenic bacteria from food handling surfaces (tabletops) from different areas with demonstration of their drug resistance properties. *Journal of Food Safety and Hygiene*, *5*(3). doi:10.18502/jfsh.v5i3.5693

Pickering, A. J., Julian, T. R., Marks, S. J., Mattioli, M. C., Boehm, A. B., Schwab, K. J., & Davis, J. (2012). Fecal Contamination and Diarrheal Pathogens on Surfaces and in Soils among Tanzanian Households with and without Improved Sanitation. *Environmental Science & Technology*, *46*(11), 5736–5743. doi:10.1021/es300022c PMID:22545817

Randall, K., Ewing, E. T., Marr, L. C., Jimenez, J. L., & Bourouiba, L. (2021). How did we get here: What are droplets and aerosols and how far do they go? A historical perspective on the transmission of respiratory infectious diseases. *Interface Focus*, *11*(6), 20210049. doi:10.1098/rsfs.2021.0049 PMID:34956601

Reynolds, K. A., Watt, P. M., Boone, S. A., & Gerba, C. P. (2005). Occurrence of bacteria and biochemical markers on public surfaces. *International Journal of Environmental Health Research*, *15*(3), 225–234. doi:10.1080/09603120500115298 PMID:16134485

Rusin, P., Maxwell, S., & Gerba, C. (2002). Comparative surface-to-hand and fingertip-to-mouth transfer efficiency of gram-positive bacteria, gram-negative bacteria, and phage. *Journal of Applied Microbiology*, *93*(4), 585–592. doi:10.1046/j.1365-2672.2002.01734.x PMID:12234341

Russotto, V., Cortegiani, A., Raineri, S. M., & Giarratano, A. (2015). Bacterial contamination of inanimate surfaces and equipment in the intensive care unit. *Journal of Intensive Care*, *3*(1), 54. doi:10.118640560-015-0120-5 PMID:26693023

Rutala, W. A., & Weber, D. J. (2013a). Disinfectants used for environmental disinfection and new room decontamination technology. *American Journal of Infection Control*, *41*(5), S36–S41. doi:10.1016/j.ajic.2012.11.006 PMID:23622746

Rutala, W. A., & Weber, D. J. (2013b). Disinfection and sterilization in healthcare facilities. Bennett & Brachman's Hospital Infections: Sixth Edition, May. doi:10.1016/j.idc.2021.04.004

Rutala, W. A., & Weber, D. J. (2018). Disinfection Chapter. *Practical Healthcare Epidemiology*, 58.

Rutala, W. A., & Weber, D. J. (2019a). Best practices for disinfection of noncritical environmental surfaces and equipment in health care facilities: A bundle approach. *American Journal of Infection Control*, *47*, A96–A105. doi:10.1016/j.ajic.2019.01.014 PMID:31146858

Rutala, W. A., & Weber, D. J. (2019b). Disinfection, sterilization, and antisepsis: An overview. *American Journal of Infection Control*, *47*, A3–A9. doi:10.1016/j.ajic.2019.01.018 PMID:31146848

Sattar, S. A., Springthorpe, S., Mani, S., Gallant, M., Nair, R. C., Scott, E., & Kain, J. (2001). Transfer of bacteria from fabrics to hands and other fabrics: Development and application of a quantitative method using Staphylococcus aureus as a model. *Journal of Applied Microbiology*, *90*(6), 962–970. doi:10.1046/j.1365-2672.2001.01347.x PMID:11412326

Scott, E., & Bloomfield, S. F. (1990). The survival and transfer of microbial contamination via cloths, hands and utensils. *The Journal of Applied Bacteriology*, *68*(3), 271–278. doi:10.1111/j.1365-2672.1990.tb02574.x PMID:2111304

Simmonds, R., Lee, D., & Hayhurst, E. (2020). Mobile phones as fomites for potential pathogens in hospitals: Microbiome analysis reveals hidden contaminants. *The Journal of Hospital Infection*, *104*(2), 207–213. doi:10.1016/j.jhin.2019.09.010 PMID:31585142

Singh, D., Joshi, K., Samuel, A., Patra, J., & Mahindroo, N. (2020). *Alcohol-based hand sanitisers as first line of defence against SARS-CoV-2 : a review of biology, chemistry and formulations.*

Sreshtaa, V. S., & Geetha, R. V. (2021). Role of handwashing and hand sanitising in preventing infection. *Annals of the Romanian Society for Cell Biology*, 724–737.

Ssekitoleko, R. T., Oshabaheebwa, S., Munabi, I. G., Tusabe, M. S., Namayega, C., Ngabirano, B. A., Matovu, B., Mugaga, J., Reichert, W. M., & Joloba, M. L. (2020). The role of medical equipment in the spread of nosocomial infections: A cross-sectional study in four tertiary public health facilities in Uganda. *BMC Public Health*, *20*(1), 1–11. doi:10.118612889-020-09662-w PMID:33066745

Stephens, B., Azimi, P., Thoemmes, M. S., Heidarinejad, M., Allen, J. G., & Gilbert, J. A. (2019). Microbial exchange via fomites and implications for human health. *Current Pollution Reports*, *5*(4), 198–213. doi:10.100740726-019-00123-6 PMID:34171005

Su, Y., Yrastorza, J. T., Matis, M., Cusick, J., Zhao, S., Wang, G., & Xie, J. (2022). Biofilms: Formation, research models, potential targets, and methods for prevention and treatment. *Advancement of Science*, *9*(29), 2203291. doi:10.1002/advs.202203291 PMID:36031384

Suleyman, G., Alangaden, G., & Bardossy, A. C. (2018). The role of environmental contamination in the transmission of nosocomial pathogens and health-associated infections. *Current Infectious Disease Reports*, *20*(6), 12. Advance online publication. doi:10.100711908-018-0620-2 PMID:29704133

Todd, E. C. D., Greig, J. D., Bartleson, C. A., & Michaels, B. S. (2009). Outbreaks where food workers have been implicated in the spread of foodborne disease. Part 6. Transmission and survival of pathogens in the food processing and preparation environment. *Journal of Food Protection*, *72*(1), 202–219. doi:10.4315/0362-028X-72.1.202 PMID:19205488

Todd, E. C. D., Michaels, B. S., Greig, J. D., Smith, D., Holah, J., & Bartleson, C. A. (2010). Outbreaks where food workers have been implicated in the spread of foodborne disease. Part 7. Barriers to reduce contamination of food by workers. *Journal of Food Protection*, *73*(8), 1552–1565. doi:10.4315/0362-028X-73.8.1552 PMID:20819372

Van Seventer, J. M., & Hochberg, N. S. (2017). Principles of infectious diseases: transmission, diagnosis, prevention, and control. International Encyclopedia of Public Health, 22.

Wißmann, J. E., Kirchhoff, L., Brüggemann, Y., Todt, D., Steinmann, J., & Steinmann, E. (2021). Persistence of pathogens on inanimate surfaces: A narrative review. *Microorganisms*, *9*(2), 1–37. doi:10.3390/microorganisms9020343 PMID:33572303

Yepiz-Gomez, M. S., Bright, K. R., & Gerba, C. P. (2006). Identity and numbers of bacteria present on tabletops and in dishcloths used to wipe down tabletops in public restaurants and bars. *Food Protection Trends*, *26*(11), 786–792.

KEY TERMS AND DEFINITIONS

Biochemical Test: These are various biochemical tests done to isolate specific bacteria species of interest.

Contamination: This describes uncleaniness, dirt, and filth caused by unwanted contaminants or microbes.

Culture: This is the medium or man-made surviving environment for growing bacteria or pathogen of interest.

Fomites: Inanimate objects or surfaces that serves to harbour potential infectious microbes.

Grams Stain: This is a stain developed by one scientist called Christian Gram that is used to stain bacteria after they have been isolated from their culture.

Infection: This describes an invasion by a microbe into one's system to compromise the immune system with mostly clinical manifestation.

Isolates: They are the microorganisms that have been isolated or grown on the artificial created environment for research and clinical purposes.

Microbes: This describes the tiny microorganisms that cannot be seen with our naked but are of medical importance.

Chapter 10
Current Trends for Using Moringa Oleifera Seed (MOS) in Water and Wastewater Treatment:
Prospects and Challenges for Developing Countries – A Review

Mansuur Husein
https://orcid.org/0000-0002-1532-5284
Tamale Technical University, Ghana

Alhassan Abdul Latif Husein
Tamale Technical University, Ghana

Niib Konwuruk
Tamale Technical University, Ghana

Hamdu Ibrahim
Tamale Technical University, Ghana

Zakaria Issaka
Tamale Technical University, Ghana

Peter Kaba
Ehime University, Japan

ABSTRACT

The unprecedented incidence of rising industrialization, population concentration, and mismanagement of water resources has contributed to water scarcity and pollution in many countries. Water pollution poses a great health threat to humans and animals, as well as the ecosystem. Researchers around the globe have brought to light different methods of using natural Moringa oleifera seed in water and wastewater treatment through coagulation, flocculation, sedimentation, biosorption, photocatalysis, and adsorption, among others. These trends toward using natural products for water purification can help avoid the side effects of synthetic chemicals

DOI: 10.4018/978-1-6684-9843-9.ch010

such as alum employed in wastewater decontamination and minimize the cost of treatment. This chapter seeks to comb through the literature and identify the available trends in using Moringa seed to treat water and wastewater, as well as highlight their prospects and challenges for developing countries.

1. INTRODUCTION

The provision of clean and safe water is a crucial aspect of public health and development. However, developing countries often face challenges in providing safe and clean water due to a lack of resources and infrastructure. Traditional water treatment methods can be expensive and require skilled personnel to operate, which further limits their accessibility in developing countries (Salgot & Folch, 2018; Velmurugan et al., 2020). The World Health Organization's (WHO) 2017 report indicated that about one-third of the world's population does not have access to safe drinking water (Water, 2017).

Also, it is estimated that about 90% of sewage in developing countries is discharged untreated directly into water bodies. Daily statistics show that 2 million metric tons of wastewater drain into the aquatic ecosystem without prior treatment (Petersen et al., 2016).

On this note, it is projected that the decline in about one-third of the global ecology is estimated to be a result of the deterioration of freshwater quality, usually caused by water resource pollution (UNICEF, 2015). Hence, pollution of water resources poses a significant socio-economic burden on people living in such areas as it affects their development. Most people living in developing countries, especially in rural communities, face the worst consequences of water pollution and scarcity (Dharwal et al., 2020). The burden of this menace lies mostly on women and girls, as it is their core responsibility to provide water for the household in most traditional settlements. This has attracted the attention of several development practitioners, including stakeholders in the water sector.

Therefore, water scarcity and gender equality are closely linked issues, particularly in developing countries where access to clean water is limited and traditional gender roles often dictate who is responsible for water collection and management (Mann et al., 2014; Owa, 2013).

Women and girls are disproportionately affected by water scarcity, as they are typically responsible for collecting water for household use. This task can be time-consuming and physically demanding, often requiring them to travel long distances and carry heavy loads, which can have negative impacts on their health, education, and economic opportunities. Furthermore, when water is scarce, girls may be unable to attend school because they need to assist with water collection or because schools lack adequate water and sanitation facilities (Safo-adu & Hanson, 2019).

In addition to the burden of water collection, women are also often excluded from decision-making processes related to water management and infrastructure development. This can limit their ability to advocate for their water needs and interests and perpetuate gender inequalities in access to resources (Hasan et al., 2019).

Therefore, potable water is seen as a necessity for life, which can propel development, improve sanitation and hygiene practices, improve girl-child school enrollment, and empower women. Nonetheless, anthropologically induced factors have led to its deterioration in both quality and quantity, causing about 80% of diseases suffered by the human population, especially in the developing world (Karaouzas et al., 2018; Landrigan et al., 2019). In spite of these, many potential methods used in water treatment (Seow et al., 2012; Trösch, 2009) are usually expensive (Zhou et al., 2019), especially in rural areas where income levels are very low to afford high and complex technology (Abubakar, 2019).

Meanwhile, many synthetic disinfectants or coagulants used for water or wastewater treatment have been proven to be harmful to human health and the ecosystem as a whole (Amagloh & Benang, 2009). This has awakened the curiosity of many researchers to find out more suitable organic or natural substitutes that are environmentally friendly (Gupta et al., 2018a, 2018b; Pawar, 2016).

Moreover, chemical and advanced technologies for wastewater treatment are not readily available in most developing countries. As a result, these may be relatively expensive for people living in rural areas in developing countries and can pose health hazards if wrongly applied (Aho & Lagasi, 2012; Shan et al., 2017). As a result, there is a need for low-cost and sustainable water treatment solutions that can be implemented in developing countries. Moringa oleifera seed (MOS) has emerged as a potential solution for water treatment in developing countries due to its coagulation and antimicrobial properties. This book chapter aims to provide an overview of the current trends, prospects, and challenges associated with the use of MOS in water and wastewater treatment, especially in developing countries.

1.1 Properties of Moringa

1.1.1 Phytochemical Properties of MOS

The phytochemical properties of MOS make it an effective natural coagulant for water and wastewater remediating pollutants, as shown in Table 1. The use of MOS as a coagulant has been shown to be effective in removing suspended particles, turbidity, and organic matter from water, making it a valuable tool for water treatment systems in developing countries where access to traditional water treatment methods is limited (Patel et al., 2014). Moringa oleifera seed has been used as a natural coagulant for water treatment due to its high content of phytochemicals such as proteins, carbohydrates, phenolic compounds, and flavonoids. These phytochemicals have unique properties that enable them to effectively remove suspended particles and pollutants from water systems (Sinha, 2012).

Proteins, which make up approximately 30% of the seed weight, are the primary coagulant agent in Moringa oleifera seeds. The proteins in the seed contain a high concentration of positively charged amino acids such as arginine and lysine, which help to destabilize negatively charged particles in water. This destabilization process results in the formation of larger, more easily settled particles that can be removed through sedimentation or filtration (Idris, Jami, Hammed, et al., 2016).

In addition to proteins, Moringa oleifera seed also contains carbohydrates, which help enhance the coagulation process by increasing the size of the flocs formed during coagulation. The carbohydrates act as a bridge between the particles in the water, resulting in the formation of larger flocs that settle more quickly. Phenolic compounds and flavonoids in MOS also contribute to the coagulation process by adsorbing onto the surface of suspended particles, increasing their size and weight, and making them more easily removable by sedimentation or filtration, as illustrated in Table 1 (Villaseor-Basulto et al., 2018).

To lessen the negative health effects of excessive turbidity and more microbiologically contaminated water, moringa seed has been approved for usage and is being used in many rural areas for water filtration. In addition, it is regarded as a natural absorbent (Echeverria et al., 2019). and antibacterial agent for treating river water (Brilhante et al., 2017). It is possible to create poly-elecrolytes from crushed seeds that act as natural flocculating agents. These agents bind suspended particles in a colloidal suspension to form bigger flocs that improve microbial adhesion (Keogh et al., 2017).

Researchers discovered a steroidal glycoside called strophantidin as a bioactive agent that was more effective in the sedimentation of inorganic and organic matter in raw water by decreasing microbial load by 55% and coliform load by 65% after 24 hours, compared to alum, which reduced microbial load by 65% and coliform load by 83% under similar circumstances (Ali, 2016; Amagloh & Benang, 2009).

MOS has natural coagulants that can be utilized to filter out turbidity, bacteria, and suspended particles from water. Studies have shown that MOS is successful at treating water for both domestic and industrial use. MOS can reduce turbidity and microbial load by up to 99%, according to a review of its usage in water treatment by Villaseor-Basulto et al. (2018).

Furthermore, pterygospermin, an antibiotic component found in Moringa oleifera seeds, is responsible for killing waterborne microorganisms (Brilhante et al., 2017; Patel et al., 2014). Moringa oleifera, as demonstrated by Shan et al. (2017),

Table 1. Phytochemicals present in MOS

	Phytochemical composition	Percentage (% w/w)	Extraction solvents			References
			Methanol	Ethanol	Water	
1	Alkaloids	0.01-0.5	x	x	x	(Patel et al., 2014; Sinha, 2012)
2	Cardiac Glycosides	0.005-0.028	x			(Sinha, 2012)
3	Crude proteins	27.1- 35.0			x	(Sinha, 2012)
4	Flavonoids	0.03-1.47	x	x	x	(Bukar et al., 2010),(Patel et al., 2014)
5	Glycosides	0.02-0.048	x			(Patel et al., 2014)
6	Saponin	1.1-1.8		x	x	(Patel et al., 2014)
7	Steroidal rings	1.3-12.9	x			(Sinha, 2012)
8	Tannins	0.4-4.0	x		x	(Patel et al., 2014)
9	Terpenoids	0.02- 0.22	x			(Sinha, 2012)
10	Volatile oil	9.3-38.2			x	(Idris, Jami, Jamal, et al., 2016)
11	Others	± 6.50				

Source: Adapted from (Idris, Jami, Hammed, et al., 2016; Patel et al., 2014).
NB: x *Means available*

includes positively charged proteins that work well as coagulants for the treatment of wastewater and water. These characteristics also support the use of MOS for water filtration. In a similar vein, as table 2 demonstrates, it has been tested and supported by numerous studies for its ability to combat organic (George et al., 2016) and inorganic pollutants (George et al., 2016; Gupta et al., 2018b) in wastewater.

Therefore, using natural coagulants derived from plants is a good alternative to using chemical coagulants such as alum. It is well known that, as compared to standard chemical coagulants, plant extracts from species like Moringa oleifera are more successful at removing suspended solids, turbidity, softening hard water, and reducing sludge production (Eman et al., 2014; Seow et al., 2012).

1.1.2 Medicinal Characteristics of MOS

Moringa seeds are a highly nutritious food source that can provide many important micronutrients (figure 2) and health benefits through Pharmacological activities as depicted in figure 1.

2. CURRENT TRENDS OF MOS APPLICATION IN WATER AND WASTEWATER TREATMENT

2.1 Wastewater Treatment

Many developing countries in Asia and Africa have been using Moringa oleifera in water treatment for several decades. It has long been discovered for its potency as a coagulant and flocculant in water and wastewater treatment (Brilhante et al., 2017). This is mainly due to the properties of the seed, such as being non-toxic, biodegradable, and acting as a coagulant and flocculant agent. The flocculation and coagulation properties have been linked to the presence of flocculating proteins and dimeric cationic proteins, respectively, which were extracted from the MOS (Biosci et al., 2014; Eman et al., 2014).

According to the studies conducted by Aho & Lagasi (2012) and Keogh et al. (2017), which compared and analyzed various natural coagulants extracted from plants including Prosopis juliflora, latifaria, and Moringa oleifera (MO) for their effectiveness in water treatment, their results showed that MO had the best removal efficiency of contaminants such as turbidity from the aqueous solution as a natural coagulant.

Copper has been discovered to be one of the most hazardous heavy chemicals in waste water, which has serious health impacts on living organisms (Gupta et al., 2018b). The sustainable removal of copper from water is very essential for plants and animals. Ali et al. (2016) (Ali, 2016) investigated the removal of copper (Cu II) from water systems using Moringa oleifera press cake (MOPC) as a natural and environmentally friendly biosorbent. The results indicated enhanced reduction of copper ion concentrations of approximately 53.49% and a turbidity removal rate of 75.45%. These further prove the efficiency of MOS in the treatment of water systems with copper concentrations. This, when highly applied in developing countries, can help mitigate water-borne disease and address water scarcity, leading to improved health conditions.

In the presence of growing industrialization and urbanization, more water is used in different stages of manufacturing and other commercial activities. Effluent from firms and sewage from domestic sources are sometimes polluted with non-biodegradable heavy metals such as lead, cadmium, zinc, and manganese, which have toxic effects on the biosphere (Biosci et al., 2014; Järup & kesson, 2009). These heavy metals accumulate in aquatic organisms and affect the food chain (Wuana & Okieimen, 2011). For these reasons, Shan et al. (2017) studied the use of Moringa oleifera seed as a natural coagulant for waste water treatment and heavy metal removal. Throughout their experiments, they finally observed about 85–94% removal of turbidity. Dissolved oxygen (DO) improved from 2.580.01 to 4.000.00

mg/L. This low-cost and convenient technology will have a bright future for many developing countries in battling water pollution while maintaining ecosystem integrity.

Bakare et al. (2016) also investigated the suitability of Moringa oleifera seed extract as a natural coagulant in the treatment of highly turbid water. The result of his experiment proves that the use of Moringa oleifera seed extract is highly effective as a natural coagulant for water treatment.

2.2 Water Purification

Vigneswaran and Sundaravadivel (2009) studied the traditional and household water purification methods of rural communities in developing countries, which have persisted for a long time. Most of the traditional methods highlighted were different kinds of filtration. Other studies have also shown that Moringa oleifera is highly cultivated in many rural areas (Aho & Lagasi, 2012; Amagloh & Benang, 2009; Bakare, 2016). This implies that promoting the use of MOS in the processes of water purification as a coagulant or flocculant to reduce turbidity, fluoride, or other impurities is of great essence. This will help improve the quality of drinking water and control its scarcity, leading to improved health conditions and a higher standard of living.

In addition, Ali (2016) conducted empirical research on the use of MOS as a natural coagulant for water treatment. Their study focused on the active constituents of MOS responsible for coagulation properties in water treatment. Their results indicated that the isolated bioactive compounds, through the coagulation/flocculation process, enhanced turbidity removal of 95.5%, 98%, and 99.3% for the purification of river water containing low, medium, and high levels of turbidity, respectively (Table 2). Since MOS cultivation is common in most developing countries, this application method can help reduce domestic wastewater treatment costs incurred by low- or middle-income countries and boost productivity (Araujo et al., 2013).

2.3 Removal of Specific Chemicals

Among the chemical elements in the periodic table, fluorine has been found to be the most electronegative and the most reactive. Its presence in water affects the growth conditions of aquatic plants and animals, as it contains some level of toxicity (Blahuskova et al., 2019). Ravikumar & Sheeja (2014) studied the removal of fluoride from aqueous systems using Moringa oleifera seed cake and its composite coagulants. The outcome of the experiment indicated that the use of MOS cake with its composite coagulants (alum and starch) reduced turbidity below 1 mg/L, which is within the international acceptable standard limit for drinking water. This puts many developing countries like India, Ghana, and Nigeria on the opportunity side to widen the scope of MOS use to remove targeted toxic chemicals like fluorine, which is viable and cost-effective.

2.4 Method of MOS Processing

Figure 5 shows MOS processing techniques for water treatment. This process for using Moringa oleifera seeds as a natural coagulant to treat water for drinking purposes involves the following procedures:

a) Collecting mature seed pods (Figure 3)
b) Dehusking or removing seeds and discarding discolored ones (Figure 3)
c) Crushing an appropriate number of seed kernels to form a fine powder
d) Mixing the powder with a small amount of clean water to form a paste
e) Adding the paste to 250 ml of clean water in a bottle, or 1 seed kernel powder, can treat 1 liter of water.
f) Filtering the solution through a muslin cloth
g) Adding the solution to the water to be treated based on the dosage
h) Stirring rapidly for at least 1 minute, then slowly for 5–10 minutes
i) letting the treated water settle undisturbed for 1-2 hours, and pouring off the clean water carefully.

2.4.1 Dosage of MOS Powder

The dosage and ratio of Moringa seed extract to water can vary depending on the quality and characteristics of the water being treated. However, in general, the recommended dosage of Moringa seed extract for water treatment is around 50–500 mg/L, with an optimal dose of around 200 mg/L. Other studies have also shown that 1 seed kernel powder can treat 1 liter of water.

To prepare the moringa seed extract, the kernels are ground into a fine powder and mixed with water at a ratio of around 1:10 (weight to volume). For instance, to prepare 1 liter of Moringa seed extract, 100 grams of Moringa seed powder would be mixed with 900 ml of water.

The clean water can be filtered or sterilized to make it safe for drinking (Figure 4). The dosage and ratio of Moringa seed cake powder to water may need to be adjusted based on factors such as the quality of the water, the pH, and the temperature.

Lea (2010) conducted research on the bioremediation of turbid surface water using seed extract from MO Lam. The study aimed at applying MOS to a traditional water purification system, which gave rise to efficient biological treatment of turbid water containing pathogens.

Also, Kavitha et al. (2012) investigated the toxicity of Moringa oleifera seed extract on some hematological and biochemical profiles in a freshwater fish, Cyprium carpio. The study focused on the acute and sublethal toxicity of MOS extract on hematological and biochemical variables of a freshwater fish cyprium under laboratory conditions. After several days of experimentation, the result finally provided baseline information about the toxicity of MOS extracts from aquatic animals like C. carpio and a standard limit for water purification and treatment. This research indicates that the use of MOS in water treatment is highly advantageous but also has some drawbacks. Therefore, caution must be exercised in its application to avoid its limitations. This can be highly achieved through massive education of the masses on the prospects and challenges of using MOS in water treatment.

Again, Nishi et al. (2012) researched the efficiency of the hybrid process of coagulation and flocculation with MOS followed by ultrafiltration to remove Microcystis sp. cells from the water supply. The research specifically aimed at evaluating the efficacy of MOS as a coagulant in removing cells like Cyanobacterium microcystis sp cells, their apparent color, and turbidity from water in the presence of ultrafiltration.

Furthermore, Camacho et al. (2017) also evaluated the impacts of MOS with a reduced amount of oil as a coagulant in conventional water treatment to remove cyanobacteria from different natural surface waters. The successful study showed chlorophyll and turbidity removals up to 85% for highly turbid waters with 50 mg/L.

A comparative study on the use of ordinary and de-oiled MO in the treatment of abattoir wastewater was also conducted by Aho & Lagasi (2012). In the study process, they compared the efficiency of using oil and de-oiled MOS extract in the decontamination of abattoir waste water. After the experiment, it came to light that de-oiled MOS seed removed turbidity up to 80%, while oiled MOS extract removed turbidity up to 68.86%.

3. CHALLENGES FACED BY DEVELOPING COUNTRIES IN IMPLEMENTING MOS-BASED WATER AND WASTEWATER TREATMENT SYSTEMS

Although MOS presents a potential solution for water treatment in developing countries, there are several challenges that must be addressed to promote its adoption. These negative impacts could hinder the large-scale application of MOS in water purification (Kumssa et al., 2017; Ueda Yamaguchi et al., 2021). The current challenges of MOS application in water treatment systems are as follows:

Table 2. Trends of MOS application in water and wastewater treatment

No.	Types of water/ waste water	Methods/process	Targeted pollutants	Composite chemical	Part of Moringa Oleifera used	Result/Rate of pollutant removal	Reference
1	Water system	Biosorption	Cu (II) Turbidity	-	MOS pressed cake	Reduced copper by 53.49% Removed turbidity by 74.45%	(Ali, 2016)
2	Waste Water	Coagulation	Turbidity DO=2.58 mg/L COD=99.5 ± 0.71 mg/L BOD=48.00 ± 0.42mg/L	-	MOS	Turbidity removed = 85-94% DO= 4.0 mg/L COD= 164.0 ± 2.83 mg/L BOD= 76.65 ± 2.33 mg/L	(Shan et al., 2017)
3	Surface water	Coagulation	Turbidity= 2.15-49.3 NTU	Alum	Processed MOS	Moringa alone= 2.7 NTU Alum alone= 1.8 NTU moringa with alum=1.4 NTU	(Muyibi & Al_Fugara, 2003)
4	Turbid water	Coagulation	Suspended particles	Alum	Fine powder of MOS	Moringa=70% Alum=75%	(Rohini & Padmini, 2016)
5	River water	Coagulation/ flocculation	Turbidity	-	MOS	Low turbidity = 95% Medium turbidity = 98% High turbidity = 99.3%	(Ali, 2016)
6	Abattoir waste water	-	Turbidity	-	Oil and De-oiled MOS	De-oiled = 80% Oiled= 68.86%s	(Aho & Lagasi, 2012)
7	Aqueous system	Coagulation	Fluoride	Alum and starch	MOS cake	Reduced turbidity below 1 mg/L	(Ravikumar & Sheeja, 2014)
8	Turbid surface water	Bioremediation-Indigenous method	Turbidity	-	MOS extract		(Lea, 2010)
9	Water supply	Coagulation/ flocculation Followed by ultrafiltration (UL)	Turbidity	-	MOS	Removed turbidity and the cells more than 99%	(Nishi et al., 2012)
10	Surface water treatment	Coagulation	Turbidity Cyanobacteria (Chlorophyll)		De-oiled MOS	Chlorophyll and turbidity were removed up to 85%	(Camacho et al., 2017)
11	Industrial dyeing wastewater	Photocatalytic water detoxification	Microbes (pathogens) Organic dye Pb metal	Silver nano particle	Seed	80% Pb metal removal > 80% organic dye Excellent anti-microbial activity against gram positive and gram negative	(Mahreen et al., 2021)
12	POME treatment	Coagulation	Turbidity Colour COD Sludge reduction		MOS extract	Turbidity =51% Colour=50% COD=65% Sludge reduction=148.0 ml/g	(Hizam et al., 2021)

continued on following page

Table 2. Continued

No.	Types of water/ waste water	Methods/process	Targeted pollutants	Composite chemical	Part of Moringa Oleifera used	Result/Rate of pollutant removal	Reference
13	Aqueous solution	Coagulation	Congo Red		MOS cake powder	Congo red removal = 97%	(Taylor et al., 2014)
14	Industrial wastewater	Biosorption	Acid Blue 9 synthetic dye		MOS	AB= 9-694.2 mg g-1 Biosorption 329.5 mg g-1	(dos Santos Escobar et al., 2021)
15	Aqueous solution	Adsorption filtration	Heavy metal (Cd, Cr, Pb)	Clay polymer composite	MOS composite	Cd-99.98% Cr-99.99% Pb-99.99%	(Ravikumar & Udayakumar, 2020)
16	Synthetic and real wastewater	Adsorption kinetics	Pb (II)		MOS	96.95% removal of Pb (II)	(Swelam et al., 2018)
17	Synthetic turbid water	Coagulation/ flocculation	Turbidity		MOS extract	Turbidity = 3.2 NTU (90%)	(Awad et al., 2013a)
18	Dairy wastewater	Coagulation	TDS, TSS, Hardness, Chlorides, COD, BOD		MOS powder	TDS(1889 mg/l), TSS(1500 mg/l), Hardness (203 mg/l), Chlorides(113 mg/l), COD(1640 mg/l), BOD (960 mg/l)	(Chonde & Raut, 2017)

3.1 Lack of Awareness and Understanding of MOS-based Treatment Systems Among Stakeholders

Awad et al. (2013b) reported that the lack of awareness and understanding of MOS-based treatment systems among stakeholders in some developing countries, such as Brazil, hindered the adoption of the technology. Another challenge is the variability in the quality of MOS seeds, which can affect the effectiveness of the treatment process. Petersen et al. (2016) also reported that the variability in the quality of MOS seeds could affect the coagulation and antimicrobial properties of MOS.

3.2 Availability and Affordability

The availability and affordability of MOS seeds may also be a challenge, particularly in areas where MOS is not grown. The moringa plant is mainly propagated through seed or stem cuttings. However, its propagation through stem cuttings has been identified to weaken the growth of the parent plant, resulting in low productivity and death in some cases (Majid et al., 2011). Also, propagation by seed causes variation in phenotypes and nutritional content as a result of changes in the genotypes of the plant species (Shan et al., 2017).

3.3 Increase Concentration of Total Organic Carbon (TOC)

In a study conducted by Awad et al. (2013a), it was observed that to achieve an acceptable level of turbidity removal from synthetic turbid water, the dosage of MOS seed powder was increased from 200 to 600 ml/L. The corresponding TOC concentration also increased from 12.43 to 34.78 mg/L. The gradual increase in TOC corresponding with the dosage increment after the coagulation/flocculation process justifies the fact that, some tiny particles of MOS powder and organic matter present in the treated solution become dispersed colloidal particles and thus increase TOC. The presence of these tiny particles can distort the three basic characteristics of water, namely odour, taste, and colour, and create a conducive microenvironmental condition for the growth of microbes (Broin et al., 2002; Okuda et al., 2001).

3.4 Demand for Large Land-Area for Vast Cultivation of MOS

There has been a steadily increasing demand for MOS products, both raw and processed. To meet the ever-increasing demands, there needs to be large-scale cultivation of MO. Mass production will also require large agricultural land areas. The prospects of large-scale cultivation may not be achieved given the limited land space in some countries coupled with the competing demand for land for various land uses owing to population concentration and the development of industries. In some cases, both the quality and quantity of MO biomass are compromised due to susceptibility to infestation by vermin and frost sensitivity. In addition, there are competing and overlapping demands within the various uses of moringa in some communities. Some communities are challenged with the choice of purpose for food (nutrition), medicine, cosmetics, and water purification as a result of the limited quantities of the available MO tree species. This may ultimately reduce biodiversity as a result of the decline in Missouri's natural tree resources (Kumssa et al., 2017).

3.5 Slow Sedimentation

Compared to conventional chemical coagulants, MOS powder takes a relatively longer time for the flocs to fully settle at the bottom of the container. Optimal pollutant removal efficiency for turbidity was achieved after 5 hours of application. Turbidity concentration decreased from 12.7 to 8.7 NTU after 1 h and 5 h, respectively, from an initial concentration of 32 NTU (Awad et al., 2013a). Moringa has been identified as being more effective in highly turbid water than in water with a low concentration. This actually prolongs the sedimentation process, thereby decreasing the rate of water production in a treatment system.

3.6 Risk of Secondary Pollution Through Bacteria Regrowth

The potency of bacteria removal by moringa as a coagulant is not totally complete. According to Awad et al. (2013b), more than 80% of total coliform (TC) was removed by using moringa seed as a coagulant to remove turbidity. This implies the remaining 20% of TC was still present in the treated water. Hence, the treated water could face the risk of bacteria regrowth, especially when kept for a relatively longer time (one day or more) (Shan et al., 2017).

3.7 Challenges of Establishing Specific Coagulation Mechanisms

Due to the availability of different species and varieties of MOS as well as different types and concentrations of water and wastewater pollutants, it is difficult to establish specific coagulation mechanisms that take place when MOS or its extracts are used for wastewater decontamination based on available literature.

4. RECOMMENDATIONS FOR FUTURE RESEARCH AND PROMOTION

To promote the adoption of MOS-based treatment systems in developing countries, there is a need for further research to address the challenges faced by developing countries. One area of research is the development of standards for MOS seed quality to ensure the effectiveness of MOS-based treatment systems. There is also a need for the development of standards for MOS quality and specific mechanisms for seed-coagulant processing to ensure the consistency and effectiveness of MOS-based treatment systems. Another area of research is the optimization of the treatment process to improve the efficiency and cost-effectiveness of MOS-based treatment systems. The authors also suggested optimizing the treatment process by exploring the effects of different parameters such as pH, temperature, and MOS dosage on treatment efficiency. Furthermore, there is a need for awareness and education programs to promote the adoption of MOS-based treatment systems among stakeholders in developing countries. Furthermore, there is a need for education and training programs to increase awareness and understanding of MOS-based treatment systems among stakeholders in developing countries. The interaction of different MOS extracts with colloids present in different effluents needs extra studies to further understand the specific coagulation mechanism. Finally, methods to help avoid bacteria regrowth after water treatment with MOS need further exploration.

5. FUTURE RESEARCH DIRECTIONS

The use of MOS in water and wastewater treatment has attracted substantial interest due to its low cost, availability, and efficacy in removing different contaminants. The following research directions can serve to promote the development of MOS-based treatment technologies for water and wastewater treatment as well as contribute to the sustainable management of water resources in poor countries:

5.1 Methods for Optimizing MOS Extraction

The extraction process used to get the active components has a large impact on the efficiency of MOS in water treatment. Future study might concentrate on improving extraction methods to increase the yield of active compounds, improve their stability, and minimize extraction time and cost.

5.2 MOS-Based Hybrid Treatment Systems

MOS may be used in conjunction with other treatment methods such as adsorption, coagulation, and membrane filtration to improve their effectiveness. A future study might look at the possibility of establishing MOS-based hybrid treatment systems that combine the advantages of MOS with other treatment technologies to achieve more effective pollutant removal.

5.3 MOS-Based Remediation for Emerging Contaminants

Emerging pollutants such as medicines and personal care items are a rising problem in water and wastewater treatment. A future study might look at the potential of MOS in removing these new pollutants and designing MOS-based treatment systems that are especially aimed at them.

5.4 Long-Term Performance of MOS-Based Treatment Systems

The long-term effectiveness and stability of MOS-based treatment systems must be assessed in order to determine their sustainability and durability. Future research could concentrate on monitoring the long-term performance of MOS-based treatment systems under various operational conditions and identifying factors that can impact their stability (Figure 6).

6. CONCLUSION

MOS can effectively remove suspended particles, turbidity, and microorganisms from water. MOS's potential activity in breaking down contaminants and purifying polluted water as a coagulant and antimicrobial agent has been justified in numerous empirical studies. The coagulation and flocculation processes of using moringa in water and wastewater treatment have proven to be excellent onsite treatment technology options for many developing countries, especially India, Ghana, Pakistan, Brazil, and Nigeria, among others. Rural households in water-stressed or polluted areas can rely on this technique to get sustainable access to potable drinking water. In the effective utilization of moringa to treat water, one of the socioeconomic implications is the bridging of the gender inequality gap. In most rural communities, the burden of collecting water from available sources lies in the hands of women and girls. Sometimes girls of school-going age would have to help their mothers fetch water instead of going to school. Therefore, by using moringa seed to improve quality and access to potable water, this can help reduce the risk and stress associated with carrying water as well as help women make time for other productive activities. In addition, MOS application technology in water treatment is simple, easy, and low-cost for both installation and maintenance. This makes it affordable for households with low incomes to access and utilize potable water, thereby reducing the risk of waterborne diseases.

ACKNOWLEDGMENTS

The authors would like to express their sincere gratitude to the Department of Water and Environmental Engineering, Tamale Technical University, for their invaluable support during the research and writing of this book chapter. Their expertise and resources have been instrumental in helping us achieve our research goals and communicate our findings effectively. We would like to specifically thank the Applied Research Conference of Technical Universities of Ghana (ARC-TU) for the opportunity given us to present our paper during the 2022 conference.

REFERENCES

Abubakar, I. R. (2019). Factors influencing household access to drinking water in Nigeria. *Utilities Policy*, *58*, 40–51. doi.org/10.1016/j.jup.2019.03.005. doi:10.1016/j.jup.2019.03.005

Aho, I. M., & Lagasi, J. E. (2012). A new water treatment system using Moringa oleifera seed. *American Journal of Scientific and Industrial Research, 3*(6), 487–492. doi:10.5251/ajsir.2012.3.6.487.492

Ali, E. N. (2016). *Removal of Copper from Water System using Moringa oleifera Press Cake*. 16–19. doi:10.17758/URUAE.AE0416214

Amagloh, F. K., & Benang, A. (2009). Effectiveness of Moringa oleifera seed as coagulant for water purification. *African Journal of Agricultural Research, 4*(2), 119–123.

Araujo, C. S. T., Carvalho, D. C., Rezende, H. C., Almeida, I. L. S., Coelho, L. M., Coelho, N. M. M., Marques, T. L., & Alves, V. N. (2013). *Bioremediation of Waters Contaminated with Heavy Metals Using Moringa oleifera Seeds as Biosorbent*. Applied Bioremediation - Active and Passive Approaches., doi:10.5772/56157

Awad, M., Wang, H., & Li, F. (2013a). Weaknesses of moringa oleifera use in water treatment. *International Journal of Current Research, 5*(5), 1165–1167.

Awad, M., Wang, H., & Li, F. (2013b). WEAKNESSES OF Moringa oleifera USE IN WATER TREATMENT. *Article in International Journal of Current Research, 5*, 1165–1167. https://www.researchgate.net/publication/281629560

Bakare, B. F. (2016). An investigation of moringa oleifera seed extract as a natural coagulant in water treatment. *Lecture Notes in Engineering and Computer Science, 2226*, 566–569.

Biosci, I. J., Alfarra, R. S., Ali, N. E., Yusoff, M. M., Razak, L. T., & Biosci, I. J. (2014). Removal of heavy metals by natural adsorbent [review] [IJB]. *International Journal of Biosciences, 6655*, 130–139. doi:10.12692/ijb/4.7.130-139

Blahuskova, V., Vlcek, J., & Jancar, D. (2019). Study connective capabilities of solid residues from the waste incineration. *Journal of Environmental Management, 231*, 1048–1055. doi:10.1016/j.jenvman.2018.10.112 PMID:30602228

Brilhante, R. S. N., Sales, J. A., Pereira, V. S., Castelo-Branco, D., Cordeiro, R. A., de Souza Sampaio, C. M., de Araújo Neto Paiva, M., Santos, J. B. F., Sidrim, J. J. C., & Rocha, M. F. G. (2017). Research advances on the multiple uses of Moringa oleifera: A sustainable alternative for socially neglected population. *Asian Pacific Journal of Tropical Medicine, 10*(7), 621–630. doi:10.1016/j.apjtm.2017.07.002 PMID:28870337

Broin, M., Santaella, C., Cuine, S., Kokou, K., Peltier, G., & Joet, T. (2002). Flocculent activity of a recombinant protein from Moringa oleifera Lam. seeds. *Applied Microbiology and Biotechnology*, *60*(1), 114–119. PMID:12382051

Bukar, A., Uba, A., & Oyeyi, T. (2010). Antimicrobial profile of Moringa oleifera Lam. extracts against some food–borne microorganisms. *Bayero Journal of Pure and Applied Sciences*, *3*(1). doi:10.4314/bajopas.v3i1.58706

Camacho, F. P., Sousa, V. S., Bergamasco, R., & Teixeira, M. R. (2017). The use of Moringa oleifera as a natural coagulant in surface water treatment. *Chemical Engineering Journal*, *313*, 226–237. doi:10.1016/j.cej.2016.12.031

Chonde, S., & Raut, P. (2017). Treatment of dairy wastewater by Moringa oleifera seeds. *World Journal of Pharmaceutical Research*, *6*(8), 1484–1493. doi:10.20959/wjpr20178-9015

Dharwal, M., Parashar, D., Shuaibu, M. S., Abdullahi, S. G., Abubakar, S., & Bala, B. B. (2020). Water pollution: Effects on health and environment of Dala LGA, Nigeria. *Materials Today: Proceedings*.

dos Santos Escobar, O., de Azevedo, C. F., Swarowsky, A., Adebayo, M. A., Netto, M. S., & Machado, F. M. (2021). Utilization of different parts of Moringa oleifera Lam. seeds as biosorbents to remove Acid Blue 9 synthetic dye. *Journal of Environmental Chemical Engineering*, *9*(4), 105553. doi:10.1016/j.jece.2021.105553

Echeverria, C. A., Handoko, W., Pahlevani, F., & Sahajwalla, V. (2019). Cascading use of textile waste for the advancement of fibre reinforced composites for building applications. *Journal of Cleaner Production*, *208*, 1524–1536. doi:10.1016/j.jclepro.2018.10.227

Eman, N. A., Tan, C. S., & Makky, E. A. (2014). Impact of Cake Residue Application on Waste Water Treatment: A Case Study. *Journal of Water Resource and Protection*, *06*(07), 677–687. doi:10.4236/jwarp.2014.67065

George, K. S., Revathi, K. B., Deepa, N., Sheregar, C. P., Ashwini, T. S., & Das, S. (2016). A Study on the Potential of Moringa Leaf and Bark Extract in Bioremediation of Heavy Metals from Water Collected from Various Lakes in Bangalore. *Procedia Environmental Sciences*, *35*, 869–880. doi:10.1016/j.proenv.2016.07.104

Gupta, S., Jain, R., Kachhwaha, S., & Kothari, S. L. (2018a). Nutritional and medicinal applications of Moringa oleifera Lam.—Review of current status and future possibilities. *Journal of Herbal Medicine*, *11*(March), 1–11. doi:10.1016/j.hermed.2017.07.003

Gupta, S., Jain, R., Kachhwaha, S., & Kothari, S. L. (2018b). Nutritional and medicinal applications of Moringa oleifera Lam.—Review of current status and future possibilities. *Journal of Herbal Medicine, 11*, 1–11. doi:10.1016/j.hermed.2017.07.003

Hasan, Md. K., Shahriar, A., & Jim, K. U. (2019). Water pollution in Bangladesh and its impact on public health. *Heliyon, 5*(8), e02145. https://doi.org/https://doi.org/10.1016/j.heliyon.2019.e02145

Hizam, M., Noor, M., Lee, W. J., Faiz, M., Azli, M., & Ngadi, N. (2021). Materials Today : Proceedings Moringa oleifera extract as green coagulant for POME Treatment : Preliminary studies and sludge evaluation. *Materials Today: Proceedings, 2020*(xxxx). doi:10.1016/j.matpr.2021.02.241

Idris, M. A., Jami, M. S., Hammed, A. M., & Jamal, P. (2016). Moringa oleifera seed extract: A review on its environmental applications. *International Journal of Applied Environmental Sciences, 11*(6), 1469–1486.

Idris, M. A., Jami, M. S., Jamal, P., & Hammed, A. M. (2016). *Moringa Oleifera Seed Extract : a Al Ā L-Friendly Alternative. October.*

Järup, L., & Åkesson, A. (2009). Current status of cadmium as an environmental health problem. *Toxicology and Applied Pharmacology, 238*(3), 201–208. doi:10.1016/j.taap.2009.04.020 PMID:19409405

Karaouzas, I., Smeti, E., Vourka, A., Vardakas, L., Mentzafou, A., Tornés, E., Sabater, S., Muñoz, I., Skoulikidis, N. Th., & Kalogianni, E. (2018). Assessing the ecological effects of water stress and pollution in a temporary river - Implications for water management. *The Science of the Total Environment, 618*, 1591–1604. doi:10.1016/j.scitotenv.2017.09.323 PMID:29054662

Kavitha, C., Ramesh, M., Kumaran, S. S., & Lakshmi, S. A. (2012). Toxicity of Moringa oleifera seed extract on some hematological and biochemical profiles in a freshwater fish, Cyprinus carpio. *Experimental and Toxicologic Pathology, 64*(7–8), 681–687. doi:10.1016/j.etp.2011.01.001 PMID:21282048

Keogh, M. B., Elmusharaf, K., Borde, P., & Mc Guigan, K. G. (2017). Evaluation of the natural coagulant Moringa oleifera as a pretreatment for SODIS in contaminated turbid water. *Solar Energy, 158*(October), 448–454. doi:10.1016/j.solener.2017.10.010

Kumssa, D. B., Joy, E. J. M., Young, S. D., Odee, D. W., Ander, E. L., Magare, C., Gitu, J., & Broadley, M. R. (2017). Challenges and opportunities for Moringa growers in southern Ethiopia and Kenya. *PLoS One, 12*(11), e0187651. doi:10.1371/journal.pone.0187651 PMID:29121079

Landrigan, P. J., Fuller, R., Fisher, S., Suk, W. A., Sly, P., Chiles, T. C., & Bose-O'Reilly, S. (2019). Pollution and children's health. *The Science of the Total Environment*, *650*, 2389–2394. doi:10.1016/j.scitotenv.2018.09.375 PMID:30292994

Lea, M. (2010). Bioremediation of turbid surface water using seed extract from Moringa oleifera Lam.(drumstick) tree. *Current Protocols in Microbiology*, *16*(1), 1G–2. doi:10.1002/9780471729259.mc01g02s16 PMID:20131221

Mahreen, H., Shahid, M., Rajoka, R., Xiong, Y., Cai, H., Muhammad, R., Mahmood, Q., He, Z., & Zhu, Q. (2021). Journal of Environmental Chemical Engineering Green synthesis of a silver nanoparticle using Moringa oleifera seed and its applications for antimicrobial and sun-light mediated photocatalytic water detoxification. *Journal of Environmental Chemical Engineering*, *9*(4), 105290. doi:10.1016/j.jece.2021.105290

Majid, N. M., Islam, M. M., Justin, V., Abdu, A., & Ahmadpour, P. (2011). *Evaluation of heavy metal uptake and translocation by Acacia mangium as a phytoremediator of copper contaminated soil. 10*(42), 8373–8379. doi:10.5897/AJB11.392

Mann, U. S., Dhingra, A., & Singh, J. (2014). Water pollution. *Causes, Effects and Remedies.*, *02*, 70–74.

Muyibi, S., & Alfugara, A. (2003). Treatment of surface water with Moringa Oleifera seed extract and alum – a comparative study using a pilot scale water treatment plant. *The International Journal of Environmental Studies*, *60*(6), 617–626. doi:10.1080/723032000087925

Nishi, L., Vieira, A. M. S., Vieira, M. F., Bongiovani, M., Camacho, F. P., & Bergamasco, R. (2012). Hybrid process of coagulation/flocculation with Moringa oleifera followed by ultrafiltration to remove Microcystis sp. cells from water supply. *Procedia Engineering*, *42*, 865–872. doi:10.1016/j.proeng.2012.07.479

Okuda, T., Baes, A. U., Nishijima, W., & Okada, M. (2001). Coagulation mechanism of salt solution-extracted active component in Moringa oleifera seeds. *Water Research*, *35*(3), 830–834. doi:10.1016/S0043-1354(00)00296-7 PMID:11228982

Owa, F. D. (2013). Water Pollution: Sources, Effects, Control and Management. *Mediterranean Journal of Social Sciences*. doi:10.5901/mjss.2013.v4n8p65

Patel, P., Patel, N., Patel, D., Desai, S., & Meshram, D. (2014). Phytochemical analysis and antifungal activity of moringa oleifera. *International Journal of Pharmacy and Pharmaceutical Sciences*, *6*(5), 144–147.

Pawar, T. A. (2016). Use of natural coagulants for the coagulation treatment of water – a review. *International Journal of Research in Engineering* [IJRESTs]. *Science and Technologies*, *2*(1), 156–159.

Petersen, H. H., Petersen, T. B., Enemark, H. L., Olsen, A., & Dalsgaard, A. (2016). Removal of Cryptosporidium parvum oocysts in low quality water using Moringa oleifera seed extract as coagulant. *Food and Waterborne Parasitology*, *3*, 1–8. doi:10.1016/j.fawpar.2016.03.002

Ravikumar, K., & Sheeja, K. A. (2014). Removal of Fluoride from Aqueous System Using Moringa oleifera Seed Cake and its Composite Coagulants, I. *International Journal of Current Engineering and Technology*, *4*(3), 1356–1360.

Ravikumar, K., & Udayakumar, J. (2020). *Moringa oleifera seed composite a novel material for hazardous heavy metals (Cd, Cr and Pb) removal from aqueous systems. 11*(1), 123–138.

Rohini, M. V., & Padmini, E. (2016). Preliminary phytochemical screening of selected medicinal plants of polyherbal formulation. *Journal of Pharmacognosy and Phytochemistry*, *5*(5), 277.

Safo-adu, G., & Hanson, R. (2019). PRACTICES AND KNOWLEDGE ABOUT ENVIRONMENTAL SANITATION AND HYGIENE AMONG URBAN DWELLERS IN GOMOA EAST DISTRICT. *GHANA*, *6*(1), 1–11.

Salgot, M., & Folch, M. (2018). Wastewater treatment and water reuse. *Current Opinion in Environmental Science & Health*, *2*, 64–74. doi:10.1016/j.coesh.2018.03.005

Seow, T. W., Lim, C. K., Nor, M. H. M., Mubarak, M. F. M., Lam, C. Y., & Yahya, A., & Zaharah Ibrahim. (2012). REVIEW ON WASTEWATER TREATMENT TECHNOLOGIES Jayashree Dhote. [IJERT]. *International Journal of Engineering Research & Technology (Ahmedabad)*, *1*(5), 1–10.

Shan, T. C., Al Matar, M., Makky, E. A., & Ali, E. N. (2017). The use of Moringa oleifera seed as a natural coagulant for wastewater treatment and heavy metals removal. *Applied Water Science*, *7*(3), 1369–1376. doi:10.100713201-016-0499-8

Sinha, S. N. (2012). Phytochemical analysis and antibacterial potential of Moringa oleifera Lam. *International Journal of Science Innovations and Discoveries*, *2*(4), 401–407.

Swelam, A. A., Sherif, S. S., & Ibrahim, A. (2018). THE ADSORPTION KINETICS AND MODELING FOR Pb (II). *REMOVAL FROM SYNTHETIC AND REAL WASTEWATER BY MORINGA OLEIFERA SEEDS.*, *29*(2), 105–124.

Taylor, P., Tie, J., Li, P., Xu, Z., Zhou, Y., Li, C., & Zhang, X. (2014). *Desalination and Water Treatment Removal of Congo red from aqueous solution using Moringa oleifera seed cake as natural coagulant.* Taylor and Francis. doi:10.1080/19443994.2014.905980

Trösch, W. (2009). Water treatment. In *Technology Guide*. Principles - Applications - Trends., doi:10.1007/978-3-540-88546-7_73

Ueda Yamaguchi, N., Cusioli, L. F., Quesada, H. B., Camargo Ferreira, M. E., Fagundes-Klen, M. R., Salcedo Vieira, A. M., Gomes, R. G., Vieira, M. F., & Bergamasco, R. (2021). A review of Moringa oleifera seeds in water treatment: Trends and future challenges. *Process Safety and Environmental Protection, 147*, 405–420. doi:10.1016/j.psep.2020.09.044

UNICEF, W. H. O. (2015). *Progress on Sanitation and Drinking Water: 2015 Update and MDG Assessment*. UN.

Velmurugan, A., Swarnam, P., Subramani, T., Meena, B., & Kaledhonkar, M. J. (2020). *Water demand and salinity*. Desalination-Challenges and Opportunities. doi:10.5772/intechopen.88095

Vigneswaran, S., & Sundaravadivel, M. (2009). Traditional and household water purification methods of rural communities in developing countries. *Wastewater Recycle. Reuse and Reclamation, 2*, 84–96.

Villaseñor-Basulto, D. L., Astudillo-Sánchez, P. D., del Real-Olvera, J., & Bandala, E. R. (2018). Wastewater treatment using Moringa oleifera Lam seeds: A review. *Journal of Water Process Engineering, 23*(February), 151–164. doi:10.1016/j.jwpe.2018.03.017

Water, U. N. (2017). *The United Nations world water development report 2017: wastewater the untapped resource*. UN Water.

Wuana, R., & Okieimen, F. (2011). Heavy Metals in Contaminated Soils: A Review of Sources, Chemistry, Risks and Best Available Strategies for Remediation. *ISRN Ecology, 2011*, 1–20. doi:10.5402/2011/402647

Zhou, Z., Ruan, D., Jiang, L.-M., Yang, Y., Ge, H., & Wang, L. (2019). Comparison on treatment strategy for chemical cleaning wastewater: Pollutants removal, process design and techno-economic analysis. *Journal of Environmental Management, 235*, 161–168. doi:10.1016/j.jenvman.2019.01.053 PMID:30682668

Chapter 11
Development of Basic Blocks for Garments Using Anthropometric Data of Pre-teen Girls

Vivian Biney-Aidoo
Accra Technical University, Ghana

William Kwesi Senayah
https://orcid.org/0000-0003-0659-2265
Accra Technical University, Ghana

Mercy Kuma-kpobee
University of Ghana, Ghana

Sherrifa Mahama
University of Ghana, Ghana

Docea Fianu
University of Ghana, Ghana

ABSTRACT

Although sizing is usually established through the development of a size chart, the real elements of fit can only be ascertained after the creation of a garment pattern, based on the measurements of the size chart. Therefore, one of most fundamental stages in the garment manufacturing business is patternmaking. This study developed basic pattern blocks for garment manufacture using anthropometric data of Ghanaian girls (7 to 12 years), measured, and processed into a size chart in a previous study. First of all, the measurements of the size chart were extracted, after which one extra size (GH 6) was created based on the intervals observed between the sizes.

DOI: 10.4018/978-1-6684-9843-9.ch011

Basic pattern blocks were developed and validated by producing trial samples and testing them on a sub-section of the target population. The major contribution of this study is the availability of a variation of ready-made standard garment blocks for use by garment manufacturers, and an enhanced fit of garments produced for the children's wear sector.

1. INTRODUCTION

The successful manufacture of garments requires the development of patterns for each design. Garment fit is directly related to the body dimensions that are built into the basic pattern blocks of the garment. Therefore, the main inputs in pattern and garment production are accurate measurements of the target population (Podbev et al., 2014; Gill, 2015). Without accurate measurements, garment manufacturers will not have the right statistics to produce good fitting garments (Adu-Gyamfi, 2006; Hsu, 2009). Body shapes and sizes, however, differ among countries, ethnicities, age, and gender (Salehi & Shahrabi, 2012; Muslim et al., 2014; Bari et al., 2015; Zakaria, 2016). Therefore, it is important to develop and use of measurement standards that fit specific target populations, in order to improve the garment manufacturing business (Hsu, 2009; Mpampa et al., 2010). Body measurements are usually standardized and classified into official sizes as is common in developed countries like the US and UK (Adu-Boakye, 2012; Forster & Ampong, 2012). In countries where official size charts are not available, most garment manufacturers use foreign size charts and varied methods of size estimation (Adu-Boakye, 2012 ; Widyanti et al., 2017). These practices lead to non-standardized measurements and subsequent fit problems.

Pattern making is one of the earliest and most critical stages in the development of a garment (Hsu, 2009; Puri, 2013; Gill, 2015). The process of garment manufacture involves the transformation of the original design into a pattern based on body measurements (Hsu, 2009). Garment samples are then sewn, fitted, and amended. The patterns which are finally obtained, become the major industrial tool for production and are taken through numerous stages such as pattern grading, marker production, spreading and cutting, in order to produce final garments (Hsu, 2009). Pattern making therefore, forms a fundamental part of sizing, since the fit of the garment is closely linked to the body dimensions incorporated in the basic pattern blocks (Podbev et al., 2014; Gill, 2015). When the body measurements are not built into a pattern, it leads to production flaws and cause numerous problems throughout the production chain (Hsu, 2009). A pattern set which bears with the right measurements therefore becomes the basic tool for garment manufacture. The use of accurate patterns helps to solve numerous industry problems like increasing customer base; demand for new products; short lead times; demand for volumes and cost management (Puri, 2013).

With the absence of accurate patterns, garment manufacturers could lose a lot of resources such as fabric and time in addition to high costs of production, rejected goods and loss of business credibility.

An anthropometric size chart has been formulated specifically for Ghanaian pre-teen girls, a demographic identified as facing challenges in clothing fit. This size chart furnishes essential data for manufacturers producing or intending to produce garments for Ghanaian pre-teen girls and analogous groups. One innovative way to present the size chart is to embed the measurements into basic pattern blocks. Additionally, the majority of Ghanaian garment manufacturers are accustomed to custom sewing through the free-hand method. Consequently, they exhibit a lack of enthusiasm towards developing basic patterns for garment production. This aversion and lack of skill in pattern-making among Ghanaian garment manufacturers may hinder their utilization of the created size chart. Thus, there is a pressing need to devise innovative strategies to encourage the adoption and utilization of the size chart.

While some studies have generated size charts for specific target demographics, there remains a dearth of research on translating these size charts into practical garment patterns. The conversion of a size chart into basic garment patterns holds the potential to become a valuable and indispensable element in garment manufacturing, specifically in addressing the distinct fit challenges prevalent in the Ghanaian children's wear segment. This study aims to develop basic pattern blocks for Ghanaian pre-teen girls, utilizing Ghanaian anthropometric data (the size chart), with the intention of streamlining and enhancing the production of well-fitting garments.

2. BACKGROUND

2.1 The Ghanaian Garment Industry

Very few manufacturers in Ghana specialize in the children's wear category. The children's wear sector of the Ghanaian clothing industry, however, has a huge potential market because of the high percentage of children in relation to the overall population (Ghana Statistical Service, 2013; Abdul-Rahman, 2018). Additionally, although the Ghanaian garment industry has a great potential in various respects (Quartey, 2006; Gavor & Ampong, 2009; Amateye, 2009), it is bedeviled by many challenges which affect its capacity to compete internationally. Some of the typical challenges are the high production costs, poor quality, massive importation especially of second-hand clothing (Fianu & Zentey, 2000; Fianu et al.,2014), as well as the use of the custom-made and free hand cutting methods (Hoefter, 2001). Furthermore, garment companies lack conformity to mass production standards, leading to inability to meet large orders (Kuma-Kpobee, 2013; Fianu et al.,2014) and subsequent collapse.

Despite Ghanaians' initial preference for custom-made over ready-to-wear (RTW) garments, the use of RTW has gradually taken over the use of custom-made garments, due to the importation of volumes of RTW as well as second-hand clothes from foreign countries (Fianu & Aryee-Atta, 2008; Mastamet-Mason et al., 2008; Denkyi-Sarpong, 2011). Moreover, most consumers derive a lot of satisfaction from RTW in terms of price and convenience (Yertutan, 2001; Mpampa et al., 2010; Alubel et al., 2017), and this has probably resulted in an increased demand for RTW among Ghanaians. This places a huge demand on the use of pre-determined size charts for Ghanaians. Most garment manufacturers who produce in large quantities for the RTW Ghanaian market, therefore use foreign size charts. Others resort to varied methods of size estimation (Adu-Boakye, 2012; Widyanti et al., 2017), resulting in non-standardized measurements and subsequent fit problems.

2.2 Fit and Size of Children's Clothing

Accurate fit important for children's development because of the rapid physical growth, (Kang et al., 2001; Zakaria, 2011; Widyanti et al., 2017). The primary activity of children is to play and learn, which affects garment fit (Armstrong, 2014). Children have a lot of clothing fit issues (Huyssteen, 2006; Chung et al., 2007), due to their peculiar growth characteristics (Aldrich, 2009; Ariadurai et al., 2009; Zakaria, 2011). Pre-teen children (7 to 12 years old) have increased growth in body girth and height (Aldrich, 2011; Zakaria, 2011; Muslim et al., 2014). They are therefore vulnerable and need extra care and attention physically and psychologically (Manna et al., 2011; Kataria, 2017; Sowah, 2020). The rapid growth trend is more rapid in girls compared to boys (Hyusteen, 2006; Aldrich, 2011; Zakaria, 2011). In the same way, Ghanaian girls also have special anthropometric features (Biney-Aidoo, 2020).

2.3 Pattern Making

Patterns are shapes of garments cut out of pieces of paper based on body measurements and transferred onto fabrics for cutting. Each of these shapes represents a component of the garment design (Puri, 2013). There are two basic and common methods of obtaining patterns: drafting and draping. Of the two, drafting requires higher theoretical inclination due the use of the flat method, thereby preventing any direct contact with the body curves, as compared to draping. Additionally, drafting is the widely used and has become the formal method of instruction in most educational institutions (Aldrich, 2009; Gill, 2015). Patterns can be produced using both manual and digital methods, because both methods offer numerous options for obtaining patterns of varied designs (Armstrong, 2014). Puri (2013) reported that pattern making has moved to an advanced level, with digital methods featuring strongly as a key aspect. However, despite the influence of digital technology, patterns are still being developed manually in most circumstances (Puri, 2013).

Gill (2015) identified seven (7) important stages of the mass production of garments. Theas are:

i. Collect anthropometric measurements of the target population.
ii. Develop patterns or adopt and/or adapt existing pattern blocks.
iii. Make Samples
iv. Evaluate Fit
v. Approve fit and grade patterns.
vi. Final fit and Amendment
vii. Make final garments for consumers.

Gill (2015) further outlined key stages in defining pattern dimensions. They are:

(1) Making individual patterns in relation to individual or standard measurements.
(2) Making standard sized pattern blocks by grading, and
(3) Incorporating the demands of the target market

3. MAIN FOCUS OF THE CHAPTER

3.1 Issues, Controversies, Problems

Patterns serve as the primary link between garment design and garment manufacture (Aldrich, 2009; Forster & Ampong, 2012; Kuma-Kpobee, 2013; Podbev et al., 2014), as they bear the measurements needed to achieve accurate fit. Most garment firms develop patterns for new designs by constructing or adapting from basic pattern blocks (Aldrich, 2004, 2009). Basic blocks can be used to produce good fitting garments, if the pattern maker is skillful and meticulous (Forster & Ampong, 2012). Yet, most Ghanaian garment manufacturers are used to the free-hand method (Fianu and Zentey, 2000; Forster & Ampong, 2012) and therefore have a very poor affinity towards the development of basic patterns for garment manufacture (Kuma-Kpobee, 2013). Connell et al. (2006), Beshah et al. (2014) and Alubel et al. (2017), recommended the development of standard patterns for garment manufacture, to enhance production efficiency. The use of patterns is time saving, economical, and gives a feeling of confidence in the management of garment manufacture, thus improves accuracy and consistency of sizing and speed to the market (Aldrich, 2004; Hsu, 2009;

Forster & Ampong, 2012; Widyanti et al., 2017). One consumer group that tends to be ignored when it comes to garment size and fit is children (Zakaria 2016). Yet children need special attention due to their special growth characteristics, leading to the physiological and psychological importance of garments with accurate size and fit. In this study, we develop basic pattern blocks for Ghanaian basic schoolgirls, using a Ghanaian size chart, to simplify and enhance the manufacture of well-fitting garments. The specific objectives are the study were to:

i. Develop basic pattern blocks using the measurements of the size chart.
ii. Conduct fit evaluation.

3.2 Materials and Methods

The procedure used to develop the basic blocks for measurement standards was executed in five (5) stages, based on Gill's (2015) model for RTW product development.

1) Measurements

The size chart developed for Ghanaian 7-12-year-old girls in an earlier study (Biney-Aidoo, 2020), was adopted and used as the initial measurements for this study. Using the size intervals generated in the size chart, the 6-year-old size was extrapolated and added to obtain a preliminary size chart for basic schoolgirls (6 to 12 years old).

2) Development of Pattern

Using the size measurements of the size chart, seven (7) sets of basic block patterns (basic bodice with basic long sleeves, and basic skirt), were drafted after adding the necessary garment ease. Ease allowances were selected based on recommendations by BSI (1990) and Otieno & Fairhurst (2000).

3) Creation of Samples

The drafted basic blocks were used to produce samples of garments for fit evaluation.

4) Fit Evaluation

The garment samples were evaluated to check how well they fit the girls. Six (6) basic schoolgirls per age group were selected randomly from two (2) of the basic schools selected for the measurements. The models wore the garment samples over undergarments and were made to move in different positions for 15 minutes. Critical areas of the body such as the bust, hip, waist, neck, shoulder-length as well as the grain and ease were assessed for fit. An evaluation form was designed to record the results.

5) Final Fit and Revision

All required corrections were made, and the blocks were revised. A fresh set of blocks were prepared, using the feedback from the evaluation. The second fit evaluation was done with 21 girls (3 from each of the 7 age groups) to validate the new measurements. The selection criteria for the 21 girls were the same as that of the first fit evaluation.

4. SOLUTIONS AND RECOMMENDATIONS

4.1 Measurements

Table 1 presents a size chart of 6- to 12-year-old basic schoolgirls, based on the size chart for 7 to 12 year old girls developed by Biney-Aidoo (2020). It also shows the size GH 6 extrapolated using the differences between the measurements of the age groups.

Table 1. Size chart of Ghanaian basic schoolgirls

Size code (Age/Height range in cm)	*6 years 112.30 to 118.29*	**7 years 118.30 to 124.29**	**8 years 124.30 to 130.29**	**9 years 130.30 to 136.29**	**10 years 136.30 to 142.29**	**11 years 142.30 to 148.29**	**12 148.30 to 154.29**	**Size Intervals**
Dimensions		**Measurements (cm)**						
Height	***115.30***	121.30	127.30	133.30	139.30	145.30	151.30	**6.0**
Bust girth	***54.60***	57.80	61.00	64.20	67.40	70.60	73.80	**3.2**
Waist girth	***51.30***	53.30	55.30	57.30	59.30	61.30	63.30	**2.0**
Hip girth	***59.20***	62.80	66.40	70. 00	73.60	77.00	80.80	**3.6**

Source: Biney-Aidoo (2020); Size GH 6 extrapolated

Table 2. Measurements and ease allowances for garment production

Age (in years)	6	7	8	9	10	11	12	Bodice	Skirt	Sleeve
Dimensions	**Measurements**							**Ease**		
Bust	54.6	57.8	61.0	64.2	67.4	70.6	73.8	**10.0**		-
Waist	51.3	53.3	55.3	57.3	59.3	61.3	63.3	**6.0**	**6.0**	-
Neck base girth	26.7	27.5	28.3	29.1	29.9	30.7	31.5	**2.0**	-	-
Nape to waist: Front	29.9	31.3	32.7	34.1	35.5	36.9	38.3	**0.0**	-	-
Shoulder length	9.0	9.4	9.8	10.2	10.6	11.0	11.0	**0.0**	-	-
Back Width	23.9	24.9	25.9	26.9	27.9	28.9	29.9	**2.0**	-	-
Front Width	21.7	22.7	23.7	24.7	25.7	26.7	27.7	**1.0**	-	-
Armhole Depth (½ Nape to waist: Front)	15.5	16.0	16.5	17.0	18.0	18.5	19.0	**1.5**	-	-
Hip girth	59.2	62.8	66.4	70.0	73.6	77.2	80.8	-	**6.0**	-
Waist to hip	16.2	17.2	18.2	19.2	20.2	21.2	22.2	-	**0.0**	-
Sleeve Length	42.6	45.1	47.6	50.1	52.6	55.1	57.6	-	-	**0.0**
Upper arm girth	17.7	18.7	19.7	20.7	21.7	22.7	23.7	-	-	**6.0**
Wrist girth	12.2	12.7	13.2	13.7	14.2	14.7	15.2	-	-	**3.0**

All measurements in cm

The size intervals as observed between the sizes were consistent and therefore could easily be calculated and applied for smaller or bigger sizes as confirmed by (Huyssteen, 2006). The measurements as well as the size intervals of the adopted chart are unique to Ghanaian girls (7 to 12 years). The intervals observed in this study (bust: 3.2 cm, waist: 2.0 cm and hip: 3.6 cm), were similar to intervals used by Shoben & Taylor (2004). Others were 3.0 cm for bust and hip, and 1.0 for waist (Aldrich, 2011) and 4.0 cm for bust, waist and hip (ISO/TR 10652, 1991).

5. DEVELOPMENT OF BASIC BLOCKS

5.1 Generation of Garment Measurements

Table 2 shows the ease allowances added to produce the basic blocks (bodice, skirt and pants. For instance, when10 cm ease allowance was added to the hip girth of 54.6 cm for size GH 6 years, the garment measurement became 64.6 cm.

The bust, waist and hip have the highest amount of ease, and this is due to fact that these are usually the key areas of rapid growth among female children. Ease allowance provides the basic ease needed, for the wearer to move comfortably in a garment, and are necessary for garment manufacture. Some garment manufacturers confuse ease allowance with seam allowance and add varied amounts indiscriminately without being sure of whether to use such measurements as garment ease or design ease. Such indecision can affect the fit of the garment.

5.2 Drafted Basic Blocks

The results of the basic blocks drafted using measurements from the size chart and following pattern drafting instructions by Winifred Aldrich (2009) is illustrated in figure 1.

Figure 1. Drafting instructions

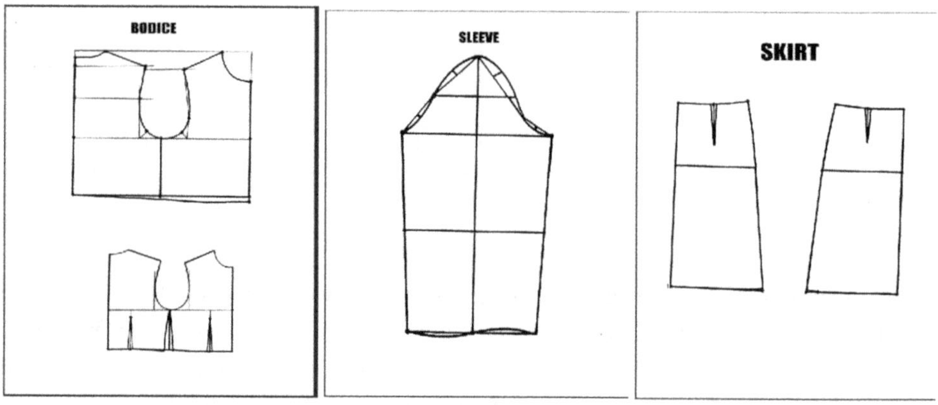

Seven (7) sets of the basic pattern blocks (illustrated in Figure 1), each bearing the standard measurements of Ghanaian basic schoolgirls were produced. The measurements used were compared with other measurements that are popularly applied in Ghanaian educational institutions, and by Ghanaian garment manufacturers.

The comparison revealed that although the basic blocks are similar in shape to other basic blocks, the measurements are different and represent the unique body sizes and shapes of Ghanaian basic schoolgirls.

Fit Evaluation of Basic Blocks

The observations made during the evaluation of the basic blocks for fit, are presented in Table 4.

Table 3. Comparison of measurements of Ghanaian girls and girls from other countries

Countries	Ages(years)	6	7	8	9	10	11	12
	Dimensionss	Measurements						
Ghana	Height	**115.30**	**121.30**	**127.30**	**133.30**	**139.30**	**145.30**	**151.30**
(Ongoing Study)	Bust	54.60	57.80	61.00	64.20	67.40	70.60	73.80
	Waist	51.30	53.30	55.30	57.30	59.30	61.30	63.30
	Hip	59.20	62.80	66.40	70. 00	73.60	77.00	80.80
South Africa (Hyusteen, 2006)	Height		**125.5**	**133.5**	**140.5**	**147.5**	**154.0**	**160.0**
	Bust		61.0	64.5	68.5	73.0	75.0	81.0
	Waist		55.3	57.8	60.2	62.5	64.1	67.1
	Hip		67.4	72.0	76.5	81.5	85.0	90.5
UK Aldrich, 2011	Height	**116.00**	**122.00**	**128.00**	**134.00**	**140.0**	**146.0**	**152.0**
	Bust	61.00	63.00	66.00	69.00	72.0	75.0	78.0
	Waist	58.00	60.00	61.00	62.00	63.0	64.0	65.0
	Hip	65.00	68.00	71.00	74.00	77.0	80.0	83.0
U. S	Height		**129 - 132**	**134 - 137**		**139 - 145**		**147 - 150**
	Bust		67.00	70.00		74.00		77.00
	Waist		58.00	60.00		62.00		65.00
	Hip		71.00	74.00		77.00		82.00
Croatia (Hrzˇenjak et al., 2013) Hrzˇenjak et al., 2013 Hrzˇenjak et al., 2013)	Height	**119.90**	**126.0**	**130.9**	**136.3**	**142.1**	**148.9**	**154.9**
	Bust	58.70	60.6	63.1	66.2	68.5	72.9	76.2
	Waist	55.60	56.9	58.5	61.1	62.2	64.7	66.4
	Hip	64.30	66.8	70.4	74.1	76.6	80.7	84.4

The ease allowances were generally sufficient for all the sample garments in each size group. A few problems were found with the key girth areas. For example, bust girth, waist girth, hip girth and upper arm girth were either too tight or too loose for many of the subjects. Therefore, the back width and front width areas were also smaller or bigger. The vertical dimensions had no major problems except for a few cases of taller subjects, making both the front and back nape to waist areas too short. Generally, there were fewer problems with 6- to 8-year-old girls compared with the 9 to 12 year old girls, which could be attributed to the extremely rapid growth of girls during puberty. As a result, two size variations (small and medium) were created within each size category, to solve problems relating to smaller or bigger girth measurements, as well as shorter or taller girls. These findings corroborate the

Table 4. Summary of observations of fit evaluation

Dimensions	Problems Identified	Frequency	(%)
Bust	Tight or loose bust and armhole	32	76
Waist	Tight or loose waist girth	33	78
Neck base girth	Neck girth too tight or too loose	36	85
Front Nape to waist	Front nape to waist too short or too long	38	90
Back Nape to waist	Back nape to waist to short or too long	38	90
Shoulder length	Shoulder length slightly smaller or bigger	36	85
Back Width	Back width smaller or bigger	34	80
Front Width	Smaller or bigger front width measurements	33	78
Depth of Armhole	Folds and tightness at occurring around armpit	37	88
Hip girth	Tight or hollow hip area	31	73
Sleeve Length	Sleeves too short or too long	38	90
Upper arm girth	Tight or hollow upper arm area	32	76
Wrist girth	Slightly tight or loose wrist girth	35	83

n = 42

advice by many authors to provide more sizes categories for pre-teens to suit the very rapid growth (Aldrich, 2009; Hrz ˇenjak et al., 2013, 2015; Zakariah, 2011; 2014).

Revision of Basic Blocks

The initial basic blocks were graded half size down, using the size intervals already established in the chart. A final measurements table (Table 4) to portray the revised measurements based on the fit evaluation was produced.

An evaluation of the revised measurements using a new set of drafted patterns and fit trials revealed that the new measurements and size intervals were suitable for the Ghanaian girls studied. Some Ghanaian manufacturers are used to the practice of applying foreign size charts and pattern blocks for garment manufacture. While some use the measurements as they are, others adapt them in different ways to obtain what they perceive as suitable for Ghanaians. The practice, according to Huyssteen (2006) and Adu-Boakye (2012) result in inconsistent measurements and fit of garments produced.

Table 5. Final measurements (cm) table for Ghanaian basic schoolgirls

Size code (Age / Height)			**GH 6 Years**		**GH 7 Years**		**GH 8 Years**	
			112.30 to 118.29 *to 118.29 to 118.29 to 118.29*		**118.30 to 124.29**		**124.30 to130.29**	
			S	**M**	**S**	**M**	**S**	**M**
Height			112.30	115.30	118.30	121.30	124.30	127.30
Bust			53.00	54.60	56.20	57.80	59.40	61.00
Waist			50.30	51.30	52.30	53.30	54.30	55.30
Hip			57.40	59.20	61.00	62.80	64.60	66.40
Size code (Age / Height)	GH 9 Years		GH 10 Years		GH 11 Years		GH 12Years	
	130.30 to 136.29		136.30 to 142.29		142.30 to 148.29		148.30 to 154.29	
	S	M	S	M	S	M	S	M
Height	130.30	133.30	136.30	139.30	142.30	145.30	148.30	151.30
Bust	62.60	64.20	65.80	67.40	69.00	70.60	72.20	73.80
Waist	56.30	57.30	58.30	59.30	60.30	61.30	62.30	63.30
Hip	68.20	70.00	71.80	73.60	75.25	77.00	79.00	80.80

All values are in cm.

CONCLUSION

In this study, we developed basic pattern blocks for garment manufacture, using a size chart of Ghanaian basic schoolgirls. We developed patterns, produced samples, conducted a fit evaluation, amended the measurements, and graded patterns. Seven (7) sets of the basic pattern blocks to represent the unique body sizes of Ghanaian basic schoolgirls were produced. A comparison of size charts that are commonly available for Ghanaian manufacturers, revealed similar but unique measurements of Ghanaian basic schoolgirls. The final body measurements and ease allowances applied for drafting the blocks after revision, were suitable for each size range. Some Ghanaians manufacturers are used to adopting or adapting foreign size charts, pattern blocks and pattern making instructions for garment manufacture. However, the application has been subjective because of the perception that those measurements are not based on Ghanaian anthropometric characteristics. The availability of a standard measurement table and standard pattern blocks for Ghanaian basic school children, although long awaited would lead to the application of consistent measurements for garment manufacture. The Ghanaian manufacturers' skill of freehand cutting can also be used to adapt the basic blocks into different garment designs. Future studies should consider replicating the same research for Ghanaian basic schoolboys. Moreover, there could be a further processing of the measurements into a computerized database, to generate patterns digitally.

ACKNOWLEDGMENTS

This work was partially supported by the Banga Africa Project under the University of Ghana, Pan African Doctorial Academy and the Accra Technical University Research and Innovation Fund (ARIF). We are grateful to everyone who in one way or another played major roles in making this paper a success.

REFERENCES

Abdul-Rahman, S. (2018, March 12). Ghana's population hits 29.6million - Statistical Service. *The Ghanaian Times.* www.ghanaiantimes.com.gh

Adu-Boakye, S. (2012). *Development of a conceptual framework relating to ready-to-wear clothing for Ghanaian women for manufacturing strategies*. [Doctoral thesis, Manchester Metropolitan University, 2012]. https://e-space.mmu.ac.uk/305090/

Adu-Gyamfi, V. (2006). *Developing size chart for Ghanaian women: A pilot study in Simpa 'A'. Project report.* Department of Home Economics Education, University of Education.

Aldrich, W. (2004). *Metric pattern cutting for women's wear*. Blackwell Science.

Aldrich, W. (2009). *Metric pattern cutting for children's wear and babywear* (4th ed.). Blackwell Publishing.

Aldrich, W. (2011). *Metric pattern cutting for children's wear and baby wear* (4th ed.). Blackwell Publishing.

Alubel, M., Yadav, M., & Singh, N. (2017). Anthropometric size chart for Ethiopian girls for better garment design. *Journal of Fashion Technology & Textile Engineering, 5*(2), 1–11. doi:10.4172/2329-9568.1000151

Amateye, L. (2009). *The role of the indigenous Ghanaian textile industry in relation to the President's Special Initiative (PSI) on textiles and garments.* [Doctoral thesis, Kwame Nkrumah University of Science and Technology]. https://ir.knust.edu.gh/xmlui/handle/123456789/288

Ampofo, V. O. (2002). *Ghana's Textiles/ Garment Industry: Industrial Development and Investment Division (Ministry of Trade and industry-Ghana).* Intracen. https://www.intracen.org/worldtradenet/docs/whatsnew/atc_lesotho_november2000/2

Ariadurai, S. A., Nilusha, T. P. G., Alwis, T., & Manori-Dissanayake, D. M. R. (2009). An Anthropometric Study on Sri Lankan School Children for Developing Clothing Sizes. *Journal of Social Sciences*, *19*(1), 51–56. doi:10.1080/09718923.2009.11892690

Armstrong, H. (2014). *Pattern making for fashion design* (5th ed.). Pearson Education Ltd.

Bari, S. B., Salleh, N. M., Sulaiman, N., & Othman, M. (2015). Development of clothing size for pre-school children based on anthropometric measurements. *Australian Journal of Sustainable Business and Society*, *1*(2), 22–32.

Beshah, B., Belay, B., Tilahun, S., Tizazu, B., & Matebu, A. (2014). Anthropometric data of Bahirdar city's adult men for clothing design. *International Journal of Vocational and Technical Education*, *6*(5), 51–57. doi:10.5897/IJVTE2014.0155.x

Biney-Aidoo, V. (2020). *Development Of Garment Size Chart Using Anthropometric Data Of Ghanaian Girls Aged 7 To 12 Years.* [Unpublished PhD thesis, Universityof Ghana Legon, 2009].

British Standard Institution. (1990). *BS 7231, Part 1, Body Measurements of Boys and Girls from Birth to 16.9 Years*. British Standard Institution, London.

Chen, C. M., LaBat, K., & Bye, E. (2010). Physical characteristics related to bra fit. *Ergonomics*, *53*(4), 514–524. doi:10.1080/00140130903490684 PMID:20309747

Chung, M., Lin, H., & Wang, M. J. W. (2007). The development of sizing systems for Taiwanese elementary and high-school students. *International Journal of Industrial Ergonomics*, *37*(8), 707–716. doi:10.1016/j.ergon.2007.05.004

Connell, L. J., Ulrich, P. V., Brannon, E. L., Alexander, M., & Presley, A. B. (2006). Body shape assessment scale : Instrument development for analyzing female figures. *Clothing & Textiles Research Journal*, *25*(2), 80–95. doi:10.1177/0887302X0602400203

Denkyi-Sarpong, G. (2011). Meeting the challenge of a competitive fashion market. *International Journal of Home Economics Research*, *2*, 157–158.

Fianu, A.G., Ayertey, A., & Francois, E.M. (2005). A pilot study of categorizing Ghanaian Female University Students into body sizes (Phase I). *ARAHE*, *12*(3).

Fianu, A.G., Ayertey, A., & Francois, E.M. (2005). Perceived and preferred body sizes of Ghanaian Female Students at University of Ghana, Legon (Phase II). *ARAHE, 12*(4), 231 - 235.

Fianu, A. G., Biney-Aidoo, V., Antiaye, E., & Oppong, J. (2014). The production and exportation of garments under the AGOA initiative: Challenges facing garment producers in Ghana. *Arts and Design Studies, 17*, 34-44. http://www.iiste.org

Fianu, A. G., & Zentey, E. A. (2000). Problems of large-scale fashion designers in Accra, Ghana. *Journal of Consumer Studies & Home Economics*, *24*(2), 128–136. doi:10.1046/j.1365-2737.2000.00153.x

Fianu, D. A. G., & Aryee-Atta, S. N. A. (2008). Ghanaian women's perception of ready to wear garments. [country_paper_ghana.pdf]. *Journal of Family and Consumer Sciences*, *100*(4), 28–33.

Forster, P., & Ampong, I. (2012). Pattern cutting skills in small scale garment industries and teacher education universities in Ghana. *International Journal of Vocational and Technical Education*, *4*(2), 14–24.

Gavor, M. E., & Ampong, I. T. (2009). Factors that affect quality of garments made by Ghanaian micro-scale manufacturers. *International Journal of Home Economics Research*, *1*, 110–121.

Ghana Statistical Service. (2013). *2010 Population & housing census: national analytical report*. Ghana Statistical Service.

Gill, S. (2015). A review of research and innovation in garment sizing, prototyping and fitting. *Textile Progress*, *47*(1), 1–85. doi:10.1080/00405167.2015.1023512

Hoefter, A. F. (2001). *The competitiveness of Ghana's industry.* [Unpublished doctoral thesis, Nr. 2494, Difo-Druck Bamberg, Germany].

Hrzˇenjak, R., Dolezal, K., & Ujevic, D. (2013). Development of sizing system for girls aged 6 to 12 years in Croatia. *Collegium Antropologicum*, *37*(4), 1095–1103. PMID:24611320

Hrženjak, R., Doležal, K., & Ujević, D. (2015). Sizing system for girls aged 13 - 20 years based on body types. *Textile Research Journal*, *85*(12), 1293–1304. doi:10.1177/0040517514559579

Hsu, C. H. (2009). Developing Accurate Industrial Standards to Facilitate Production in Apparel Manufacturing Based on Anthropometric Data. *Human Factors and Ergonomics in Manufacturing*, *19*(3), 199–211. doi:10.1002/hfm.20148

Huyssteen, S. V. (2006). *Development of standardized sizing systems for the South African children's wear market.* [Doctoral thesis, University of Stellenbosch]. https://scholar.sun.ac.za/handle/10019.1/4646

ISO/TR 10652. (1991). *Standard Sizing Systems for Clothes.* International Organization for Standardization, Geneva.

Kang, Y., Choi, H. S., & Hee, D. W. (2001). A study of the apparel sizing of children's wear - An analysis of the size increments utilized in children's wear based on an anthropometric survey. *Journal of Korean Home Economics Association English Edition, 2*(1), 95–110. doi:10.1533/9780857096890.1.95

Kataria, M. V. (2017). Tween consumers: A study of the impact of social media on tween's buying decisions. *International Journal of Marketing & Financial Management, 5*(1), 2348–3954. doi:10.5281/zenodo.238698

Kuma-Kpobee, M. A. (2009). *Determination of a Sizing System for Mass Customisation of Ghanaian Women's Traditional Dress: and a Conceptual Framework for Small and Medium Scale Enterprises.* [Unpublished PhD thesis, Manchester Metropolitan University, Manchester].

Kuma-Kpobee, M. A. (2013). The evolution and current manufacturing practice applied to the traditional dress of women in Ghana. *International Journal of Technology and Management Research, 2*, 56–66. www.koforiduapoly.edu.gh

Manna, P. K., De, D., Bera, T. K., Chatterjee, K., & Ghosh, D. (2011). Anthropometric assessment of physical growth and nutritional status among school children of North Bengal. *Anthropologist, 13*(4), 299–305. doi:10.1080/09720073.2011.11891211

Mastamet-Mason, A., De Klerk, H. M., & Ashdown, S. (2008). Sizing and fit research at grassroots level: A methodology for the identification of unique body shapes in African developing countries. *Journal of Family Ecology and Consumer Sciences, 36*, 9–21. doi:10.4314/jfecs.v36i1.47110.x

Mastamet Mason, A., De Klerk, H. M., Sommervile, J., & Ashdown, S. P. (2008). Consumers' knowledge on sizing and fit issues : A solution to successful apparel selection in developing countries. *International Journal of Consumer Studies, 32*(3), 276–284. doi:10.1111/j.1470-6431.2008.00679.x

Mpampa, M. L., Azariadis, P. N., & Sapidis, N. S. (2010). A new methodology for the development of sizing systems for the mass customization of garments. *International Journal of Clothing Science and Technology, 22*(1), 49–68. doi:10.1108/09556221011008802

Muslim, E., Moch, B. N., Fileinti, N. D., Puspasari, M. A., Sibarani, T. M. L., & Laskana, D. G. N. (2014). The Development of standard size for clothes of Indonesian boys based on anthropometric data as a reference to formulate RSNI 0555. *International Journal of Ergonomics*, *4*(2), 15–23.

Otieno, R. B., & Fairhurst, C. (2000). The development of new clothing size charts for female Kenyan children, part 2: Size-range Categorization and Presentation of Final Size Charts. *Journal of the Textile Institute*, *91*(2), 153–162. doi:10.1080/00405000008659535

Pheasant, S. (2003). *Bodyspace anthropometry, ergonomics and the design of work*. Taylor & Francis Ltd.

Podbevek, T. (2014). For a good fitted skirt, the waist-to-hip distance should be measured. *Anthropological Notebooks, 20*(2), 77-88.

Puri, A. (2013). Efficacy of Pattern Making Software in Product Development. *International Journal of Advanced Quality Management*, *1*(1), 21–39.

Quartey, P. (2006). The textile and clothing industry in Ghana. In H. Jauch & R. Traub- Merz (Eds.), the future of the textile and clothing industry in Sub-Saharan Africa, (pp. 35-146). Bonn: Friedrich-Ebert-Stiftung.

Salehi, M., & Shahrabi, J. (2012). Upgrading fitness in the production of garments. *Research Journal of Textile and Apparel*, *16*(4), 100–104. doi:10.1108/RJTA-16-04-2012-B011.x

Shoben, M. M., & Taylor, P. J. (2004). *Grading for the fashion industry, the theory and practice* (3rd ed.). LCFS Fashion Media.

Widyanti, A., Mahachandra, M., Soetisna, H. R., & Sutalaksan, I. Z. (2017). Anthropometry of Indonesian Sundanese children and the development of clothing size system for Indonesian Sundanese children aged 6 to10 years. *International Journal of Industrial Ergonomics*, *6*, 37–46. doi:10.1016/j.ergon.2017.05.008.x

Yertutan, C. (2001). Problems of young people related to anthropometric measures of ready-made clothes. *Journal of Qafqaz University, Fall(8),* 27–32.

Zakaria, N. (2011). Sizing system for functional clothing- Uniforms for school children. *Indian Journal of Fibre and Textile Research*, *36*, 348–357.

Zakaria, N. (2014). Body shape analysis and identification of key dimensions for apparel sizing systems. Woodhead Publishing Limited. https://dx.doi.org/10.1533/9780857096890.1.95.

Zakaria, N. (2016). Size and fit of clothing for children and teenagers. In *Clothing for children and teenagers* (pp. 15–32). Elsevier Ltd. doi:10.1016/B978-0-08-100226-1.00002-3

Zakaria, N. (2016b). Conducting anthropometric surveys of children and teenagers. In *Clothing or children and teenagers* (pp. 33–51). Elsevier Ltd. doi:10.1016/B978-0-08-100226-1.00003-5

Zakaria, N., & Gupta, D. (2014). *Apparel sizing: existing sizing systems and the development of new sizing systems*. Woodhead Publishing Limited. doi:10.1533/9780857096890.1.3.x

Chapter 12
Design and Construction of a Drying Chamber for Millinery and Accessories

Beauty Awusi Nyavor Akpoyo
Accra Technical University, Ghana

Benjamin Kofi Asinyo
Kwame Nkrumah University of Science and Technology, Ghana

Daniel Kwabena Danso
Akenten Appiah-Menka University of Skills Training and Entrepreneurial Development, Ghana

ABSTRACT

The study was conducted to design and construct a drying chamber for millinery and accessories production to reduce the risks involved in the old ways of drying hat. The design of the study is descriptive while the population includes fashion schools. The sample size for the study was 100 respondents while a researcher's self-developed questionnaire was used as the instrument for data collection. The blocked hat used the sun drying method, rather than the drying chamber, and considered the drying chamber more effective than sun drying method. The dryer constructed could dry the hats effectively and efficiently.

INTRODUCTION

According to Harrison (1960), Millinery is the design, manufacture and sale of hats and head-ware. A person engaged in this trade is called a milliner or hatter.

DOI: 10.4018/978-1-6684-9843-9.ch012

Millinery is sold to women, men and children, though some definitions limit the term to women's hats. Historically, milliners, typically female shopkeepers, produced or imported an inventory of garments for men, women, and children, including hats, shirts, cloaks, shifts, caps, neckerchiefs, and undergarments, and sold these garments in their millinery shop. Millinery and Accessories such as shoes, handbags, belts, hats, and jewelry are designed to coordinate with apparel to create a total fashion look. The colorful variety of these accessories creates a whole new sense of fashion in the industry. But we cannot overlook how these millinery accessories are made. In recent times, modern ways of making accessories have been introduced but the millinery producers still stick to the old ways of making the products.

PROBLEM STATEMENT

In Ghana, the traditional way of drying the hat is sun drying in the open space. Drying in the sun comes with its attendant problems such as difficulty to regulate the temperature hence exposure to outsider interference, changes in the weather such as rain and wind/dust.

Also, the construction of millinery accessory is a tedious and labor-intensive process. Therefore, this study is conducted to design and construct a drying chamber for millinery and accessories production which may intend to help reduce the risks involved in the old ways of drying.

LITERATURE REVIEW AND RELATED WORK

Historical Background of Millinery

According to Bailey (1998) Headwear for women began in earnest during the Middle Ages when the church decreed that their hair must be covered. During the 18th century, milliners took the hat-making art out of the home and established the millinery profession. Today, a "milliner" defines a person associated with the profession of hat making. In the 18th century however, a milliner was more of a stylist. Traditionally a woman's occupation, the milliner not only created hats or bonnets to go with costumes but also chose the laces, trims and accessories to complete an ensemble. The term „milliner" comes from the Italian city of Milan, where in the 1700"s, the finest straws were braided, and the best quality hat forms were made. Pancake style „shepherdess" hats were popular throughout most of the 18th century, in varying brim widths. These hats were considered necessary to keep the sun away from fair complexions, especially as the parasol was not a fashionable accessory during this period (https://vintagefashonguild.orgsthehistoryofwomenshats).

According to Casper (2002), more recently, the term milliner has evolved to describe a person who designs, makes, sells or trims hats primarily for a female clientele. The origin of the term is probably the Middle English milliner, meaning an inhabitant of the city of Milan or one who deals in items from Milan, known for its fashion and clothing (https://www.revolvy.com/main).

WAYS OF DRYING

Drying Under the Sun

In Ghana, our traditional way of drying hats is by drying in the sun. However, the finishing of this hat barely comes out pretty well finished. The sun also comes with its problems, such as difficulty to regulate the temperature, exposure to outside interference, changes in the weather such as rain, wind and dust.

According to Pakowski & Mujumdar (1995), drying of textiles outdoors, however, depends on the weather conditions. In today's society, a lack of space to hang the millinery and the need to reduce the drying time have made artificial drying more common. Drying cabinets and tumble dryers offer a fast and convenient way of drying millinery. Drying chambers offer the gentleness of drying hats, shoes and fascinators. The drying chamber keeps the hats neat instead of taking up the wet work to dry in the sun at any open place outside. Many drying chambers have special hooks that lets you dry items such as shoes, fascinators and hats (Stienen, 2007).

MATERIALS AND METHODS / RESEARCH METHODOLOGY

Data Collection, Source, and Nature of Data

The data for this study was taken from a primary source. The primary data for the research was taken from some fashion designers in some fashion school. The data is on millinery and how they are dried during the production process. Also, a drying chamber will be designed and constructed.

Descriptive Statistics

Descriptive analysis of data is presented in text for easy interpretation. Descriptive statistics refers to the collection, organizing, presenting and analyzing of data in a significant way by using analytical tools such as tables, graphs, frequencies and percentages (Kaur et al. 2018).

Inferential Statistics

This section reviews the theory of inferential statistics which was used in the data analysis. It is mostly termed as the further analysis of the study. Inferential Statistics is a statistical method which is used to make inferences on a population based on a sample collected. It is also a body of methods for drawing conclusions about characteristics of a population based on information available in a sample from the population (Sutanapong & Louangrath, 2015).

DATA ANALYSIS

The data collected were analyzed using excel to draw graphs to present the data. AutoCAD was used to design the structure of the chamber going to be built.

DISCUSSION OF FINDING AND RESULTS

The purpose of this study is to build a millinery drying chamber which will serve the functions of an oven dryer used to dry millinery hats after preparation. This section of the study presents the discussion of results from the questionnaire used and the designing and construction stages for the drying chamber.

LENGTH OF TIME OF DRYING HAT

Finally, respondents were asked whether they had blocked a hat before. This was done in order to know whether respondents had personally engaged in the art of millinery making to which 95% responded in the affirmative. The remaining 5% responded negatively. This implies that an overwhelming majority of respondents had actually had hands-on experience in millinery making and hence are suitable for this study. Since respondents claimed they had adequate knowledge about the millinery process and also had personally engaged themselves in blocking hats and millinery accessories, the researcher sought their response on how long it took them.

The result shows that it took 45% of respondents one day to dry a hat whilst for 36% of respondents, it took them two to five days. However, 19% of respondents indicated that it took them about a week in order to fully dry a hat that had been

Figure 1. Length of line for drying hat

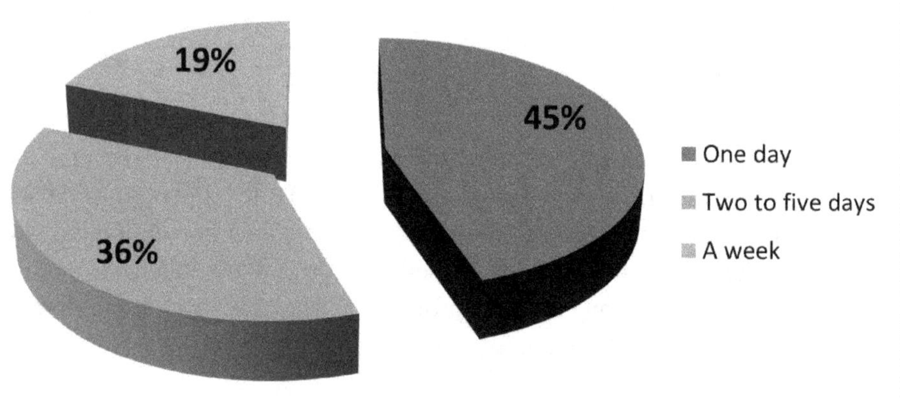

blocked. This result is indicative of the fact that different respondents may be using different methods to dry their hats, hence, the differences in their drying periods.

DRYING METHODS USED BY MILLINERS

As a follow up, the researcher sought to know the drying methods employed by respondents when drying their hats.

Figure 2. Types of drying

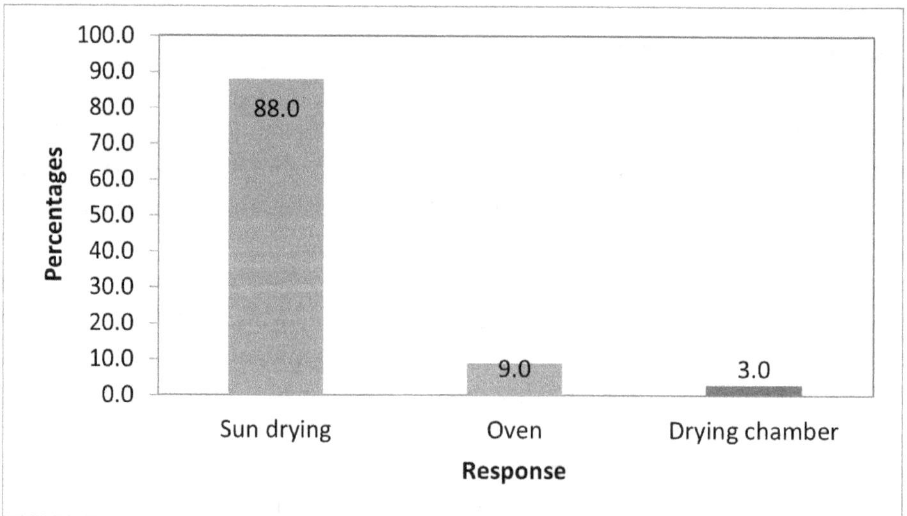

From Figure 2, it is evident that the most commonly used method of drying millinery hats and accessories by respondents was sun drying method. This method was used by a sound majority, 88% of respondents whilst only 9% responded indicated that they used ovens to dry their hats. Further, 3% of respondents indicated they used specialized drying chamber for millinery to dry their hats.

COMPARISM OF USE OF DRYING CHAMBER AND SUN DRYING METHODS

Respondents were then asked to compare the drying chamber and the sun regarding their effectiveness in drying hats. To this, 87% said the drying chamber was more effective whilst only 13% said the sun was effective.

Figure 3. Comparing the effectiveness of drying hats with drying chamber and with the sun

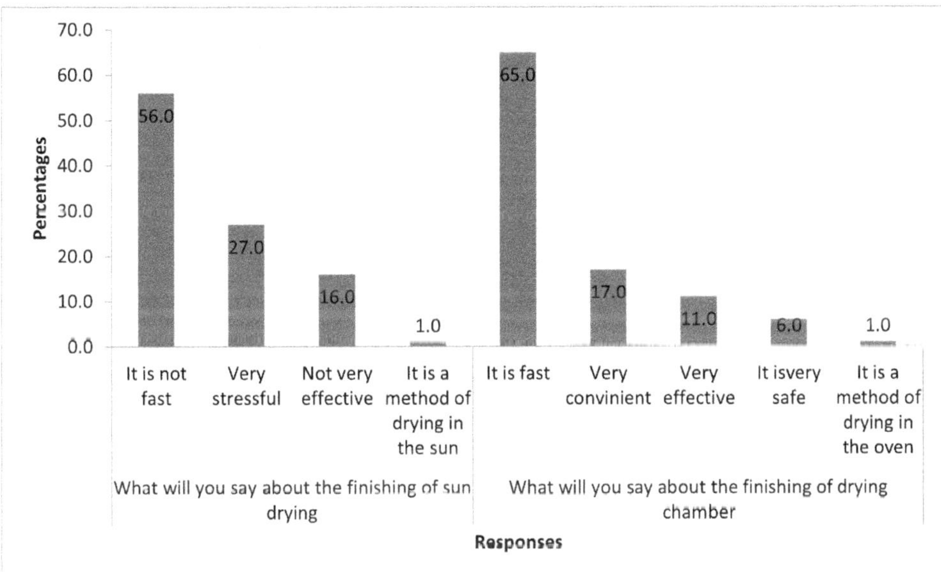

Figure 3 represents the comparison between the responses from drying under the sun and drying in a drying chamber. From the responses, 56.0% responded that the sun drying was not fast; 27.0% said it was very stressful; 16.0% said sun drying was not effective; and 1.0% gave answers that were not relevant to this thesis. On the other hand, 65.0% responded that the drying chamber method was a fast way of

Table 1. Respondents opinion on best method for drying millinery products

Which method is the best method for drying hats? Give reasons (N=100)	Frequency	Percentage
Drying in chamber because it is safe to use	12	12.0
Drying in chamber because it is fast	71	71.0
Drying in chamber because it is effective	14	14.0
Sun drying because it is faster	3	3.0

drying; 17.0% said the drying chamber was very convenient; 11.0% said the drying chamber was very effective; 6.0% said the drying chamber was very safe; and 1.0% of the response was not relevant to this thesis. The results clearly show that the drying chamber is the best method for drying hats.

BEST METHOD FOR DRYING MILLINERY PRODUCTS

As a means of endorsing the best way for drying millinery products, the researcher sought respondents' opinions and presented their views in Table 1. According to the data, a total of 97% of respondents said that the drying chamber was the best method for drying millinery products citing reasons as is safe to use (12%), drying chamber is faster (71%) and drying chamber is effective (14%). However, 3% of respondents were of the view that the sun drying was faster and that to them, drying of millinery

Figure 4. Drying rate for senamney in drying chamber

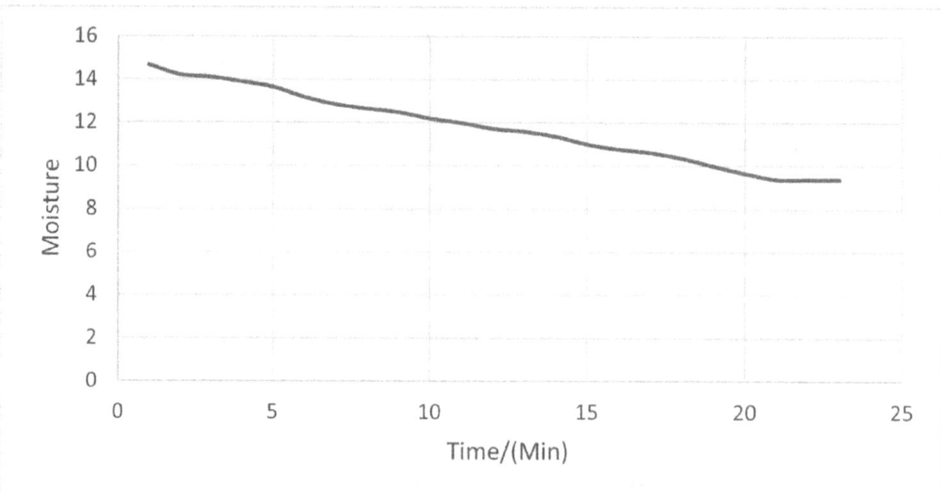

products using the sun was the best. To ascertain the effectiveness of the drying chamber, tests were conducted, and the drying rate of a blocked hat recorded and plotted on a graph. This is presented in Figure 4.

According to the data presented in the graph above, the moisture content of a blocked millinery hat placed in the drying chamber dried fast and significantly. From the data, there was an actual reduction of moisture content which was reflected in the weight of the hat from 15g to 9g in just under 25minutes. This is a result that can never be achieved using the conventional method of sun drying even on the hottest or sunniest days in Ghana.

STRUCTURAL DESIGN

The structure going to be put up was designed using AutoCAD software to develop a nice and accurate structure. The structure is in eight different faces. The designs have been well structured to suit its purpose. These faces are the cross-section areas showing clearly how the design of the structure will be.

MATERIALS NEED FOR THE CONSTRUCTION OF THE DRYING CHAMBER

The materials needed are:

Dryer: a drying equipment, especially for thick material such as felt.

Galvanized sheet/Logging sheet: it is fiber glasses that will be used to fabricate the body of the dryer.

Screw: there are small nail like metal, which will be used to join the pieces together.

Spray: A paint that will be used to cover the body of the dryer.

Heater: It will be connected to the heat as welding electrodes with materials together.

Insulator: They are going to serve as a thermal barrier to the heat getting out.

Figure 5. Galvanized structure showing the back and the front with the fixed insulators

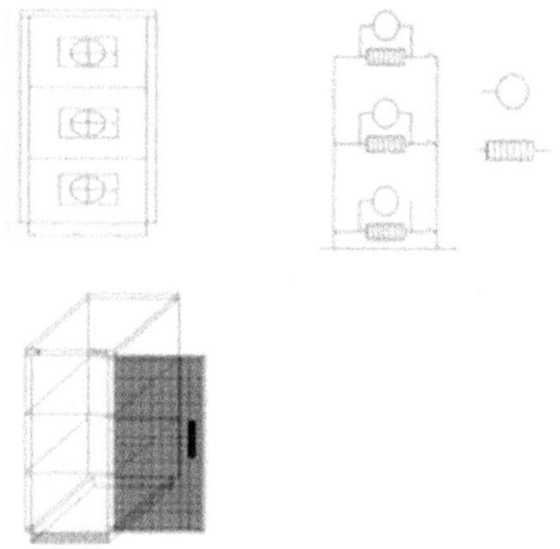

Figure 6. Blocked hat being tested in the new drying chamber

CONCLUSION

The study sort to design and construct a drying chamber for drying hats. The study was to manufacture millinery and accessories drying chamber. The result of the study indicated that oven or the drying chamber is the most effective equipment for drying hats. The research also proves that sun drying could be one of the ways but without any doubt, suggests that chamber drying is most effective and quickest way to drying hats. The dryer constructed could dry the hats effectively and efficiently. It is recommended that the drying chamber be used in other institutions. It is also evident that the most used method of drying millinery hats and accessories is the sun drying method. This study also compared the sun drying with the drying chamber method. The design and construction of millinery and accessories drying chamber is aimed at the rate of drying of hat and fascinators. Also, to produce efficient drying chamber to reduce the length of time it takes for blocked hats and accessories to dry.

In conclusion, the drying chamber can perform better and dry hats faster. The study revealed that the chamber drying is better than the sun drying, this because the chamber can perform multiple tasks in that, it can be used for different purposes in terms of drying. The demographic information for the study revealed that the field of millinery and accessories is mostly dominated by females. Secondly, the study shows that respondents also have moderate experience in the fashion industry judging from the time span they have been in it.

In the light of the above factors the following recommendations were made to address the issue of sun drying and chamber drying. That there should be a policy to make all fashion schools, vocational and technical school to have at least one chamber dryer in the schools. The researcher recommends that policy makers should review the curriculum to suit the current demand for millinery and accessories business. Though some tertiary offer fashion in their institutions, this research has revealed that they do not factor in millinery and accessories in the curriculum. The researcher recommends that all tertiary institutions that offer fashion must include millinery and accessories in the curriculum. Finally, the researcher recommends that there should be a policy from the government agencies (NCTE) to improve upon the curriculum in millinery and accessories.

The following are recommendations for further research: The researcher also recommended for further studies that solar power can be a source of energy for a new model drying chamber. Finally, it is recommended that for further study, a timer could be fixed inside the drying chamber as well as the switch in front of the chamber.

ACKNOWLEDGMENT

This work appreciates all who supported us in writing this paper.

REFERENCES

Feldman, E. (1992). Fashions of a 1990s. New York: Facts on File.

Grass, M. N. (1955). *History of hosiery. 2*. Fairchild Publishing Company.

Haghi, A. K. (2005). *A study of heat and mass transfer in porous material under equilibrium conditions.*

Hekmat, D., & Fisk, W. J. (1984). Improving the energy performance of residential clothes dryers. The *35th Annual International Appliance Technical Conference, Ohio State University. History of Millinery Hat styles of the Past*. Millinery Techniques. http://www.millinerytechniques.com/ history-of-millinery.html

Kaur, P., Stoltzfus, J., & Yellapu, V. (2018). Descriptive statistics. *International Journal of Academic Medicine*, *4*(1), 60–63. doi:10.4103/IJAM.IJAM_7_18

Merceron, D. L. (2007). *Millinery in Vogue.* New York: Rizzoli International Publications, Inc.

Mujumdar, A. S., & Menon, A. S. (1995). *Drying of Solids: Principles*. Classification, and Selection of Dryers.

Pakowski, Z., & Mujumdar, A. S. (1995). Basic process calculations in drying. In A.S. Mujumdar (Ed.), Handbook of Industrial Drying. Marcel Dekker.

Sutanapong, C. & Louangrath, P.I. (2015). Descriptive and Inferential Statistics. *Inter. J. Res. Methodol. Soc. Sci.*

Millinery Techniques. (n.d.). *The Art of Millinery - A Guide to making a hat*. Millinery Techniques. http://www.millinerytechniques.com/ history-of millinery.html

Torb & Reiner. (2014). *A Brief History of Millinery*. Torb & Reiner. http://www. torbandreiner. com/Millinery Materials/history-of-millinery

Chapter 13
Fabrication of a Prosthetic Socket for a Transtibial Amputee:
The Sculptor's Contribution

Evans Kwadwo Donkor
https://orcid.org/0000-0001-5855-8718
Takoradi Technical University, Ghana

Fredrick Boakye-Yiadom
Takoradi Technical University, Ghana

Rex Ifelaja Akinruntan
Takoradi Technical University, Ghana

ABSTRACT

This chapter focused on sculptural fabricated prosthetic socket for a transtibial (below-knee) amputee. As part of this study, an in-depth examination of the special needs for lower limb prostheses with quality and attractiveness was conducted in Sekondi-Takoradi Metropolis. Qualitative inquiry employing studio-based and descriptive research designs was drawn from collecting data from ten (10) purposively selected samples consisting of two (2) Prosthetists, one (1) Orthotist, two (2) Laboratory technicians, one (1) Amputee, one (1) STMA official and three (3) Sculptors. Research instruments comprising personal interviews and direct observations were used for the on-site data collection for the study. Data were analysed using thematic content and input-output transformation model analysis tools. The results revealed that producing a prosthetic socket for the amputee increased the active workforce of the amputee. Prosthetic socket fabrication should be made a course-related programme in the sculpture technology department at Takoradi Technical University together with Supomu-Dunkwa health centre and STMA.

DOI: 10.4018/978-1-6684-9843-9.ch013

INTRODUCTION

The amputation of a limb is an irreversible intervention in the natural physiological makeup of a human being. Physical loss of a body part is caused by underlying disease, comorbidities, and concurrent injuries as well as the psychological instability of the victim. Cardiovascular complications and diabetes cause limb loss, increasing obesity, an ageing population and trauma (Dal Canto et al., 2019). Amputation makes ambulation extremely difficult for amputees (Finch, 2011). Much of amputee's ambulation is associated with transtibial prostheses of the upper and lower extremities (Sewell, Noroozi, Vinney & Andrews, 2000). The prosthesis is categorised into upper and lower prostheses (Bates, Fergason & Pierrie, 2020). The upper-extremity prostheses are used at varying levels of amputation: forequarter, shoulder disarticulation, trans-humeral prosthesis, elbow disarticulation, trans-radial prosthesis, wrist disarticulation, full hand, finger, partial finger and hand, whereas the lower-extremity prostheses provide replacements at varying levels of amputation (Bates et al., 2020; Chadwell et al., 2020). These include hip disarticulation, trans-femoral prosthesis, knee disarticulation, transtibial prosthesis, Syme's amputation, foot, partial foot, and toe (Bates et al., 2020; Chadwell et al., 2020; Krebs, Edelstein & Thornby, 1991).

Grosz (2003) posits that there are two main subcategories of lower extremity prosthetic devices, namely trans-tibial (any amputation transecting the tibia bone or a congenital anomaly resulting in a tibial deficiency) and trans-femoral (any amputation transecting the femur bone or a congenital anomaly resulting in a femoral deficiency). Lower limb amputations are considerable among 805,426,000 people with locomotor disability (Oosterhoff, Geertzen & Dijkstra, 2022; Spoden, Nimptsch & Mansky, 2019). Most members of this group are young, active-earning males (Finch, 2011). Ambulation with prostheses is considered a basic need and a significant means of survival (Lee, Winson, Zhang & Ming, 2005). Even though passive prostheses replace missing body parts to a high degree and improve patients' independence and mobility, those in the lower-income bracket in developing countries, especially Ghana, still have strong feelings about their inability to procure prostheses for their livelihoods. Hence, in Ghana, especially those with transtibial cases (legs missing below the knee), amputees are selective as to what they consider appropriate to assist them in ambulation (Pollock, 2015). Amputees experience activities of limitation and participation restrictions. Typical activity limitations and participation restrictions for lower extremity amputees relate to self-care activities and mobility (Gallagher et al., 2011). These affect the person's ability to return to and maintain work, maintain social relationships, participate in leisure activities and be active members of the community (Brigham and Women's Hospital (BWH), 2011). Environmental factors such as barriers in the community related to physical/structural environments and personal factors such as age, gender, level of education, and ability to adjust restrict participation in normal social roles for lower extremity amputation (Hawkins et al., 2016; BWH, 2011).

Although there is a growing interest in research into transtibial prostheses to support amputees in Ghana, evidence suggests that research on transtibial prostheses in Ghana is scarce. Romm (1989) even asserts that transtibial prosthesis designs remain virtually unexplored in sculpture, notwithstanding their obvious centrality to the sculptural experience. Research on transtibial prosthetic socket preferences, however, focuses on Western countries with prosthetic studies. Africa, including Ghana, produces few transtibial prostheses socket preferences (Stevens, DePalma & Wurdeman, 2019; Davenport et al., 2017). Despite the central role of transtibial prostheses sockets in ambulation experiences, the interface between prostheses and sculpture in Ghana has received scant research attention. Existing studies, including Aduayom-Ahego (2020) and Aduayom-Ahego and Ehara (2016), focus on the position and challenges in prosthetics and orthotics education in Ghana. Specifically, Mock, Quansah, Addae-Mensah and Donkor's (2005) study investigated the guidelines and importance of prostheses for amputees in Ghana, whilst Tabur et al. (2015) examined and tested limb prostheses. None of these studies focused on the fabrication of prosthetic sockets for a transtibial amputee and the sculptural processes involved in the fabrication. These constitute knowledge gaps that need to be filled.

The study's motivations are threefold: first, the study addresses one of the gaps in sculpture, thus the production of transtibial prostheses in Ghana. Although Ghana has promoted sculpture in international exhibitions for many decades, there is no detailed information on the production of transtibial prostheses in sculpture. This study makes valuable contributions to the amputee situation by highlighting transtibial prosthetic sockets made from this study in Ghana. Second, as the first description of the locally made transtibial socket by sculptors' experience among Ghana's amputees, this study is intended to provide a strong point for more in-depth studies of the unique qualities and attractiveness of transtibial prosthesis sockets at Takoradi Technical University, Ghana. Finally, an insight into the production of transtibial prosthesis sockets by Takoradi Technical University is vital to maximising the potential and job avenues for sculpture students at the university. As a result, this study sought to sculpturally fabricate a prosthetic socket for transtibial (below-knee) amputees for ambulation. As a case study intervention, the study investigates an amputee situation in the Sekondi Takoradi Metropolis and fabricates a quality, attractive and fit-for-purpose lower limb prosthesis for one of the amputees.

Conceptual Framework

The study adopted Faustini et al.'s (2006) socket design and fabrication framework conceptualised from five stages of socket design. The surface model and residual limb geometry are developed through laser scanning. Faustini et al.'s conceptual framework also embodies computer-aided design model development using appropriate design constraints, Finite Element Modelling (FEM) analysis of the socket related to boundary conditions, Selective Laser Sintering (SLS) fabrication, and experimental validation of the FEM analysis results (Faustini et al., 2006). As a workability test for this study, the socket design and fabrication framework by Faustini et al. (2006) was integrated in the prosthetic design, measuring the residual limb geometry, production of the socket, alignment and prosthesis ready to be used for the study. Post-surgical management and Thomas test analysis were used as a case study for a transtibial (below-knee) amputee. Several factors were considered in producing a socket, such as scientific evaluations, the possibility of economic production, actual living conditions of a person with disabilities, and disability (Poonekar, 1992; Gh. Pirouzi et al., 2014; Aydın & Okur, 2018).

Using Faustini et al.'s (2006) conceptual framework, the amputee's history such as the cause of his amputation, weight, and age, was established. The data on the amputee's ambulation time and distance covered are factors in the design of the prosthetic socket for the amputee. In this sense, the amputee needs a prosthesis to ambulate. This statement supports Grosz's (2003) and Romm's (1989) assertions that it is essential to ascertain the ambulation of an amputee before any prescription is administered. The researchers, before the design consideration stage, discussed with the amputee the following parameters:

1. Does the person with an amputation want to walk?
2. Will it be possible for the person with an amputation to walk again?
3. Does the amputee have any underlying medical condition that accrued to the amputation?
4. Where does the amputee want to walk?
5. Will prosthetic rehabilitation improve the quality of life of an amputee?

After the amputee's assessment, the researchers considered the design parameters shown in Figure 1. It describes the general perspective of the study and ends by stating the study's objectives.

Figure 1. Conceptual framework for prosthesis socket design, analysis and fabrication (Faustini et al., 2006)

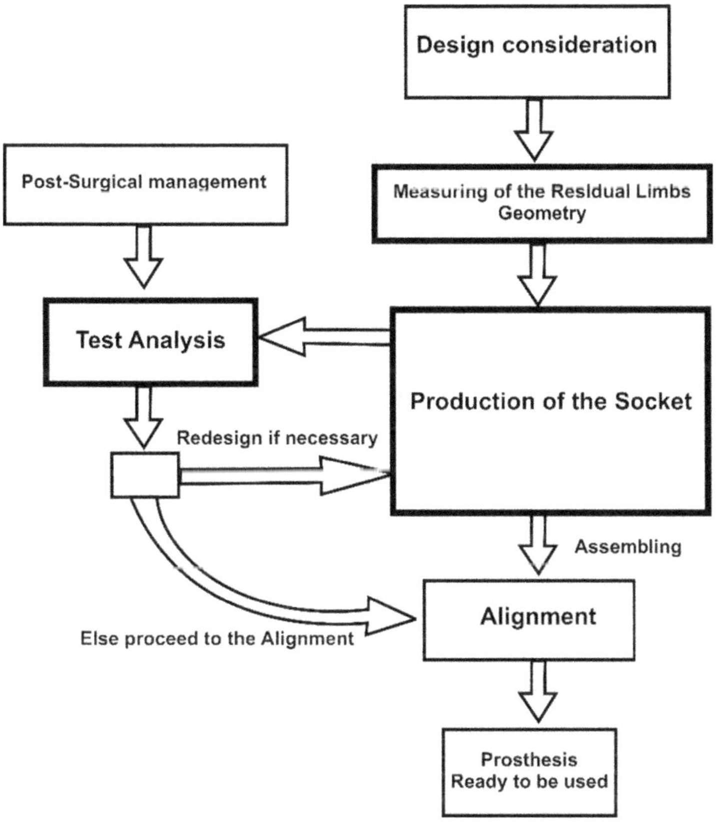

MATERIALS AND METHODS

The qualitative approach was adopted for the study where the researchers elicited opinions from the research participants. The research population was made up of prosthetists, orthotists, laboratory technicians, an amputee, sculptors and an official from Sekondi-Takoradi Metropolis Assembly (STMA) in the Western region of Ghana. The sample size for the study was 10, comprising two (2) Prosthetists, one (1) Orthotist, two (2) Laboratory technicians, one (1) Amputee, one (1) STMA official and three (3) Sculptors. These groups of people were purposely sampled due to their expertise and leadership in the association. These experts have first-hand information about their group. The study employed the expert-type purposive sampling technique to select respondents, as it is most likely to yield appropriate information (Kelly, 2010). Respondents' opinions were solicited through personal interviews. It was conducted face-to-face because the researchers wanted respondents to freely express their opinions on issues concerning the study. Another reason for doing the face-to-face interview was that aside from its added advantage of providing a physical encounter between the interviewer and the interviewee, it was also flexible, interactive, and generative, resulting in an in-depth understanding of the situations under study. The researchers used direct observation to observe the amputee's ambulation. The visual analysis tool provided detailed descriptions of some specific visual displays at the studio stage on fabrication processes of prosthetic sockets for transtibial amputee. In doing so, the study analyses the visuals and describes the elements in detail by making the visuals or images stand out (Leeuwen & Jewitt, 2004). The Thomas test ensured that the researchers assessed the length of the hip flexor muscles that differentiated between tightness in single-joint and two-joint muscles in the amputee. The length of the hip flexors was directly associated with the range of motion achieved at the hip and knee joints (Physiopedia, 2023; Kendall et al., 2005).

Research Designs

The study employed studio-based (Aesthetico-action) and descriptive research designs where the study focused on techniques, tools, materials and processes for the execution of a prosthetic socket for a transtibial amputee. The descriptive research design was chosen to describe the systematic and sequential way of executing the work (Salkind, 2010). It was also used to describe the documentation of studio results where the studio-based (Aesthetico-action) research design was employed as a detailed instructional explanation of studio ideas, thoughts, techniques and processes. It was used in the construction of the prosthesis as a cyclic, open-ended, active process that involves observing, reflecting, planning, acting, producing and evaluating the finished product (Marshall, 2010). All these processes of studio-based inquiry were captured in the pre-intervention, intervention and post-intervention phases of the study's findings and discussion.

Technique

The study employed modelling and casting techniques in sculpture to produce the prosthetic socket. The information recorded about the materials used in this prosthesis work was systematic and sequential. Identifying techniques for the materials chosen by the artists or researchers was critical to understanding the work and how it was created. For example, it was imperative to know the name of the material and how it was used to create the prosthesis for the amputee. The description of the technique provided an additional clue to understanding the process of fabricating a transtibial prosthesis and identifying the researchers' materials. Sculptors employ four primary techniques in their craft. They encompass carving, casting, modeling, and assembling (Getty, n.d; Walliman, 2001). Generally, processes are classified as subtractive (involves removing/carving out material) or additive (involves adding material). Casting is a technique where sculptures are created by pouring melted or mixed material, typically metal or cement, into a mold. The mold is then allowed to cool and harden, making casting an additive process. On the other hand, modeling involves building up and shaping a soft or malleable material, such as clay to form a sculpture. It is often supported by an armature. Modeling is also an additive process (Getty, n.d). In this particular study, the researchers chose modeling and casting techniques as additive processes to create a prosthetic socket for the transtibial amputee in their studio-based research.

Tools and Materials

The sculptural fabrication process of the prosthetic socket involved the utilisation of various tools and materials. These included examination bed, bench, knife, router, calliper, measuring tape, wire mesh, file, Allen or hex keys, screwdrivers, pencil, grinder, sandpaper, air compressor, box filled with sand, plastic bucket, measuring can, metallic pipe, heat gun, cutter, hand drills, Plaster of Paris powder (P.O.P), Plaster of Paris bandage, resin, fiberglass, hardener, stockinet, talcum powder, thermoplastic sheet, crepe bandage, blue colour, half-inch nail, liner, P.V.A plastic, pylon, zodee (colour), distal padding, plaster zote, glue, perlon, thinner, ML gauge, stump sock, and cotton sleeve. These tools and materials were utilised throughout the process to create the prosthetic socket.

Ethical Considerations

Research participants received verbal information regarding their consent and the purpose of the study. It was necessary to obtain the amputee's consent before using him in the study. The study sought permission to capture and incorporate his photographs at all stages of the prosthetic socket fabrication process. For the study, the researchers obtained permission from the administration office of the Supomu-Dunkwa Health Centre. The Supomu-Dunkwa Health Centre is located near the villages of Beposo and Badukrom in the Western Region. The Centre is located 5 kilometers northwest of nearby Daboase, a town in the Western Region of Ghana and the capital of Wassa East District.

FINDINGS AND DISCUSSION

1. Pre-Intervention Stage (Pre-Studio)

An Amputee Situation in the Sekondi-Takoradi Metropolis

In the amputee situation, trauma was rated the highest against peripheral vascular disease (P.V.D.), tumour, infections, and congenital limb deficiency (J. Kwofie, personal communication, September 11, 2016). This finding was at variance with Sewell, Noroozi, Vinney, Andrews' (2000) assertion that most female amputees were a result of trauma as against other causes such as a tumour, infections, congenital limb and peripheral vascular disease (P.V.D.) but consistent with Lee & Zhang (2005) report that 90% of amputees in Africans are as a result of trauma. It was found that the amputee lost his job. He was stigmatised and not considered for societal positions. Some amputees in developing countries, especially West Africans have no jobs (Aduayom-Ahego, 2020; Aduayom-Ahego & Ehara, 2016; Mock et al., 2005). This situation is an affront to section 29 subsections (1) and (3) of Ghana's 1992 constitution, and section 14 subsection "E" of the labour act of Ghana (2003). This demands equal treatment regardless of one's deed, social status, religion and political affiliations as far as job acquisition and societal positions are concerned.

In dealing with the amputee, he used a crutch to ambulate. Again, it was found that the participant had less knowledge of the transtibial prosthesis. The need for a prosthetic socket was discussed with the amputee. The participant was still afraid of losing his entire affected leg. This result does not support Romm's (1989) comments about easy ambulation; the amputee fits into prostheses rather than any other means such as sticks and crutches.

Post-Surgical Management

In fabricating a prosthesis for the amputee, the history such as the causes of his amputation, weight and age was established. The cause of the participant's amputation was severe infection (diabetes), 66 years of age and 78kg in weight. He was made to ambulate for five (5) minutes, and the distance covered was 20 metres without a prosthesis. The result indicates that an amputee needs a prosthesis to ambulate. This result supports Grosz's (2003) and Romm's (1989) assertions that it is essential to ascertain the ambulation of an amputee before any prescription is administered. This finding agrees with Torburn et al. (1995), that the amputee's medical history be known because it helps avoid any unnecessary injury that might aggravate the patient's condition.

2. Intervention Stage (Studio)

Sculptural Fabrication Processes of Prosthetic Socket for a Transtibial (Below-Knee) Amputee

Test Analysis

Thomas' test was run on the patient to examine how the amputation was done. The test was done to palpate the bones (tibia and fibula) (Physiopedia, 2023; Kendall et al., 2005). The bone had been cut at the fibula lower (0.3cm) than the tibia in the lower extremity below the knee amputation. The test aligns with Lee & Zhang's (2005) statement that the fibula grows with time, especially in young amputees. If they are cut equally, the patient will find it difficult to wear the prosthesis because it will grow taller than the tibia with time. The Thomas test was used to determine the abduction and adduction of the residual limb by establishing the strength of the muscles and how the socket should be made (Physiopedia, 2023; Kendall et al., 2005). This test confirmed Finch's (2011) statement that the Thomas Test is used to determine the strength and weakness of muscles. The researchers examined the amputee with their hands on the examination bed.

Measurement of the Body to Determine the Size Required for the Artificial Limb

The production of a prosthetic socket begins with capturing the geometry of the residual limb, which is called shape capture. This process accurately represents the residual limb, which is critical to achieving a good socket fit. Measuring the stump and the active leg was done using tape and a calliper. This measurement was done to determine the difference in the size required for the artificial limb LAL-LRL=PF.

Key: L.A.L. means the length of the active limb, i.e., from the floor to knee height, L.R.L. is the length of the residual limb, and P.F. is prosthesis fabrication. LAL=52 cm. LRL=12 cm. 52 cm-12 cm= PF=40 cm. The circumference of the residual limb (CRL=16 cm) was taken, and the active calf was 17 cm. The identification of the patient's patella, cordials and medial laterals was secured and marked for build-ups formation of the cast to ease ambulation with an indelible pencil. The researchers further traced the patient's active foot, which was measured 10 cm to select the prosthesis foot. The active small ankle measured 5 cm. This measurement was done to check the size of the prosthesis ankle.

Modelling of the Main Socket

56 cm of P.O.P. bandages were cut with scissors. It was soaked in a bucket containing 2000 ml of water (H20). This activity was done to make the bandage soak to model

Figure 2. Measuring the stump

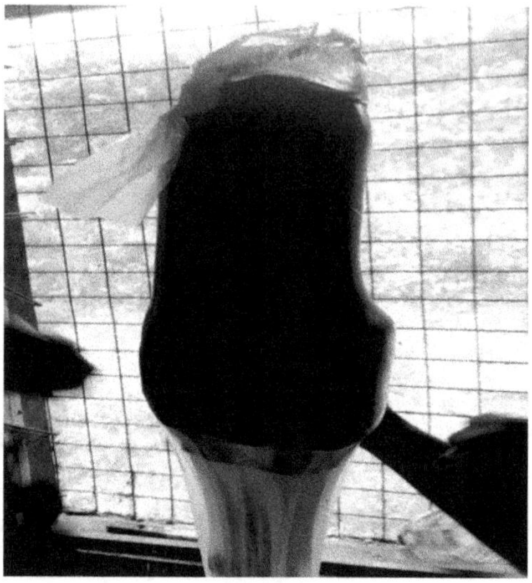

Figure 3. Negative cast from the stump

the socket. The P.O.P. bandage was kept in water for 5 seconds. It was then worn over the amputee's stump, pressed with the researchers' fingers to identify all the possible bony areas marked. This process was done to sit for 5 minutes after which it was taken from the residual limb. The mould took the shape of the residual limb or stump with all the markings on the patient's patella, condyles and medial laterals.

The mould or negative cast flare was then modified with a knife to arrive at the actual socket shape. Figures 2 and 3 show the P.O.P. plaster cast from the stump.

Casting From the Negative Stump

The mould was covered with an X-ray sheet to extend the height of the socket for the positive cast. 1600ml of powdered P.O.P. was dissolved in 2500 ml of H2O in a bucket. The mixture was stirred with a stick until all the P.O.P. dissolved completely without any lump seen. A metallic pipe of a height of 18 cm known as a mandrill was fixed onto the receiver mandrill to suck out the air, was placed into the mould and kept in a box of sand to stand firm to receive the mixture. The mixture was poured into the mould to the brim. A 5 cm extension was made to cater to the proximal cast. The mixture was left for 15 minutes to set. The cast was then taken out of the negative cast by dissecting the mould with a knife. The cast had all the markings for build-ups indicated in indelible ink. The markings were deepened for visibility with the indelible pencil. The cast was shaped and smoothed to the actual size of the measurement taken previously. The ends of the tibia bone, medial laterals and condyles on the cast were raised for build-ups with 4½ nails sent through the cast with a hammer. A mixture of P.O.P paste and two (2) teaspoons full of blue was used for the build-ups. The colour again showed where the build-ups have to be made. The build-ups were done to create space for the tibia bone, medial, lateral and condyles without hurting the stump by creating sourness on the patient's flesh if he wears the permanent socket. The distal and proximal build-ups were done with the surform blade or filed to appreciable levels with rough, smooth files and wire mesh.

Figure 4. Shaping of the cast from the negative

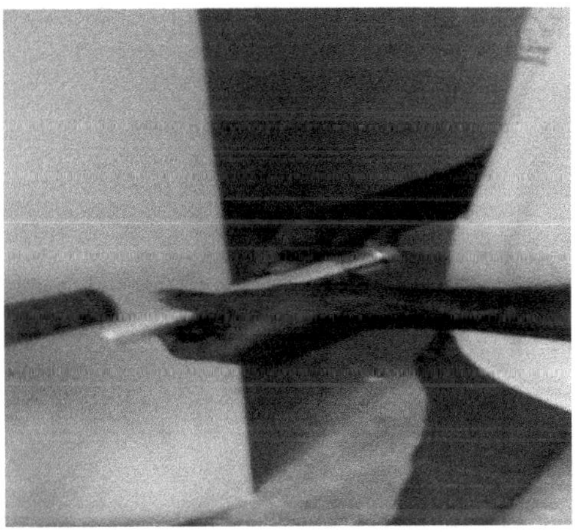

Figure 5. Cast build-up mould

Figure 6. Shaping of the cast with the router

Figure 7. Shaped plaster zote

The back of the positive cast also received a build-up where P.O.P paste was used to fill pores in the cast. It was then covered with P.V.A rubber to remove air trapped in the cast. This was done to free the semi-tendinosis, semi-menbrenosis and the biceps femora. If it is not done, the permanent socket will be rough, making it difficult for the patient to wear. The distal circumference was used to prepare the soft liner for the socket. The plaster zote was prepared and shaped with a router.

Formation of the Thermoplastic Sheet Around the Cast

The casts' proximal and distal circumferences and diameters were taken with measuring tape and a veneer calliper. They were 42 cm and 31 cm, respectively. These measurements were taken to prepare the thermoplastic sheet to fit the cast. The 42 cm was meant for the distal approximately 31 cm. The two long ends of the thermoplastic sheets were flattened by the thickness of the material of half an inch (½"). This activity was done by achieving a cylindrical shaped of the two long ends of the thermoplastic sheet with a brush and glue. The thermoplastic sheet mould was heated with a heat gun at 450 Degrees Celsius to soften it to fit onto the cast. Figure 8 shows the heated thermoplastic sheet. Talcum powder was applied to the cast as a releasing agent for the thermoplastic sheet mould. Again, P.V.A. plastic was fitted onto the cast to serve as another lubricant to remove the cast from the thermoplastic sheet mold socket easily. A thermoplastic sheet mould was fitted onto the cast, which was mounted onto the receiver mandrill. It was left to harden before the cast was taken and shaped for 4 minutes. The cast's thermoplastic sheet

Figure 8. Heating of thermoplastic sheet

Figure 9. Initial formation of the sockets

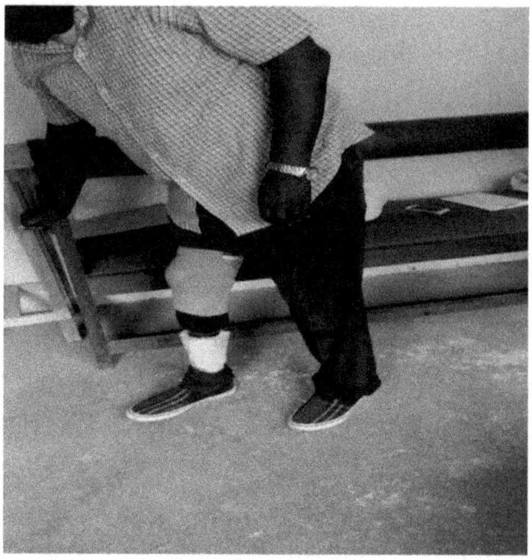

mould was tightened with a crepe bandage to ensure airtight. In the distal area of the flares, the thermoplastic sheet mould were shaped to receive the distal padding. Knife and router, as shaping tools, were used for the distal padding. The distal padding was mounted onto the thermoplastic sheet mould with glue application. At this point, a thermoplastic sheet liner was formed and put aside. The whole process was repeated, and this time, the thermoplastic sheet liner was left on the cast to form a permanent socket.

Formation of the Permanent Socket

A five (5) layer of perlon was added to the liner of the thermoplastic sheet mould to take the shape of the cast. Stockinet was also added to the perlon to give a permanent cast colour. This was closer to the body and skin colour. Another P.V.A. plastic height of 55 cm in cylindrical form was added to the perlon, the stockinet mould, and the socket with polyester resin. 2 teaspoons full of hardener and 7 ml of resin were added and stirred to form a mixture of resin and hardener. The mixture was then poured into the P.V.A. plastic mould to solidify to form the socket. The P.V.A, the perlon and the stockinet were knotted with a rope around the receiver mandrill. This method does not allow air to enter and leave. The distal padding area was tied, and the top was also tied together. The air compressor drained air from the socket. The whole experiment was left to set for 15 minutes whilst the compressor was on.

Removal of the Cast From the Socket

The distal area of the socket was filled with the usage of a router. The power drill was used to create a space to reach the cast in the distal area of the socket. The compressor tube was inserted into the space of the socket. The powered compressor was added to push the cast out of the socket. The cast is now a piece of cast that can be used at any given time. Specification for the socket was measured and shaped with a powered cutter and router. P.V.A. plastics were removed from in and out of the socket. A shuttle lock space was created in the center of the distal area.

Figure 10. Removal of the cast

Selection of Metal Parts of the Artificial Limb (Pylon, Shank, and Connectors)

The metal part creates distance and support between the knee joint and the foot (lower-leg prosthesis) or between the socket and the foot. The type of connector used between the Shank and the knee/foot determines the modularity of the prosthesis. Modular means that the foot's angle and displacement concerning the socket can be changed after fitting. In developing countries, prostheses are non-modular to reduce costs. Modularity for adults' age and height is essential because of their average growth of 1.9 cm annually (J. Kwofie, personal communication, September 12, 2016).

Figure 11. Shaping the socket

The modular adaptor and pylon with a single cramp were selected for the amputee. The patient wants the foot's angle and socket replacement after fitting over time.

Consideration of Type of Foot

Concerning the selection of a foot, an amputee was factored in. The patient's daily activities were considered, therefore, a soft ankle cushion heel (SACH) foot was selected for him.

Figure 12. Selection of the metal parts

Figure 13. Selection of the SACH Foot

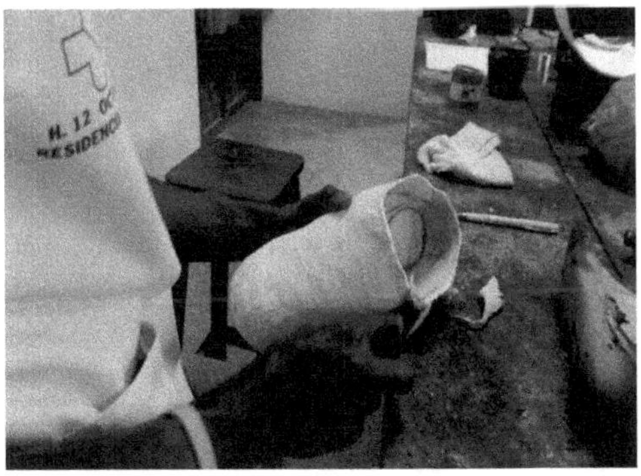

Assemblage of the Entire Limb With Bench Alignment

The consideration of three types of alignment was done at this point to factored in the assembly of the entire limb. The first type was bench alignment. The lateral side of the socket was equally shared between two parts. The ML gauge was used to measure the proximal and distal sockets. The foot was divided into three parts on the lateral side. The divisions were marked with a marker. The foot's posterior part was chosen and marked. The socket line was in line with the posterior mark. The flexion socket was tilled posteriorly to arrive at 5 degrees Celsius. Again, the posterior part of the foot was equally shared between the two parts. A 1cm gap was created from the posterior mark to help the toe out of the foot. The anterior part of the foot was divided into two equal parts. The division line passed through the toe break. Divisions were done with a marker. After bench alignment, the bolts were tightened together with an alien wrench as heights were considered. A thorough

Figure 14. Assembly of entire limb

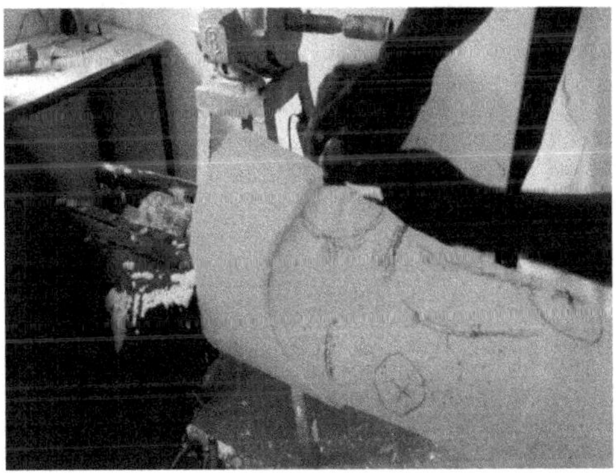

Figure 15. Bench alignment

examination of the residual limb was made to ensure its status. The examination showed that the stump was in good shape. The patient wore a stump sock. The liner around the stump sock was similar to the cotton sleeves on the liner. The stump sock held the residual limb in good shape.

Static Alignment

Static alignment was followed. This was where the patient wore the prosthesis. Observations of the amputee's demeanor were his facial expressions, standing position and the positioning of the prosthesis leg. The researchers engaged in a face-to-face interview with the amputee in line with how he felt about the prosthesis.

Figure 16. Static alignment

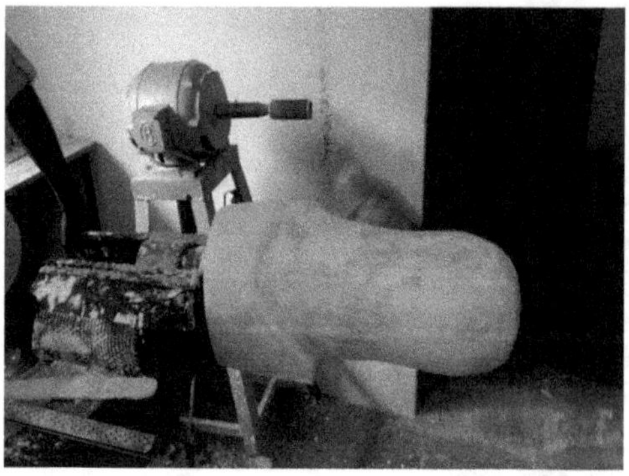

Figure 17. Static alignment

Dynamic Alignment

This was the stage at which the amputee began to ambulate. Again, the researchers observed gate deviations (such as lateral trunk bending, lateral thrust and medial thrust). This section was done for an hour. After the residual limp was examined, a check on the tire and skin bruises was done to ensure comfortability. After the final alignment, the shuttle, pylon and the upper part of the SACH foot were moulded using plaster zote and glue. It was covered again with a stockinet to hide the metal portions of the prosthesis.

Figure 18. Dynamic alignment

Figure 19. Dynamic alignment

Reinkensmeyer (2009) asserts that air pockets can allow sweat to accumulate which softens the skin. Over time, this can lead to skin breakdown. Ultimately, this is caused by frequent itching and skin rashes. On the other hand, a very tight fit may increase excessively the interface pressures that may also lead to skin breakdown after prolonged use. It can be very painful for an amputee. The researchers ensured that the fit between the residual limb and socket attachment was not too loose or tight. The socket checks were spot on at the contact of the residual limb and socket or a liner that supported easy ambulation. With this in the minds of the researchers, knowing that the optimal socket fit between the residual limb and socket is critical to the function and usage of the entire prosthesis, care was ensured.

3. Post-Intervention Stage (Post-Studio)

Responses From the Audience

Psychologically, the transtibial socket provides the amputee with functionally designed prostheses that fit below-knee and above-knee amputations (Gh. Pirouzi et al., 2014; Pollock, 2015). The design makeup of the prosthesis and residual limb interface provides comfort, appropriate load transmission, and efficient movement control (Aydın & Çağlar Okur, 2018). Based on the responses shared by some research participants including the test case transtibial amputee, the following responses were revealed:

Conversation 1:

I know it is expensive to buy both local and foreign prostheses depending on the type of artificial limb. I would prefer the prosthetic socket than wood sticks or metal crutches since the study makes it affordable with its fabrication processes.

Conversation 2:

I get scared any time I see fabricated or artificial limbs. How is it possible for someone to walk on this fabricated leg or arms?

Amputee:

It is better than none. My life is back. I feel comfortable walking on this artificial limb created for me. Now, I can work and generate money for my family and myself. Now, I can be with my peers, family and everybody. God bless you all. (Implied to researchers).

CONCLUSION

This study fabricates a sculptural prosthetic socket for a transtibial (below-knee) amputee. In an attempt, the study investigated an amputee situation in the Sekondi-Takoradi Metropolis and explored in-depth studies on the special need for lower limb prostheses of quality and attractiveness. It was evident from the studio-based inquiry that the amputee's amputation was caused by diabetes. Other causes of amputation were peripheral vascular disease (P.V.D.), trauma, tumours, infections, and congenital limb deficiency. It was envisaged that the research participant (amputee) was discriminated against in all social activities in the Sekondi-Takoradi Metropolis due to his challenge. The amputee used a wood stick for ambulation.

In addition, the study revealed that the prosthetic socket became one of the significant factors that reduced the amputee's burden. The fabrication of prosthetic sockets and assembling them for the amputee in the study provided a medium through which the amputee test case connected with his family, friends and everyone in Sekondi-Takoradi Metropolis and Ghana as a whole. Unlike other ambulation devices such as metal crutches and wood sticks, the Thomas test and a prosthetic socket for the transtibial amputee provided a robust connection between the amputee and his whole being. As it was established in the studio-based production processes of the fabrication of prosthetic sockets for transtibial amputee, the realisation of prosthesis has caused the transtibial amputee to pursue his dreams by working to make ends meet instead of begging for arms on the streets of Sekondi-Takoradi metropolis.

FUTURE RESEARCH DIRECTIONS

Implications to Supomu-Dunkwa Health Center, STMA, and Sculpture Technology Department, TTU

The psychology support unit at the health centre should restructure its programmes with contents that incorporate counselling of amputees on the streets on the importance of prostheses in ambulation. STMA must enforce disability-friendly building measures, while the human rights commission in the region should protect amputees' basic right to a decent job. STMA must move all amputees on the streets of the Metropolis and acquire prostheses for them through the disability fund from the Assembly's common fund from the central government. The disability fund would help procure locally fabricated prosthetic sockets as established in this study for the rest of the amputees. This effort would bring joy and hope to them and their families.

Additionally, the sculpture technology department, Takoradi Technical University together with Supomu-Dunkwa Health Centre should mount a programme to fabricate transtibial prostheses for amputees in the university's catchment area and beyond. A prosthesis programme would be a perfect fit for TTU's strategic plan and the Government of Ghana's TVET policy.

Funding Information

This research received no funding.

ACKNOWLEDGMENTS

The researchers are grateful to the management of these wonderful institutions namely; Supomu-Dunkwa Health Center, Department of Sculpture Technology, Takoradi Technical University, STMA-Sekondi-Takoradi-Western Region of Ghana and all the research participants involved in this study.

REFERENCES

Aduayom-Ahego, A. (2020). Prosthetic rehabilitation of multiple-digit amputations using silicone material in sub-Saharan African country Ghana. *The Pan African Medical Journal*, *36*, 357. doi:10.11604/pamj.2020.36.357.25361 PMID:33224423

Aduayom-Ahego, A., & Ehara, Y. (2016). Current position and challenges in prosthetics and orthotics education in Ghana. *Niigata Journal of Health and Welfare, 16*(1), 26–34.

Aydın, A., & Okur, S. (2018). Effects of test socket on pain, prosthesis satisfaction, and functionality in patients with transfemoral and transtibial amputations. *Medical Science Monitor, 24*, 4031–4037. doi:10.12659/MSM.910858 PMID:29897893

Bates, T. J., Fergason, J. R., & Pierrie, S. N. (2020). Technological advances in prosthesis design and rehabilitation following upper extremity limb loss. *Current Reviews in Musculoskeletal Medicine, 13*(4), 485–493. doi:10.100712178-020-09656-6 PMID:32488625

Brigham and Women's Hospital (BWH). (2011). *Standard of care: Lower extremity amputation*. The Brigham and Women's Hospital, Inc., Department of Rehabilitation Services.

Cairo, A. (2005). *Orthopaedic project in Afghanistan*. The International Committee of the Red Cross (ICRC).

Carroll, K., & Edelstein, J. E. (2006). *Prosthetics and patient management: A comprehensive clinical approach*. SLACK Incorporated.

Chadwell, A., Diment, L., Micó-Amigo, M., Morgado Ramírez, D. Z., Dickinson, A., Granat, M., Kenney, L., Kheng, S., Sobuh, M., Ssekitoleko, R., & Worsley, P. (2020). Technology for monitoring everyday prosthesis use: A systematic review. *Journal of Neuroengineering and Rehabilitation, 17*(1), 93. doi:10.118612984-020-00711-4 PMID:32665020

Dal Canto, E., Ceriello, A., Rydén, L., Ferrini, M., Hansen, T. B., Schnell, O., Standl, E., & Beulens, J. W. (2019). Diabetes as a cardiovascular risk factor: An overview of global trends of macro and micro vascular complications. *European Journal of Preventive Cardiology, 26*(2, suppl), 25–32. doi:10.1177/2047487319878371 PMID:31722562

Davenport, P., Noroozi, S., Sewell, P., & Zahedi, S. (2017). Systematic review of studies examining transtibial prosthetic socket pressures with changes in device alignment. *Journal of Medical and Biological Engineering, 37*(1), 1–17. doi:10.100740846-017-0217-5 PMID:28286462

Faustini, M. C., Neptune, R. R., Crawford, R. H., Rogers, W. E., & Bosker, G. (2014). An experimental and theoretical framework for manufacturing prosthetic sockets for transtibial amputees. *IEEE Transactions on Neural Systems and Rehabilitation Engineering, 14*(3), 304–310. doi:10.1109/TNSRE.2006.881570 PMID:17009490

Finch, J. (2011). The art of medicine: The ancient origins of prosthetic medicine. *Lancet, 377*(9765), 348–349. doi:10.1016/S0140-6736(11)60190-6 PMID:21341402

Gallagher, P., O'Donovan, M. A., Doyle, A., & Desmond, D. (2011). Environmental barriers, activity limitations and participation restrictions experienced by people with major limb amputation. *Prosthetics and Orthotics International, 35*(3), 278–284. doi:10.1177/0309364611407108 PMID:21937573

Getty. (n.d). 4 basic sculpture techniques. *The J. Paul Getty Museum.* http://www.getty.edu/education/teachers/classroom_resources/curricula/sculpture/background2.html

Greenberg, C. (2012, August 10). Oscar Pistorius, South African 4×400m relay team finish 8th as Bahamas wins gold. *Huffington Post.* https://www.huffpost.com/entry/oscar-pistorius-south-africa-relay-4x400-olympics_n_1765596

Grosz, E. (2003). Prosthetic objects. In I. Cheng, & B. Tschumi, The state of architecture at the beginning of the 21st Century (pp. 96-97). Columbia Books of Architecture.

Hawkins, A. T., Pallangyo, A. J., Herman, A. M., Schaumeier, M. J., Smith, A. D., Hevelone, N. D., Crandell, D. M., & Nguyen, L. L. (2016). The effect of social integration on outcomes after major lower extremity amputation. *Journal of Vascular Surgery, 63*(1), 154–162. doi:10.1016/j.jvs.2015.07.100 PMID:26474508

Kelly, S. (2010). Qualitative interviewing techniques and styles. In I. Bourgeault, R. Dingwall, & R. de Vries (Eds.), *The Sage handbook of qualitative methods in health research.* SAGE Publications. doi:10.4135/9781446268247.n17

Kendall, F. P., McCreary, E., Provance, P., Rodgers, M., & Romani, W. (2005). *Muscles: Testing and function with posture and pain* (5th ed.). Lippincott Williams & Wilkins.

Krebs, D. E., Edelstein, J. E., & Thornby, M. A. (1991). Prosthetic management of children with limb deficiencies. *Physical Therapy, 71*(12), 920–934. doi:10.1093/ptj/71.12.920 PMID:1946626

Lee, W. C. C., & Zhang, M. (2005). Design of monolimb using finite element modelling and statistics-based Taguchi method. *Clinical Biomechanics (Bristol, Avon), 20*(7), 759–766. doi:10.1016/j.clinbiomech.2005.03.015 PMID:15963612

Marshall, C. (2010). *A research design for studio-based research in art. Taylor and Francis Group*. L.L.C.

Mock, C. N., Quansah, R. E., Addae-Mensah, L., & Donkor, P. (2005). The development of continuing education for trauma care in an African nation. *Injury*, *36*(6), 725–732. doi:10.1016/j.injury.2004.12.044 PMID:15910824

Murdoch, G., Wilson, A. & Bennett, Jr. (1997). *A primer on amputations and artificial limbs*. Charles C Thomas Publisher, Ltd.

Oosterhoff, M., Geertzen, J. H. B., & Dijkstra, P. U. (2022). More than half of persons with lower limb amputation suffer from chronic back pain or residual limb pain: A systematic review with meta-analysis. *Disability and Rehabilitation*, *44*(6), 834–854. doi:10.1080/09638288.2020.1783377 PMID:32603198

Physiopedia. (2023). Thomas test. *Physiopedia*. https://www.physio-pedia.com/Thomas_Test

Pirouzi, G., Abu Osman, N. A., Eshraghi, A., Ali, S., Gholizadeh, H., & Wan Abas, W. A. B. (2014). Review of the socket design and interface pressure measurement for transtibial prosthesis. *The Scientific World Journal*, *2014*, 1–9. doi:10.1155/2014/849073 PMID:25197716

Pollock, A. (2015, May 20). *Amputees control bionic legs with their thoughts*. Reuters. https://www.reuters.com/article/us-iceland-mind-controlled-limb-idUSKBN0O51EQ20150520

Poonekar, P. (1992). *Prosthetics and orthotics in India, In prosthetic and orthotic research for the twenty-first century*. National Institute of Child Health and Human Development.

Reinkensmeyer, D. J. (2009). Robotic assistance for upper extremity training after stroke. *Studies in Health Technology and Informatics*, *145*, 25–39. PMID:19592784

Romm, S. (1989). Arms by design. *Plastic and Reconstructive Surgery*, *84*(1), 158–163. doi:10.1097/00006534-198907000-00029 PMID:2660173

Salkind, N. J. (Ed.). (2010). *Encyclopedia of research design* (Vol. 1). SAGE Publications, Inc. doi:10.4135/9781412961288

Sewell, P., Noroozi, S., Vinney, J., & Andrews, S. (2000). Developments in the transtibial prosthetic socket fitting process: A review of past and present research. *Prosthetics and Orthotics International*, *24*(2), 97–107. doi:10.1080/03093640008726532 PMID:11061196

Spoden, M., Nimptsch, U., & Mansky, T. (2019). Amputation rates of the lower limb by amputation level – observational study using German national hospital discharge data from 2005 to 2015. *BMC Health Services Research, 19*(1), 8. doi:10.118612913-018-3759-5 PMID:30612550

Stevens, P. M., DePalma, R. R., & Wurdeman, S. R. (2019). Transtibial socket design, interface, and suspension: A clinical practice guideline. *Journal of Prosthetics and Orthotics, 31*(3), 172–178. doi:10.1097/JPO.0000000000000219

Tabur, S., Eren, M. A., Celik, Y., Dag, O. F., Sabunca, Z. A., & Savas, E. (2015). The major predictors of amputation and length of stay in diabetic patients with acute foot ulceration. *Wiener Klinische Wochenschrift, 127*(1-2), 45–50. doi:10.100700508-014-0630-5 PMID:25398288

Torburn, L., Powers, C. M., Guiterrez, R., & Perry, J. (1995). Energy expenditure during ambulation in dysvascular and traumatic below-knee amputees: A comparison of five prosthetic feet. *Journal of Rehabilitation Research and Development, 32*(2), 111–119. PMID:7562650

Van Leeuwen, T., & Jewitt, C. (2004). *The handbook of visual analysis*. SAGE Publications., doi:10.4135/9780857020062

Walliman, N. S. R. (2001). *Your research project: A step-by-step guide for the first-time researcher*. SAGE Publications.

Wigley, M. L. (1991). Prosthetic theory: The disciplining of architecture. *Assemblage*, (15), 6–29. doi:10.2307/3171122

Wire, C. (2003). Creative metal clay jewelry: Techniques, projects, inspirations. Lark Books, Division of Sterling Publishing Co., Inc.

Chapter 14
Sustaining Small Enterprises Through Digital Technologies in the Post-COVID-19 Era

Elikem Chosniel Ocloo
https://orcid.org/0000-0003-4729-3093
Accra Technical University, Ghana

Isaac Sewornu Coffie
Accra Technical University, Ghana

Gretel Rejoice Mawunyo Asigbe-Tsriku
Accra Technical University, Ghana

Frank Frimpong Opuni
https://orcid.org/0000-0001-8794-6541
Accra Technical University, Ghana

ABSTRACT

Though studies have recently examined the effect of the coronavirus on micro and small enterprises (MSEs) and the role digital technologies could play in mitigating those effects, little attention is paid to which digital tool is likely to have the most significant impact on MSEs' performance. This notwithstanding previous findings that most MSEs are reluctant to embrace digitization due to the need for more evidence on what works best for them or digital tool to adopt first. From the resource-based view theory's perspective, the study sought to examine the most effective digital technologies for MSEs' survival post-Covid-19 era. Quantitative data from 375 MSEs in Ghana was analysed using PLS-SEM. The results show that though all four digital technologies examined in the study significantly influence the survival of MSEs in the post-Covid-19 era, the adoption of social media as a digital resource will have the most significant influence on their survival.

DOI: 10.4018/978-1-6684-9843-9.ch014

INTRODUCTION

The outbreak of the Covid-19 pandemic has affected the well-being and health of individuals, businesses and economies across the globe. Due to the devastating nature of the pandemic, governments around the world implemented stringent policies like lockdowns, closures of physical operations of businesses, and reductions in working hours in order to contain or slow the spread of the virus (Bai et al., 2021; Coffie & Hinson, 2021; Papadopoulos et al., 2020). These measures have negatively affected economies and businesses worldwide to the extent that their long-term viability and survival are in question. For instance, according to Eurostat (2021)'s, the Eurozone has recorded the sharpest fall in GDP since 1995 to 3.8% due to the pandemic. The US economy also contracted by 19.2% as of the second quarter of 2020, the lowest decline since 1946. Economies in less developed countries were also not spared. Ghana's economy, for instance, grew by 1.7%, the lowest since 1992 (World Bank, 2021). Despite several economic recovery programmes, many economies are yet to recover from the devastating effects of the pandemic.

A critical analysis of events by scholars also shows that, though the outbreak of the pandemic has severely impacted several MSEs, those in developing economies are the major victims of the pandemic and its containment measures (Bai et al., 2021; Kimuli et al., 2021; Tweneboah-Koduah & Coffie, 2022). In other words, the containment measures had a more severe negative impact on MSEs compared to larger and global businesses. This has been attributed to the limited attention these enterprises give to digital technologies (Kimuli et al., 2021; Papadopoulos et al., 2020). Thus, businesses that primarily operate on a face-to-face contact basis with customers, usually in dilapidated structures with little or no attention to digital technologies, are the businesses that were mostly affected by the Covid-19 pandemic. The World Bank (2021) report on the devastating impact of Covid-19, for instance, shows that about 85% of MSEs (mainly in the informal sectors) in most emerging economies have either closed down or are struggling to survive in the post-Covid-19 world. For instance, the findings of Oduntan and Isere (2022) provide evidence to show that businesses operating online were less affected by the pandemic and its containment measures. These findings suggest that digitisation is a significant way businesses, particularly MSEs in developing economies, could be sustained or protected against natural disasters like Covid-19. As stated by past scholars (e.g., Bai et al., 2021), the challenges presented by the Covid-19 pandemic have taught an important lesson that adopting technological tools, particularly by micro and small enterprises, is critical for their survival and growth. Papadopoulos *et al.* (2020), for instance, argue that digitisation is simply a matter of survival in today's competitive marketplace and the post-Covid-19 world. Moreover, businesses that are hesitant to embrace digitisation may struggle for survival. This is because, with digitisation, firms can conduct most of their business transactions with customers, suppliers or intermediaries online with little or no physical contact.

In addition to contributing to the survival of businesses, digitisation also brings many significant benefits to firms, such as reducing transaction costs by providing better and quicker access to information and communication between staff, suppliers and networks (Kimuli et al., 2021). It also makes it easier for businesses to access resources like government services, which are rapidly being provided online, as well as finance (like peer-to-peer lending), training, and recruitment channels. Additionally, it encourages innovation and allows businesses to produce data and analyse their operations in novel ways to boost performance (OECD, 2021). The other benefits of digitisation include the potential of expanding beyond geographical reach and acquiring new customers, better engagement with customers, and efficient customer credit management.

Despite the benefits and opportunities digital technologies bring, the review of the literature shows that though managers of MSEs recognise the importance of adopting digital technologies in this era of rapid technological change, only small number of these enterprises have considered or taken effort to digitise their operations, even after Covid-19 (OECD, 2021). For instance, a survey by Micro Save (MSC, 2021) in India, Kenya, Indonesia, and the Philippines shows that only 28% of small businesses have reported increasing their digitisation efforts, or considering digitising their operations despite the devastating effects of Covid-19 on their businesses. In other words, many micro and small enterprises continue to lag in digitisation or adoption of digital tools for their business operations. The digital adoption gap for MSEs compared to larger firms has widened over the last decade. The report by OECD (2021), for instance, shows that while digital adoption among small firms remains below 30%, large firms, especially in frontier countries, have shown rapid progress of about 80% or more. The report also shows that one of the major reasons why many MSEs are not embracing digitisation is the lack of the first bold step to adopt their first digital platform. Research has shown that most small enterprises are reluctant to adopt digital technologies due to lack of resources and evidence on what works best for them (Alvarenga et al., 2023; Bai et al., 2021). The work of Oduntan and Isere (2022) for instance, shows that about 75% of business owners in Nigeria reported a lack of knowledge about which digital technology would yield maximum benefit or which tool to adopt first since they lack the financial capacity to embrace all forms of digital technologies at a go. These findings, for instance, confirm OECD's (2021) report that most MSEs would be willing to embrace digitisation if they know which digital tool is best for them and which tool to start with. Thus, identifying what works best for MSEs, given their unique characteristic of financial constraints, is likely to encourage the adoption of digital technologies by MSEs. Scholars (e.g., Cueto et al., 2022; Oduntan and Isere, 2022) have explicitly called for empirical evidence on digital tools that are likely to have the most significant influence on the survival, growth and performance of small businesses. Others (e.g., Bai et al., 2021) have suggested that given the resource constraints of most MSEs, there should be a gradual adoption of digital tools by these small enterprises and called for studies to identify which tool to start with.

Despite these calls, studies examining the impact of the pandemic on MSEs and the role of digital technologies in mitigating this impact have been given limited attention, especially in the identification of which digital tools work best for MSEs (Bai et al., 2021; Papadopoulos et al., 2020; Alvarenga et al., 2023). In other words, the question about which digital tool or platform MSEs should adopt first to encourage or drive their digitisation process largely remains unanswered. As Raghavan *et al.* (2018) noted, the initial adoption of a digital tool is likely to speed up the digitisation of MSEs, as it becomes a catalyst or a strong motivator for further adoption of other digital tools. Additionally, given the resource constraints of most MSEs, identifying what works best will also save managers of these businesses scarce resources and increase their willingness to adopt other digital tools. Therefore, from the resource-based view theory's perspective, the current study sought to determine the most effective digital technology for MSEs' survival during and after crises like the Covid-19 pandemic and should be adopted first by MSEs. The survival of MSEs is crucial, given their contributions to national development and job creation. For instance, in most African countries, MSEs are a major driving force for sustainable social development as they constitute over 90% of businesses that provide income for a large number of vulnerable and disadvantaged groups (UNCTAD, 2021; Bai et al., 2021). In Ghana, MSEs contribute about 37% of GDP and employ about 65% of the workable population (Bai et al., 2021). Thus, the survival and continuity of these businesses are critical to national development and growth.

LITERATURE REVIEW AND RELATED WORK

According to the Ghana Enterprise Agency (GEA), micro and small enterprises are businesses with ten or fewer employees and working capital, excluding land and building not exceeding $25,000. About 90% of the private sector's workforce is employed by these businesses and contributes about 37% of Ghana's GDP (Zumanu, 2019). Despite these contributions, MSEs are the most vulnerable to uncertain and unforeseen situations such as Covid-19. They require targeted sustainable efforts such as investment in digital technologies to drive their growth and survival (Zumanu, 2019).

The Covid-19 pandemic has altered market conditions and consumer behaviour, forcing businesses to innovate and adopt strategic approaches to survive in this crisis era and beyond (Alvarenga et al., 2023; Papadopoulos et al., 2020). In other words, as shown by past studies, Covid-19 has altered consumers' habits and tastes for convenience (Zumanu, 2019). Therefore, businesses need to be market-oriented to adjust to the changing habits of consumers in recent times to remain competitive (Coffie & Hinson, 2021). The adoption of digital technologies has become a critical survival strategy for businesses to catch up with these changing habits, especially MSEs operating in less developed economies (Alvarenga et al., 2023). Prior to the COVID-19 pandemic, reports indicate that online presence, social media presence, or e-commerce actions were not considered a core business priority for MSEs, particularly those operating in less developed economies (UNCTAD, 2019). It has however, become a necessity for survival in the new post- Covid-19 era and steps are needed to encourage MSEs, particularly those operating in less developed economies like Ghana to digitise their operations.

In the new post-Covid-19 world, taking advantage of digital technologies could help MSEs access markets, advertise their businesses, and reach out to customers locally and globally (Papadopoulos et al., 2020). The use of digital technologies has been identified as a critical tool for limiting the impact of the natural crisis as presented by the Covid-19 pandemic (Papadopoulos et al., 2020). For instance, Alvarenga et al. (2023) have shown that using digital technologies limited the impact of the pandemic on large organisations that do business online compared to those with a limited online presence. For MSEs in developing countries like Ghana, to survive in the uncertain business environment that has constrained global business activities, digital technologies must be integrated into their operations.

THE RELATIONSHIP BETWEEN DIGITALIZATION AND SUSTAINABILITY OF BUSINESSES

In the context of this study, sustainability defined as the ability of businesses to adapt, thrive, and endure in a constantly evolving and challenging environment (Alvarenga et al., 2023; Bai et al., 2021). Digitalization plays a crucial role in enhancing the sustainability and survival of businesses, particularly micro and small enterprises. Firstly, digitalization enables operational agility and efficiency, allowing businesses to respond quickly to market changes and optimize their resources. By automating processes, implementing digital tools for inventory management and supply chain optimization, and utilizing data analytics, businesses can streamline operations, reduce costs, and improve productivity, ultimately increasing their chances of survival (Zumanu, 2019).

Secondly, digitalization expands market opportunities and customer reach. Through websites, online stores, and digital marketing channels, businesses can access a broader customer base and target specific market segments. This expanded market reach opens doors to new revenue streams, diversifies customer bases, and reduces reliance on traditional physical locations, which is particularly valuable during times of disruption or restricted physical access (UNCTAD, 2021; Bai et al., 2021). Thirdly, digitalization enables businesses to adapt to changing consumer behaviours and preferences. By embracing e-commerce platforms, businesses can offer convenient and contactless purchasing options, personalized customer experiences, and efficient delivery services. This adaptability to digital channels is crucial for businesses to remain competitive and meet customer expectations in an increasingly digital-centric world (Raghavan *et al.,* 2018).

Lastly, digitalization facilitates data-driven decision-making and strategic planning. By leveraging data analytics, businesses can gain insights into market trends, customer behaviour, and operational performance. These insights inform informed decision-making, allowing businesses to identify new growth opportunities, mitigate risks, and adjust their strategies to align with market demands (Bai et al., 2021).

Digitalization is a vital enabler of business survival and sustainability. By enhancing operational efficiency, expanding market reach, adapting to changing consumer behaviours, and leveraging data-driven insights, businesses can navigate challenges, seize opportunities, and ensure their long-term survival in an increasingly digital landscape. Embracing digital technologies is essential for businesses, especially micro and small enterprises, to remain agile, competitive, and resilient in the face of uncertainties and disruptions.

THEORETICAL FOUNDATION AND HYPOTHESIS DEVELOPMENT

This study draws on the firm's resource-based view (RBV) theory to explain the importance of MSEs' adoption of digital technologies as a strategic tool for survival during and after the pandemic (Barney, 1991; Raghavan et al., 2018). According to the RBV, businesses can gain an advantage over their competitors by utilising valuable, rare, unique, and non-replaceable resources. Assets, knowledge, capabilities, skills, and relations are examples of these resources (Barney, 1991; Khayer et al., 2021). As explained in the resource-based view theory, the resources of MSEs could play an essential role in generating competitive advantage and superior performance.

The RBV's central concept is that the availability and creative utilisation of a firm's resources and competencies that are valuable, rare, perfectly inimitable, and non-substitutable will enable it to transform into excellence (Raghavan et al.,

2018). In other words, the RBV emphasises the relationship between strategy and a firm's internal resources. The resources of a business can be internal (or inside-out) or external (or outside-in). Internal resources are those that the firm owns and controls, such as financial, human, physical, and technological resources. In contrast, external resources are those that can be earned and controlled, to a degree depending on factors such as industry attractiveness and structural autonomy, but are not necessarily owned by the firm, such as customers, competitors, and suppliers (Raghavan *et al.,* 2018; Khayer et al., 2021). Thus, from the RBV's perspective, MSEs' adoption and utilisation of digital technologies is an internal resource that could provide valuable, rare, inimitable resources that could sustain or lessen the devastating effect of natural crises like Covid-19 on their survival.

In general, digital tools are regarded as a more robust internal firm resource than an external firm resource for establishing competitive advantage (Raghavan *et al.,* 2018; Khayer et al., 2021). This could be attributed to the chain or multiplier effect of having an information advantage, which facilitates the creation of other competitive advantages, such as cost and differentiation (Khayer et al., 2020). More importantly, maintaining digital systems as a competitive advantage is challenging but rewarding. It enables firms to detect and respond to market opportunities and threats (Raghavan et al., 2018).

Though past studies have used Capability Theory and Technology, Organisation and Environment to explain the importance and adoption intention of digital technologies by small businesses and found significant results, the current study seeks to explain the phenomenon from the perspective of RBV. As shown by past studies, we argue that digital technologies in this era of 'new normal' could become an important resource for gaining competitive advantage and survival of small businesses during and beyond crises, as presented by Covid-19. The following section discusses how the various digital technologies could be a resource for MSEs' survival.

ONLINE STORE/WEBSITES AND SURVIVAL OF MSES

An online store is an internet platform or website that allows visitors or consumers to shop without human interaction and make payments online or upon delivery. On the other hand, a website is a simple platform that only allows visitors to access information and make requests, such as subscribing to a newsletter or making contacts. The adoption of websites and online stores, in particular, has become a critical component in the survival of businesses, especially MSEs in less developed economies due to the pandemic. The findings of Gu et al. (2021) provide evidence of the shifting influence of online consumer purchasing behaviour as consumers have experienced more online buying during the pandemic. For instance, a survey by UNCTAD (2021) in nine emerging and developed economies shows that consumers in emerging economies have made a significant shift in purchase behaviour by shifting to online shopping. The survey also shows that “during the pandemic, online consumption habits of consumers have changed significantly, with a greater proportion of internet users buying essential products, such as food and beverages, cosmetics and medicines online.” The International Telecommunications Union’s (ITU) report on internet penetration and e-commerce shows that as of March 2020, the percentage of online shoppers in Ghana has increased from 18.3% in 2018 to 27.5% out of the 14.76 million internet users in Ghana (ITU, 2020). The findings of Khayer et al. (2021) also show that Nigerian consumers have also embraced online shopping with a 44% increase in online shopping among retailers, especially among consumers living in urban areas. This shows that the Covid-19 pandemic has accelerated the shift toward a more digital world. These changes are more likely to have a lasting effect as economies worldwide begin to recover. These reports and studies have shown that due to Covid-19, the importance of adopting digital technologies should be treated as a necessity with website creation and hosting as no exception. Based on this discussion, the following hypothesis is proposed.

H1: Online store/website will positively and significantly influence MSEs’ survival during and after the pandemic.

Social Media Platforms and Survival of MSEs

Known as “user-generated communication,” social media have reformed the tools and strategies employed by several organisations in their communications with their customers, who wield power in terms of information control (Etim & Daramola et al., 2020). In the face of challenges presented by the Covid-19 pandemic, scholars (e.g., Gu et al., 2021) have argued that embracing social media marketing is critical to the survival of businesses, especially MSEs operating in developing economies. Embracing social media channels with billions of users offers MSEs more opportunities for growth by enabling them to reach more customers than traditional mass media.

The unique characteristics of social media, such as compatibility with most devices with internet connectivity and cost-effectiveness in reaching large customers, its adoption has become a critical tool for MSEs to establish close relationships with customers (Gu et al., 2021). In other words, social media marketing has emerged as a cost-effective tool for businesses to enhance their brand image and customer relationships. Thus, digital media such as social media platforms has become one of the most effective and efficient choices for communicating with communities and customers. Given this, it seems unfathomable that MSEs would not use this technological advancement, which has greater advantages than traditional mass media in terms of cost, reach, and interaction (Gu et al., 2021). The findings of Etim and Daramola et al. (2020) have shown that with limited resources, SMEs could circumvent challenges that impede growth and survival by using social media platforms. In addition, Odoom *et al.* (2017) show that social media provide a good marketing platform for small and medium enterprises (SMEs) in Ghana by providing access to huge audiences at a very low cost.

Facebook, LinkedIn, Twitter, Instagram, WhatsApp, Viber, WordPress, YouTube, Flikr, Google+, Pinterest, Quora, and Snapchat are among the most popular networking sites (Gu et al., 2021; Etim & Daramola et al., 2020). For entrepreneurs, these platforms give them a wider audience, a better understanding of what they want and need, and innovatively collapse time and space between creating products or services and greater customer engagement (Odoom et al., 2017; Etim & Daramola et al., 2020). Market intelligence from social media could fill the MSEs' resource gaps and reduce uncertainty by using their market knowledge, resource-matching abilities, and learning and marketing capabilities in real-time to fill the gap and reduce uncertainty.

As a result of the new normal which is triggered by the Covid-19 pandemic, the buying attitudes of customers are increasingly shifting in recent times from the traditional "brick and mortar" to digital platforms, and customers are choosing the convenience, assertiveness and control that digital buying provides (Bai et al., 2021). It is, therefore, important for owners/managers to critically consider social networking platforms as valuable tools for meeting the convenient demands of customers and as a means of opportunity recognition.

H2: Social media platforms will serve as a significant means for MSEs' survival during and after the Covid-19 pandemic

Electronic Payment Systems and Survival of MSEs

Electronic payment (E-payment) is the electronic transfer of money between two parties using computer networks, the internet, and digital systems (Alumasa & Muathe, 2021). E-payment also refers to electronic payments for goods and services made by businesses, individuals, and governments (Khayer et al., 2021; Alumasa & Muathe, 2021; Pham et al., 2018). Examples of such networks include the internet and modern devices like credit/debit cards, EFT, POS, and mobile phones. Besides promoting economic growth, e-payment saves time and is user-friendly (Pham et al., 2018).

In today's corporate environment, the electronic payment system is quickly becoming a revolutionary payment method due to its efficiency, convenience and timeliness (Gu et al., 2021; Raghavan et al., 2018). The electronic payment system is continuously being embraced and adopted by businesses in developed and developing countries to simplify and ease payments in business transactions. The most prevalent electronic payment method in Ghana is mobile payment, which automates the merchant and the client experience by removing the need for paper money (Zumanu, 2019). This is followed by Point-of-Sale devices such as the E-zwich (Zumanu, 2019).

Based on the above discussion, we argue that when MSEs in less developed economies like Ghana adopt these electronic means of payment, it is most likely to lessen the cost of transacting business and the effect of the Covid-19 pandemic and similar natural disasters that might occur in the future. For instance, the analysis of Alumasa and Muathe (2021) has shown that large and international firms operating in Nigeria that encourage online buying and electronic payment systems were less affected by the containment measures imposed as a result of the outbreak of the Covid-19 pandemic.

H3: Using E-payment systems will significantly contribute to the survival of MSEs during and after the Covid-19 pandemic.

Cloud Computing and Survival of MSEs

In response to the increased demand for better business integration, flexibility, and agility, cloud computing uses modern information technology in a new business environment. Using Internet technologies in conjunction with on-demand IT services in a pay-per-use model makes cloud computing revolutionary (Khayer et al., 2021; Priyadarshinee et al., 2017). Cloud computing is a new IT approach to strategically creating company value.

Cloud computing is a cutting-edge architecture that employs distributive and parallel computing techniques to enable widespread, accessible, and on-demand access to virtualised and distributed computing resources such as hardware, software, networks, servers, storage, and databases (Yeboah-Boateng & Essandoh, 2014). It allows organisations to increase capacity or add capabilities without investing in new infrastructure, training new employees, or licensing new software because it delivers rental services, which allows the degree of consumption to be adjusted based on current demands (Khayer et al., 2021; Priyadarshinee et al., 2017). Because MSEs lack adequate technical capabilities to attain their desired goals, cloud services can provide a level playing field for them by adopting cutting-edge technologies that were previously unaffordable (Khozani et al., 2023).

Cloud computing's unlimited flexibility, improved reliability, and high security enable SMEs to customise services and data for increased availability (Khayer et al., 2021). Additionally, cloud computing delivers value at strategic and operational levels, such as boosting annual revenue and lowering operating costs (Khozani et al., 2023). Thus, cloud computing may improve business operations and has brought substantial benefits to SMEs, allowing them to compete in today's dynamic business environment (Khozani et al., 2023). Past studies have shown the effectiveness of cloud computing in influencing organisational performance. For instance, Khayer et al. (2021) have shown a positive and significant association between cloud computing and firm performance among SMEs in Nigeria. From the South African service firms' perspective, Johnston et al. (2016) have also shown cloud computing as an effective tool for providing speed and agility as a value driver to organisations.

Even though cloud computing has many advantages, the adoption rate and use are still low, particularly among micro and small enterprises (Khozani et al., 2023). Based on the above discussion on the benefits of adopting cloud computing, we argue that its adoption by MSEs operating in Ghana will be a critical avenue for the survival of these MSEs during and after the pandemic. The study thus, hypothesised as follows:

H4: Adopting Cloud computing will have the most significant influence on MSEs' survival during and after Covid-19

METHODOLOGY

The quantitative research method was employed in analysing the study's hypothesised relations. Structured questionnaires were used in collecting data from the respondents. The questionnaire was divided into two parts. Part one consists of the following information: number of years in operation, technological tools currently employed by the business, number of employees, industry, and type of social media users. The second part comprises questions on the study's constructs (online store/website, social media, electronic payment systems, and cloud computing). The statements measuring these constructs were measured on a five-point Likert scale anchored with "1 strongly disagree" to "5 strongly agree" with "3 being neutral." The items measuring these constructs were adapted from the literature. For instance, the items measuring online/website were adapted from Gu et al. (2021) and Raghavan et al. (2018). The items measuring social media were adapted from Odoom et al. (2017) and Bai et al. (2021). All the items measuring electronic payment were adapted from Pham et al. (2018). Concerning cloud computing, the measurement items were adapted from Khayer et al. (2020) and Johnston *et al.* (2016).

The population for the study comprises MSEs that are duly registered with the GEA as of the time of data collection. The GEA is a governmental agency under the Ministry of Trade and Industry with the mandate to identify, nature, promote and develop micro, small and medium-sized enterprises in Ghana (Zumanu, 2019). Based on the list obtained from GEA, the total number of MSEs registered with the agency as of May 2022 was 6,235. Data was, however, collected from MSEs located within the Accra-Tema metropolis since the list from GEA shows that about 51.2% of registered MSEs with the institution are located in these areas.

A probabilistic sampling technique (precisely, a systematic sampling technique) was used in selecting the MSEs for the study. The sample size formula by Bartlett et al. (2001) determined a designated sample size of 485. However, when the selected MSEs were contacted via telephone, and the purpose of the survey explained to them, 25 declined to partake in the survey, and 55 were no longer in operation, making the total number available and willing to participate to be 405. The structured questionnaires were administered offline by scheduling an appointment with the owners or operations managers and visiting their premises. However, 375 valid responses (representing a 77% response rate) were used for data analysis after data collection and data cleansing. The data was analysed using the PLS-SEM technique.

RESULTS OF THE STUDY

Profile

Table 1 summarises the profile of the MSEs in the study. The table shows that, regarding the distribution of the businesses in terms of industry, the majority of the MSEs surveyed (60%) are in the services sector, followed by manufacturing (26%) and agriculture (14%). This confirms the work of past scholars who found that Ghana's service sector has overtaken the agriculture sector and become the largest contributor to economic activities (Odoom et al., 2017; Coffie & Hinson, 2021). The table further shows that concerning the usage of digital technologies, the majority of the MSEs surveyed (75%) mostly employs two digital technologies (social media and electronic payment systems). 16% uses only one form of digital technology: the electronic payment system.

Additionally, the profile result shows that only 9% of MSEs have a combination of website, social media presence and electronic payment systems. This is not surprising, as past studies have shown that MSEs, particularly those in developing economies like Ghana, are reluctant to establish an online presence (Gu et al., 2021). Regarding social media usage, the profile result has shown that most MSEs in Ghana use WhatsApp and Facebook, followed by only WhatsApp and only Facebook. This finding is in tandem with the findings of Odoom et al. (2017), who found these social media platforms as the most common types of social media used by Ghanaian SMEs.

Table 1. Profile of MSEs sampled

Industry			**Social media platforms**		
Agriculture	53	14%	Only WhatsApp	78	35%
Manufacturing	97	26%	Only Facebook	51	23%
Service	225	60%	Only Instagram	-	-
Current digital technology (Only one)			WhatsApp + Instagram	35	16%
Website			WhatsApp + Facebook	44	20%
Only E-payment	58	16%	Facebook + Instagram	-	-
Only Social media	-	-	Facebook + Twitter	-	-
Only Cloud computing	-	-	Twitter + Instagram	-	-
No. using two digital technologies			WhatsApp, Facebook, Twitter	-	-
Website + E-pay	-	-	WhatsApp, Facebook, Instagram	14	6
Website + SM	-	-	Facebook, Instagram, Twitter	-	-
Website + Cloud	-	-	Instagram, Twitter, WhatsApp	-	-
SM + E-pay	282	75%	WhatsApp, Facebook, Instagram, Twitter	-	-
SM + Cloud	-	-	**TOTAL**	**224**	**100**
E-pay + Cloud	-	-			
No. using three digital technologies			**Electronic payment systems**		
Website, E-pay, SM	35	9%	Only mobile money (Momo)	251	67
Website, E-pay, Cloud	-	-	Only E-zwich	31	9
Website, SM, Cloud	-	-	Momo + E-zwich	58	24
E-pay, SM, Cloud	-	-	**TOTAL**	**340**	**100**
Cloud, SM, E-pay	-	-	**Size (Number of employees)**		
No. using four digital technologies			4	156	42
Website, SM, E-pay, Cloud **TOTAL 375**		**100**	3	122	33
			2	68	18
			1	29	7

Source: Survey Data (2022)

THE MEASUREMENT MODEL

Following the two-step approach to structural equation modelling (Anderson and Gerbing, 1988), the measurement model was first evaluated to determine the

Table 2. Assessment of validity and reliability of the measurement model

Construct	Code	Loading	Alpha (α)	CR	AVE
Online/Website	OWEB1	0.945	0.943	0.957	0.816
	OWEB2	0.938			
	OWEB3	0.815			
	OWEB4	0.940			
	OWEB5	0.871			
	OWEB6	0.945			
Social Media	SM1	0.937	0.963	0.970	0.844
	SM4	0.944			
	SM5	0.916			
	SM6	0.876			
	SM7	0.924			
	SM8	0.912			
E-payment	E-PAY1	0.902	0.951	0.961	0.804
	E-PAY2	0.910			
	E-PAY3	0.920			
	E-PAY4	0.918			
	E-PAY5	0.884			
	E-PAY6	0.845			
Cloud Computing	CC1	0.912	0.969	0.975	0.867
	CC2	0.941			
	CC3	0.938			
	CC4	0.935			
	CC5	0.926			
	CC6	0.934			
MSEs' Survival	SUSTAIN1	0.945	0.948	0.958	0.794
	SUSTAIN2	0.935			
	SUSTAIN3	0.883			
	SUSTAIN4	0.908			
	SUSTAIN5	0.839			
	SUSTAIN6	0.830			

Table 3. Discriminant validity

	CC	E-PAY	OWEB	SM	SUSTAIN
CC	0.931				
E-PAY	0.622	0.897			
OWEB	0.818	0.587	0.903		
SM	0.837	0.585	0.858	0.918	
SUSTAIN	0.845	0.661	0.876	0.908	0.891

Discriminant validity was also achieved since the square root of the AVEs (diagonal values) is greater than the constructs' correlation coefficients.

reliability and validity of the measurement items. The reliability of the items was assessed using factor loadings, Cronbach Alpha (α) and Composite Reliability (CR). Items (SM2, SM3) with factor loadings of less than 0.6 were deleted (Sarstedt et al., 2017). As shown in **table 2**, the values of both α and CR are above the recommended threshold value of 0.7 (Sarstedt et al., 2017). The convergent validity of the items was assessed using Average Variance Extracted (AVE). The AVE values are also above the recommended threshold of 0.5. On the other hand, discriminant validity was evaluated using the Fornell and Larcker (1981) criterion. As shown in **table 3**, discriminant validity was also achieved since the square root of the AVEs (diagonal values) is greater than the constructs' correlation coefficients.

ASSESSMENT OF THE STRUCTURAL MODEL

The measurement model assessing the structural paths and relationships was tested using PLS-SEM after achieving an acceptable measurement model (see figure 1 and table 4). As displayed in figure 2 and table 4, the results show that hypothesis H1, which suggests that the adoption of online/websites will have a significant favourable influence on MSEs survival during and beyond the pandemic, was supported ($t = 4.46$, $p < .001$). Additionally, the hypothesis that social media will be a significant factor in the survival of MSEs during and after the Covid-19 pandemic was also found to be statistically significant ($t = 7.78$, $p < .001$), thereby providing support for hypothesis H2. The path between electronic payment systems and the survival of MSEs during and after a natural crisis such as Covid-19 was also statistically significant ($t = 5.26$, $p < .001$), providing support for H3. Finally, H4, which hypothesised that cloud computing would be a significant factor in the survival of MSEs during and after the Covid-19 pandemic, was also supported ($t = 2.95$, $p < .003$).

Table 4. Analysis of hypothesized paths

Hypothesis	Path relationship	β	T-stats	p-value	Decision
H1	OWEB -> SUSTAIN	0.270	4.456	0.000	Supported
H2	SM -> SUSTAIN	0.485	7.775	0.000	Supported
H3	E-PAY -> SUSTAIN	0.135	5.261	0.000	Supported
H4	CC -> SUSTAIN	0.134	2.945	0.003	Supported

Figure 1.

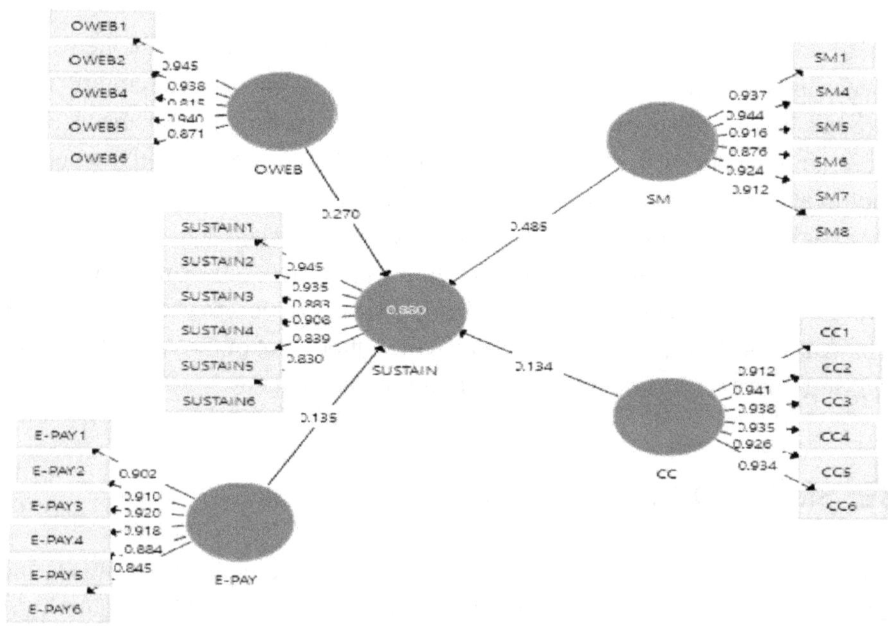

DISCUSSION

The study's objective was to determine which digital technology, when adopted by MSEs, particularly those from a developing country context like Ghana, will have the most significant influence on their survival during and after a crisis and should be adopted first by MSEs. The study further sought to determine the digital tool that small business should adopt first. The study's results have shown that all four digital technologies examined in the study (online/website, social media, electronic payment and cloud computing) significantly influence the survival of MSEs during and after a natural crisis, as presented by Covid-19. However, with regards to which of these digital technologies have the most significant influence on MSEs' survival, the results have shown that the adoption of social media will have the most significant influence on the survival of Ghanaian MSEs during and beyond the pandemic. This is followed by electronic payment systems, online store or website and cloud computing. These results suggest that although adopting these digital technologies will significantly influence MSEs' survival, adoption of social media will have the most significant impact. Given the resource constraints characterizing these MSEs, the study's outcome is important as it shows the most important digital technology to adopt first. Thus, as their resource capability improves, they could continue adding up until they fully adopt all these digital technologies to improve their survival, agility and growth during crisis as presented by Covid-19.

Contrary to Bai et al. (2021), who found electronic payment as a critical tool for the digital transformation of micro and small enterprises, this study shows social media adoption as the most critical factor for MSEs' survival and likely to encourage the digital transformation process among MSEs if adopted first. As shown by previous studies, social media being the most significant influencer of MSEs' survival could be attributed to the large number of consumers using social media and its interactive nature (Etim & Daramola et al., 2020). Additionally, the highly interactive nature of social media, coupled with their high degree of compatibility with any device on the internet, makes social media adoption and deployment easy for most MSEs, given their chronic resource-constrained context. Thus, with internet connectivity, most social media platforms such as Facebook, WhatsApp, and Twitter are very simple and easily adoptable by any firm (Odoom et al., 2017).

The study also supports the findings of Gu et al. (2021) who found online store as an important variable for success during the pandemic as their study has proved the shifting interest of consumers towards online purchasing during the pandemic. The study's finding is also in tandem with the work of Pham et al. (2018) who found a positive and significant association between online store and performance of MSEs.

Consistent with past findings, the study found a positive significant relationship between electronic payment and survival of MSEs in Ghana (Bai et al., 2021; Khayer et al., 2021; Zumanu, 2019). The result partially agrees with the findings of Bai et al. (2021) who found electroninc payment as the most significant digital tool for survival of businesses. The current study specifically found electronic payment systems as the second most significant variable for MSEs survival during and beyond the pandemic. The significant relationship between electronic payment and survival of MSEs could be explained by the increasing demand for safety and convenience in making transactions by consumers in recent times (Khayer et al., 2021). Thus, given the numerous advantages associated with electronic payment systems, its adoption is a critical factor for the development, growth and survival of MSEs in developing economies like Ghana. For instance, as a result of the rising incidents of armed robbery attacks on people carrying cash or those who have gone to the bank to withdraw money, more customers now prefer electronic cash as it is safer. In the times of the COVID-19 pandemic where people have to regularly sanitize their hand after handling money, customers are increasingly finding ways of avoiding physical cash and human touch to save themselves from getting affected by the coronavirus. Due to this, the need for contactless payment has increased and this means that small businesses need to start using the latest technology to prevent the risk of being overtaken by their competitors. This finding for instance, corroborates the work of Raghavan et al. (2018) who found electronic payment as most singular factor for SME growth and performance in India.

Similarly, the study's result corroborates the findings of past studies that found positive and significant association between cloud computing and performance of businesses (Khayer et al., 2021; Priyadarshinee et al., 2017; Yeboah-Boateng & Essandoh, 2014). This suggests that the adoption of cloud computing is most likely to have a positive significant influence on the survival of MSEs during and after crisis such as the Covid-19 pandemic. The study is for instance, consistent with the findings of Khayer *et al.* (2021) which confirms the positive impact of the adoption of cloud computing on firm performance. The result also agrees with the work of Raghavan et al. (2018) which shows that adoption of cloud computing positively increases market value and performance of firms.

THEORETICAL AND PRACTICAL IMPLICATIONS OF THE STUDY

The study makes some theoretical contributions to the digitization literature of micro and small enterprises. Firstly, in the Ghanaian context, the study, employs the resource-based view theory (RBV) to identify digital technologies that are likely to have the most significant influence on the survival of micro and small enterprises during the post Covid-19 crisis. The result shows that though all four digital technologies examined by the study have a significant influence on survival of MSEs, social media adoption has the most significant influence on their survival during and post Covid-19. Secondly, given the resource constraints characterizing most MSEs and the call for their gradual adoption of digital tools, our study contributes to the literature by showing that MSEs could start their digitization efforts from social media adoption, followed by electronic payment, online/website and cloud computing. As suggested by Bai et al. (2021) this is likely to encourage and speed up the adoption rate of digital technologies among MSEs, especially in developing countries like Ghana. Thus, from the RBV theory's perspective, the study has demonstrated that small enterprises could gain a competitive advantage and other benefits of digitization through the gradual adoption of digital tools starting from social media, followed by electronic payment, online/website and cloud computing. It is worth mentioning that although the RBV has been extensively used to explain the importance of resources in the performance of businesses, its application to identify significant digital resources for MSEs has received limited attention. Thus, the study contributes to theory-building efforts in the micro and small enterprise literature.

However, owners/managers are at liberty to adopt all these digital technologies at a go if they have the ability to do so, since the result has shown that all these technologies have a significant impact on the survival of MSEs in Ghana.

For policymakers, we recommend that targeted policies such as tax incentives could be used to influence micro and small enterprises to encourage them to incorporate electronic payment systems into their operations. This is most likely to improve on government's effort to achieve a cashless economy and reduce challenges associated with cash transactions. Since micro and small enterprises are primarily in the informal sector, encouraging electronic payment systems among them is also likely to reduce the non-bankable population and help formalize their operations which could also increase tax revenue collection. Additionally, based on the findings of the study, we recommend that state institutions such as the Ministry of Trade and Industry, GEA should embark on specialized training for MSEs on social media marketing, electronic payment and the benefits of adopting cloud computing and online/websites. This will also help speed up the government's agenda of digitizing the Ghanaian economy.

CONCLUSION

In conclusion, this research study focused on exploring the role of digital technologies in sustaining small enterprises in the post-Covid-19 era. The study examined four digital tools, namely websites/online stores, cloud computing, social media, and electronic payment, and their impact on the survival of MSEs.

The findings of the study indicate that all four digital technologies significantly influence the survival of MSEs in the post-Covid-19 era. However, it was revealed that the adoption of social media as a digital resource holds the most significant influence on their survival.

Social media platforms provide MSEs with an effective means to engage with customers, promote their products or services, and expand their reach. By leveraging social media, MSEs can build brand awareness, foster customer relationships, and drive sales. The interactive and viral nature of social media allows MSEs to connect with a wide audience, even on a limited budget.

While websites/online stores, cloud computing, and electronic payment systems also play important roles in supporting MSEs, the study emphasizes the distinctive impact of social media in enhancing their survival. Thus, MSEs should prioritize integrating social media strategies into their digital transformation efforts.

These findings have significant implications for MSEs, policymakers, and practitioners. MSEs should recognize the value of social media and invest in developing an effective social media presence. Policymakers should support initiatives that promote digital literacy and provide resources to facilitate the adoption of digital technologies. Practitioners and consultants can guide MSEs in utilizing social media effectively and maximizing its benefits for business growth.

Overall, the study highlights the importance of embracing digital technologies, with particular emphasis on social media, as a means to sustain small enterprises in the post-Covid-19 era. By leveraging these digital tools, MSEs can adapt to the changing business landscape, remain competitive, and thrive in the digital age.

DIRECTION FOR FUTURE RESEARCH

This study has some limitations that may lead to future lines of research. Although the study has found social media as having the most significant influence on survival of MSEs, how the specific types of social media platforms influence survival has not been addressed by the current study. Further study is therefore needed to show the various types of social media and their influence on the survival of MSEs. Additionally, further study is also required to identify how the various forms of electronic payment systems influence survival of MSEs.

REFERENCES

Alumasa, S., & Muathe, S. (2021). Mobile Credit and Performance: Experience and Lessons from Micro and Small Enterprises in Kenya. *Journal of Applied Finance & Banking*, *11*(4), 135–161. doi:10.47260/jafb/1146

Alvarenga, M. Z., Oliveira, M. P. V. D., & Oliveira, T. A. G. F. D. (2023). The impact of using digital technologies on supply chain resilience and robustness: The role of memory under the covid-19 outbreak. *Supply Chain Management*, *28*(5), 825–842. doi:10.1108/SCM-06-2022-0217

Anderson, J. C., & Gerbing, D. W. (1988). Structural equation modeling in practice: A review and recommended two-step approach. *Psychological Bulletin*, *103*(3), 411–423. doi:10.1037/0033-2909.103.3.411

Bai, C., Quayson, M., & Sarkis, J. (2021). COVID-19 pandemic digitization lessons for sustainable development of micro-and small- enterprises. *Sustainable Production and Consumption*, *27*(5), 1989–2001. doi:10.1016/j.spc.2021.04.035 PMID:34722843

Barney, J. (1991). Firm resources and sustained competitive advantage. *Journal of Management*, *17*(1), 99–120. doi:10.1177/014920639101700108

Bartlett, J. E., Kotrlik, J. W., & Higgins, C. C. (2001). Organizational Research : Determining Appropriate Sample Size in Survey Research. *Information Technology, Learning and Performance Journal*, *19*(1), 43–50.

Coffie, I. S., & Hinson, R. E. (2021). Types of Corporate Social Responsibility Initiatives as Response to COVID-19 Pandemic in Emerging Economies. In *Marketing Communications in Emerging Economies* (Vol. I, pp. 161–192). Palgrave Macmillan. doi:10.1007/978-3-030-81329-1_7

Cueto, L. J., Frisnedi, A. F. D., Collera, R. B., Batac, K. I. T., & Agaton, C. B. (2022). Digital Innovations in MSMEs during Economic Disruptions : Experiences and Challenges of Young Entrepreneurs. *Administrative Sciences*, *12*(1), 8–22. doi:10.3390/admsci12010008

Etim, E., & Daramola, O. (2020). The informal sector and economic growth of South Africa and Nigeria: A comparative systematic review. *Journal of Open Innovation*, *6*(4), 134. doi:10.3390/joitmc6040134

Eurostat. (2021). *Euro zone GDP contracts by a record 3.8% in the first quarter*. https://ec.europa.eu/eurostat/web/covid-19/economy

Fornell, C., & Larcker, D. F. (1981). *Structural equation models with unobservable variables and measurement error: Algebra and statistics*.

Gu, S., Ślusarczyk, B., Hajizada, S., Kovalyova, I., & Sakhbieva, A. (2021). Impact of the covid-19 pandemic on online consumer purchasing behavior. *Journal of Theoretical and Applied Electronic Commerce Research*, *16*(6), 2263–2281. doi:10.3390/jtaer16060125

Khayer, A., Talukder, M. S., Bao, Y., & Hossain, M. N. (2020). Cloud computing adoption and its impact on SMEs' performance for cloud supported operations: A dual-stage analytical approach. *Technology in Society*, *60*(3), 101225. doi:10.1016/j.techsoc.2019.101225

Khozani, M. L., Behzadi, H., Nowkarizi, M., & Neizar, F. S. (2021). We live in cloud computing world, without using it in our libraries. *Library Hi Tech*, *40*(6), 1916–1929. doi:10.1108/LHT-03-2021-0107

Kimuli, S. N. L., Sendawula, K., & Nagujja, S. (2021). Digital technologies in micro and small enterprise: Evidence from Uganda's informal sector during the COVID-19 pandemic. *World Journal of Science. Technology and Sustainable Development, 18*(2), 93–108.

Odoom, R., Anning-Dorson, T., & Acheampong, G. (2017). Antecedents of social media usage and performance benefits in small-and medium-sized enterprises (SMEs). *Journal of Enterprise Information Management, 30*(3), 383–399. doi:10.1108/JEIM-04-2016-0088

Oduntan, E. B., & Isere, V. O. M. (2022). Digital Technologies for Business Sustainability in Nigeria : An Empirical Analysis. *African Journal of Management and Business Research, 5*(1), 12–22.

OECD. (2021). *The Frontiers of Digital Learning: bridging the SME digital skills gap*. OECD. https://www.oecd.org/digital/sme/events/Frontiers of Digital Learning - Key Highlights - June 2021.pdf

Papadopoulos, T., Baltas, K. N., & Balta, M. E. (2020). The use of digital technologies by small and medium enterprises during COVID-19: Implications for theory and practice. *International Journal of Information Management, 55*(3), 102–132. doi:10.1016/j.ijinfomgt.2020.102192 PMID:32836646

Pham, Q. T., Tran, X. P., Misra, S., Maskeliunas, R., & Damaševičius, R. (2018). Relationship between convenience, perceived value, and repurchase intention in online shopping in Vietnam. Sustainability (Switzerland), 10(1).

Priyadarshinee, P., Raut, R. D., Jha, M. K., & Kamble, S. S. (2017). A cloud computing adoption in Indian SMEs: Scale development and validation approach. *The Journal of High Technology Management Research, 28*(2), 221–245. doi:10.1016/j.hitech.2017.10.010

Raghavan, V., Wani, M., & Abraham, D. M. (2018). Exploring E-Business in Indian SMEs: Adoption, Trends and the Way Forward. In *Emerging Markets from a Multidisciplinary Perspective* (pp. 95–106). Springer. doi:10.1007/978-3-319-75013-2_9

Sarstedt, M., Ringle, C. M., & Hair, J. F. (2021). Partial least squares structural equation modeling. In *Handbook of market research* (pp. 587–632). Springer International Publishing. doi:10.1007/978-3-319-05542-8_15-2

Tweneboah-Koduah, E. Y., & Coffie, I. S. (2022). Social Distancing, Hand Washing and Handshaking Behaviour During and Beyond Coronavirus Pandemic: A Social Marketing Perspective. *Social Marketing Quarterly*, *28*(4), 288–306. doi:10.1177/15245004221134208

UNCTAD. (2021). *COVID-19 and E-commerce : findings from a survey of online consumers in 9 countries*. UNCTAD.

World Bank. (2021). *Unmasking the Impact of COVID-19 on Businesses : Firm Level Evidence from Across the World*. World Bank. https://openknowledge.worldbank.org/handle/10986/34626

Yeboah-Boateng, E. O., & Essandoh, K. A. (2014). Factors Influencing the Adoption of Cloud Computing by Small and Medium Enterprises in Developing Economies. *International Journal of Emerging Science and Engineering*, *2*(4), 631–642.

Zumanu, V. O. (2019). Performance of SMEs in Ghana: Do mobile money payments matter? University of Ghana.

KEY TERMS AND DEFINITIONS

Cloud Computing: The delivery of computing services, including storage, processing power, and software applications, over the internet. Instead of relying on local servers or personal computers to store and process data, cloud computing utilizes remote servers and networks to provide on-demand access to resources and services.

Digitalization: This involves using digital technologies to transform a business model, creating new revenue and value production opportunities. Achieving this requires integrating digital tools and systems into different aspects of a business's operations, such as management, communication, production, and customer service.

Digitization: Converting data to a digital format. It describes the pure analog-to-digital conversion of existing data and documents.

Electronic Payment System: An electronic payment system, also known as a digital payment system or e-payment system, refers to a method of making financial transactions electronically, without the need for physical currency or traditional paper-based methods such as checks or cash. It involves the exchange of funds between buyers and sellers or between parties involved in a transaction through electronic means.

Online Store: An online store is an internet platform or website that allows visitors or consumers to shop without human interaction and make payments online or upon delivery.

Social Media: Social media refers to a collection of online platforms and applications that allow users to create, share, and exchange content in the form of text, images, videos, and other multimedia formats. It enables individuals and organizations to connect, communicate, and interact with others virtually, regardless of geographical distances.

Chapter 15
The Moderating Role of State Regulation on Stakeholder Market Orientation and Sustainability Performance at Tourism Destinations

Ernest Kafui Kwasi Tsetse
https://orcid.org/0000-0001-9043-3907
Ho Technical University, Ghana

Elikem Chosniel Ocloo
https://orcid.org/0000-0003-4729-3093
Accra Technical University, Ghana

Mahmoud Abdulai Mahmoud
University of Ghana, Ghana

Emmanuel Kofi Agbesi
E.P. University College, Ho, Ghana

ABSTRACT

Using a survey method, the study used responses from 313 Ghanaian tourism businesses in selected regions and the PLS-SEM in analysing the data. The study established that ecological regulation significantly moderates the StMO-SP relationship while political regulation negatively moderates the StMO-SP nexus. State regulations significantly influence Stakeholder market orientation and sustainability performance at tourism destinations. The study recommends that tourism managers and planners be cognisant with state regulation requirements in fashioning and implementing their StMO-SP strategies. Tourism destination managers and planners who are involved in using StMO in improving sustainability management practices to take into account the role of key stakeholders in fashioning and implementing their sustainability strategies. Future research directions have been discussed.

DOI: 10.4018/978-1-6684-9843-9.ch015

INTRODUCTION

Tourism as an industry has grown tremendously in recent years and its importance to nations, regions, communities, cities, towns and specific destination cannot be underestimated. The travel and tourism sector has contributed 10.3% to the world's Gross Domestic Product (GDP) and contributed 330 million jobs globally {World Travel & Tourism Council (WTTC, 2019)}. In Ghana, the travel and tourism sector is the third contributor to GDP. The sector contributed in total GHS12, 573m to Ghana's GDP in 2017, or 6% to total GDP and 5.5% in 2018 (WTTC, 2018). There are a lot of potentials in this sector and governments and the private sector are investing heavily in infrastructure such roads, power, pipe borne water as well as coming out with good regulatory frameworks to regulate tourism and its related industry. When governments are able to effectively and efficiently regulate the stakeholders then the industry can reap the huge investments in the sector. Tourism destinations (TD) are subsets of destinations that need special attention due their peculiar nature in terms of the role they, can play in improving the status of the rural populations, contributing to sustainable income of nations and conserving natural and social environment of the TDs (Peña, Jamilena, & Molina, 2013). TDs offer unique offerings that give visitors memorable experiences in order to build competitive advantage (Musa, Kayat, & Thirumoorthi, 2010). However, tourism destinations are saddled with some challenges, for instance, low investments in infrastructure such as roads, water, power, and as a result make these destinations somewhat unattractive. Improvement in tourism destinations do not translate only into the physical landscape of the destination but also the social life of the community in general (Hwang, Stewart, & Ko, 2012).

The above benefits to the nation might be cut short due to human activities that threaten the continuous survival of natural vegetative cover, animals, landmarks and water bodies which are used as attractors to the tourism destinations. Bad mining practices, bad farming practices, excessive bushfires and global climate change are destroying the green vegetative cover, the animals and water bodies of the destinations. These strategic natural resources serve as the tourist destinations for the country. The forest vegetation in Ghana, for instance, has seen massive deterioration with 80% of it disappearing in the past 100 years. The forest reduced from 8.2 million hectares to 1.6 million hectares with an annual depletion rate of 65000 hectares (Forestry Commission (FC), 2018). The country has lost the quality and quantity of its water bodies as a result of illegal mining called "galamsey" and other human activities like farming and bush burning. These unsustainability human activities pose a lot of threat to the natural resources of the country hence demand immediate action to stop the continuous destruction of these strategic natural resources. This calls for all stakeholders to come on board to put in sustainability policies and practices to prevent such deterioration to these natural resources. One option that will help in tackling this deterioration is through stakeholder management of these natural resources where stakeholders would commit to putting in resources to prevent further destruction by way of behavioural change which will support the existence of these strategic natural resources that form the tourism destinations of Ghana.

One business strategy that can be adopted by TDs is Stakeholder Market Orientation (StMO). The researcher therefore conceptualized StMO as a broad network of destination's stakeholders that establish relationships among themselves with the aim of gathering market information and other resources, exchange them among the stakeholders to be used to make strategic marketing decisions that benefit all stakeholders at the destination to achieve long term survivor of the resources of the destinations that translate into providing sustainability superior services to their customers and other stakeholders (Narver & Slater, 1990; Jaworski & Kohli, 1993; Polo Peña et al., 2013; Mahmoud, Blankson, & Hinson, 2017).

State regulation is the kind of regulation that uses coercive force to command tourism businesses to comply with sustainability management practices (Zhu & Sarkis, 2007; Li, Tang, & Zhang, 2020). According to this study, state regulation has been classified into political and ecological regulations. Political regulation involves the use of state institutions such as Ghana Tourism Authority (GTA), to control management of tourism businesses to charge appropriate fees and provide safe business environments for tourists (Line and Wang, 2017a). Ecological regulation on the other hand is explained as state institutions such as the Environmental Protection Authority (EPA), Minerals Commission, Water Resource Commission and all other state institutions that are empowered by an Act of Parliament to regulate environmental management practices and obtain compliance from tourism and other businesses that use the natural environment as their basic resource for production (Line and Wang, 2017a). Extant literature has observed that, sustainability practices add to the cost of production of businesses and as a result, most businesses on their own might not put in place sustainability management practices, hence, calls for regulatory frameworks to regulate businesses to comply with environmental sustainability practices (Crittenden, Crittenden, Ferrell, Ferrell, & Pinney, 2011b; Nkamnebe, 2011; Akram, Abrar-ul-Haq, & Raza 2019). Moderators absent between StMO and SP. No work to the best of our knowledge has used both stakeholder and institutional theories to explain the relationship between StMO and SP and with state regulation as a moderator. MO's application to destination marketing has received little attention by scholars (Mitchell, Wooliscroft, & Higham, 2010; Line & Wang, 2017a). Scholars have called for further research in tourism destinations in other geographical locations to stimulate debate and to raise questions for studying and monitoring stakeholder participation (Longart et al. 2017)

This study seeks to contribute to sustainability regulations, moderation literature, sustainability marketing and development of comprehensive conceptual framework for measuring StMO-SP relationship with state regulations as moderating variable. The main objective of the chapter is to moderate state regulations with stakeholder market orientation with sustainability performance to determine the strength or otherwise of the StMO-SP nexus.

THEORETICAL FOUNDATION AND HYPOTHESIS DEVELOPMENT

Macro marketing scholars have called for constructs that are able to measure sustainability (Line & Wang, 2017b). Green et al., (2015) assessed the impact of MO on green supply chain management practices (GSCMP) and environmental performance (EP) and established positive impact of MO on GSCMP and EP. But the relationship between MO, StMO and social and economic sustainability is yet to be established in the macro marketing literature to the best of my knowledge. Our inspiration for this study is grounded in this new macro marketing thought where the business adopts the system approach to its stakeholders by generating, disseminating and responding to market information from multiple stakeholders to enable the business to adopt and practice sustainability behaviours that give superior value to its stakeholders as well as meeting the environmental, economic and societal needs.

For sustainability marketing to be effective in emerging markets, there is a need for proper regulatory framework. This is because there is a trade-off between sustainability performance and profit motives of businesses (Nkamnebe, 2011; Crittenden, Crittenden, Ferrell, Ferrell, & Pinney, 2011a; Line & Wang, 2017b). It believed that when businesses are left to embark on sustainability performance practices most might not do due to the cost implications for them. The study would, therefore, moderate, StMO sustainability performance relationship with state regulations to determine the impact on the relationship.

This study is guided by the stakeholder and institutional theories. The stakeholder theory which explains the relationship between the business and its stakeholders (Donaldson & Preston, 1995; Freeman, 1999; Mitchell, Wooliscroft, & Higham, 2010; Hult, Mena, Ferrell, & Ferrell, 2011; Line & Wang 2017a). Freeman (1999) explains stakeholders as individuals or groups who can affect and can be affected by the attainment of the organisational goals. It is assumed that if firms manage their stakeholders efficiently, it would result in improved sustainability practice and performance of the tourism destination.

A stakeholder has been explained by Freeman, (1999) as a group or individual who can affect or is affected by the achievement of organization's objectives. This explanation gives a very broad perspective of a stakeholder to mean that anybody or institution that the business interacts with directly or indirectly can be termed as a stakeholder and have some level of interest in the business. In a narrow view, stakeholders of a business are key in providing resources, internal regulations, time and attention to that business and in return, the business tends to prioritize those stakeholders that give more resources as being most important in its relationship.

Stakeholders possess at least three of the following attributes, "power, legitimacy and urgency". According to Mitchell et al. (2010), business managers or corporation managers give low priorities to those stakeholders that possess one of these attributes, moderate priorities to those who possess two of these attributes and high priorities to those who possess all three attributes. The power of a stakeholder refers to the degree to which that stakeholder can command the direction of the relationship by using coercive utilitarian or normative means. Legitimacy refers to an acceptable position or an assumption that an action or actions of an entity are presumed desirable, proper or appropriate within some socially accepted system that is regulated by norms, values and beliefs. Urgency refers to the extent to which an actor or a stakeholder demands urgent attention. This is defined by how far management of an entity can delay a response to an important demand of a stakeholder. Businesses that give prompt response to stakeholders have the urgency, power and legitimacy attributes.

Furthermore, Lawrence and Shadnam, (2008) argued that institutional theory is a theory used in examining organizational situations which views the social world as significantly comprising of institutions. Institutional theory is very important in analysing organizational behaviour and was developed to understand the behaviour of organizations (Najeeb, 2014). Institutional theory highlights the contextual, historical and processual facets in which organizational activities occur.

Isomorphic is an outcome of both institutional and competitive pressures. Isomorphic stands on three pillars such as coercive, mimetic and normative (Najeeb, 2014). For the relevance of this study, the coercive pillar was highlighted. Coercive is the situation where institutions or organizations take certain actions or implement an activity as a result of pressure by regulatory stakeholders. This is where institutions and/or organization's activities are guided by regulations.

The integration of the stakeholder and the institutional theories in explaining the Stakeholder Market Orientation and Sustainability Performance (StMO–SP) relationship is that the state regulators; the political and ecological regulators, who use coercive force to compel tourism businesses to comply with sustainability practices are also the key stakeholders (primary stakeholders) of the tourism businesses. In other words, the state institutions have a direct interest in the tourism businesses. This is because, if the tourism businesses do well, the state will eventually benefit from them through the payment of taxes, employment of its citizens, among others.

Government or state institutions provide regulatory frameworks for the businesses and require them to pay taxes and as well as to meet standards for certification and the communities within which the firm or business is located give land and labour. The firm is dependent on these actors in one way or the other and this gives them power over the firm. As a result, primary stakeholders are the most important stakeholders to the business and any disregard for them can trigger serious consequences for the business (Mitchell et al. 2010). Normally, the primary stakeholder has a contractual relationship with the business thus has a direct influence in the decisions of the business.

The conceptual framework for the study describes the relationship between the variables that form the basis for the study. This study interrogates the key dimensions of StMO in the area of competitor orientation (CompO), visitor orientation (VTO), community orientation (ComO), industry orientation (IndO) on sustainability performance dimensions of environmental sustainability performance (EnvSP), social sustainability performance (SocSP) and economic sustainability performance (EcoSP). The study also moderates the StMO-SP relationship with state regulation (SR) in terms of political regulation (PR) and ecological regulation (ER). The outcome variables are controlled by macro level factors such as energy and infrastructure policy which are beyond the control of the tourism businesses. On the other hand, there are market level control variables such as technology turbulence, competitive intensity and buyer power; which are within the control of these tourism businesses.

Figure 1. Conceptual framework

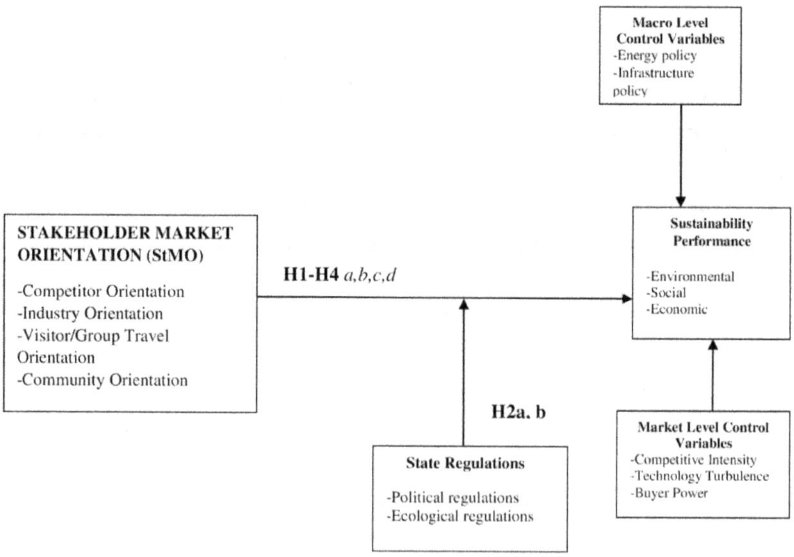

HYPOTHESES DEVELOPMENT

Below are the direct hypotheses for the study:

i. Competitor orientation (CompO) (H1a, H1b, H1c) positively predict environmental sustainability performance (EnvSP), social sustainability performance (SocSP) and economic performance (EcoSP)
ii. Industry orientation (IndO) (H2a, H2b, H2c) positively predicts EnvSP, SocSP and EcoSP
iii. Visitor Orientation (VTO) (H3a, H3b, H3c) positively predicts EnvSP, SocSP and EcoSP.
iv. Community Orientation (ComO) (H4a, H4b, H4c) positively predicts EnvSP, SocSP and EcoSP.

Below are indirect hypotheses for the study:

Extant literature has demonstrated a positive impact when corporate environmental responsibility and sustainability performance are moderated by government or state regulations (Zhu & Sarkis, 2007; Ortiz-de-Mandojana et al., 2016; Li et al., 2017; Akram et al., 2019). According to a study by Akram et al., (2019) on the role of corporate governance and firm's environmental performance: a moderating role of institutional regulations, it was established that, corporate governance has a positive and significant impact on green performance. The results have indicated positive and significant impact on the moderating role of institutional regulation for corporate governance and firm performance.

Prior studies by Li et al., (2017) on the effects of corporate environmental responsibility (CEP) on corporate financial performance (CFP): the moderating role of government regulation established that, the moderating effect of government regulation is significantly positive, i.e., stringent government regulation significantly reinforces the positive influence of CER on CFP. Another study by Ortiz-de-Mandojana et al., (2016) on corporate governance and environmental sustainability: the moderating role of the national institutional context established that, there is a positive moderating effect in the relationship between corporate governance mechanisms and environmental sustainability.

However, there is no known study yet that has moderated state regulation on StMO-SP nexus. In order to close this gap, the study therefore moderates state regulation on StMO-SP to determine whether the moderator will strengthen or weaken the relationship. The study therefore hypothesises that:

H5a: Political regulation positively predicts StMO-SP relationship.

H5b: Ecological regulation positively predicts StMO-SP relationship.

RESULTS AND DISCUSSION

The hypothesis begins with an assessment of the measurement and structural models (Hair et al., 2019; Shmueli et al., 2019). Thus, the SmartPLS 3.2.8 (Ringle, Wende, & Becker, 2015) was run to generate the pictorial and calculation results of reliability, convergent and discriminant validity. Considering suggestions from Henseler et al. (2015), Fornell and Larcker (1981), Nunnally and Bernstein (1994) and Hair et al. (2019), an iterative process is adopted for elimination of the indicators. Thus, through an iterative process, weak items were discarded from the developed model. As shown in Appendix A (Table 1)

Scores for all variables were recorded on a Likert scale ranging from 1 (strongly disagree) to 5 (strongly agree). From the results in Table 1, the visitor/group travel orientation subscale had the highest mean (M = 3.92; SD = .621) and the industry orientation subscale had the lowest mean (M = 3.41; SD = .770). The mean for the overall StMO was 3.61(SD = .561), with a highest and lowest score of 5 and 1, respectively. The results show that the scores for the StMO were generally above average and that the data was spread close to the mean. The skewness and kurtosis values for the StMO were -0.362 and 0.442, respectively, indicating that the scores were normally distributed around the mean. However, the scores for the visitor/group travel orientation subscale of the StMO were non-normal and negatively skewed and therefore not normally distributed around the mean, emanating from the skewness value of -1.085. See Table 2 in Appendix A

For sustainability performance (Table 2), the environmental sustainability performance subscale had the highest mean (M = 3.84; SD = .687) and the social sustainability performance subscale had the lowest mean (M = 3.72; SD = .656). The mean for the overall SP was 3.80(SD = .564), with a highest and lowest score of 5 and 1, respectively. The results show that the scores for the SP were generally above average and that the data was spread close to the mean. Both skewness and kurtosis values indicate that the scores were normally distributed around the mean. See Table 3 Appendix 3

Regarding state regulations (Table 3), the political regulations subscale had the highest mean (M = 3.66; SD = .768) and the ecological regulations subscale had the lowest mean (M = 3.57; SD = .722). The mean for the overall SR was 3.62(SD = .678), with a highest and lowest score of 5 and 1, respectively. The results show that the scores for the SR were generally above average and that the data was spread close to the mean. Both skewness and kurtosis values indicate that the scores were normally distributed around the mean.

In summary, using the overall mean scores of the three main constructs where all variables and their dimensions had mean scores above the cut-off point of 3.2, it is evident that the sample had positive perception about stakeholder market orientation, state regulations and sustainability performance. However, the levels of stakeholder market orientation, state regulations and sustainability performance cannot be described as high but as average.

Table 4 above and Figure 2 below show the final iteration results. Finally, the results show that all the reflectively measured constructs' measures have adequate construct reliability and convergent validity. More specifically, all the loadings exceed the threshold value of 0.708, the average variance extracted (AVE) is higher than the critical value of 0.5, and all the construct reliabilities (i.e. Cronbach's alpha, the coefficients rhoA, and the composite reliability) have values above 0.7 (Hair et al., 2019; Shmueli et al., 2019). Once the iteration process is completed, the final model is checked for discriminant validity based on Fornell and Larcker, the cross loadings and the Heterotrait-Monotrait (HTMT) approaches. The details are provided in the next section.

Figure 2. Results

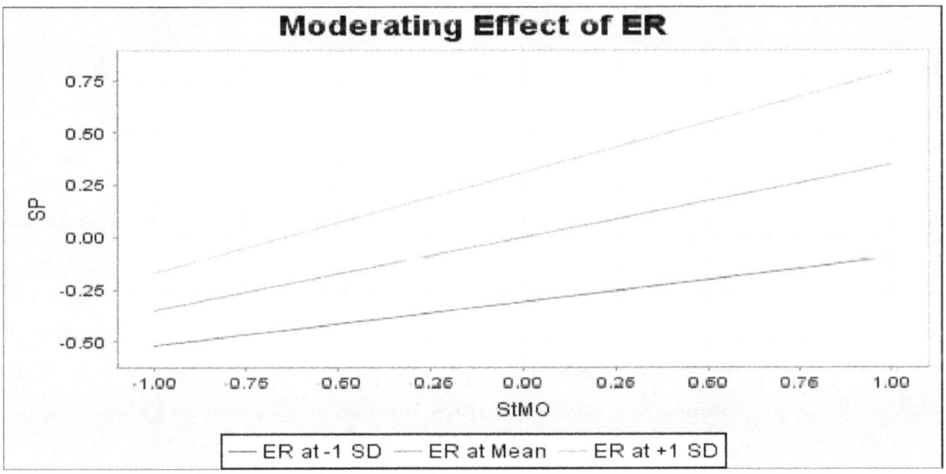

Furthermore, the discriminant validity of all dimensions of the three constructs was examined. Three approaches to assessing the discriminant validity of the constructs (Fornell and Larcker criterion, the cross loadings and the Heterotrait-Monotrait (HTMT) approach) were applied. The results are depicted in Tables 3.

For discriminant validity, the latest criteria for evaluating discriminant validity, the Heterotrait-Monotrait (HTMT) ratios were considered. All HTMT values (Table 3) were below the threshold of 0.85 (Henseler et al., 2015) or 0.90 (Hair et al., 2014; 2019), and this also confirms that discriminant validity was attained.

Figure 3. Moderating effect of PR

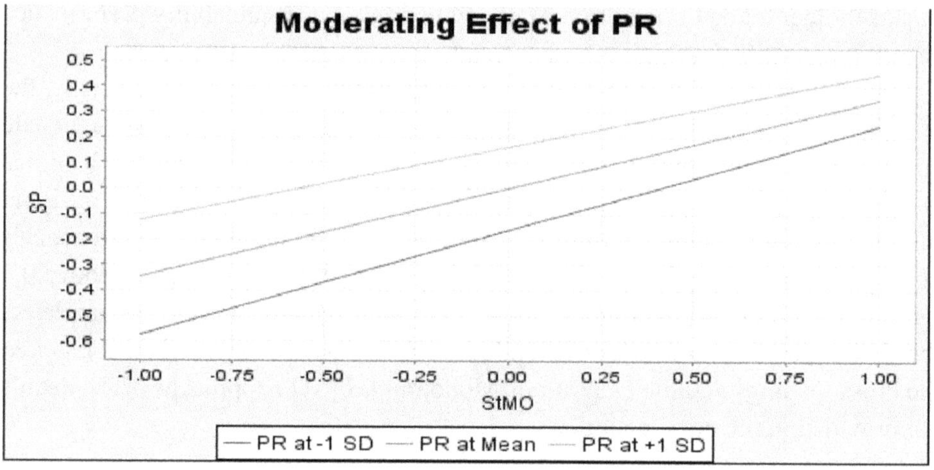

Moderation Analysis

As presented in Table 4, StMO and SR explains 52.6% of the variance in predicting SP ($R^2 = 0.526$). Ecological regulation (ER) significantly moderated the effect StMO had on SP ($\beta = 0.136$; $t = 2.719$; $p = 0.007$), hence supporting H_{5b}. This suggests that ER strengthens the effect of StMO on SP. Surprisingly, political regulation (PR) was found to insignificantly moderate (i.e., weaken) the effect StMO had on SP ($\beta = -0.062$; $t = 1.191$; $p = 0.234$).

Figure 4 and 5 above show the slope results for SP predicted by StMO for each category of state regulations. In summary, the results indicate that only ecological regulation aspect of state regulations strengthened the effect of StMO on SP. The final results are presented in Figure 5 to show the path relationships in the model.

Review of extant literature has assisted in developing a comprehensive conceptual framework that is appropriate in measuring StMO and sustainability performance at tourism destinations. and hence became the framework that was used to measure StMO and SP as well as the moderating variables. This conceptual framework could also be used to measure SP in other services sectors such as education, health, and finance and so on in Ghana and in other countries.

The main objective is to moderate ecological and political regulation on StMO-SP relationship, as revealed and tested in these hypotheses.

H5a: *Moderating* effect of *political regulation positively predicts the relationship between StMO and SP.*

H5b: *Moderating* effect of *ecological regulation positively predicts the relationship between StMO and SP.*

Figure 4. Slope analysis of the moderation effect of ER
Note: *Blue line = Less ecological regulation; Green line = High ecological regulation; Red line = No ecological regulation*

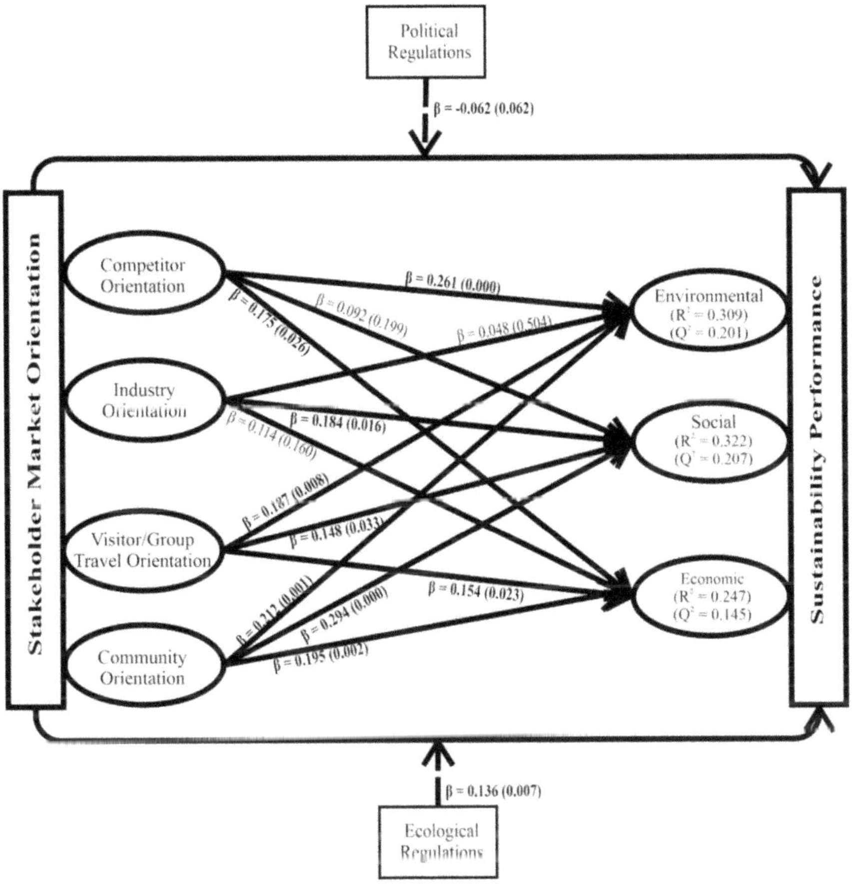

Hypothesis 5a: ***Moderating*** **effect of** ***political regulation positively predicts the relationship between StMO and SP.*** To determine the moderating effects of political regulation on SP, findings from the analysis show a negative impact of political regulations on StMO-SP nexus, hence hypothesis 5a was rejected *(See Appendix A Table 1)*. This finding contradicts findings from extant literature which reveal that, government regulations significantly impacts positively on corporate environmental responsibility and corporate financial performance (Li et al., 2017). The above finding proves that, political regulation negatively affects StMO-SP relationship hence, government institutions such as GTA must relook at its regulations in relation tourism business sustainability performance by engaging tourism destination

Figure 5. Slope analysis of the moderation effect of PR

Note: *Blue line = Less political regulation; Green line = High political regulation; Red line = No political regulation*

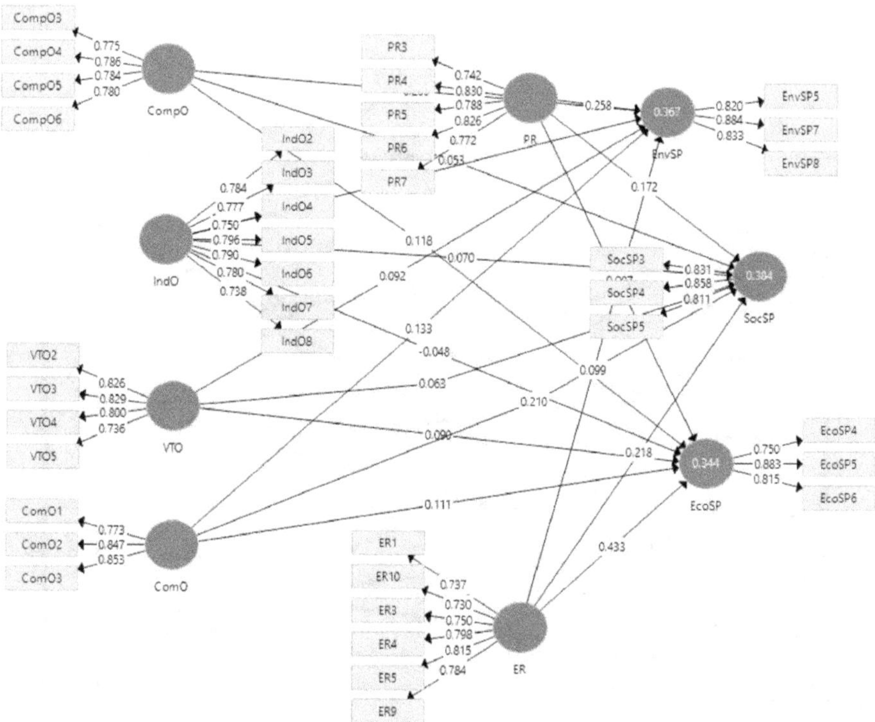

stakeholders to come out with appropriate rules and regulations that will improve economic sustainability performance. For instance, for GTA to determine the fees charged at tourism destination by tourism businesses, it must consult customers, tourism businesses, and other essential stakeholders so that the fees charged will be affordable for visitors so as to improve patronage at the tourism destination, thereby leading to an improvement in revenue generation. It is also possible that, officials of GTA might be compromised hence, do not enforce the regulations to its latter which might result in poor sustainability performance at the tourism destinations.

Hypothesis 5b: ***Moderating* effect of *ecological regulation positively predicts the relationship between StMO and SP.*** In testing hypothesis *5b*, findings from this analysis established that, ecological regulation positively and significantly impacts StMO-SP relationship hence, the hypothesis *5b* is supported. This finding is in agreement with studies by Akram et al., (2019), where it concluded that, there is a positive and significant effect of the moderating role of institutional regulation on corporate governance and environmental performance. The plausible reason for this finding is that, ecological institutions like the EPA, Water Resource Commission, and the Fire Service have been strict in regulating and enforcing tourism businesses at the southern part of Ghana to comply with environmental management practice.

RESEARCH IMPLICATIONS

The study has both theoretical and managerial implication. Moderating the StMO-SP relationship with state regulations also contributes to the moderation literature, as moderators help to determine the strength or otherwise of the relationship. In this study, it has been noticed that, while political regulation moderated the StMO-SP negatively, the ecological regulation is significantly positive on the relationship. This therefore means that ecological regulatory institutions should continue to enforce the rules to improve environmental sustainability or compliance while political regulatory institutions revisit their rules and mode of regulation. The study also contributes to theory by developing a comprehensive conceptual framework which will guide future researchers in their study. The framework involves state regulation as a moderating variable. Introduction of state regulations (political and ecological) as a moderation variable is a modest contribution to theory. The study was able to integrate the stakeholder and the institutional theory in the study of StMO-SP nexus and this assisted in further explanation of the variables.

The study proposes some managerial implications some of which include aiding tourism destination managers and planners who are involved in using StMO in improving sustainability management practices to take into account the role of key stakeholders like political and ecological regulations in fashioning and implementing their sustainability strategies. The study will also assist state regulators in their oversight responsibilities on tourism businesses in complying with sustainability practices at the tourism destinations. The study also reveals that political regulators must relook at their role in sustainability practices, as it is now, they are negatively impacting sustainability performance at the tourism destinations. The study will therefore, inform them to review their regulations to better improve sustainability performance at the tourism destinations.

FUTURE RESEARCH DIRECTIONS

Despite the above contributions, our results should be limited to the tourism destinations. One of the first limitations of the study is that it was a cross sectional survey and so longitudinal study should be undertaken in the future. The study is based on moderation and so future study should undertake mediation. The study is restricted to the tourism sector and a result generalisation was limited to the tourism destinations. Future research can examine other sectors such as banking, telecommunication, higher education, and manufacturing sectors especially small and medium sized enterprises. Furthermore, future study should consider the segregation of each StMO dimensions to SP, as well as moderating state regulation on each of the StMO dimensions on SP.

CONCLUSION

It must be noted that there are very key managerial implications that can be derived from these research findings. To management of tourism destinations, this research comes in handy as it will help tourism business managers to identify key stakeholders from whom information can be generated and disseminated to fashion out sustainability management strategies that will help improve sustainability performance at the TDs. If this is done, it will help TD businesses to be more competitive and attract more visitors to their destinations, and hence become more profitable. The last but not least conclusion is that ecological regulation significantly impacts positively on sustainability performance as has been asserted other scholarly findings (Akram et al., 2019). However, political regulation does not impact sustainability performance. This implies that, StMO alone is not enough to support sustainability. However, ecological regulation by ecological institutions is needed to strengthen SP. However, political regulations that are applied to improve sustainability performance needs to be revisited as they are not providing effective regulation such as institutions like GTA must revisit its regulations that are used for these tourism businesses with regards to SP.

REFERENCES

Akram, F., Abrar-ul-Haq, M., & Raza, S. (2019). a Role a Corporate Governance and Firm'S Environmental Performance: A Moderating Role of Institutional Regulations. *International Journal of Management Studies*, *25*(2), 19–37. doi:10.32890/ijms.25.2.2018.10498

Crittenden, V. L., Crittenden, W. F., Ferrell, L. K., Ferrell, O. C., and Pinney, C. C. (2011b). *Market-oriented sustainability : a conceptual framework and propositions*, 71–85. doi:10.1007/s11747-010-0217-2

Dabija, D. C., Postelnicu, C., Dinu, V., & Mihăilă, A. (2017). Stakeholders' perception of sustainability orientation within a major Romanian University. *International Journal of Sustainability in Higher Education*, *18*(4), 533–553. doi:10.1108/IJSHE-10-2015-0169

Donaldson, T., & Preston, L. E. (1995). The Stakeholder Theory of the Corporation: Concepts, Evidence, and Implications. *Academy of Management Review*, *20*(1), 65–91. doi:10.2307/258887

FC. (2018). *National Forest Plantation Development Programme. Annual Report 2018*. Forest Services Division, Forestry Commission.

Fornell, C., and Larcker, D. F. (1981). *Structural equation models with unobservable variables and measurement error: Algebra and statistics*.

Freeman, R. E. (1999). Response : Divergent Stakeholder Theory. *Academy of Management, 24*(2), 233–236.

Green, K. W., Toms, L. C., & Clark, J. (2015). Impact of market orientation on environmental sustainability strategy. *Management Research Review*, *38*(2), 217–238. doi:10.1108/MRR-10-2013-0240

Hair, J. F., Hult, G. T. M., & Ringle, C. M. (2014). Partial Least Squares Structural Equation Modeling (Pls-Sem) (pp. 1–329).

Hair, J. F., Risher, J. J., Sarstedt, M., & Ringle, C. M. (2019). When to use and how to report the results of PLS-SEM. *European Business Review*, *31*(1), 2–24. doi:10.1108/EBR-11-2018-0203

Henseler, J., Ringle, C. M., & Sarstedt, M. (2015). A new criterion for assessing discriminant validity in variance-based structural equation modeling. *Journal of the Academy of Marketing Science*, *43*(1), 115–135. doi:10.100711747-014-0403-8

Hult, G. T. M., Mena, J. A., Ferrell, O. C., & Ferrell, L. (2011). Stakeholder marketing: A definition and conceptual framework. *AMS Review*, *1*(1), 44–65. doi:10.100713162-011-0002-5

Hwang, D., Stewart, W. P., & Ko, D. (2012). Community Behavior and Sustainable Rural Tourism Development. *Journal of Travel Research*, *51*(3), 328–341. doi:10.1177/0047287511410350

Jaworski, B. J., & Kohli, A. K. (1993). Market Orientation : Antecedents and Consequences. *Journal of Marketing*, *57*(3), 53–70. doi:10.1177/002224299305700304

Kohli, A. K., Jaworski, B. J., & Kumar, A. (1993). MARKOR : A Measure of Market Orientation, *30*(4), 467–477.

Lawrence, T. B., and Shadnam, M. (2008). Institutional theory. *The international encyclopedia of communication.*

Li, C., Cao, C., Zhang, L., Chen, X., Ren, S., & Zhao, Y. (2017). Effects of corporate environmental responsibility on financial performance: The moderating role of government regulation and organizational slack. *Journal of Cleaner Production*, *166*, 1323–1334. doi:10.1016/j.jclepro.2017.08.129

Li, T., Tang, F., & Zhang, L. (2020). Differential effects of voluntary environmental programs and mandatory regulations on corporate green innovation. *Natural Hazards*, *103*(3), 3437–3456. doi:10.100711069-020-04137-y

Line, N. D., & Wang, Y. (2017a). A multi-stakeholder market oriented approach to destination marketing. *Journal of Destination Marketing & Management*, *6*(1), 84–93. doi:10.1016/j.jdmm.2016.03.003

Mahmoud, B. & Hinson. (2017). *Market orientation and corporate social responsibility : towards an integrated conceptual framework,* 1–15. doi:10.1186/s40991-017-0020-1

Mitchell, W., Wooliscroft, B., & Higham, J. (2010). Sustainable Market Orientation : A New Approach to Managing Marketing Strategy. *Journal of Macromarketing*, *30*(2), 160–170. doi:10.1177/0276146710361928

Musa, G., Kayat, K., & Thirumoorthi, T. (2010). The experiential aspect of rural home-stay among Chinese and Malay students using diary method. *Tourism and Hospitality Research*, *10*(1), 25–41. doi:10.1057/thr.2009.26

Najeeb, A. (2014). Institutional theory and human resource management. *Being Practical with Theory: A Window into Business Research*, 25–30. http://eurekaconnection.files.wordpress.com/2014/02/p-25-30-institutional-theory-theori-ebook_finaljan2014-v3.pdf%0Aresearch-pubs@uow.edu.au

Narver, J. C., & Slater, S. F. (1990). The Effect of a Market Orientation on Business Profitability. *Journal of Marketing*, *54*(4), 20–35. doi:10.1177/002224299005400403

Nkamnebe, A. D. (2011). Sustainability marketing in the emerging markets: Imperatives, challenges, and agenda setting. *International Journal of Emerging Markets*, *6*(3), 217–232. doi:10.1108/17468801111144058

Nunnally, J. C., & Bernstein, I. H. (1994). *Psychometric Theory* (3rd ed.). McGraw-Hill, Inc.

Ortiz-de-Mandojana, N., Aguilera-Caracuel, J., & Morales-Raya, M. (2016). Corporate Governance and Environmental Sustainability: The Moderating Role of the National Institutional Context. *Corporate Social Responsibility and Environmental Management*, *23*(3), 150–164. doi:10.1002/csr.1367

Polo Pena, A. I., Frias Jamilena, D. M., & Rodriguez Molina, M. A. (2013). Market orientation as a strategy for the rural tourism sector: Its effect on tourist behavior and the performance of enterprises. *Journal of Travel Research*, *52*(2), 225–239. doi:10.1177/0047287512459108

Reuter, C., Goebel, P., & Foerstl, K. (2012). The impact of stakeholder orientation on sustainability and cost prevalence in supplier selection decisions. *Journal of Purchasing and Supply Management*, *18*(4), 270–281. doi:10.1016/j.pursup.2012.06.004

Roberts, S., & Tribe, J. (2008). Sustainability indicators for small tourism enterprises — An exploratory perspective. *Journal of Sustainable Tourism*, *16*(5), 575–594. doi:10.1080/09669580802159644

Shmueli, G., Sarstedt, M., Hair, J. F., Cheah, J.-H., Ting, H., Vaithilingam, S., & Ringle, C. M. (2019). Predictive model assessment in PLS-SEM: Guidelines for using PLSpredict. *European Journal of Marketing*, *53*(11), 2322–2347. doi:10.1108/EJM-02-2019-0189

WTTC. (2018). *Travel & Tourism Economic Impact 2018 World*. London: WTTC. https://www.wttc.org/-/media/files/reports/economic-impactresearch/regions-2018/world2018.pdf

WTTC. (2019). *Home*. World Travel & Tourism Council. http://www.wttc.org

Zhu, Q., & Sarkis, J. (2007). The moderating effects of institutional pressures on emergent green supply chain practices and performance. *International Journal of Production Research*, *45*(18–19), 4333–4355. doi:10.1080/00207540701440345

APPENDIX A

Table 1. Moderating effect of ER and PR

Hypotheses	Path	β	T Statistics	P Values	R^2	Decision
H_{5a}	PR*StMO => SP	-0.062	1.191	0.234	0.526	Rejected
H_{5b}	ER*StMO => SP	0.136	2.719	0.007		Supported

Table 2. Descriptive statistics for state regulations (SR)

Code	Statements	Mean	Std. Deviation	Skewness	Kurtosis
Political Regulations (PR)		***3.66***	***0.768***	***-0.422***	***0.371***
PR1	The state regulates our business in terms of the fees charged	3.59	1.078	-0.681	0.007
PR2	The state provides standards within which our business operates	3.80	0.994	-0.743	0.377
PR3	There are state institutions that provide training for our staff regarding tourism issues	3.45	1.163	-0.450	-0.502
PR4	Our business inputs concern from tourism regulators in making business decisions.	3.58	1.029	-0.545	-0.005
PR5	There are laws regarding compliance with the minimum wage payments for employees	3.65	1.023	-0.518	0.048
PR6	There are laws regulating the provision of certain services at destinations	3.71	0.956	-0.491	0.101
PR7	There are laws regarding safety and security standards at the destination	3.85	0.948	-0.638	0.374
Ecological Regulations (ER)		***3.57***	***0.722***	***-0.344***	***0.074***
ER1	Our business engages the services of the Fire Service to educate the local community about causes and effects of bush fires	3.31	1.156	-0.296	-0.654
ER2	Our destination engages the services of the Forestry Commission to improve and protect the vegetation	3.18	1.111	-0.254	-0.623
ER3	There are laws regarding the protection of water bodies at the destination	3.63	0.966	-0.468	0.047
ER4	There are laws regarding forest preservation at the destination	3.37	1.056	-0.495	-0.099
ER5	There are laws regarding the protection of animals at destinations	3.44	1.060	-0.418	-0.187
ER6	There are laws regulating the disposal of litter at destinations	3.81	1.020	-0.978	0.919
ER7	There are laws governing the disposal of liquid & solid waste at destinations in Ghana	3.85	0.929	-0.833	0.804
ER8	There are regulations governing the separation of waste at destinations in Ghana	3.66	0.923	-0.628	0.507
ER9	There are laws regarding air pollution at destinations in Ghana	3.77	0.928	-0.737	0.724
ER10	There are laws regulating noise pollution at destinations in Ghana	3.72	0.994	-0.558	0.022
ER11	There are laws preventing mining and sand winning from the destination	3.54	1.141	-0.565	-0.302
State Regulations (SR)		**3.62**	**0.678**	**-0.337**	**-0.021**

Source: Fieldwork, 2022

Table 3. Results on reliability and convergent validity: final iteration

Variable	Indicator	Loading	CA	rho_A	CR	AVE
ComO	ComO1	0.773	0.768	0.787	0.864	0.680
	ComO2	0.847				
	ComO3	0.853				
	ComO4	Omitted				
	ComO5	Omitted				
	ComO6	Omitted				
IndO	IndO1	Omitted	0.889	0.893	0.913	0.599
	IndO2	0.784				
	IndO3	0.777				
	IndO4	0.750				
	IndO5	0.796				
	IndO6	0.790				
	IndO7	0.780				
	IndO8	0.738				
	IndO9	Omitted				
VTO	VTO1	Omitted	0.810	0.810	0.876	0.638
	VTO2	0.826				
	VTO3	0.829				
	VTO4	0.800				
	VTO5	0.736				
	VTO6	Omitted				
	VTO7	Omitted				
CompO	CompO1	Omitted	0.788	0.791	0.862	0.610
	CompO2	Omitted				
	CompO3	0.775				
	CompO4	0.786				
	CompO5	0.784				
	CompO6	0.780				
	CompO7	Omitted				

continued on following page

Table 3. Continued

Variable	Indicator	Loading	CA	rho_A	CR	AVE
ER	ER1	0.737	0.862	0.862	0.897	0.593
	ER2	Omitted				
	ER3	0.750				
	ER4	0.798				
	ER5	0.815				
	ER6	Omitted				
	ER7	Omitted				
	ER8	Omitted				
	ER9	0.784				
	ER10	0.730				
	ER11	Omitted				
PR	PR1	Omitted	0.851	0.854	0.894	0.628
	PR2	Omitted				
	PR3	0.742				
	PR4	0.830				
	PR5	0.788				
	PR6	0.826				
	PR7	0.772				
EcoSP	EcoSP1	Omitted	0.751	0.767	0.857	0.668
	EcoSP2	Omitted				
	EcoSP3	Omitted				
	EcoSP4	0.750				
	EcoSP5	0.883				
	EcoSP6	0.815				
EnvSP	EnvSP1	Omitted	0.801	0.806	0.883	0.716
	EnvSP2	Omitted				
	EnvSP3	Omitted				
	EnvSP4	Omitted				
	EnvSP5	0.820				
	EnvSP6	Omitted				
	EnvSP7	0.884				
	EnvSP8	0.833				

continued on following page

Table 3. Continued

Variable	Indicator	Loading	CA	rho_A	CR	AVE
SocSP	SocSP1	Omitted	0.781	0.784	0.872	0.695
	SocSP2	Omitted				
	SocSP3	0.831				
	SocSP4	0.858				
	SocSP5	0.811				
	SocSP6	Omitted				

Table 4. Results of Heterotrait-Monotrait ratio (HTMT) analysis criterion of discriminant validity

Variable	1	2	3	4	5	6	7	8	9
1. ComO									
2. CompO	0.545								
3. ER	0.611	0.623							
4. EcoSP	0.498	0.508	0.690						
5. EnvSP	0.526	0.582	0.570	0.610					
6. IndO	0.589	0.739	0.716	0.468	0.473				
7. PR	0.589	0.507	0.764	0.504	0.616	0.559			
8. SocSP	0.619	0.501	0.652	0.691	0.687	0.537	0.615		
9. VTO	0.503	0.574	0.562	0.468	0.512	0.555	0.617	0.500	-

Chapter 16
Assessment of Environmental Knowledge:
The Case of the Ghanaian Contractor

Nana Benyi Ansah
https://orcid.org/0000-0003-2400-1406
Accra Technical University, Ghana

Isaac Aidoo
https://orcid.org/0000-0003-1081-2814
Accra Technical University, Ghana

Justice Agbevade
Accra Technical University, Ghana

ABSTRACT

While developed countries continue to progress in environmental impact mitigation, global south countries have immature environmental mitigation and performance strategies. The need for formidable knowledge of various environmental themes and categories is thus essential to educate construction contractors on the awareness of their environmental challenges. Knowledge of sustainable standards provides the pathway to achieving the U.N.'s sustainable development goals (SDGs) in providing construction sustainability. This study assessed the awareness level of the Ghanaian construction contractor using climate change as the environmental theme. A survey was carried out on the theme using a 5-point Likert scale. The survey was conducted using purposive and snowball sampling methods. The contractor's environmental consciousness level was estimated using linear regression analysis to determine their grasp of climate change. According to the findings, the Ghanaian construction contractor is adequately informed of climate change.

DOI: 10.4018/978-1-6684-9843-9.ch016

INTRODUCTION

Global warming and climate change have become synonymous in the environmental research lexicon. While developed countries continue to progress in environmental impact mitigation, global south countries have immature environmental mitigation and performance strategies (Durdyev et al., 2018). According to Maepa et al. (2017), only 31 research articles on environmental impacts from Nigeria, Ghana, and Ivory Coast, published between 2000 and 2016, were available on databases; fifteen were found on Google Scholar, and twelve on Elsevier and four on Springer. The need for formidable knowledge of various environmental themes and categories is thus essential to educate construction contractors on the awareness of their environmental challenges. Knowledge of sustainable standards provides the pathway to achieving the U.N.'s sustainable development goals (SDGs) in providing construction sustainability. Construction projects are well-acknowledged to have a significant environmental impact (Elmasoudi et al., 2019; Durdyev et al., 2018; Sakr et al., 2010; Liyin & Hong, 2006). Alsanad (2015) claims that more effort is needed to improve stakeholder understanding of sustainable development because general awareness and implementation techniques are lacking.

Environmental knowledge and management are essential motivators for meeting project management objectives in environmental performance. According to Hossain et al. (2020), governments must work together to combat climate change due to the global environmental problem. Therefore, improving contractor motivations at the environmental knowledge and management level is thus seen as a proactive strategy for achieving the desired environmental results (Liyin & Hong, 2006). The knowledge needed to illuminate and reduce the effects of environmental constraints comes from understanding the environmental footprint of construction activities. Undoubtedly, contractors' awareness of their environmental responsibilities is guaranteed to help identify the source of the most extensive damage, allowing mitigation targets to be developed. However, according to Quitmann et al. (2021), most organisations are unaware of their environmental footprints and consequently unable to track and minimise their effects. The aim of this study is thus to assess the Ghanaian construction contractor's understanding of environmental issues by focusing on climate change. To achieve this aim, the contractors' level of environmental awareness is estimated by analysing their level of environmental knowledge. Ansah et al. (2022) identified informed construction decisions as a forerunner to boost construction professionals' confidence and demonstrate in-depth accounts of leading alternative environmental mitigation solutions. Again, Rahman et al. (2020) demonstrated the understanding and use of environmental activities as instruments for managing relationships with stakeholders, who benefit from green performance in terms of market share and customer awareness to accrue valuable performance rewards against environmentally focused activities. Therefore, despite the significant contributions to creating awareness to mitigate the environmental burdens of projects, the assessment of environmental knowledge and subsequent integration into sustainable development still lacks the attention it deserves.

BACKGROUND

The sustainability paradigm affirms that buildings affect climate change, and this change is now a global reality.

In the building and construction sector, all aspects of a building's life cycle – design, raw material extraction, processing, manufacturing, construction, usage, and maintenance impact the environment (Bilec et al., 2009). There are growing demands on identifying the environmental attributes of construction items linked to their impact on the environment to explain the worth of a material's or product's sustainability qualities in the commercial competition. Also, poor construction practices and high environmental material impact (especially in the production phase) contribute to the challenge of climate change (Piccardo et al., 2019; Hong et al., 2019); and, more specifically, contribute to acid rain and eutrophication, especially within the catchment area of the project. As a result, a good grasp of environmental knowledge reduction measures is required for the correct construction configuration of various building elements contributing to the project, including their selection and use to mitigate unacceptable environmental impacts.

CLIMATE CHANGE

Greenhouse gas emissions, the most significant of which is carbon dioxide, are the principal cause of climate change and are usually the primary component for calculating global warming potential (GWP). According to Araujo et al. (2020), the reaction to climate change is divided into mitigation and adaptation strategies. Mitigation strategies help minimise emissions of heat-trapping gases, whereas adaptation entails improving one's ability to avoid or manage negative consequences while taking advantage of positive consequences for the present and future. The Global warming potential is usually measured using a hundred-year baseline (GWP100). The GWP100 is expressed in kg of CO2 equivalents (kgCO_2e). Ritchie & Roser (2020) state that manufacturing and construction processes account for nearly one-third of world emissions. Several countries have either developed or vowed climate-reduction plans, with evidence to support their claims. Despite little documentation on Ghana's climate change policies, evidence backing their implementation is scarce (Ritchie

& Roser, 2020). Climate change is a cause for concern because Lindsey (2020) claims that the growth of the quantities of carbon dioxide is responsible for nearly 67% of the overall energy imbalance, causing the rise in temperature on the Earth. Debrah et al. (2021) acknowledged the need to create environmental awareness to control emissions and decrease climate change. Their study identified behavioural and attitudinal change as significant factors in pushing the call for a decrease in climate change. Hossain et al. (2020) demonstrated that awareness does not translate immediately into implementations and that practical application of environmental knowledge and practices are needed. Whilst the awareness of organisations' carbon reduction strategies is improving, there is still a missing link between awareness and behaviours (Lu et al., 2012).

ENVIRONMENTAL KNOWLEDGE

Organisations have been more cautious in managing knowledge at the organisational level to favourably effectuate project decisions since global access to information was implemented. However, Severo et al. (2019) raised concerns about the management of environmental awareness because environmental challenges have global implications. Environmental awareness inspires new consumption behaviour, advancing the green and sustainability goals of the contractor. Nonetheless, there is a paucity of knowledge in integrating environmental information into the contractors' project sustainability performance (Pham & Pham, 2021). Their research concluded that environmental knowledge must be developed and integrated into an organisation's environmental management processes. According to Rahman et al. (2020), having an environmental knowledge repository improves the efficiency of legal and environmental reporting. Being environmentally conscious involves expressing gratitude for nature and, as a result, the ability to take responsibility for it. Mensah et al. (2021) reiterated Du Plessis & Brandon's (2015) proposal for attitudinal changes to promote regenerative sustainability (to adapt to approaching changes in the planetary system into the globe) in light of these environmental knowledge requirements. Environmental awareness and knowledge will decouple the shortfall in knowledge management, act as a prerequisite for fighting environmental concerns, and boost policymakers' attempts to deal with environmental strains. Table 1. provides a study on a brief overview of environmental awareness in developing countries to illuminate the subject to reflect the current status of the study area in the global south.

Table 1. A summary of environmental awareness in developing countries

Author and Date	Purpose	Finding	Background
On and Techapeer-aparnich, (2021)	To identify the barriers to the successful implementation of sustainable construction in Cambodia's construction industry.	Government entities have a common understanding of sustainable construction; however, there are no government-supported procedures for developing building regulations and financial incentive schemes for sustainable construction projects.	Google Scholar
Hossain et al., (2020)	To examine the impact of environmental awareness and green technology usage on sustainable green practices in Bangladesh.	Technology usage significantly influences environmental awareness and green practices.	Research Gate
Chen et al. (2019)	To study the impacts of environmental quality and income-pollution relationship.	Environmental awareness has a positive impact on environmental policy, environmental quality, and the relationship between income and pollution. Environmental awareness has an impact on how much money is spent on the environment.	Economic Modelling
Durdyev et al., (2018)	To investigate stakeholder awareness and the factors hindering the adoption of sustainable construction in Cambodia.	Due to a lack of education, knowledge and awareness, stakeholder participation and the adoption of green building practices are poor.	Research Gate
Darko &Chan (2018)	To identify the important strategies to promote Green Building Technology (GBT) adaptation with reference to the Ghanaian Construction Market	More publicity, education and training programmes relevant to key stakeholders and awareness through research and development must be considered in GBT adaptation.	Building and Environment
Vyas & Jha (2016)	Integrating the green building approach to achieve environmental, social and economic impacts associated construction	The need to increase environmental awareness among stakeholders to reduce environmental impacts.	Sustainable cities And societies
Younis (2015)	To identify the effect of the thought processes of leaders who make environmental decisions using Liberia as an example.	Environmental awareness should be incorporated into any project's short and long-term impact analysis to provide the required validation.	Routledge (Taylor and Francis)

METHODOLOGY

The Ghanaian construction industry classifies construction companies as small, medium and large contractors. These firms are identified by the codes D4K4, D3K3, D2K2 and D1K1, respectively. According to Moo & Eyiah (2020), large-scale contractors can undertake projects of unlimited value. Climate change as a theme was assessed to reflect the environmental consciousness of the Ghanaian construction contractor. Data was collected through a survey of professionals from the contractor classifications using purposive and snowball sampling via the Google form platform (Saunders et al., 2019). In all, 16 responses were received. The questionnaire was adapted from Akadiri's (2011) existing literature using a 5-point Likert scale. The scaling ranged from least important to most important. Quantitative methodology and descriptive research using the mean values and linear regression tests were adopted to predict the awareness level of contractors on climate change. The study used SPSS version 26 for the analysis. In the analysis, Eutrophication Potential, Global warming potential, Ozone Depletion Potential, and Acidification Potential were predictors. Climate change was set as the response variable.

RESULTS

This study assessed the Ghanaian construction contractor's understanding of environmental issues by assessing their knowledge of climate change. It was hypothesised that Ghanaian construction contractors have low knowledge of environmental issues regarding climate change. Table 2. shows the demographic data of the respondents. The mean values of the study's five climate change predictors of the awareness level of the various respondents are illustrated in Table 3. The minimum mean awareness value for carbon emissions (global warming potential) was 3.25. Ozone depletion and acidification are two predictors that indicate promising knowledge awareness. Although awareness may not always convert into action, it is a valuable tool for understanding various topics. The survey values for environmental criteria were analysed for their Correlation. From Table 4. Pearson's correlation coefficient of 99% and confidence interval of 95% were greater than zero. Since a strong correlation was established, a linear regression model was created to test the hypothesis.

The R Square value is 0.933, which means the predictors can explain 93.3% of the variance.

From Table 6. (ANOVA table), the p-value (0.000014) is less than the alpha-value(0.05). We reject our null hypothesis and conclude that the model is significant. Collectively, $(F(5,10) = 27.936)$, $p < 0.05$.

From Table 7. the linear model would be Climate Change = 6.587 + 2.505 (Environmental impact) + 2.007 (Global warming potential) + 2.698 (Ozone depletion potential) + 3.493 (Acidification Potential) + 2.727 (Environmental impact) + 2.505 (Environmental impact) + 2.007 (Global warming potential) + 2.698 (Ozone depletion potential) + 3.493 (Acidification Potential) + 2.727 (Environmental (Eutrophication Potential). The model in Table 6. shows that each

Table 2. Demographic data of respondents

Variable	Group	Frequency	Per cent
Type of organisation	Architectural Design	1	6.3
	Real Estate	1	6.3
	Contractor	3	18.8
	Engineering	5	31.3
	Q.S	1	6.3
	PM	2	12.5
	Others	3	18.8
	Total	16	100
Size of Company	<10 staff	5	31.3
	11-30 Staff	4	25
	31-49 Staff	2	12.5
	50-69staff	1	6.3
	70-99 Staff	1	6.3
	> 100 staff	3	18.8
	Total	16	100
Type of Building speciality	Residential	3	18.8
	Industrial	1	6.3
	Commercial	6	37.5
	Institutional	5	31.3
	Others	1	6.3
	Total	16	100
Experience	Less than 6 Years	1	6.3
	6-10 Years	5	31.3
	11-20 Years	6	37.5
	31-40 Years	1	6.3
	>40 Years	3	18.8
	Total	16	100

Table 3. Level of awareness on climate change

Descriptive Statistics	Mean	Std. D	N
Environmental impact during material harvest or extraction	3.94	1.3	16
Global warming Potential	3.25	1.4	
Ozone Depletion Potential	4.19	1.4	16
Acidification Potential	4.19	1.2	16
Eutrophication Potential	3.75	1.4	16

Table 4. Correlation of environmental criteria

Pearson Correlations for main study variables	Environmental impact during material harvest or extraction	Ozone Depletion Potential	Acidification Potential	Eutrophication Potential
Ozone Depletion Potential	.796**			
Acidification Potential	.811**	.928**		
Eutrophication Potential	.586*	.618*	.696**	
Climate_Change	.807**	.887**	.911**	.795**
** Correlation is significant at the 0.01 level (2-tailed).				
* Correlation is significant at the 0.05 level (2-tailed).				

Table 5. Model Summary

Model	R	R^2	Adj R^2	RMSE
1	.966a	0.933	0.9	4.74623

Table 6. ANOVA[a]

Model		SS	df	M S	F	Sig.
1	Regression	3146.483	5	629.297	27.936	.000b
	Residual	225.267	10	22.527		
	Total	3371.75	15			

Table 7. Coefficients

Model		B	S.E.	β	t	p
1	(Constant)	6.587	4.925		1.338	0.211
	Environmental impact during material harvest or extraction	2.505	1.716	0.215	1.46	0.175
	Global warming potential	2.007	0.974	0.186	2.06	0.066
	Ozone Depletion Potential	2.698	2.479	0.248	1.088	0.302
	Acidification Potential	3.493	3.071	0.285	1.138	0.282
	Eutrophication Potential	2.727	1.29	0.253	2.114	0.061

β: Standardised regression coefficient

variable's coefficients are positive. Environmental impact during material harvest or extraction (= 0.215, t = 1.46, p = 0.175), Global warming potential (=0.186, t=2.06, p=0.066), Ozone Depletion Potential (= 0.248, t=1.088, p=0.302), Acidification Potential (=0.285, t=0.1138, p= 0.282), and Eutrophication Potential (= 0.253, t= 2.114, p= 0.0061). The study concludes that the Ghanaian contractor is sufficiently aware of environmental issues (climate change predictors) and is thus knowledgeable and conscious of climate change.

DISCUSSION

The construction industry greatly contributes to global warming (Estokova et al., 2022; Ansah et al., 2022; Feng, 2020). A building is a large and complex unique design. The manufacturer is the contractor, and the construction typically occurs at the site (factory). Therefore, the manufacturer of buildings (contractor) must possess a high knowledge of climate change to help build sustainably. This study also contributes to developing knowledge in the built environment sustainability paradigm by determining whether the construction contractor, as a critical stakeholder in the industry, has sufficient knowledge to understand the global warming discourse and the 'fight' against climate change. According to Shengju et al. (2019), implementing environmental awareness can speed up the innovation and development of construction technology to reduce carbon dioxide emissions, the primary contributor to global warming. The high environmental awareness can also translate into constructing more environmentally friendly buildings. Addy et al. (2020) explained the lack of awareness as the absence of public knowledge on sustainable construction. Therefore the challenge to improve the climate change paradigm may be extended to all construction stakeholders to ensure the spread of knowledge since construction is a teamwork

FUTURE RESEARCH DIRECTIONS

Research on the relationship between the awareness of environmental performance and economic growth of the construction industry can help create the required sustainability metrics in construction project success requirements. Future research can focus on the relationship between the environment, economy, and society. This way, stakeholder knowledge may focus on material performance to create the most sustainable structures.

CONCLUSION

It is widely acknowledged that the building industry has significant potential to lessen its negative environmental impact. As a result, the construction contractor requires awareness to help reduce the harmful consequences of construction activities on the environment. This study assessed the environmental awareness of the Ghanaian building construction contractor by focusing on climate change. The survey found that Ghanaian contractors understand climate change and are knowledgeable in climate-related topics. This understanding suggests that the Ghanaian construction contractor is informed about climate change. Though the contractor is knowledgeable on climate change issues, strengthening the understanding of the Ghanaian contractor's knowledge through workshops, seminars, conferences, and training will go a long way towards achieving climate change consciousness. Further research is required to look at a broader range of indicators to have a complete picture of environmental awareness and determine whether this awareness has been translated into action to provide the consciousness within the construction industry to build sustainably. Because the construction industry is a significant environmental polluter, the Ghanaian construction contractor can lead in mitigating climate change by translating awareness into action.

ACKNOWLEDGEMENT

This research received no specific grant from any funding agency in the public, commercial, or not-for-profit sectors.

REFERENCES

Addy, M., Adinyira, E., Danku, J. C., & Dadzoe, F. (2020). Impediments to the Development of the Green Building Market in Sub-Saharan Africa: The case of Ghana. *Smart and Sustainable Built Environment*, *10*(2), 193–207. doi:10.1108/SASBE-12-2019-0170

Akadiri, O. P. (2011). *Development of a Multi-Criteria Approach for Selecting Sustainable Materials for Building Projects*. PhD Thesis - The University of Wolverhampton. http://wlv.openrepository.com/wlv/bitstream/2436/129918/1/Akadiri_PhD thesis.pdf

Alsanad, S. (2015). Awareness, Drivers, Actions, and Barriers of Sustainable Construction in Kuwait. *Procedia Engineering, 118*, 969–983. doi:10.1016/j.proeng.2015.08.538

Ansah, M. K., Chen, X., & Yang, H. (2022). A Holistic Environmental and Economical Design Optimisation of Low Carbon Buildings Considering Climate Change and Confounding Factors. *The Science of the Total Environment, 821*, 153442. doi:10.1016/j.scitotenv.2022.153442 PMID:35092783

Ansah, N. B., Adinyira, E., Agyekum, K., & Aidoo, I. (2022). Bibliometric Study on Particle Emissions of Natural and Alternative Building Materials. In M. Gorse, C., Scott, L., Booth, C., Dastbaz (Ed.), *Managing, Building, and Delivering the Sustainable Development Goals* (pp. 495–507). Springer International Publishing. doi.org/10.1007/978-3-030-79450-7_37

Araujo, G. A., Carneiro, A. M. P., & Palha, P. R. (2020). Predictive Methodology for the Quantification of Environmental Aspects in Urban Infrastructures. *Sustainability (Basel), 12*(18), 7636. doi:10.3390u12187636

Bilec, M. M., Ries, R. J., & Matthews, H. S. (2009). Life-Cycle Assessment Modeling of Construction Processes for Buildings. *Journal of Infrastructure Systems, 16*(3), 199–205. doi:10.1061/(ASCE)IS.1943-555X.0000022

Chen, X., Huang, B., & Te Lin, C. (2019). Environmental Awareness and Environmental Kuznets Curve. *Economic Modelling, 77*, 2–11. doi:10.1016/j.econmod.2019.02.003

Darko, A., & Chan, A. P. C. (2018). Strategies to Promote Green Building Technologies Adoption In Developing Countries: The Case of Ghana. *Building and Environment, 130*, 74–84. doi:10.1016/j.buildenv.2017.12.022

Debrah, J. K., Vidal, D. G., & Dinis, M. A. P. (2021). Raising Awareness on Solid Waste Management Through Formal Education for Sustainability: A Developing Countries Evidence Review. *Recycling, 6*(1), 1–21. doi:10.3390/recycling6010006

Du Plessis, C., & Brandon, P. (2015). An Ecological Worldview as the Basis for a Regenerative Sustainability Paradigm for the Built Environment. *Journal of Cleaner Production, 109*, 53–61. doi:10.1016/j.jclepro.2014.09.098

Durdyev, S., Zavadskas, E. K., Thurnell, D., Banaitis, A., & Ihtiyar, A. (2018). Sustainable Construction Industry in Cambodia: Awareness, Drivers and Barriers. *Sustainability (Basel), 10*(2), 1–19. doi:10.3390u10020392

Elmasoudi, I., Ibrahim, M. G., Tokimatsu, K., & Elbeltagi, E. (2019). Environmental Impact Assessment Model For Buildings' Construction Activities. *International Journal of Construction Management*, *0*(0), 1–12. doi:10.1080/15623599.2019.1657256

Estokova, A., Fabianova, W. M., & Marcela, O. (2022). Concrete Structures and Their impacts on Climate Change and Water and Raw Material Resource Depletion. *International Journal of Civil Engineering*, *13*. doi:10.1007/s40999-022-0701-8

Hong, J., Hong, T., Kang, H., & Lee, M. (2019). A Framework for Reducing Dust Emissions and Energy Consumption on Construction Sites. *Energy Procedia*, *158*, 5092–5096. doi:10.1016/j.egypro.2019.01.637

Hossain, M. I., Slow, M. L., Ong, S. T., & Said, R. M. (2020). The Role of Environmental Awareness and Green technological Usage to Foster Sustainable Green Practices in Bangladeshi Manufacturing SMEs. *International Journal of Advanced Science and Technology*, *29*(7s), 3115-3124. https://www.researchgate.net/publication/341576178

Liyin, S., Hong, Y., & Griffith, A. (2006). Improving Environmental Performance by Means of Empowerment of Contractors. *Management of Environmental Quality*, *17*(3), 242–257. doi:10.1108/14777830610658674

Lu, Y., Zhu, X., & Cui, Q. (2012). Effectiveness and Equity Implications of Carbon Policies in the United States Construction Industry. *Building and Environment*, *49*(1), 259–269. doi:10.1016/j.buildenv.2011.10.002

Maepa, M., Bodunrin, M. O., Burman, N. W., Croft, J., Engelbrecht, S., Ladenika, A. O., MacGregor, O. S., & Harding, K. G. (2017). Review: Life Cycle Assessments in Nigeria, Ghana, and Ivory Coast. *The International Journal of Life Cycle Assessment*, *22*(7), 1159–1164. doi:10.100711367-017-1292-0

Mensah, S., Nani, G., & Ayarkwa, J. (2021). *Contractors' Adaptation to Environmentally Sustainable Construction: A Micro-Level Implementation Framework*. Smart and Sustainable Built Environment. doi:10.1108/SASBE-11-2020-0174

Moo, F., & Eyiah, A. (2020). Factors Influencing the Growth of Small and Medium Construction Firms in Northern Ghana. *Journal of African Business*, *21*(3), 416–431. doi:10.1080/15228916.2019.1641306

On, K., & Techapeeraparnich, W. (2021). Barriers to Implementation of Sustainable Construction in Cambodia Construction Industry. *ACM International Conference Proceeding Series*, (pp. 213–219). ACM. 10.1145/3473141.3473251

Pham, T., & Pham, H. (2021). Improving the Green Performance Of Construction Projects Through Supply Chain Integration: The Role of Environmental Knowledge. *Sustainable Production and Consumption*, *26*, 933–942. doi:10.1016/j.spc.2021.01.004

Piccardo, C., Dodoo, A., Gustavsson, L., & Tettey, Y. A. (2019). Comparative Life-Cycle Analysis of Building Materials for the Thermal Upgrade of an Existing Building. In *IOP Conference Series: Earth and Environmental Science*. IOP Science. 10.1088/1755-1315/225/1/012044

Quitmann, C., Sauerborn, R., & Herrmann, A. (2021). *Gaps in Reporting Greenhouse Gas Emissions by German Hospitals: A Systematic Grey Literature Review*.

Rahman, M., Aziz, S., & Huges, M. (2020). The product-market performance benefits of environmental policy. Why customer awareness? *Business Strategy and Environment*. WILEY. https://doi.org/DOI:10.1002/bse.2484

Ritchie, H., & Roser, M. (2020). CO_2 and Greenhouse Gas Emissions. *OurWorldInData.Org*.

Sakr, D. A., Sherif, A., & El-Haggar, S. M. (2010). Environmental Management Systems' Awareness: An Investigation of Top 50 Contractors In Egypt. *Journal of Cleaner Production*, *18*(3), 210–218. doi:10.1016/j.jclepro.2009.09.021

Saunders, M. N. K., Lewis, P., & Thornhill, A. (2019). *Research Methods for Business Students* (Eighth Edi). Pearson Education Limited.. doi:10.1108/qmr.2000.3.4.215.2

Severo, E. A., De Guimarães, J. C. F., Dellarmelin, M. L., & Ribeiro, R. P. (2019). The Influence of Social Networks on Environmental Awareness and the Social Responsibility Of Generations. *Brazilian Business Review*, *16*(5), 500–518. doi:10.15728/bbr.2019.16.5.5

Shengju, Z., Rui, W., & Xingang, W. (2019). Study the Influencing Factors of Carbon Emission in the Production Stage of Building Materials and the Countermeasures for Reducing Carbon Emission. *IOP Conference Series. Earth and Environmental Science*, *371*(3), 032082. doi:10.1088/1755-1315/371/3/032082

Vyas, G. S., & Jha, K. N. (2016). Benchmarking Green Building Attributes to Achieve Cost Effectiveness Using a Data Envelopment Analysis. *Sustainable Cities and Society*. doi:10.1016/j.scs.2016.08.028

Younis, K. M. (2015). Views on Potential Methods For Raising Environmental Awareness in Developing Countries: A Study on Social Responsibility Engagement in Liberia. *Global Bioethics*, *26*(2), 128–144. doi:10.1080/11287462.2015.1031469

KEY TERMS AND DEFINITIONS

Attitudinal Change: The constructive behavioural change towards a particular phenomenon due to self-awareness.

Contractor Motivation: The construction contractor's desire to achieve positive environmental footprints due to the awareness of the adverse effects of poor construction practices.

Environmental Theme: The main heading of an environmental footprint as a result of the description of a specific environmental indicator.

Formidable Knowledge: The capacity to have a thorough comprehension of discourse-related concerns.

Global South Countries: Poor and least developed countries which struggle with economic and human development, unemployment, and other issues.

Stakeholder: Parties interested in lessening the effects, in which their actions or inactions may have on the environment.

Sustainable Standard: The appropriate documentation and methods used in regular construction practices to meet the current generation's needs without compromising the following generation's capacity to meet their own needs.

Chapter 17
Assessment of Fresh/Dried Makuea Poo-Ung Consumption and Its Benefit on Recommended Daily Allowance

Frances Betty Fraikue
Takoradi Technical University, Ghana

ABSTRACT

Makuea poo-ung is a seasonal crop which grows abundantly during the rainy periods. The purpose of the chapter was to assess the relation between 50μg/g/ml of nutrients present in fresh/dried makuea poo-ung compared to RDA of children, teenagers, and adults. Laboratory analysis was undertaken to assess nutrients in fresh/dried makuea poo-ung, by the use of proximate analysis, mineral analysis, atomic absorption spectrometer, and liquid chromatography-mass spectrometry. Techniques used to analyse data were SPSS, Independent T. Test, and test for hypothesis. Twenty nutrients were present in fresh/dried makuea poo-ung. Comparing with RDA, it affirmed that a consumption of at least 50 μg/g of makuea poo-ung each day could provide humankind with 12-50% of the dietary allowance needed then serve as a natural food supplement. Assessment of nutrient in makuea poo-ung provided unique results. Hence, commended that consumers patronize makuea poo-ung as natural food supplement.

DOI: 10.4018/978-1-6684-9843-9.ch017

INTRODUCTION

Evolution of eggplant began in the 3rd century in India, in China around the 4th and 5th century, then in Africa around the 9th century (Bhaskar & Kumar, 2015; Sękara, Cebula, & Kunicki, 2007). Historically, it was evidently revealed that production of eggplant was highly concentrated in China - 58% and India - 25%, as the leading producers worldwide (Bhaskar & Kumar, 2015; Li, Chen, Zhuang, & Chen, 2010; Chong, 2005; Matsubara, Kaneyuki, Miyake, & Mori, 2005). Similarly, India was then labeled as the center of diversity and primary growth of varietal eggplant (Sękara et al., 2007). In Africa however, discovery of makuea poo-ung described as semi-domesticated species were gathered mostly in farmlands and backyard gardens (Cassidy, Mukamal, Liu, Franz, Eliassen, & Rimm, 2013; Putra, 2011). The growth of makuea poo-ung commenced as a green plant which had white and yellow flower and with small/tiny fruits. It was much like unripe cherry tomatoes or garden peas that appear in clusters of about 10-15 (Bhaskar and Kumar, 2015; Kwon, Apostolidis, & Shetty, 2004). It has a diameter of about 0.4 inch or 1 cm amidst thorns on the stem (Bliss & Elstein, 2004; Whitaker & Stommel, 2003). The stem appears jade green in colour and leaves appear broad and bright green (Caguiat & Hautea, 2014; Whitaker & Stommel, 2003). Makuea poo-ung is a seasonal fruit vegetable available on the market throughout the year, they grow best and in abundance from June to August then drastically reduce in October due to the onset of the dry season (Bhaskar & Kumar, 2015; Bliss & Elstein, 2004).

Makuea poo-ung is very nutritious, could be consumed fresh or dry and is termed as one of the ten world's best healthiest food (Caguiat & Hautea, 2014). As an eggplant, it is highly presumed as iron chelator (Krisban, 2013; Putra, 2011), yet, consumers patronize it when it is convenient unlike the exotic vegetables. Makuea poo-ung is branded as indigenous, cheap and has cumbersome method of preparation (Buzby, Hyman, Stewart, & Wells, 2011; Habwe, Waling, & Onyango, 2008). Similarly, consumers are not privy to the nutritional percentage gained when makuea poo-ung is consumed daily as food supplement. Still on the gap, the assessment of nutrient presents in fresh and dried makuea poo-ung had neither been reviewed (Umayal Sundari, Neelamegam, & Subramanian, 2013; Omar, Abdullah, Humeid, & Yamani, 2012) nor compared to recommended daily allowances (RDA) to ascertain the utility of this product. In the light of these, there was the need to assess and disseminate knowledge on the benefit derived after consumption of makuea poo-ung in relation to RDA. The main objective of the study was to assess the benefits of consuming at least 50µg/g/l of fresh and dried makuea poo-ung daily, in relation to RDA.

There were numerous nutrients present in all the species of eggplant and the body predominantly needs them for growth and repair and also protection against diseases. They contain a multitude of vitamins and minerals, dietary fibre, proteins, antioxidants, plus phytochemicals which were known to possess antioxidant activity (Concellòn, An˜o´n, & Chaves, 2004; Whitaker & Stommel, 2003; Noda, Kneyuki, Igarashi, & Mori 2000). One chief phytochemical present in eggplant was glucoside, it also contained phenolic compounds (caffeic plus chlorogenic) then flavonoids (nasunin and delphindin). In addition, nasunin was a pigment present in the rind of eggplant (Bhaskar & Kumar, 2015; Cassidy et al., 2013; Matsubara et al., 2005).

Table 1. Main nutrients present in fresh eggplant per 100mg/g

Food Nutrients	Nutrients Present in Raw Eggplant
Energy	100 - 104 kJ
Carbohydrates	4.70 - 5.88 g
Sugars	2.35 - 3.53 g
Dietary fiber	2.80 - 3.40 g
Fat	0.18 - 0.20 g
Protein	0.80 - 1.01 g
Retinol (A)	0.80 - 1.00 mg
Thiamine (B1)	0.040 mg
Riboflavin (B2)	0.04 - 0.11 mg
Niacin (B3)	0.59 - 0.65 mg
Pantothenic acid (B5)	0.28 mg
Vitamin B6	0.09 - 0.10 mg
Vitamin K	2.90 - 3.50 mg
Folate (B9)	18 - 22 mg
Vitamin C	1.80 - 2.20 mg
Vitamin E	0.20 - 0.40 mg
Calcium	7.40 - 9.00 mg
Iron	0.20 - 0.24 mg
Magnesium	13.50 - 14 mg
Manganese	0.20 - 0.25 mg
Phosphorus	22.50 - 25 mg
Potassium	129 - 130 mg
Zinc	0.10 - 0.16 mg
Sodium	1.60 – 2.00 mg
Copper	0.10 - 0.12 mg

Source: (Ashworth, 2000; Azevedo, Alves de Lima, Gomes, Stringheta, Ribeiro, & Salvadori, 2007; Bhaskar & Ramesh Kumar, 2015; Biology of Brinjal, 2015; Bliss & Elstein, 2004; Cassidy et al., 2013; Dias, 2012; Noda et al., 2000; Sharma, Sharma, & Rana 2013; USDA database, 2012)

Nutrients present in eggplant were generalized by several authors (Ashworth, 2000; Azevedo, Alves de Lima, Gomes, Stringheta, Ribeiro, & Salvadori, 2007; Bhaskar & Kumar, 2015; Biology of Brinjal, 2015; Bliss & Elstein, 2004; Cassidy et al., 2013; Dias, 2012; Noda et al., 2000; Sharma, Sharma, & Rana 2013; USDA database, 2012). The nutritional value per 100mg/g of fresh eggplant were lengthily calculated with closely related quantities for each nutrient.

The presence of carbohydrate provided energy needed for sustenance, helped the body with good digestion and protects the digestive tract due to the high fibre and low soluble carbohydrate contents. Eggplants contained low saturated fat, and sodium content. The essential fatty acids present in eggplants promoted proper brain functioning and prevented cardiovascular diseases and cancer. All vitamins present protected the body against diseases. Vitamin K prevents blood clots and strengthens capillaries. However, eggplants did not contain vitamin D. Mineral content also promoted the metabolism of fat and carbohydrates and ensured the proper brain functioning of the body (Axe, 2015; Cassidy et al., 2013; Krisban, 2013; Sharma et al., 2013; Dias, 2012).

Eggplants were dried when they underwent numerous processes in order to expend moisture content. It was assumed that dried eggplants encountered some loss of nutrients, and moisture content left after processing was less than 10% (Biology of Brinjal, 2015; Muyandar & Devahastin, 2014; Santacatalina, Ozuna, Cárcel, García-Pérez, & Mulet 2011). Other authors affirmed that there were either nutrient gain or loss after processing eggplant into powder (Çağlarırmak & Hepçimen, 2013; Nyadanu, Aboagye, Akromah, Osei, & Dordoe, 2011, Putra, 2011; Skip ThePie link, 2015; Small, 2009; USDA database, 2012). Notwithstanding, powdered or dried eggplant served as convenient food and stored at room temperature for about three months.

Recommended dietary allowance are designed to depict the quantity of food nutrients required by each category of humankind. It is the average daily level of intake sufficient to meet nutritional requirements needed for healthy living. These levels are assumed to ensure nutritional adequacy sufficient to meet requirement of about 97/98% of daily consumption. Since nutrients present in makuea poo-ung was the focus of study, only RDA of nutrients presents in makuea poo ung was significant and reviewed for the study (Dietary Reference Intake (DRI), 2020, RDA, 2020).

Table 3, depicted the age range of male and females RDA. This ranged from less than 11 years, 11 to 14, 15 to 18, 19 to 24, 25 to 50 and above 50 years. It also exhibited the RDA required by expectant and lactating mothers. On the whole, average RDA for male is different from that of female, although some of their requirements were the same. An average compilation of RDA needed by men/ women respectively were carbohydrate 2900/2200g, protein 58/46g, fat 95/70g, vitamin A 1000/800µg,

Table 2. Nutrients present in 100mg/g of dried eggplant

Food Nutrients	Dried Eggplant
Carbohydrates	9.00 g
Sugars	3.10 g
Dietary fiber	2.50 g
Fat	0.22 g
Protein	0.90 g
Moisture content	8.88
Retinol (A)	1.0 mg
Thiamine (B1)	0.08 mg
Riboflavin (B2)	0.2 mg
Niacin (B3)	0.6 mg
Vitamin k	2.9 mg
Vitamin C	1.3 mg
Vitamin E	0.41 mg
Calcium	6.0 mg
Iron	0.23 mg
Magnesium	11 mg
Manganese	0.11 mg
Phosphorus	15 mg
Potassium	123 mg
Zinc	0.12 mg
Sodium	1.00 mg
Copper	0.06 mg

Source: (Biology of Brinjal, 2015; Çağlarırmak and Hepçimen, 2013, Nyadanu *et al.*, 2011, Putra, 2011; Skip ThePie link, 2015; Small, 2009; USDA database, 2012)

vitamin B1 1.5/1.1µg, vitamin B2 1.7/1.3µg and vitamin B3 19/15µg. Others were, magnesium 350/280µg, iron 10/15µg, manganese 2.3/1.8µg and zinc 15/12µg. On the contrary, some other nutrients required same quantities for both men and women. These were, fibre 24g, vitamin C 60µg, calcium 1200µg, potassium 4.7µg, sodium 1.5µg, copper 900µg, and phosphorus 1200µg. Similarly, expectant and lactating mothers virtually need the same amount for themselves and the developing embryos/ babies. The only exception is that, these mothers need a bit more of carbohydrate, protein, vitamin A, vitamin B1. Vitamin B2, Vitamin B3, vitamin C, magnesium, potassium, copper, iron, manganese and zinc (Dietary Reference Intake (DRI), 2020; Recommended Daily Intake, 2020).

Table 3. Recommended daily allowance

Type of Food nutrients	Age of Children	Age Range of Males					Age Range of Females					Expectant mothers	Lactating mothers	
	5 – 10	11 - 14	15 - 18	19 - 24	25 - 50	50+	11 - 14	15 - 18	19 - 24	25 - 50	50 +	0 to 9 months	1st 6 months	2nd 6 months
Carbohydrate	1800	2500	3000	2900	2900	2300	2200	2200	2200	2200	1900	2500	2700	2700
Fibre	15	24	24	24	24	24	24	24	24	24	24	24	24	24
Protein	24	45	59	58	63	63	46	44	46	50	50	60	65	62
Fat	70	95	95	95	95	95	70	70	70	70	70	70	70	70
Vitamin A	400	1000	1000	1000	1000	1000	800	800	800	800	800	800	800	800
Vitamin B_1	0.9	1.3	1.5	1.5	1.5	1.2	1.1	1.1	1.1	1.1	1.0	1.5	1.6	1.6
Vitamin B_2	1.0	1.5	1.8	1.7	1.7	1.4	1.3	1.3	1.3	1.3	1.2	1.6	1.8	1.7
Vitamin B_3	11	17	20	19	19	15	15	15	15	15	13	17	20	20
Vitamin C	40	50	60	60	60	60	50	60	60	60	60	70	90	95
Magnesium	170	270	400	350	350	350	280	300	280	280	280	300	355	340
Calcium	400	1200	1200	1200	800	800	1200	1200	1200	800	800	1200	1200	1200
Potassium	3.0	3.8	4.5	4.7	4.7	4.7	3.8	4.5	4.7	4.7	4.7	4.7	5.1	5.1
Sodium	1.0	1.2	1.5	1.5	1.5	1.3	1.2	1.5	1.5	1.5	1.3	1.5	1.5	1.5
Copper	700	890	900	900	900	900	890	900	900	900	900	1000	1200	1200
Iron	12	12	12	10	10	10	15	15	15	15	10	30	15	15
Manganese	1.5	1.9	2.2	2.3	2.3	2.3	1.6	1.6	1.8	1.8	1.8	2.0	2.6	2.6
Phosphorus	500	1200	1200	1200	800	800	1200	1200	1200	800	800	1200	1200	1200
Zinc	10	15	15	15	15	15	12	12	12	12	12	15	19	16

Sources: (Dietary Reference Intake (DRI), 2020; Recommended Daily Intake, 2020).

METHODOLOGY

This experimental laboratory survey of the nutritional value of 100µg/g/l of fresh and dried makuea poo-ung took place at two places. The Soil Science Laboratory, Kwame Nkrumah University of Science and Technology (KNUST), Kumasi, Ghana and Advance Technical Laboratory, Ramaiah University of Applied Sciences (RUAS), Bangalore, India. The nutrients present in eggplants in general were calculated per g/100g and mg/100g. With this reference, g/100g and mg/100g of raw makuea poo-ung fruits were weighed and nutrients assessed, likewise, nutrients present in g/100g and mg/100g of dried makuea poo-ung. Assessment of the nutrients were done twice for both the fresh and dried makuea poo-ung and the average was taken. Once the results were obtained, comparison was made to evaluate their nutritional differences. The principle underlying this comparison was to assess the relevant importance in the consumption of more fresh or dried makuea poo-ung, or both depending on the resources (time, energy, money, ability) available to the homemaker.

At KNUST, the details of the techniques to which the input data were subjected to came in various forms. The term used to determine the protein, fat, fibre, ash and moisture was called Proximate analysis. Percolation was used to identify carbohydrate present. Mineral analysis was used to ascertain macro nutrients present in some minerals (calcium, magnesium, potassium and phosphorus). Atomic absorption spectrometer was also used to find out the quantity of micro nutrients present in iron, zinc, manganese and copper of fresh and processed makuea poo-ung. In general, wet digestion method was used for fresh makuea poo-ung whilst dry digestion was used for the dried one. At RUAS, the details of the techniques to which the input data were subjected to were in two forms. The vitamins were divided into two; water-soluble and fat-soluble. The main instruments used for the analysis were Liquid Chromatography-Mass Spectrometry (termed as LC-MS) waters acquity; this was a diagnostic chemistry process that conglomerates physical separation capabilities

Table 4. Nutritional value of 100µg/g/l fresh/dried Makuea poo-ung

Type of Nutrient	Fresh Makuea poo-ung	Dried Makuea poo-ung
Calcium	0.99	1.02
Magnesium	0.41	0.58
Potassium	2.29	3.05
Sodium	0.12	0.11
Phosphorus	0.28	0.73
Copper	11.70	6.20
Iron	1.21	12.43
Manganese	0.50	7.60
Zinc	0.47	4.48
Fat	0.88	1.08
Fibre	7.60	7.89
Ash	3.01	0.98
Moisture	80.55	8.68
Protein	2.01	14.57
Carbohydrate	78.92	67.60
Vitamin A	0.36	0.76
Vitamin B1	0.10	0.08
Vitamin B2	0.56	0.47
Vitamin B3	0.11	0.22
Vitamin C	0.20	0.12
Vitamin D	None	None

Source: Fieldwork, 2020

and transfers the separated components from the LC column into the MS ion source. Ultra-Performance Liquid Chromatography (UPLC), was used in analytical techniques in order to improve chromatographic resolution, data processing and control, speed and sensitivity, High Class (H-CLASS) and Tandem Quadrupole Detector (TQD) for high-sensitivity quantitative analysis. In the assessment of nutrients present in the mobile phase of fat and water-soluble vitamins, different reagents and chemicals amidst standards and Sound Pressure Level (SPL) were utilized.

One kilogram of freshly harvested makuea poo-ung was split into two. The quantity for fresh analysis was pre-treated and analysed in order to ascertain the nutrients present. Just after the pre-treatment analysis of nutrients commenced on that same day with 100g/mg. Dried makuea poo-ung was also prepared from the same source and nutrients were analysed from 100g/mg. In order to assess the variance between fresh and dried makuea poo-ung, Independent Sample T-Test was used to investigate their differences after the use of frequency tables. This Independent T-Test was used to analyse the nutritional differences between fresh and dried makuea poo-ung. This was undertaken because there is a general assumption that consumption of fresh products is highly nutritious than processed ones. Additionally, these were compared to the RDA of children, teenagers and adult's consumption of 50g/mg intake on daily basis to assess the nutritional benefits to humankind.

DISCUSSION OF FINDINGS AND RESULTS

It is highly unhealthy for an individual to consume more than 50µg/g makuea poo-ung daily since it can basically lead to diarrhea and stomach upset. Notwithstanding, the review of RDA for children, teenagers and adults were compared with 50µg/g of makuea poo-ung to assess the nutritional benefits to humankind. Generally, the comparative study reviewed that the RDA for children was lower than that of teenagers and adults. However, for teenagers and adults some RDA were the same. These included carbohydrates, fat, all vitamins and all minerals present except magnesium and copper, as well as protein and fibre.

The comparative study revealed that most of the nutrients only served as a food supplement since their nutritional content could not be relied upon in meeting the RDA of humankind. These were carbohydrate, protein, fat, vitamin A, B1, B3, and C, together with magnesium, calcium, sodium, copper and potassium. It meant that out of the 20 nutrients present in makuea poo-ung, 12 of them could not be consumed mainly for the provision those required nutrients. The reason being that a consumption of 50µg/g (about 3 tablespoons full) on daily basis does not offer any significant impact when compared to RDA.

Table 5. Nutrients present in 50mg/g of fresh/dried Makuea poo-ung in relation to RDA

Food Nutrients	50mg/g of Fresh & Dried Makuea poo-ung	Males/Females RDA	Adolescents Boys/ Girls RDA	RDA for Children
Carbohydrate	39.46/33.80 g	2900/2200 g	2900/2200 g	1410 g
Fibre	3.80/3.95 g	24 g	15 g	15 g
Protein	2.01/7.28 g	58/46 g	56/44 g	38 g
Fat	0.44/0.54 g	95/70 g	70 g	70 g
Vitamin A	0.18/0.38 mg	1000/800 mg	1000/800 mg	400 mg
Vitamin B_1	0.05/0.04 mg	1.5/1.1 mg	1.5/1.1 mg	0.6 mg
Vitamin B_2	0.28/0.24 mg	1.7/1.3 mg	1.7/1.3	0.6 mg
Vitamin B_3	0.6/0.11 mg	19/15 mg	19/15 mg	7 mg
Vitamin C	0.10/0.06 mg	60 mg	60 mg	36 mg
Magnesium	0.21/0.29mg	350/280 mg	400/300 mg	240 mg
Calcium	0.50/0.51 mg	1200 mg	1200 mg	550 mg
Potassium	1.15/1.53 mg	4.7 mg	4.7 mg	4.5 mg
Sodium	0.06/0.06 mg	1.5 mg	1.5 mg	1.2 mg
Copper	5.85/3.10 mg	900 mg	700 mg	500 mg
Iron	0.61/6.22 mg	10/15 mg	10/15 mg	9 mg
Manganese	0.25/3.80 mg	2.3/1.8 mg	2.3/1.8 mg	2.0 mg
Phosphorus	0.14/0.37 mg	1200 mg	1200 mg	800 mg
Zinc	0.24/2.24 mg	15/12 mg	15/12 mg	11 mg

Source: Fieldwork 2020

Nonetheless, 6 out of the 20 nutrients assessed (zinc, manganese, potassium, iron, fibre and vitamin B2) provided between half to one-twelfth of the RDA when 50µg/g is consumed daily. Specifically, the daily consumption of 50µg/g makuea poo-ung provided children and teenagers with 25% fibre and 17% for adults. As children attain 50% of vitamin B2 from the consumption of makuea poo-ung, 20% and 17% were gained by teenage/adult females and teenage/adult males respectively. The consumption of makuea poo-ung provided all people with 25% potassium. As dried makuea poo-ung provided humankind with more than 100% manganese as compared to RDA, consumption of the fresh provided all people with between 10-13%. Zinc present in dried makuea poo-ung was able to provide 20% nutrients for children, 17% for teenage/adult females and 14% for teenage/adult males. The mineral iron, which is the assumed primary reason for people consuming makuea poo-ung does not provide much nutrients when fresh makuea poo-ung is consumed.

Table 6. Summary of six nutrients in fresh/dried Makuea poo-ung compared to RDA

Nutrients	Adult male	Adult female	Teenage boy	Teenage girl	Children
Fibre	17%		25%		
Vitamin B_2	17%	20%	17%	20%	40-50%
Potassium			25%		
Manganese			10 13%		
Zinc (Dried)	14%	17%	14%	17%	20
Iron (Dried)	50%	47%	50%	47%	50%

Note: Manganese (dried) provides more than 100% RDA
Copper (fresh and dried) is so abundant in makuea poo-ung, therefore anyone who needs so much of it can conveniently consume makuea poo-ung, and vice versa
Source: Fieldwork, 2020

However, the daily consumption of dried makuea poo-ung provided children and teenage/adult males with around 50% iron and then for teenage/adult females 47% iron. Finally, results emanating from RDA compared to 50μg/g of fresh and processed makuea poo-ung revealed that this fruity vegetable serves as a food supplement for all. Results from other nutrients compared with RDA implied that consumers attained 25% fibre when makuea poo-ung was consumed. Again, a person who needed less fat and protein could also consume fresh makuea poo-ung. As for carbohydrate the content present in makuea poo-ung had no bearing on the RDA of consumers. Finally, moisture and ash had no significant link with humankind RDA.

CONCLUSION

Makuea poo-ung is one of the perfect foods available, since it contains all the food nutrients needed by humankind. The presence of nine minerals, five vitamins as well as carbohydrate, moisture and protein amongst others are all needed by man for nutritional benefit. Collectively, all the 20 nutrients present in fresh makuea poo-ung were also present in dried makuea poo-ung. An amount of 50μg/g of fresh and dried makuea poo-ung compared to RDA revealed that most of the nutrients only served as a food supplement because their nutritional content could not be relied upon in meeting the RDA of humankind. These were carbohydrate, protein, fat, vitamin A, B_1, B_3, and C together with magnesium, calcium, sodium, copper and phosphorus.

Nonetheless, 6 out of the 20 nutrients provides between half to one-twelfth of the RDA when 50μg/g is consumed daily. These were Vitamin B_2, potassium, iron, manganese, zinc and fibre. Additionally, 2 nutrients namely moisture and ash had no significant link with RDA required. Finally, iron which is the assumed primary reason for people consuming makuea poo-ung does not provide much nutrients when fresh makuea poo-ung is consumed. However, the daily consumption of dried makuea poo-ung provided consumers with between 47% to 50% iron. It is highly commended that, makuea poo-ung should be consumed by all people, that is young or old, expectant or lactating mothers and even babies. People should try and consume at least 50µg/g of fresh or dried makuea poo-ung each day for nutritional benefits.

Section title should be "Conclusion", not "Conclusions". Provide discussion of the overall coverage of the chapter and concluding remarks.

REFERENCES

Ashworth, S. (2000). *Growing Eggplants Successfully, Tauton's Kitchen Gardener, 2*, 24–28.

Axe, D. (2015). Eggplant nutrition, benefits & recipes. *Food is medicine*. draxe.com/eggplant-nutrition/

Azevedo, L., Alves de Lima, P. L., Gomes, J. C., Stringheta, P. C., Ribeiro, D. A., & Salvadori, D. M. (2007). Differential response related to genotoxicity between eggplants (Solanum Melanogena) skin aqueous extract and its main purified anthocyanin (delphinidin) in vivo. *Food and Chemical Toxicology, 45*(5), 852–858. doi:10.1016/j.fct.2006.11.004 PMID:17194516

Bhaskar, B., & Kumar, P. (2015). Genetically modified (GM) crop face an uncertain future in India: Bt Brinjal Appraisal – A perspective. *Annals of Plant Sciences, 4*, 960–975.

Biology of brinjal. (2015). General description of Brinjal or eggplant (Solanum melongena L.) Government of India Department of biotechnology, Ministry of science and technology. dbtbiosafety.nic.in/guidelines/brinjal

Bliss, R. M., & Elstein, D. (2004) Scientists get under eggplants skin. *ARS magazine*. http://wwwars.usda.gov/is/AR/archive/jan04/skin0104.htm.

Buzby, J. C., Hyman, J., Stewart, H., & Wells, H. F. (2011). The value of retail - and consumer-level fruit and vegetable losses in the United States. *The Journal of Consumer Affairs, 5*(3), 492–515. doi:10.1111/j.1745-6606.2011.01214.x

Çağlarırmak, N., & Hepçimen, A. Z. (2013). An investigation of nutritional values of dried vegetables. *Gida, 36*, 327–333.

Caguiat, X. G. I., & Hautea, D. M. (2014). Genetic diversity analysis of eggplant (Solanum melongena L.) and related wild species in the Philippines using morphological and SSR markers. *SABRAO Journal of Breeding and Genetics, 46*, 183–201.

Cassidy, A., Mukamal, K. J., Liu, L., Franz, M., Eliassen, A. H., & Rimm, E. B. (2013). High anthocyanin intake is associated with a reduced risk of myocardial infarction in young and middle-aged women. *Circulation*, *127*(2), 188–196. doi:10.1161/CIRCULATIONAHA.112.122408 PMID:23319811

Chong, M. (2005). Perception of the visit and benefits of Bt eggplant by Indian farmers. *Journal of Risk Research*, *8*(7-8), 617–634. doi:10.1080/1366987050004299O

Concellòn, A., An ˜o'n, M. C., & Chaves, A. R. (2005). Effect of chilling on ethylene production in eggplant fruit. *Food Chemistry*, *92*(1), 63–69. doi:10.1016/j.foodchem.2004.04.048

Dias, J. S. (2012). Nutritional quality and health benefits of vegetables: A Review. *Food and Nutrition Sciences*, *3*(10), 1354–1374. doi:10.4236/fns.2012.310179

Dietary reference intake – DRI. (2020). Recommended intakes for individuals, elements. Food and nutrition board, Institute of medicine and national academics. *DaddyBlog*. daddyblog.wordpress.com

Habwe, F. O., Waling, K. M. & Onyango, M. O. A. (2008). Food Processing and Preparation Technologies for Sustainable Utilization of African Indigenous Vegetables for Nutrition Security and Wealth Creation in Kenya. In *using Food Science and Technology to Improve Nutrition and Promote National Development*.

Krisban, S. (2013). *8 excellent reasons to eat more eggplant*. Care 2 healthy living. www.care2.com/greenliving/8-great-reasons-to-eat-eggplant-html

Kwon, Y. I., Apostolidis, E., & Shetty, K. (2008). In vitro studies of eggplant (Solanum Melongena) phenolics as inhibitors of key enzymes relevant for type 2 diabetes and hypertension. *Bioresource Technology*, *99*(8), 2981–2988. doi:10.1016/j.biortech.2007.06.035 PMID:17706416

Li, H., Chen, H., Zhuang, T., & Chen, J. (2010). Analysis of genetic variation in eggplant and related Solanum species using sequence-related amplified polymorphism markers. *Scientia Horticulturae*, *125*(1), 19–24. doi:10.1016/j.scienta.2010.02.023

Matsubara, K., Kaneyuki, T., Miyake, T., & Mori, M. (2005). Antiangiogenic activity of nasunin, antioxidant anthocyanin, in eggplant peels. *Journal of Agricultural and Food Chemistry*, *53*(16), 6272–6275. doi:10.1021/jf050796r PMID:16076105

Muyandar, A. S., & Devahastin, S. (2014). *Fundamental principles of drying, Handbook of industrial drying* (4th ed.). CRC Press.

Noda, Y., Kneyuki, T., Igarashi, K., Mori, A., & Packer, L. (2000). Antioxidant activity of nasunin, an anthocyanin in eggplant peels. *Toxicology, 148*(2-3), 119–123. doi:10.1016/S0300-483X(00)00202-X PMID:10962130

Nyadanu, D., Aboagye, L. M., Akromah, M. K., Osei, M. K., & Dordoe, M. B. (2014). Agromorphological characterization of gboma eggplant, an indigenous fruit and leafyvegetable in Ghana. *African Crop Science Journal, 22*, 281–289.

Omar, S. S., Abdullah, Z. A., Humeid, M. A., & Yamani, M. I. (2012). Optimal composition and heat processing requirements for canning of eggplant dip (Motabbal Al-bathinjan). *Czech Journal of Food Sciences, 30*(1), 35–44. doi:10.17221/62/2010-CJFS

Putra, G. (2011). Nutrition is a key to health. *Nurtient.* www.nutrient.net

Santacatalina, J. V., Ozuna, C., Cárcel, J. A., García-Pérez, J. V., & Mulet, A. (2011) Quality assessment of dried eggplant using different drying methods: hot air drying, vacuum freeze drying and atmospheric freeze drying. *International Congress on Engineering and Food Conference* (pp. 1-6). ICEF.

Sękara, A., Cebula, S., & Kunicki, E. (2007). Cultivated eggplants – origin, breeding objectives and genetic resources, A Review'. *Folia Horticulture Ann, 19*, 97–114.

Sharma, D. K., Sharma, N., & Rana, S. (2013). Seed-borne diseases of brinjal (Solanum Melongena L.) and their control measures: A Review. *Review Article International Journal of Bioassays, 2*, 1428–1433.

Skip The Pie. (2015) USDA Record"11210" Nutritional Comparison: Kale, raw vs Eggplant, cooked, boiled. *Skip the Pie.*

Small, E. (2009). *Top 100 food plants; the world's most important culinary crops.* NRC Research Press.

Umayal Sundari, A. R., Neelamegam, P., & Subramanian, C. V. (2013). An experimental study and analysis on solar drying of bitter gourd using an evacuated tube air collector in Thanjavur, Tamil Nadu, India. *Conference Papers in Energy, 2*, 125–128. doi:10.1155/2013/125628

USDA database. (2012). *USDA National Nutrient Database for Standard Reference Release,* 28, 1028-1035. USDA.

Whitaker, B. D., & Stommel, J. R. (2003). Distribution of Hydroxycinnamic acid conjugates in fruit of commercial eggplant (Solanum melongena L.) cultivars. *Journal of Agricultural and Food Chemistry, 51*(11), 3448–3454. doi:10.1021/jf026250b PMID:12744682

Compilation of References

Abbas, S., Ozal, G., Shahbaz, S. H., & Shahbaz, M. Q. (2019). A new generalized weighted Weibull distribution. *Pakistan Journal of Statistics and Operation Research, 15*(1), 161-178.

Abdul-Rahman, S. (2018, March 12). Ghana's population hits 29.6million - Statistical Service. *The Ghanaian Times*. www.ghanaiantimes.com.gh

Abeens, M., Muruganandhan, R., & Thirumavalavan, K. (2019). Comparative analysis of different surface modification processes on AA 7075 T651. *IOP Conference Series. Materials Science and Engineering, 574*(1), 012018. Advance online publication. doi:10.1088/1757-899X/574/1/012018

Abeens, Q., Rajasekaran, E., Premnath, M., Sarukasan, D., Muruganandhan, R., & Thirumavalavan, K. (2021). Experimental and numerical investigation of residual stresses in low pulsed laser shock peening on AA7075 T651. *IOP Conference Series. Materials Science and Engineering, 1126*(1), 012086. doi:10.1088/1757-899X/1126/1/012086

Abid, S. A., Aziz, S. N., Saeed, N. A.-H. A. A., Mizil, S. N., Al-Kadmy, I. M. S., Hussein, N. H., Al-Saryi, N., Ibrahim, S. A., & Hussein, J. D. (2023). Investigation of virulence factors in microbial organisms that are associated with public health risk isolates from different regions. *Al-Mustansiriyah Journal of Science, 33*(5), 1–7. doi:10.23851/mjs.v33i5.1303

Abrokwah, F. K., Asamoah, K. A., & Esubonteng, P. K. A. (2009). Effects of the intake of natural cocoa powder on some biochemical and haematological indices in the rat. *Ghana Medical Journal, 43*(4), 164–168. PMID:21326996

Abubakar, I. R. (2019). Factors influencing household access to drinking water in Nigeria. *Utilities Policy, 58*, 40–51. doi.org/10.1016/j.jup.2019.03.005. doi:10.1016/j.jup.2019.03.005

Abujaber, A., Fadlalla, A., Gammoh, D., Abdelrahman, H., Mollazehi, M., & El-Menyar, A. (2020). Prediction of in-hospital mortality in patients with post traumatic brain injury using National Trauma Registry and Machine Learning Approach. *Scandinavian Journal of Trauma, Resuscitation and Emergency Medicine, 28*(1), 44. doi:10.118613049-020-00738-5 PMID:32460867

Acheampong, F., Larbie, C., Arthur, F. K. N., Appiah-Opong, R., & Tuffour, I. (2015). Antioxidant and anticnacer study of ageratum conyzoides aqueous extracts. *Journal of Global Biosciences, 4*(1), 1804–1815.

Addy, M., Adinyira, E., Danku, J. C., & Dadzoe, F. (2020). Impediments to the Development of the Green Building Market in Sub-Saharan Africa: The case of Ghana. *Smart and Sustainable Built Environment, 10*(2), 193–207. doi:10.1108/SASBE-12-2019-0170

Adisakwattana, S., & Chanathong, B. (2011). Alpha-glucosidase inhibitory activity and lipid-lowering mechanisms of Moringa oleifera leaf extract. *European Review for Medical and Pharmacological Sciences, 15*(7), 803–808. PMID:21780550

Aduayom-Ahego, A. (2020). Prosthetic rehabilitation of multiple-digit amputations using silicone material in sub-Saharan African country Ghana. *The Pan African Medical Journal, 36*, 357. doi:10.11604/pamj.2020.36.357.25361 PMID:33224423

Aduayom-Ahego, A., & Ehara, Y. (2016). Current position and challenges in prosthetics and orthotics education in Ghana. *Niigata Journal of Health and Welfare, 16*(1), 26–34.

Adu-Boakye, S. (2012). *Development of a conceptual framework relating to ready-to-wear clothing for Ghanaian women for manufacturing strategies.* [Doctoral thesis, Manchester Metropolitan University, 2012]. https://e-space.mmu.ac.uk/305090/

Adu-Gyamfi, S. (2018). Effects of Laser Shock Peening on Mechanical Properties and Surface Morphology of AA2024 Alloy. *American Journal of Mechanical and Materials Engineering, 2*(2), 15. doi:10.11648/j.ajmme.20180202.11

Adu-Gyamfi, S., Ren, X. D., Larson, E. A., Ren, Y., & Tong, Z. (2018). The effects of laser shock peening scanning patterns on residual stress distribution and fatigue life of AA2024 aluminium alloy. *Optics & Laser Technology, 108*, 177–185. doi:10.1016/j.optlastec.2018.06.036

Adu-Gyamfi, V. (2006). *Developing size chart for Ghanaian women: A pilot study in Simpa 'A'. Project report.* Department of Home Economics Education, University of Education.

Afewerky, H. K. (2020). Pathology and pathogenicity of severe acute respiratory syndrome coronavirus 2 (SARS-CoV-2). *Experimental Biology and Medicine, 245*(15), 1299–1307. doi:10.1177/1535370220942126 PMID:32635753

Afriyie, E. O., Edusei, A. K., Mensah, K. A., Nkoom, B., Okyere, P., Adjei, R. O., & Appiah-Brempong, E. (2015). Antibiotics Availability and Usage in Health Facilities: A Case of the Offinso-South Municipality of Ghana. *Journal of Biology, 5*(2), 132–138.

Ågerstrand, M., Berg, C., Björlenius, B., Breitholtz, M., Brunström, B., Fick, J., Gunnarsson, L., Larsson, D. G. J., Sumpter, J. P., Tysklind, M., & Rudén, C. (2015). Improving environmental risk assessment of human pharmaceuticals. *Environmental Science & Technology, 49*(9), 5336–5345. doi:10.1021/acs.est.5b00302 PMID:25844810

Agrawal, N., Pandey, P., Mishra, D., & Maheshwaris, K. (2017). *Anticarcinogenic and Chemopreventive effect of AndrographisPaniculata extract in Swiss albino mice. 7*(6).

Ahmadini, A. A. H., Naeem, M., Aamir, M., Dewan, R., Alshqaq, S. S. A., & Mashwani, W. K. (2021). Analysis and Forecast of the Number of Deaths, Recovered Cases, and Confirmed Cases From COVID-19 for the Top Four Affected Countries Using Kalman Filter. *Frontiers in Physics (Lausanne)*, *9*, 629320. doi:10.3389/fphy.2021.629320

Aho, I. M., & Lagasi, J. E. (2012). A new water treatment system using Moringa oleifera seed. *American Journal of Scientific and Industrial Research*, *3*(6), 487–492. doi:10.5251/ajsir.2012.3.6.487.492

Ahuja, S., Shelke, N. A., & Singh, P. K. (2022). A deep learning framework using CNN and stacked Bi-GRU for COVID-19 predictions in India. *Signal, Image and Video Processing*, *16*(3), 579–586. doi:10.100711760-021-01988-1 PMID:34335985

Akadiri, O. P. (2011). *Development of a Multi-Criteria Approach for Selecting Sustainable Materials for Building Projects*. PhD Thesis - The University of Wolverhampton. http://wlv.openrepository.com/wlv/bitstream/2436/129918/1/Akadiri_PhD thesis.pdf

Akaike, A. (1974). A new look at the statistical model identification. *IEEE Transactions on Automatic Control*, *19*(6), 716–723. doi:10.1109/TAC.1974.1100705

Akin, B. S. (2016). Contaminant properties of hospital clinical laboratory wastewater: A physiochemical and microbiological assessment. *Journal of Environmental Protection*, *7*(5), 635–642. doi:10.4236/jep.2016.75057

Akinyemi, F. O. (2007). Spatial data needs for poverty management. *Research and Theory in Advancing Spatial Data Infrastructure Concepts, 5*.

Akonor, P. T., & Akonor, M. A. (2013). *Food Safety Knowledge: The case of domestic food handlers in Accra*.

Akram, F., Abrar-ul-Haq, M., & Raza, S. (2019). a Role a Corporate Governance and Firm'S Environmental Performance: A Moderating Role of Institutional Regulations. *International Journal of Management Studies*, *25*(2), 19–37. doi:10.32890/ijms.25.2.2018.10498

Al Aukidy, M., Al Chalabi, S., & Verlicchi, P. (2017). Hospital wastewater treatments adopted in Asia, Africa, and Australia. In P. Verlicchi (Ed.), *Hospital Wastewaters. The Handbook of Environmental Chemistry* (Vol. 60). Springer., doi:10.1007/698_2017_5

Alberts, B., Johnson, A., Lewis, J., Raff, M., Roberts, K., & Walter, P. (2002). The plant cell wall. In *Molecular Biology of the Cell* (4th ed.). Garland Science.

Aldahdooh, M. A. A., Jamrah, A., Alnuaimi, A., Martini, M. I., Ahmed, M. S. R., & Ahmed, A. S. R. (2018). Influence of various plastics-waste aggregates on properties of normal concrete. *Journal of Building Engineering*, *17*, 13–22. doi:10.1016/j.jobe.2018.01.014

Aldhshan, S. R. S., Abdul Maulud, K. N., Wan Mohd Jaafar, W. S., Karim, O. A., & Pradhan, B. (2021). Energy consumption and spatial assessment of renewable energy penetration and building energy efficiency in Malaysia: A review. *Sustainability (Basel), 13*(16), 9244. doi:10.3390u13169244

Aldrich, W. (2004). *Metric pattern cutting for women's wear.* Blackwell Science.

Aldrich, W. (2009). *Metric pattern cutting for children's wear and babywear* (4th ed.). Blackwell Publishing.

Aldrich, W. (2011). *Metric pattern cutting for children's wear and baby wear* (4th ed.). Blackwell Publishing.

Al-Fakih, A., Mohammed, B. S., Wahab, M. M. A., Liew, M. S., Amran, Y. M., Alyousef, R., & Alabduljabbar, H. (2020, August). Characteristic compressive strength correlation of rubberized concrete interlocking masonry wall. In *Structures* (Vol. 26, pp. 169–184). Elsevier.

Alfonso-Muniozguren, P., Serna-Galvis, E. A., Bussemaker, M., Torres-Palma, R. A., & Lee, J. (2021). A review on pharmaceuticals removal from waters by single and combined biological, membrane filtration and ultrasound systems. *Ultrasonics Sonochemistry, 76*, 105656. doi:10.1016/j.ultsonch.2021.105656 PMID:34274706

Al-Ghamdi, A. K., Abdelmalek, S. M. A., Ashshi, A. M., Faidah, H., Shukri, H., & Jiman-Fatani, A. A. (2011). Bacterial contamination of computer keyboards and mice, elevator buttons and shopping carts. *African Journal of Microbiological Research, 5*(23), 3998–4003. doi:10.5897/AJMR11.770

Ali, E. N. (2016). *Removal of Copper from Water System using Moringa oleifera Press Cake.* 16–19. doi:10.17758/URUAE.AE0416214

Ali, A., Imran, M., Jabeen, M., Ali, Z., & Mahmood, S. A. (2021). Factors influencing integrated information management: Spatial data infrastructure in Pakistan. *Information Development.* doi:10.1177/02666669211048483

Almalki, S. J., & Yuan, J. (2013). A new modified Weibull distribution. *Reliability Engineering & System Safety, 111*, 164–170. doi:10.1016/j.ress.2012.10.018

Al-Rashedi, A., & Al-Hagery, M. A. (2023). Deep Learning Algorithms for Forecasting COVID-19 Cases in Saudi Arabia. *Applied Sciences (Basel, Switzerland), 13*(3), 1816. doi:10.3390/app13031816

Alsanad, S. (2015). Awareness, Drivers, Actions, and Barriers of Sustainable Construction in Kuwait. *Procedia Engineering, 118*, 969–983. doi:10.1016/j.proeng.2015.08.538

Alubel, M., Yadav, M., & Singh, N. (2017). Anthropometric size chart for Ethiopian girls for better garment design. *Journal of Fashion Technology & Textile Engineering, 5*(2), 1–11. doi:10.4172/2329-9568.1000151

Alumasa, S., & Muathe, S. (2021). Mobile Credit and Performance: Experience and Lessons from Micro and Small Enterprises in Kenya. *Journal of Applied Finance & Banking, 11*(4), 135–161. doi:10.47260/jafb/1146

Alvarenga, M. Z., Oliveira, M. P. V. D., & Oliveira, T. A. G. F. D. (2023). The impact of using digital technologies on supply chain resilience and robustness: The role of memory under the covid-19 outbreak. *Supply Chain Management, 28*(5), 825–842. doi:10.1108/SCM-06-2022-0217

Amagloh, F. K., & Benang, A. (2009). Effectiveness of Moringa oleifera seed as coagulant for water purification. *African Journal of Agricultural Research, 4*(2), 119–123.

Amateye, L. (2009). *The role of the indigenous Ghanaian textile industry in relation to the President's Special Initiative (PSI) on textiles and garments.* [Doctoral thesis, Kwame Nkrumah University of Science and Technology]. https://ir.knust.edu.gh/xmlui/handle/123456789/288

Amiard, J., & Amiard-Triquet, C. (2015). Conventional risk assessment of environmental contaminants. In: Advancing tools for dealing with emerging risks. Aquatic Ecotoxicology, 2, 25-49. doi:10.1016/B978-0-12-800949-9.00002-4

Amoaei, A., Asgharnia, H., Fallah, H., Faraji, H., Barari, R., & Naghipour, D. (2015). Characteristics of effluent wastewater in hospitals of Babol University of Medical Sciences, Babol, Iran. *Health Scope, 4*(2), e23222. doi:10.17795/jhealthscope-23222

Ampofo, V. O. (2002). *Ghana's Textiles/ Garment Industry: Industrial Development and Investment Division (Ministry of Trade and industry-Ghana).* Intracen. https://www.intracen.org/worldtradenet/docs/whatsnew/atc_lesotho_november2000/2

Anderson, E., Hughes, G., & Patterson, E. (2020). *Inactivation of SARS-CoV-2 on surfaces and in solution with Virusend (TX-10), a novel disinfectant.*

Anderson, L. A. (2019). Antibiotics guide. *Drugs.com.* https://www.drugs.com/article/antibiotics.html. Accessed on 25/12/2021.

Anderson, J. C., & Gerbing, D. W. (1988). Structural equation modeling in practice: A review and recommended two-step approach. *Psychological Bulletin, 103*(3), 411–423. doi:10.1037/0033-2909.103.3.411

Andrade, S. N., Veloso, C. M., Fontan, R. C. I., Bonomo, R. C. F., Santos, L. S., Brito, M. J. P., & Diniz, G. A. (2018). Chemical-activated carbon from coconut (Cocos nucifera) endocarp waste and its application in the adsorption of ß-lactoglobulin protein. *Revista Mexicana de Ingeniería Química, 17*(2), 463–475. doi:10.24275/10.24275/uam/izt/dcbi/revmexingquim/2018v17n2/Andrade

Ansah, N. B., Adinyira, E., Agyekum, K., & Aidoo, I. (2022). Bibliometric Study on Particle Emissions of Natural and Alternative Building Materials. In M. Gorse, C., Scott, L., Booth, C., Dastbaz (Ed.), *Managing, Building, and Delivering the Sustainable Development Goals* (pp. 495–507). Springer International Publishing. doi.org/10.1007/978-3-030-79450-7_37

Ansah, M. K., Chen, X., & Yang, H. (2022). A Holistic Environmental and Economical Design Optimisation of Low Carbon Buildings Considering Climate Change and Confounding Factors. *The Science of the Total Environment*, *821*, 153442. doi:10.1016/j.scitotenv.2022.153442 PMID:35092783

Anwar, F., Kalsoom, U., Sultana, B., Mushtaq, M., Mehmood, T., & Arshad, H. A. (2013). Effect of drying method and extraction solvent on the total phenolics and antioxidant activity of cauliflower (Brassica oleracea L.) Extracts. *International Food Research Journal*, *20*(2), 653–659.

Appiah-Opong, R., Asante, I. K., Safo, D. O., Tuffour, I., Ofori-Attah, E., Uto, T., & Nyarko, A. K. (2016). Cytotoxic effects of Albizia zygia (DC) J. F. Macbr, a Ghanaian medicinal plant, against human T-lymphoblast-like leukemia, prostate and breast cancer cell lines. *International Journal of Pharmacy and Pharmaceutical Sciences*, *8*(5), 392–396.

Appiah-Opong, R., Tuffour, I., Annor, K. G., Blankson-Darku, A. D., Cramer, P., Kissi-Twum, A. A., & Uto, T. O. A. (2016). Antiproliferative, antioxidant activities and apoptosis induction by Morinda lucida and Taraxacum officiale in Human HL-60 leukemia cells. *J. Global Biosci*, *5*(7), 4281–4291.

Araujo, C. S. T., Carvalho, D. C., Rezende, H. C., Almeida, I. L. S., Coelho, L. M., Coelho, N. M. M., Marques, T. L., & Alves, V. N. (2013). *Bioremediation of Waters Contaminated with Heavy Metals Using Moringa oleifera Seeds as Biosorbent.* Applied Bioremediation - Active and Passive Approaches., doi:10.5772/56157

Araujo, G. A., Carneiro, A. M. P., & Palha, P. R. (2020). Predictive Methodology for the Quantification of Environmental Aspects in Urban Infrastructures. *Sustainability (Basel)*, *12*(18), 7636. doi:10.3390u12187636

Argiolas, M. (2014). Surveying housing market supply affordability using a spatial data mining approach. *WIT Transactions on Ecology and the Environment*, *191*, 125–137. doi:10.2495/SC140111

Ariadurai, S. A., Nilusha, T. P. G., Alwis, T., & Manori-Dissanayake, D. M. R. (2009). An Anthropometric Study on Sri Lankan School Children for Developing Clothing Sizes. *Journal of Social Sciences*, *19*(1), 51–56. doi:10.1080/09718923.2009.11892690

Armenakis, C., & Nirupama, N. (2013). Prioritization of disaster risk in a community using GIS. *Natural Hazards*, *66*(1), 15–29. Advance online publication. doi:10.100711069-012-0167-8

Armstrong, H. (2014). *Pattern making for fashion design* (5th ed.). Pearson Education Ltd.

Ashworth, S. (2000). *Growing Eggplants Successfully, Tauton's Kitchen Gardener*, *2*, 24–28.

Asrat, S., de Jesús, D. A., Hempstead, A. D., Ramabhadran, V., & Isberg, R. R. (2014). Bacterial pathogen manipulation of host membrane trafficking. *Annual Review of Cell and Developmental Biology*, *30*(1), 79–109. doi:10.1146/annurev-cellbio-100913-013439 PMID:25103867

Atjanasuppat, K., Wongkham, W., Meepowpan, P., Kittakoop, P., Sobhon, P., Bartlett, A., & Whitfield, P. J. (2009). In vitro screening for anthelmintic and antitumour activity of ethnomedicinal plants from Thailand. *Journal of Ethnopharmacology*, *123*(3), 475–482. doi:10.1016/j.jep.2009.03.010 PMID:19473794

aus der Beek, T., Weber, F. A., Bergmann, A., Hickmann, S., Ebert, I., Hein, A., & Küster, A. (2016). Pharmaceuticals in the environment—Global occurrences and perspectives. *Environmental Toxicology and Chemistry*, *35*(4), 823–835. doi:10.1002/etc.3339 PMID:26666847

Avila-Aguero, M. L., German, G., Paris, M. M., & Herrera, J. F., (2004). Toys in a pediatric hospital: Are they a bacterial source? *American Journal of Infection Control*, *32*(5), 287–290. doi:10.1016/j.ajic.2003.10.018 PMID:15292894

Awad, M., Wang, H., & Li, F. (2013b). WEAKNESSES OF Moringa oleifera USE IN WATER TREATMENT. *Article in International Journal of Current Research*, *5*, 1165–1167. https://www.researchgate.net/publication/281629560

Awad, M., Wang, H., & Li, F. (2013a). Weaknesses of moringa oleifera use in water treatment. *International Journal of Current Research*, *5*(5), 1165–1167.

Axe, D. (2015). Eggplant nutrition, benefits & recipes. *Food is medicine*. draxe.com/eggplant nutrition/

Aydın, A., & Okur, S. (2018). Effects of test socket on pain, prosthesis satisfaction, and functionality in patients with transfemoral and transtibial amputations. *Medical Science Monitor*, *24*, 4031–4037. doi:10.12659/MSM.910858 PMID:29897893

Ayoub, N., & Musharavati, F. (2014). A Future Prospect for Domestic Waste Management in Qatar. *International Conference on Earth, Environment and Life Sciences (EELS-2014)*. 10.15242/IICBE.C1214080

Azanu, D., Styrishave, B., Darko, G., Weisser, J. J., & Abaidoo, R. C. (2018). Occurrence and risk assessment of antibiotics in water and lettuce in Ghana. *The Science of the Total Environment*, *622*, 293–305. doi:10.1016/j.scitotenv.2017.11.287 PMID:29216470

Azeko, S. T., Arthur, E. K., Danyuo, Y., & Babagana, M. (2018). Mechanical and physical properties of laterite bricks reinforced with reprocessed polyethylene waste for building applications. *Journal of Materials in Civil Engineering*, *30*(4), 04018039. doi:10.1061/(ASCE)MT.1943-5533.0002205

Azeko, S. T., Arthur, E. K., Minh, D. P., Lyczko, N., Nzihou, A., & Soboyejo, W. O. (2022). Mechanical and Thermal Properties of Sustainable Composite Building Materials Produced by the Reprocessing of Low-Density Polyethylene, Biochar, Calcium Phosphate, and Phosphogypsum Wastes. *Journal of Materials in Civil Engineering*, *34*(2), 04021457. doi:10.1061/(ASCE)MT.1943-5533.0004021

Azeko, S. T., Etuk-Udo, G. A., Odusanya, O. S., Malatesta, K., Anuku, N., & Soboyejo, W. O. (2015). Biodegradation of linear low-density polyethylene by Serratia marcescens subsp. marcescens and its cell free extracts. *Waste and Biomass Valorization*, *6*(6), 1047–1057. doi:10.100712649-015-9421-0

Azeko, S. T., Mustapha, K., Annan, E., Odusanya, O. S., & Soboyejo, W. O. (2016). Recycling of polyethylene into strong and tough earth-based composite building materials. *Journal of Materials in Civil Engineering*, *28*(2), 04015104. doi:10.1061/(ASCE)MT.1943-5533.0001385

Azevedo, L., Alves de Lima, P. L., Gomes, J. C., Stringheta, P. C., Ribeiro, D. A., & Salvadori, D. M. (2007). Differential response related to genotoxicity between eggplants (Solanum Melanogena) skin aqueous extract and its main purified anthocyanin (delphinidin) in vivo. *Food and Chemical Toxicology*, *45*(5), 852–858. doi:10.1016/j.fct.2006.11.004 PMID:17194516

Azuma, T., Arima, N., Tsukada, A., Hirami, S., Matsuoka, R., Moriwake, R., Ishiuchi, H., Inoyama, T., Teranishi, Y., Yamoako, M., Mino, Y., Hayasshi, T., Fujita, Y., & Masada, M. (2016). Detection of pharmaceuticals and phytochemicals together with their metabolites in hospital effluents in Japan, and their contribution to sewage treatment plant influents. *The Science of the Total Environment*, *548-549*, 189–197. doi:10.1016/j.scitotenv.2015.12.157 PMID:26802347

Azzalini, A. (1985). A class of distributions which includes the normal ones. *Scandinavian Journal of Statistics*, *12*(2), 171–178.

Badisa, R. B., Darling-Reed, S. F., Joseph, P., Cooperwood, J. S., Latinwo, L. M., & Goodman, C. B. (2009). Selective cytotoxic activities of two novel synthetic drugs on human breast carcinoma MCF-7 cells. *Anticancer Research*, *29*(8), 2993–2996. PMID:19661306

Baharum, Z., Akim, A. M., Taufiq-Yap, Y. H., Hamid, R. A., & Kasran, R. (2014). In vitro antioxidant and antiproliferative activities of methanolic plant part extracts of Theobroma cacao. *Molecules (Basel, Switzerland)*, *19*(11), 18317–18331. doi:10.3390/molecules191118317 PMID:25389662

Baharum, Z., Md Akim, A., Hin, T. Y. Y., Hamid, R. A., & Kasran, R. (2016). Theobroma cacao: Review of the extraction, isolation, and bioassay of its potential anti-cancer compounds. *Tropical Life Sciences Research*, *27*(1), 21–42. PMID:27019680

Bai, C., Quayson, M., & Sarkis, J. (2021). COVID-19 pandemic digitization lessons for sustainable development of micro-and small- enterprises. *Sustainable Production and Consumption*, *27*(5), 1989–2001. doi:10.1016/j.spc.2021.04.035 PMID:34722843

Bakare, B. F. (2016). An investigation of moringa oleifera seed extract as a natural coagulant in water treatment. *Lecture Notes in Engineering and Computer Science*, *2226*, 566–569.

Baker, R. E., Mahmud, A. S., Miller, I. F., Rajeev, M., Rasambainarivo, F., Rice, B. L., Takahashi, S., Tatem, A. J., Wagner, C. E., Wang, L. F., Wesolowski, A., & Metcalf, C. J. E. (2022). Infectious disease in an era of global change. *Nature Reviews. Microbiology*, *20*(4), 193–205. doi:10.103841579-021-00639-z PMID:34646006

Balakrishna, K., Rath, A., Praveenkumarreddy, Y., Guruge, K. S., & Subedi, B. (2017). A review of the occurrence of pharmaceuticals and personal care products in Indian water bodies. *Ecotoxicology and Environmental Safety*, *137*, 113–120. doi:10.1016/j.ecoenv.2016.11.014 PMID:27915141

Ballard, P., Fournier, J., Fabbro, R., & Frelat, J. (1991). Residual stresses induced by laser-shocks. *Journal de Physique. IV*, *1*(C3), 1. doi:10.1051/jp4:1991369

Bari, S. B., Salleh, N. M., Sulaiman, N., & Othman, M. (2015). Development of clothing size for pre-school children based on anthropometric measurements. *Australian Journal of Sustainable Business and Society*, *1*(2), 22–32.

Barney, J. (1991). Firm resources and sustained competitive advantage. *Journal of Management*, *17*(1), 99–120. doi:10.1177/014920639101700108

Bartlett, J. E., Kotrlik, J. W., & Higgins, C. C. (2001). Organizational Research : Determining Appropriate Sample Size in Survey Research. *Information Technology, Learning and Performance Journal*, *19*(1), 43–50.

Basiri, A., Haklay, M., Foody, G., & Mooney, P. (2019). Crowdsourced geospatial data quality: challenges and future directions. In International Journal of Geographical Information Science, 33(8). doi:10.1080/13658816.2019.1593422

Bates, T. J., Fergason, J. R., & Pierrie, S. N. (2020). Technological advances in prosthesis design and rehabilitation following upper extremity limb loss. *Current Reviews in Musculoskeletal Medicine*, *13*(4), 485–493. doi:10.100712178-020-09656-6 PMID:32488625

Bauer, D., De Abreu, J. P., Oliveira, H. S. S., Goes-Neto, A., Koblitz, M. G. B., & Teodoro, A. J. (2016). Antioxidant Activity and Cytotoxicity Effect of Cocoa Beans Subjected to Different Processing Conditions in Human Lung Carcinoma Cells. *Oxidative Medicine and Cellular Longevity*, *2016*, 1–11. Advance online publication. doi:10.1155/2016/7428515 PMID:27034742

Bazrafshan, E., Mahvi, A. H., & Naseri, S. (2015). Evaluation of Electrocoagulation Process Efficiency for Direct and Indirect Dyes Removal from Synthetic Wastewater. *Journal of Environmental Health Science & Engineering*, *13*(1), 43. doi:10.118640201-015-0191-6

Beaugerie, L., & Petit, J.-C. (2004). Antibiotic-associated diarrhoea. *Best Practice & Research. Clinical Gastroenterology*, *18*(2), 337–352. doi:10.1016/j.bpg.2003.10.002 PMID:15123074

Bebianno, M. J., & Gonzalez-Rey, M. (2015). Ecotoxicological risk of personal care products and pharmaceuticals. In *Advancing Tools for Dealing with Emerging Risks 16* (pp. 383–416). Aquatic Ecotoxicology., doi:10.1016/B978-0-12-800949-9.00016-4

Belden, J. (2020). Introduction to ecotoxicology. In: An Introduction to Interdisciplinary Toxicology: From Molecules to Man, 5, 381-393. doi:10.1016/B978-0-12-813602-7.00028-4

Bera, S. P., Godhaniya, M., & Kothari, C. (2022). Emerging and advanced membrane technology for wastewater treatment: A review. *Journal of Basic Microbiology*, *62*(3-4), 245–259. doi:10.1002/jobm.202100259 PMID:34496068

Berkner, S., Schwirn, K., & Voelker, D. (2016). Nanopharmaceuticals: Tiny challenges for the environmental risk assessment of pharmaceuticals. *Environmental Toxicology and Chemistry, 35*(4), 780–787. doi:10.1002/etc.3039 PMID:25931425

Berthe, L., Fabbro, R., Peyre, P., Tollier, L., & Bartnicki, E. (1997). Shock waves from a water-confined laser-generated plasma. *Journal of Applied Physics, 82*(6), 2826–2832. doi:10.1063/1.366113

Beshah, B., Belay, B., Tilahun, S., Tizazu, B., & Matebu, A. (2014). Anthropometric data of Bahirdar city's adult men for clothing design. *International Journal of Vocational and Technical Education, 6*(5), 51–57. doi:10.5897/IJVTE2014.0155.x

Bhaskar, B., & Kumar, P. (2015). Genetically modified (GM) crop face an uncertain future in India: Bt Brinjal Appraisal – A perspective. *Annals of Plant Sciences, 4*, 960–975.

Bilec, M. M., Ries, R. J., & Matthews, H. S. (2009). Life-Cycle Assessment Modeling of Construction Processes for Buildings. *Journal of Infrastructure Systems, 16*(3), 199–205. doi:10.1061/(ASCE)IS.1943-555X.0000022

Biney-Aidoo, V. (2020). *Development Of Garment Size Chart Using Anthropometric Data Of Ghanaian Girls Aged 7 To 12 Years.* [Unpublished PhD thesis, Universityof Ghana Legon, 2009].

BIO Intelligence Service. (2013). *Study on the environmental risks of medicinal products, Final Report.* Executive Agency for Health and Consumers. https://ec.europa.eu>files > files >study_environment.

Biology of brinjal. (2015). General description of Brinjal or eggplant (Solanum melongena L.) Government of India Department of biotechnology, Ministry of science and technology. dbtbiosafety.nic.in/guidelines/brinjal

Biosci, I. J., Alfarra, R. S., Ali, N. E., Yusoff, M. M., Razak, L. T., & Biosci, I. J. (2014). Removal of heavy metals by natural adsorbent [review] [IJB]. *International Journal of Biosciences, 6655*, 130–139. doi:10.12692/ijb/4.7.130-139

Birkin, M. (2019). Spatial data analytics of mobility with consumer data. *Journal of Transport Geography, 76*, 245–253. doi:10.1016/j.jtrangeo.2018.04.012

Blahuskova, V., Vlcek, J., & Jancar, D. (2019). Study connective capabilities of solid residues from the waste incineration. *Journal of Environmental Management, 231*, 1048–1055. doi:10.1016/j.jenvman.2018.10.112 PMID:30602228

Bliss, R. M., & Elstein, D. (2004) Scientists get under eggplants skin. *ARS magazine.* http://wwwars.usda.gov/is/AR/archive/jan04/skin0104.htm.

Bloom, D. E., & Cadarette, D. (2019). Infectious disease threats in the twenty-first century: Strengthening the global response. *Frontiers in Immunology, 10*(MAR), 1–12. doi:10.3389/fimmu.2019.00549 PMID:30984169

Bonadonna, L., Briancesco, R., & Coccia, A. M. (2017). Analysis of microorganisms in hospital environments and potential risks. *SpringerBriefs in Public Health, 9783319491592*, 53–62. doi:10.1007/978-3-319-49160-8_5

Boone, S. A., & Gerba, C. P. (2007). Significance of fomites in the spread of respiratory and enteric viral disease. *Applied and Environmental Microbiology, 73*(6), 1687–1696. doi:10.1128/AEM.02051-06 PMID:17220247

Bordogna, G., Kliment, T., Frigerio, L., Brivio, P. A., Crema, A., Stroppiana, D., Boschetti, M., & Sterlacchini, S. (2016). A spatial data infrastructure integrating multisource heterogeneous geospatial data and time series: A study case in agriculture. *ISPRS International Journal of Geo-Information, 5*(5), 73. Advance online publication. doi:10.3390/ijgi5050073

Borquaye, L. S., Ekuadzi, E., Darko, G., Ahor, H. S., Nsiah, S. T., Lartey, J. A., Mutala, A. H., Boamah, V. E., & Woode, E. (2019). Occurrence of antibiotics and antibiotic-resistant bacteria in landfill sites in Kumasi, Ghana. *Journal of Chemistry, 2019*, 2019. doi:10.1155/2019/6934507

Bradford, M. M. (1976). A rapid and sensitive method for the quantitation of microgram quantities of protein utilizing the principle of protein-dye binding. *Analytical Biochemistry, 72*(1-2), 248–254. doi:10.1016/0003-2697(76)90527-3 PMID:942051

Braisted, W., & Brockman, R. (1999). Finite Element Simulation of Laser Shock Peening. *International Journal of Fatigue, 21*(7), 719–724. doi:10.1016/S0142-1123(99)00035-3

Brandt, D., Alnyme, O., & Heldt, T. (2022). Building a spatial decision support system for tourism and infrastructure planning: Technical solution and data integration challenges. *European Journal of Geography, 13*(1), 94–108. doi:10.48088/ejg.d.bra.13.1.094.108

Brand-Williams, W., Cuvelier, E., & Berset, C. (1995). Use of a free radical method to evaluate antioxidant activity. *Lebensmittel-Wissenschaft + Technologie, 28*(1), 25–30. doi:10.1016/S0023-6438(95)80008-5

Brigham and Women's Hospital (BWH). (2011). *Standard of care: Lower extremity amputation*. The Brigham and Women's Hospital, Inc., Department of Rehabilitation Services.

Bright, K. R., Boone, S. A., & Gerba, C. P. (2010). Occurrence of bacteria and viruses on elementary classroom surfaces and the potential role of classroom hygiene in the spread of infectious diseases. *The Journal of School Nursing: the Official Publication of the National Association of School Nurses, 26*(1), 33–41. doi:10.1177/1059840509354383 PMID:19903773

Brilhante, R. S. N., Sales, J. A., Pereira, V. S., Castelo-Branco, D., Cordeiro, R. A., de Souza Sampaio, C. M., de Araújo Neto Paiva, M., Santos, J. B. F., Sidrim, J. J. C., & Rocha, M. F. G. (2017). Research advances on the multiple uses of Moringa oleifera: A sustainable alternative for socially neglected population. *Asian Pacific Journal of Tropical Medicine, 10*(7), 621–630. doi:10.1016/j.apjtm.2017.07.002 PMID:28870337

British Standard Institution. (1990). *BS 7231, Part 1, Body Measurements of Boys and Girls from Birth to 16.9 Years*. British Standard Institution, London.

Broin, M., Santaella, C., Cuine, S., Kokou, K., Peltier, G., & Joet, T. (2002). Flocculent activity of a recombinant protein from Moringa oleifera Lam. seeds. *Applied Microbiology and Biotechnology*, *60*(1), 114–119. PMID:12382051

Brown, G. (2005). Mapping spatial attributes in survey research for natural resource management: Methods and applications. In Society and Natural Resources, 18(1). doi:10.1080/08941920590881853

Brownlee, J. (2014, December 18). How To Work Through A Problem Like A Data Scientist. *Machine Learning Mastery*. https://machinelearningmastery.com/how-to-work-through-a-problem-like-a-data-scientist/

Bukar, A., Uba, A., & Oyeyi, T. (2010). Antimicrobial profile of Moringa oleifera Lam. extracts against some food–borne microorganisms. *Bayero Journal of Pure and Applied Sciences*, *3*(1). doi:10.4314/bajopas.v3i1.58706

Bunce, R. G. H., Metzger, M. J., Jongman, R. H. G., Brandt, J., De Blust, G., Elena-Rossello, R., Groom, G. B., Halada, L., Hofer, G., Howard, D. C., Kovář, P., Mücher, C. A., Padoa-Schioppa, E., Paelinx, D., Palo, A., Perez-Soba, M., Ramos, I. L., Roche, P., Skånes, H., & Wrbka, T. (2008). A standardized procedure for surveillance and monitoring European habitats and provision of spatial data. *Landscape Ecology*, *23*(1), 11–25. doi:10.100710980-007-9173-8

Burgos-Castillo, R. C., Sirés, I., Sillanpää, M., & Brillas, E. (2018). Application of electrochemical advanced oxidation to bisphenol A degradation in water. Effect of sulfate and chloride ions. *Chemosphere*, *194*, 812–820. doi:10.1016/j.chemosphere.2017.12.014 PMID:29268102

Burton, M., Cobb, E., Donachie, P., Judah, G., Curtis, V., & Schmidt, W.-P. (2011). The effect of handwashing with water or soap on bacterial contamination of hands. *International Journal of Environmental Research and Public Health*, *8*(1), 97–104. doi:10.3390/ijerph8010097 PMID:21318017

Buzby, J. C., Hyman, J., Stewart, H., & Wells, H. F. (2011). The value of retail - and consumer-level fruit and vegetable losses in the United States. *The Journal of Consumer Affairs*, *5*(3), 492–515. doi:10.1111/j.1745-6606.2011.01214.x

Çağlarırmak, N., & Hepçimen, A. Z. (2013). An investigation of nutritional values of dried vegetables. *Gida*, *36*, 327–333.

Caguiat, X. G. I., & Hautea, D. M. (2014). Genetic diversity analysis of eggplant (Solanum melongena L.) and related wild species in the Philippines using morphological and SSR markers. *SABRAO Journal of Breeding and Genetics*, *46*, 183–201.

Cairo, A. (2005). *Orthopaedic project in Afghanistan*. The International Committee of the Red Cross (ICRC).

Cai, W., Xie, L., Chen, Y., & Zhang, H. (2013). Purification, characterization and anticoagulant activity of the polysaccharides from green tea. *Carbohydrate Polymers*, *92*(2), 1086–1090. doi:10.1016/j.carbpol.2012.10.057 PMID:23399132

Calheiros, C. S. C., Rangel, A. O. S. S., & Castro, P. M. L. (2009). The effect of water and wastewater composition on the adsorption of pharmaceuticals onto activated carbons. *The Science of the Total Environment*, *407*(12), 3793–3802. PMID:19345984

Camacho, F. P., Sousa, V. S., Bergamasco, R., & Teixeira, M. R. (2017). The use of Moringa oleifera as a natural coagulant in surface water treatment. *Chemical Engineering Journal*, *313*, 226–237. doi:10.1016/j.cej.2016.12.031

Cao, Y., Shin, Y. C., & Wu, B. (2010). Parametric Study on Single Shot and Overlapping Laser Shock Peening on Various Metals via Modeling and Experiments. *Journal of Manufacturing Science and Engineering*, *132*(6), 061010. doi:10.1115/1.4002850

Cao, Y., Xie, X., Antonaglia, J., Winiarski, B., Wang, G., Shin, Y. C., Withers, P. J., Dahmen, K. A., & Liaw, P. K. (2015). Laser shock peening on Zr-based bulk metallic glass and its effects on plasticity: Experiment and modeling. *Scientific Reports*, *5*(1), 10789. doi:10.1038rep10789 PMID:25991412

Carnut, L., Mendes, Á., & Guerra, L. (2020). Coronavirus, Capitalism in Crisis and the Perversity of Public Health in Bolsonaro's Brazil. *International Journal of Health Services*, *0020731420965137*. doi:10.1177/0020731420965137 PMID:33070682

Carra, I., Ortega-Clemente, L. A., Perales, J. F., & Giménez, J. (2013). Optimization and validation of a photoreactor for the degradation of diclofenac in wastewater treatment plant effluents. *Chemical Engineering Journal*, *230*, 79–88.

Carraro, E., Bonetta, S., Bertine, C., Lorenzi, E., Bonetta, S., & Gilli, G. (2016). Hospital effluents management: Chemical, physical, microbiological risks and legislation in different countries. *Journal of Environmental Management*, *168*, 185–199. doi:10.1016/j.jenvman.2015.11.021 PMID:26708649

Carroll, K., & Edelstein, J. E. (2006). *Prosthetics and patient management: A comprehensive clinical approach*. SLACK Incorporated.

Carvalho, R. N., Arukwe, A., Ait-Aissa, S., Bado-Nilles, A., Balzamo, S., Baun, A., Belkin, S., Blaha, L., Brion, F., Conti, D., Creusot, N., Essig, Y., Ferrero, V. E. V., Flander-Putrle, V., Fürhacker, M., Grillari-Voglauer, R., Hogstrand, C., Jonáš, A., Kharlyngdoh, J. B., Loos, R., Lundebye, A., Modig, C., Olsson, P., Pillai, S., Polak, N., Potalivo, M., Sanchez, W., Schifferli, A., Schirmer, K., Sforzini, S., Stürzenbaum, S. R., Søfteland, L., Turk, V., Viarengo, A., Werner, I., Yagur-Kroll, S., Zounková, R. & Lettieri, T. (2014). Mixtures of chemical pollutants at European Legislation Safety Concentrations: How safe are they? *Toxicological Sciences*, *141*(1), 218–233, https://doi.org/10.1093/toxsci/kfu118 doi:.

Cassidy, A., Mukamal, K. J., Liu, L., Franz, M., Eliassen, A. H., & Rimm, E. B. (2013). High anthocyanin intake is associated with a reduced risk of myocardial infarction in young and middle-aged women. *Circulation*, *127*(2), 188–196. doi:10.1161/CIRCULATIONAHA.112.122408 PMID:23319811

Castaño, N., Cordts, S. C., Kurosu Jalil, M., Zhang, K. S., Koppaka, S., Bick, A. D., Paul, R., & Tang, S. K. Y. (2021). Fomite transmission, physicochemical origin of virus–surface interactions, and disinfection strategies for enveloped viruses with applications to SARS-CoV-2. *ACS Omega*, *6*(10), 6509–6527. doi:10.1021/acsomega.0c06335 PMID:33748563

Cave, R., Cole, J., & Mkrtchyan, H. V. (2021). Surveillance and prevalence of antimicrobial resistant bacteria from public settings within urban built environments: Challenges and opportunities for hygiene and infection control. *Environment International*, *157*, 106836. doi:10.1016/j.envint.2021.106836 PMID:34479136

Cedergreen, N. (2014). Quantifying synergy: A systematic review of mixture toxicity studies within environmental toxicology. *PLoS One*, *9*(5), e96580. doi:10.1371/journal.pone.0096580 PMID:24794244

Centers for Disease Control and Prevention. (2021). *Antibiotic resistance questions and answers*. CDC. https://www.cdc.gov/antibiotic-use/antibiotic-resistance.html

Cevallos-Mendoza, J., Amorim, C. G., Rodríguez-Díaz, J. M., & Montenegro, M. D. C. B. (2022). Removal of contaminants from water by membrane filtration: A review. *Membranes*, *12*(6), 570. doi:10.3390/membranes12060570 PMID:35736277

Chadwell, A., Diment, L., Micó-Amigo, M., Morgado Ramírez, D. Z., Dickinson, A., Granat, M., Kenney, L., Kheng, S., Sobuh, M., Ssekitoleko, R., & Worsley, P. (2020). Technology for monitoring everyday prosthesis use: A systematic review. *Journal of Neuroengineering and Rehabilitation*, *17*(1), 93. doi:10.118612984-020-00711-4 PMID:32665020

Chao, Y., Zhu, W., Wu, X., Hou, F., Xun, S., Wu, P., Ji, H., Xu, H., & Li, H. (2014). Application of graphene-like layered molybdenum disulfide and its excellent adsorption behavior for doxycycline antibiotic. *Chemical Engineering Journal*, *243*, 60–67. doi:10.1016/j.cej.2013.12.048

Chaula, J. (2019). Opportunities and Challenges for Integrating Statistical and Spatial Data in East African Countries. *Technology, and Sciences (ASRJETS) American Scientific Research Journal for Engineering*, *59*(1).

Chawla, S., Saxena, A., & Seshadri, S. (1988). In-vitro availability of iron in various green leafy vegetables. *Journal of the Science of Food and Agriculture*, *46*(1), 125–127. doi:10.1002/jsfa.2740460112

Chen, C. M., LaBat, K., & Bye, E. (2010). Physical characteristics related to bra fit. *Ergonomics*, *53*(4), 514–524. doi:10.1080/00140130903490684 PMID:20309747

Chen, C., Zhang, B., Huang, Q., Fu, X., & Liu, R. H. (2017). Microwave-assisted extraction of polysaccharides from Moringa oleifera Lam. leaves: Characterization and hypoglycemic activity. *Industrial Crops and Products*, *100*, 1–11. doi:10.1016/j.indcrop.2017.01.042

Chen, F., Li, D., Shen, H., Wang, C., Li, E., Xing, H., & Nguyen, H. (2017). Polysaccharides from Trichosanthes Fructus via Ultrasound-Assisted Enzymatic Extraction Using Response Surface Methodology. *BioMed Research International*. PMID:29147656

Cheng, J., Turkstra, J., Peng, M., Du, N., & Ho, P. (2006). Urban land administration and planning in China: Opportunities and constraints of spatial data models. *Land Use Policy*, *23*(4), 604–616. doi:10.1016/j.landusepol.2005.05.010

Chen, M., Yu, Q., & Sun, H. (2013). Novel strategies for the prevention and treatment of biofilm related infections. *International Journal of Molecular Sciences*, *14*(9), 18488–18501. doi:10.3390/ijms140918488 PMID:24018891

Chen, R., Li, S., Liu, C., Yang, S., & Li, X. (2012). Ultrasound complex enzymes assisted extraction and biochemical activities of polysaccharides from Epimedium leaves. *Process Biochemistry (Barking, London, England)*, *47*(12), 2040–2050. doi:10.1016/j.procbio.2012.07.022

Chen, X., Huang, B., & Te Lin, C. (2019). Environmental Awareness and Environmental Kuznets Curve. *Economic Modelling*, *77*, 2–11. doi:10.1016/j.econmod.2019.02.003

Chen, Y., Jackson, K. M., Chea, F. P., & Schaffner, D. W. (2001). Quantification and variability analysis of bacterial cross-contamination rates in common food service tasks. *Journal of Food Protection*, *64*(1), 72–80. doi:10.4315/0362-028X-64.1.72 PMID:11198444

Chen, Y., Vymazal, J., Březinová, T., Koželuh, M., Kule, L., Huang, J., & Chen, Z. (2016). Occurrence, removal and environmental risk assessment of pharmaceuticals and personal care products in rural wastewater and treatment wetlands. *The Science of the Total Environment*, *566-567*, 1660–1669. doi:10.1016/j.scitotenv.2016.06.069 PMID:27342641

Chhikara, B. S., & Parang, K. (2022). Global Cancer Statistics 2022: The trends projection analysis. In *Chemical Biology Letters Chem.* []. https://pubs.thesciencein.org/cbl]. *Biol. Lett*, *2023*(1).

Chikere, C. B., Omoni, V. T., & Chikere, B. O. (2008). Distribution of potential nosocomial pathogens in a hospital environment. *African Journal of Biotechnology*, *7*(20).

Chonde, S., & Raut, P. (2017). Treatment of dairy wastewater by Moringa oleifera seeds. *World Journal of Pharmaceutical Research*, *6*(8), 1484–1493. doi:10.20959/wjpr20178-9015

Chong, M. (2005). Perception of the visit and benefits of Bt eggplant by Indian farmers. *Journal of Risk Research*, *8*(7-8), 617–634. doi:10.1080/13669870500042990

Chonova, T., Keck, F., Labanowski, J., Montuelle, B., Rimet, F., & Bouchez, A. (2016). Separate treatment of hospital and urban wastewaters: A real scale comparison of effluents and their effect on microbial communities. *The Science of the Total Environment*, *542*, 965–975. doi:10.1016/j.scitotenv.2015.10.161 PMID:26562343

Chung, M., Lin, H., & Wang, M. J. W. (2007). The development of sizing systems for Taiwanese elementary and high-school students. *International Journal of Industrial Ergonomics*, *37*(8), 707–716. doi:10.1016/j.ergon.2007.05.004

Ciampi, F., Marzi, G., Demi, S., & Faraoni, M. (2020). The big data-business strategy interconnection: a grand challenge for knowledge management. A review and future perspectives. In Journal of Knowledge Management, 24(5). doi:10.1108/JKM-02-2020-0156

Claessens, M., Vanhaecke, L., Wille, K., & Janssen, C. R. (2013). Emerging contaminants in Belgian marine waters: Single toxicant and mixture risks of pharmaceuticals. *Marine Pollution Bulletin, 71*(1-2), 41–50. doi:10.1016/j.marpolbul.2013.03.039 PMID:23643408

Cliquet, G. (2020). From Geomarketing to Spatial Marketing. In *Spatial Economics* (Vol. II). Applications., doi:10.1007/978-3-030-40094-1_10

Coffie, I. S., & Hinson, R. E. (2021). Types of Corporate Social Responsibility Initiatives as Response to COVID-19 Pandemic in Emerging Economies. In *Marketing Communications in Emerging Economies* (Vol. I, pp. 161–192). Palgrave Macmillan. doi:10.1007/978-3-030-81329-1_7

Concellòn, A., An ˜o'n, M. C., & Chaves, A. R. (2005). Effect of chilling on ethylene production in eggplant fruit. *Food Chemistry, 92*(1), 63–69. doi:10.1016/j.foodchem.2004.04.048

Connell, L. J., Ulrich, P. V., Brannon, E. L., Alexander, M., & Presley, A. B. (2006). Body shape assessment scale : Instrument development for analyzing female figures. *Clothing & Textiles Research Journal, 25*(2), 80–95. doi:10.1177/0887302X0602400203

Correa, C., Peral, D., Porro, J. A., Díaz, M., Ruiz de Lara, L., García-Beltrán, A., & Ocaña, J. L. (2015). Random-type scanning patterns in laser shock peening without absorbing coating in 2024-T351 Al alloy: A solution to reduce residual stress anisotropy. *Optics & Laser Technology, 73*, 179–187. doi:10.1016/j.optlastec.2015.04.027

Coviello, T., Matricardi, P., Marianecci, C., & Alhaique, F. (2007). Polysaccharide hydrogels for modified release formulations. *Journal of Controlled Release, 119*(1), 5–24. doi:10.1016/j.jconrel.2007.01.004 PMID:17382422

Criscuolo, E., Signorelli, C., Mancini, N., & Clementi, M. (2021). *Fast inactivation of SARS-CoV-2 by UV-C and ozone exposure on di ff erent materials. 10*, 2–5.

Crittenden, V. L., Crittenden, W. F., Ferrell, L. K., Ferrell, O. C., and Pinney, C. C. (2011b). *Market-oriented sustainability : a conceptual framework and propositions*, 71–85. doi:10.1007/s11747-010-0217-2

Cudjoe, D., Balali, G. I., Titus, O. O., Osafo, R., & Taufiq, M. (2022). Food Safety in Sub-Sahara Africa, An insight into Ghana and Nigeria. *Environmental Health Insights, 16*. doi:10.1177/11786302221142484 PMID:36530486

Cueto, L. J., Frisnedi, A. F. D., Collera, R. B., Batac, K. I. T., & Agaton, C. B. (2022). Digital Innovations in MSMEs during Economic Disruptions : Experiences and Challenges of Young Entrepreneurs. *Administrative Sciences, 12*(1), 8–22. doi:10.3390/admsci12010008

Cunningham, B., O'Neill, V., Devereux, M., McGann, D., & O'Hora, J. (2022). Use of a door handle disinfection system to reduce the risks associated with microbial loads on fomites in a healthcare setting. *The Journal of Hospital Infection, 130*, 104–107. doi:10.1016/j.jhin.2022.09.003 PMID:36113634 da Costa Luciano, C., Olson, N., Tipple, A. F. V., & Alfa, M. (2016). Evaluation of the ability of different detergents and disinfectants to remove and kill organisms in traditional biofilm. *American Journal of Infection Control, 44*(11), e243–e249. doi:10.1016/j.ajic.2016.03.040 PMID:27234012

Dabija, D. C., Postelnicu, C., Dinu, V., & Mihăilă, A. (2017). Stakeholders' perception of sustainability orientation within a major Romanian University. *International Journal of Sustainability in Higher Education, 18*(4), 533–553. doi:10.1108/IJSHE-10-2015-0169

Dal Canto, E., Ceriello, A., Rydén, L., Ferrini, M., Hansen, T. B., Schnell, O., Standl, E., & Beulens, J. W. (2019). Diabetes as a cardiovascular risk factor: An overview of global trends of macro and micro vascular complications. *European Journal of Preventive Cardiology, 26*(2, suppl), 25–32. doi:10.1177/2047487319878371 PMID:31722562

Darko, A., & Chan, A. P. C. (2018). Strategies to Promote Green Building Technologies Adoption In Developing Countries: The Case of Ghana. *Building and Environment, 130*, 74–84. doi:10.1016/j.buildenv.2017.12.022

Davenport, P., Noroozi, S., Sewell, P., & Zahedi, S. (2017). Systematic review of studies examining transtibial prosthetic socket pressures with changes in device alignment. *Journal of Medical and Biological Engineering, 37*(1), 1–17. doi:10.100740846-017-0217-5 PMID:28286462

Dawodu, O. G., & Akanbi, R. B. (2021). Isolation and identification of microorganisms associated with automated teller machines on Federal Polytechnic Ede campus. *PLoS ONE, 16*(8), 1–15. doi:10.1371/journal.pone.0254658

de Ilurdoz, M. S., Sadhwani, J. J., & Reboso, J. V. (2022). Antibiotic removal processes from water & wastewater for the protection of the aquatic environment-a review. *Journal of Water Process Engineering, 45*, 102474. doi:10.1016/j.jwpe.2021.102474

Debrah, J. K., Vidal, D. G., & Dinis, M. A. P. (2021). Raising Awareness on Solid Waste Management Through Formal Education for Sustainability: A Developing Countries Evidence Review. *Recycling, 6*(1), 1–21. doi:10.3390/recycling6010006

Denkyi-Sarpong, G. (2011). Meeting the challenge of a competitive fashion market. *International Journal of Home Economics Research, 2*, 157–158.

Derakhshan, F., & Yousefi, S. (2019). A review on the applications of multiagent systems in wireless sensor networks. *International Journal of Distributed Sensor Networks, 15*(5), 1550147719850767. doi:10.1177/1550147719850767

Dhakal, B., & Swaroop, S. (2018). Review: Laser shock peening as post welding treatment technique. In Journal of Manufacturing Processes, 32. doi:10.1016/j.jmapro.2018.04.006

Dharwal, M., Parashar, D., Shuaibu, M. S., Abdullahi, S. G., Abubakar, S., & Bala, B. B. (2020). Water pollution: Effects on health and environment of Dala LGA, Nigeria. *Materials Today: Proceedings*.

Dia, N., Azanu, D., Darmey, J., & Sanogo, A. (2023). Adsorption of ciprofloxacin using composite microplastics. *Research, Education and Emerging Trends for Technological and Sustainable Development, 7th International Conference on Applied Sciences and Technology 2023-KsTU.*

Dias, J. S. (2012). Nutritional quality and health benefits of vegetables: A Review. *Food and Nutrition Sciences, 3*(10), 1354–1374. doi:10.4236/fns.2012.310179

Dibua, U. M. E., Onyemerela, I. S., & Nweze, E. I. (2014). Frequency, urinalysis and susceptibility profile of pathogens causing urinary tract infections in Enugu State, southeast Nigeria. *Revista do Instituto de Medicina Tropical de São Paulo, 56*(1), 55–59. doi:10.1590/S0036-46652014000100008 PMID:24553609

Dietary reference intake – DRI. (2020). Recommended intakes for individuals, elements. Food and nutrition board, Institute of medicine and national academics. *DaddyBlog*. daddyblog.wordpress.com

Diggle, P. (2005). Applied Spatial Statistics for Public Health Data. *Journal of the American Statistical Association, 100*(470), 702–703. doi:10.1198/jasa.2005.s15

Ding, K. (2003). FEM simulation of two sided laser shock peening thin sections of Ti-6Al-4V alloy. *Surface Engineering, 19*(2), 127–133. doi:10.1179/026708403225002568

Ding, K., & Ye, L. (2006). Simulation of multiple laser shock peening of a 35CD4 steel alloy. *Journal of Materials Processing Technology, 178*(1–3), 162–169. doi:10.1016/j.jmatprotec.2006.03.170

Dolar, D., Košutić, K., Pavlović, D. M., & Kunst, B. (2009). Removal of emerging contaminants of industrial origin by NF/RO-A pilot scale study. *Desalination and Water Treatment, 6*(1-3), 197–203. doi:10.5004/dwt.2009.636

Donaldson, T., & Preston, L. E. (1995). The Stakeholder Theory of the Corporation: Concepts, Evidence, and Implications. *Academy of Management Review, 20*(1), 65–91. doi:10.2307/258887

Dopico, A. M., & Tigyi, G. J. (2007). *A glance at the structural and functional diversity of membrane lipids Methods in Membrane Lipids*. Springer.

Dorman, M., Toparli, M. B., Smyth, N., Cini, A., Fitzpatrick, M. E., & Irving, P. E. (2012). Effect of laser shock peening on residual stress and fatigue life of clad 2024 aluminium sheet containing scribe defects. *Materials Science and Engineering A, 548*, 142–151. doi:10.1016/j.msea.2012.04.002

dos Santos Escobar, O., de Azevedo, C. F., Swarowsky, A., Adebayo, M. A., Netto, M. S., & Machado, F. M. (2021). Utilization of different parts of Moringa oleifera Lam. seeds as biosorbents to remove Acid Blue 9 synthetic dye. *Journal of Environmental Chemical Engineering, 9*(4), 105553. doi:10.1016/j.jece.2021.105553

Drankiewicz, D., & Dundes, L. (2003). Handwashing among female college students. *American Journal of Infection Control, 31*(2), 67–71. doi:10.1067/mic.2003.6 PMID:12665738

Dsani, E., Afari, E. A., Danso-Appiah, A., Kenu, E., Kaburi, B. B., & Egyir, B. (2020). Antimicrobial resistance and molecular detection of extended spectrum β-lactamase producing Escherichia coli isolates from raw meat in Greater Accra region, Ghana. *BMC Microbiology*, *20*(1), 1–8. doi:10.118612866-020-01935-z PMID:32795260

Du Plessis, C., & Brandon, P. (2015). An Ecological Worldview as the Basis for a Regenerative Sustainability Paradigm for the Built Environment. *Journal of Cleaner Production*, *109*, 53–61. doi:10.1016/j.jclepro.2014.09.098

Dubman, R. (2019). The digital governance of data-driven smart cities: Sustainable urban development, big data management, and the cognitive internet of things. *Geopolitics, History, and International Relations*, *11*(2), 34. doi:10.22381/GHIR11220195

Dubois, M., Gilles, K. A., Hamilton, J. K., Rebers, P., & Smith, F. (1956). Colorimetric method for determination of sugars and related substances. *Analytical Chemistry*, *28*(3), 350–356. doi:10.1021/ac60111a017

Ducel, G., Fabry, J., & Nicolle, L. (2002). Prevention of hospital acquired infections: a practical guide. Prevention of Hospital Acquired Infections: A Practical Guide., Ed. 2.

Durdyev, S., Zavadskas, E. K., Thurnell, D., Banaitis, A., & Ihtiyar, A. (2018). Sustainable Construction Industry in Cambodia: Awareness, Drivers and Barriers. *Sustainability (Basel)*, *10*(2), 1–19. doi:10.3390u10020392

Echeverria, C. A., Handoko, W., Pahlevani, F., & Sahajwalla, V. (2019). Cascading use of textile waste for the advancement of fibre reinforced composites for building applications. *Journal of Cleaner Production*, *208*, 1524–1536. doi:10.1016/j.jclepro.2018.10.227

Edinburgh Centre for Toxicology. (2015). Environmental Risk Assessment. *UNEP/IPCS Training Module No. 3. SectionB*. UNEP. http://www.chem.unep.ch/irptc/Publications/riskasse/B2text.pdf.

Editors of Encyclopaedia. (2023, May 21). antibiotic. *Encyclopedia Britannica*. https://www.britannica.com/science/antibiotic

Eid, S. Y., El-Readi, M. Z., Fatani, S. H., Mohamed Nour Eldin, E. E., & Wink, M. (2015). Natural Products Modulate the Multifactorial Multidrug Resistance of Cancer. *Pharmacology &amp. Pharmacology & Pharmacy*, *06*(03), 146–176. doi:10.4236/pp.2015.63017

Elmasoudi, I., Ibrahim, M. G., Tokimatsu, K., & Elbeltagi, E. (2019). Environmental Impact Assessment Model For Buildings' Construction Activities. *International Journal of Construction Management*, *0*(0), 1–12. doi:10.1080/15623599.2019.1657256

Elsergany, M., Ben Moussa, M., Ahsan, A., Khalfan, A., & Eissa, A. (2015). Exploratory study of bacterial contamination of different surfaces in four shopping malls in Sharjah, UAE. *Journal of Environmental and Occupational Science*, *4*(2), 101–105. doi:10.5455/jeos.20150611030039

Eman, N. A., Tan, C. S., & Makky, E. A. (2014). Impact of Cake Residue Application on Waste Water Treatment: A Case Study. *Journal of Water Resource and Protection*, *06*(07), 677–687. doi:10.4236/jwarp.2014.67065

Estokova, A., Fabianova, W. M., & Marcela, O. (2022). Concrete Structures and Their impacts on Climate Change and Water and Raw Material Resource Depletion. *International Journal of Civil Engineering*, *13*. doi:10.1007/s40999-022-0701-8

Etim, E., & Daramola, O. (2020). The informal sector and economic growth of South Africa and Nigeria: A comparative systematic review. *Journal of Open Innovation*, *6*(4), 134. doi:10.3390/joitmc6040134

European Chemicals Agency. (2008). Guidance on information requirements and chemical safetyassessment. *Chapter R.10: Characterisation of dose [concentration]-response for environment*. Europea. https://echa.europa.eu/documents/10162/13632/information_requirements_r10_en.pdf.

European Medicines Agency. (2015). Environmental risk-assessment of medicines. *Science Health Medicines*. www.ema.europa.eu.

Eurostat. (2021). *Euro zone GDP contracts by a record 3.8% in the first quarter*. https://ec.europa.eu/eurostat/web/covid-19/economy

Ezugbe, E. O., & Rathilal, S. (2020). Membrane technologies in wastewater treatment: A review. *Membranes*, *10*(5), 89. doi:10.3390/membranes10050089 PMID:32365810

Fabbro, R., Fournier, J., Ballard, P., Devaux, D., & Virmont, J. (1990). Physical study of laser-produced plasma in confined geometry. *Journal of Applied Physics*, *68*(2), 775–784. doi:10.1063/1.346783

Fadaei, A. (2021). *Viral Inactivation with Emphasis on SARS-CoV-2 Using Physical. 2021.*

Fairand, B. P., Wilcox, B. A., Gallagher, W. J., & Williams, D. N. (1972). Laser shock-induced microstructural and mechanical property changes in 7075 aluminum. *Journal of Applied Physics*, *43*(9), 3893–3895. doi:10.1063/1.1661837

Fane, A. G. T., Wang, R., & Jia, Y. (2011). *Membrane and Desalination Technologies*. Membrane and Desalination Technologies., doi:10.1007/978-1-59745-278-6

Fatemi, A., Molaei, R., Simsiriwong, J., Sanaei, N., Pegues, J., Torries, B., Phan, N., & Shamsaei, N. (2019). Fatigue behaviour of additive manufactured materials: An overview of some recent experimental studies on Ti-6Al-4V considering various processing and loading direction effects. In Fatigue and Fracture of Engineering Materials and Structures, 42(5). doi:10.1111/ffe.13000

Faustini, M. C., Neptune, R. R., Crawford, R. H., Rogers, W. E., & Bosker, G. (2014). An experimental and theoretical framework for manufacturing prosthetic sockets for transtibial amputees. *IEEE Transactions on Neural Systems and Rehabilitation Engineering*, *14*(3), 304–310. doi:10.1109/TNSRE.2006.881570 PMID:17009490

FC. (2018). *National Forest Plantation Development Programme. Annual Report 2018.* Forest Services Division, Forestry Commission.

Fedaku, S., Alemayehu, E., Dewil, R., & Van der Bruggen, B. (2019). Pharmaceuticals in freshwater aquatic environments. In: A comparison of the African and European challenge. *The Science of the Total Environment*, *654*, 324–337. doi:10.1016/j.scitotenv.2018.11.072 PMID:30448654

Feldman, E. (1992). Fashions of a 1990s. New York: Facts on File.

Fianu, A. G., Biney-Aidoo, V., Antiaye, E., & Oppong, J. (2014). The production and exportation of garments under the AGOA initiative: Challenges facing garment producers in Ghana. *Arts and Design Studies, 17*, 34-44. http://www.iiste.org

Fianu, A.G., Ayertey, A., & Francois, E.M. (2005). A pilot study of categorizing Ghanaian Female University Students into body sizes (Phase I). *ARAHE, 12*(3).

Fianu, A.G., Ayertey, A., & Francois, E.M. (2005). Perceived and preferred body sizes of Ghanaian Female Students at University of Ghana, Legon (Phase II). *ARAHE, 12*(4), 231 - 235.

Fianu, A. G., & Zentey, E. A. (2000). Problems of large-scale fashion designers in Accra, Ghana. *Journal of Consumer Studies & Home Economics*, *24*(2), 128–136. doi:10.1046/j.1365-2737.2000.00153.x

Fianu, D. A. G., & Aryee-Atta, S. N. A. (2008). Ghanaian women's perception of ready to wear garments. [country_paper_ghana.pdf]. *Journal of Family and Consumer Sciences*, *100*(4), 28–33.

Finch, J. (2011). The art of medicine: The ancient origins of prosthetic medicine. *Lancet*, *377*(9765), 348–349. doi:10.1016/S0140-6736(11)60190-6 PMID:21341402

Firehock, K., & Walker, R. A. (2019). *Green infrastructure: map and plan the natural world with GIS*. Esri Press Redlands.

Fisher, R. A. (1922). On the mathematical foundations of theoretical statistics. *Philosophical Transactions of the Royal Society of London. Series A, Mathematical and Physical Sciences*, 309–368.

Flomo, M. K., Azeko, S. T., Arthur, E. K., Kukurah, J. D., Mustapha, K., Annan, E., & Agyei-Tuffour, B. (2021). Reinforcement of cement mortar with recycled polyethylene waste for construction applications. *Journal of Composite Materials*, *55*(21), 2867–2875. doi:10.1177/00219983211000249

Flores, G. E., Bates, S. T., Knights, D., Lauber, C. L., Stombaugh, J., Knight, R., & Fierer, N. (2011). Microbial biogeography of public restroom surfaces. *PLoS One*, *6*(11), e28132. doi:10.1371/journal.pone.0028132 PMID:22132229

Fornell, C., & Larcker, D. F. (1981). *Structural equation models with unobservable variables and measurement error: Algebra and statistics.*

Fornell, C., and Larcker, D. F. (1981). *Structural equation models with unobservable variables and measurement error: Algebra and statistics.*

Forster, P., & Ampong, I. (2012). Pattern cutting skills in small scale garment industries and teacher education universities in Ghana. *International Journal of Vocational and Technical Education, 4*(2), 14–24.

Franch-Pardo, I., Napoletano, B. M., Rosete-Verges, F., & Billa, L. (2020). Spatial analysis and GIS in the study of COVID-19. A review. *The Science of the Total Environment, 739*, 140033. doi:10.1016/j.scitotenv.2020.140033 PMID:32534320

Freeman, R. E. (1999). Response : Divergent Stakeholder Theory. *Academy of Management, 24*(2), 233–236.

Friess, W., & Schlapp, M. (2001). Effects of processing conditions on the rheological behavior of collagen dispersions. *European Journal of Pharmaceutics and Biopharmaceutics, 51*(3), 259–265. doi:10.1016/S0939-6411(01)00136-9 PMID:11343891

Fuglie, L. J. (2001). *The miracle tree: the multiple attributes of Moringa.*

Fu, J., Zhu, Y., Zheng, C., Liu, R., & Ji, Z. (2014). Effect of laser shock peening on mechanical properties of Zr-based bulk metallic glass. *Applied Surface Science, 313*, 692–697. doi:10.1016/j.apsusc.2014.06.056

Fu, J., Zhu, Y., Zheng, C., Liu, R., & Ji, Z. (2015). Evaluate the effect of laser shock peening on plasticity of Zr-based bulk metallic glass. *Optics & Laser Technology, 73*, 94–100. doi:10.1016/j.optlastec.2015.04.009

Fu, J., Zhu, Y., Zheng, C., Liu, R., & Ji, Z. (2016). Enhanced plasticity of bulk metallic glass in different aspect ratios via laser shock peening with multiple impacts. *Optics & Laser Technology, 83*, 43–48. https://doi.org/https://doi.org/10.1016/j.optlastec.2016.03.028. doi:10.1016/j.optlastec.2016.03.028

Fu, W., Huang, Y., Sun, J., & Ngan, A. H. W. (2022). Strengthening CrFeCoNiMn0.75Cu0.25 high entropy alloy via laser shock peening. *International Journal of Plasticity, 154*, 103296. doi:10.1016/j.ijplas.2022.103296

Gallagher, P., O'Donovan, M. A., Doyle, A., & Desmond, D. (2011). Environmental barriers, activity limitations and participation restrictions experienced by people with major limb amputation. *Prosthetics and Orthotics International, 35*(3), 278–284. doi:10.1177/0309364611407108 PMID:21937573

Ganguly, A. (2021). *IBM WATSON SOLUTIONS FOR MACHINE LEARNING achieving successful results across computer... Vision, natural language processing and ai project.* BPB Publications.

Ganiyu, S. O., Van Hullebusch, E. D., Cretin, M., Esposito, G., & Oturan, M. A. (2015). Coupling of membrane filtration and advanced oxidation processes for removal of pharmaceutical residues: A critical review. *Separation and Purification Technology, 156*, 891–914. doi:10.1016/j.seppur.2015.09.059

Gao, Y., Yang, W., Huang, Z., & Lu, Z. (2021). Effects of residual stress and surface roughness on the fatigue life of nickel aluminium bronze alloy under laser shock peening. *Engineering Fracture Mechanics, 244*, 107524. Advance online publication. doi:10.1016/j.engfracmech.2021.107524

García, J., García-Galán, M. J., Day, J. W., Boopathy, R., White, J., Wallace, S., & Hunter, R. G. (2020). A review of emerging organic contaminants (EOCs), antibiotic resistant bacteria (ARB) and antibiotic resistance genes (ARGs) in the environment: Increasing removal with wetlands and reducing environmental impacts. *Bioresource Technology, 307*, 123228. doi:10.1016/j.biortech.2020.123228 PMID:32247686

Gavor, M. E., & Ampong, I. T. (2009). Factors that affect quality of garments made by Ghanaian micro-scale manufacturers. *International Journal of Home Economics Research, 1*, 110–121.

Geddes, C. D., & Lakowicz, J. R. (2007). *Glucose sensing* (Vol. 11). Springer Science & Business Media.

Ge, M. Z., Xiang, J. Y., Fan, Z., Lu, Y. L., & Lei, W. N. (2018). Effect of laser energy on microstructure of Mg-3Al-1Zn alloy treated by LSP. *Journal of Alloys and Compounds, 734*, 266–274. doi:10.1016/j.jallcom.2017.11.032

George, K. S., Revathi, K. B., Deepa, N., Sheregar, C. P., Ashwini, T. S., & Das, S. (2016). A Study on the Potential of Moringa Leaf and Bark Extract in Bioremediation of Heavy Metals from Water Collected from Various Lakes in Bangalore. *Procedia Environmental Sciences, 35*, 869–880. doi:10.1016/j.proenv.2016.07.104

Getty. (n.d). 4 basic sculpture techniques. *The J. Paul Getty Museum*. http://www.getty.edu/education/teachers/classroom_resources/curricula/sculpture/background2.html

Ghana Statistical Service. (2013). *2010 Population & housing census: national analytical report*. Ghana Statistical Service.

Gidari, A., Sabbatini, S., Bastianelli, S., Pierucci, S., Busti, C., Stabile, A. M., Monari, C., Galli, F., Rende, M., Cruciani, G., & Francisci, D. (2021). *SARS-CoV-2 Survival on Surfaces and the Effect of UV-C Light*. 2–9.

Gill, C. O., & Jones, T. (2002). Effects of wearing knitted or rubber gloves on the transfer of Escherichia coli between hands and meat. *Journal of Food Protection, 65*(6), 1045–1048. doi:10.4315/0362-028X-65.6.1045 PMID:12092720

Gill, S. (2015). A review of research and innovation in garment sizing, prototyping and fitting. *Textile Progress, 47*(1), 1–85. doi:10.1080/00405167.2015.1023512

Giovinazzo, R., Caradonna, L., Giaquinta, G., Guerrera, E., Mameli, M., Mansi, A., Marena, G., Mastromartino, T., Sarto, D., & Tomao, P. (2017). guidance values for microbiological monitoring on surfaces: A literature overview. *Biomedicine & Prevention, 4*(135), 174–180.

Girona, T. (2020). Confinement Time Required to Avoid a Quick Rebound of COVID-19: Predictions From a Monte Carlo Stochastic Model. *Frontiers in Physics (Lausanne)*, *8*, 186. doi:10.3389/fphy.2020.00186

Gogate, A., & Pandit, A. S. (2004). A review of imperative technologies for wastewater treatment II: Hybrid methods. *Advances in Environmental Research*, *8*(3-4), 553–597. doi:10.1016/S1093-0191(03)00031-5

Goodchild, M., Haining, R., & Wise, S. (1992). Integrating GIS and spatial data analysis: Problems and possibilities]. *International Journal of Geographical Information Systems*, *6*(5), 407–423. doi:10.1080/02693799208901923

Graham, P. L. III, Lin, S. X., & Larson, E. L. (2006). A US population-based survey of Staphylococcus aureus colonization. *Annals of Internal Medicine*, *144*(5), 318–325. doi:10.7326/0003-4819-144-5-200603070-00006 PMID:16520472

Grassi, M., Kaykioglu, G., Belgiorno, V., & Lofrano, G. (2012). Removal of Emerging Contaminants from Water and Wastewater by Adsorption Process. In G. Lofrano (Ed.), *Emerging Compounds Removal from Wastewater. SpringerBriefs in Molecular Science(). Springer*. doi:10.1007/978-94-007-3916-1_2

Grass, M. N. (1955). *History of hosiery. 2*. Fairchild Publishing Company.

Greenberg, C. (2012, August 10). Oscar Pistorius, South African 4×400m relay team finish 8th as Bahamas wins gold. *Huffington Post*. https://www.huffpost.com/entry/oscar-pistorius-south-africa-relay-4x400-olympics_n_1765596

Green, K. W., Toms, L. C., & Clark, J. (2015). Impact of market orientation on environmental sustainability strategy. *Management Research Review*, *38*(2), 217–238. doi:10.1108/MRR-10-2013-0240

Grimshaw, D. J. (2001). Harnessing the power of geographical knowledge: The potential for data integration in an SME. *International Journal of Information Management*, *21*(3), 183–191. . doi:10.1016/S0268-4012(01)00004-4

Grosz, E. (2003). Prosthetic objects. In I. Cheng, & B. Tschumi, The state of architecture at the beginning of the 21st Century (pp. 96-97). Columbia Books of Architecture.

Gu, H., Yan, P., Jiao, L., Chen, S., Song, Y., Zou, S., & Wang, X. (2023). Effect of laser shock peening on boring hole surface integrity and conformal contact fretting fatigue life of Ti-6Al-4 V alloy. *International Journal of Fatigue*, *166*, 107241. doi:10.1016/j.ijfatigue.2022.107241

Gu, J., Chen, C., Huang, X., Mo, J., Xie, Q., & Zeng, Q. (2021). Occurrence and risk assessment of tetracycline antibiotics in soils and vegetables from vegetable fields in Pearl River Delta, South China. *The Science of the Total Environment*, *776*, 145959. doi:10.1016/j.scitotenv.2021.145959

Gu, L., & Ozbakkaloglu, T. (2016). Use of recycled plastics in concrete: A critical review. *Waste Management (New York, N.Y.)*, *51*, 19–42. doi:10.1016/j.wasman.2016.03.005 PMID:26970843

Gunnarsson, L., Snape, J. R., Verbruggen, B., Owen, S. F., Kristiansson, E., Margiotta-Casaluci, L., Österlund, T., Hutchinson, K., Leverett, D., Marks, B., & Tyler, C. (2019). Pharmacology beyond the patient – the environmental risks of human drugs. *Environment International*, *129*, 320–332. doi:10.1016/j.envint.2019.04.075 PMID:31150974

Guo, D., Wu, H., Wang, S., Xin, H., Wang, J., Dai, Y., Xia, X., Zhang, Y., Zhang, H., Hong, Y., & Yu, C. (1999). Study of mechanism of strengthening materials with laser shock processing. *Science in China, Series E. Technological Sciences*, *42*(3), 288–293. Advance online publication. doi:10.1007/BF02916775

Gupta, R. C., Gupta, P. I., & Gupta, R. D. (1998). Modelling failure time data by Lehmann alternatives. *Communications in Statistics. Theory and Methods*, *27*(4), 887–904. doi:10.1080/03610929808832134

Gupta, R. D., & Kundu, D. (2009). A new class of weighted exponential distributions. *Statistics*, *43*(6), 621–634. doi:10.1080/02331880802605346

Gupta, S., Jain, R., Kachhwaha, S., & Kothari, S. L. (2018a). Nutritional and medicinal applications of Moringa oleifera Lam.—Review of current status and future possibilities. *Journal of Herbal Medicine*, *11*(March), 1–11. doi:10.1016/j.hermed.2017.07.003

Gu, S., Ślusarczyk, B., Hajizada, S., Kovalyova, I., & Sakhbieva, A. (2021). Impact of the covid-19 pandemic on online consumer purchasing behavior. *Journal of Theoretical and Applied Electronic Commerce Research*, *16*(6), 2263–2281. doi:10.3390/jtaer16060125

Gutiérrez, D., Delgado, S., Vázquez-Sánchez, D., Martínez, B., Cabo, M. L., Rodríguez, A., Herrera, J. J., & García, P. (2012). Incidence of Staphylococcus aureus and analysis of associated bacterial communities on food industry surfaces. *Applied and Environmental Microbiology*, *78*(24), 8547–8554. doi:10.1128/AEM.02045-12 PMID:23023749

Habwe, F. O., Waling, K. M. & Onyango, M. O. A. (2008). Food Processing and Preparation Technologies for Sustainable Utilization of African Indigenous Vegetables for Nutrition Security and Wealth Creation in Kenya. In *using Food Science and Technology to Improve Nutrition and Promote National Development*.

Hackel, L., Rankin, J. R., Rubenchik, A., King, W. E., & Matthews, M. (2018). Laser peening: A tool for additive manufacturing post-processing. *Additive Manufacturing*, *24*, 67–75. doi:10.1016/j.addma.2018.09.013

Haddaway, N. R., Page, M. J., Pritchard, C. C., & McGuinness, L. A. (2022). PRISMA2020: An R package and Shiny app for producing PRISMA 2020-compliant flow diagrams, with interactivity for optimised digital transparency and Open Synthesis. *Campbell Systematic Reviews*, *18*(2), e1230. doi:10.1002/cl2.1230 PMID:36911350

Haghi, A. K. (2005). *A study of heat and mass transfer in porous material under equilibrium conditions*.

Hair, J. F., Hult, G. T. M., & Ringle, C. M. (2014). Partial Least Squares Structural Equation Modeling (Pls-Sem) (pp. 1–329).

Hair, J. F., Risher, J. J., Sarstedt, M., & Ringle, C. M. (2019). When to use and how to report the results of PLS-SEM. *European Business Review, 31*(1), 2–24. doi:10.1108/EBR-11-2018-0203

Hammi, K. M., Jdey, A., Abdelly, C., Majdoub, H., & Ksouri, R. (2015). Optimization of ultrasound-assisted extraction of antioxidant compounds from Tunisian Zizyphus lotus fruits using response surface methodology. *Food Chemistry, 184*, 80–89. doi:10.1016/j.foodchem.2015.03.047 PMID:25872429

Han, E. J., & Lee, D. S. (2017). Significance of metabolites in the environmental risk assessment of pharmaceuticals consumed by human. *The Science of the Total Environment, 592*, 600–607. doi:10.1016/j.scitotenv.2017.03.044 PMID:28318699

Han, X., Liang, C. F., Li, T. Q., Wang, K., Huang, H. G., & Yang, X. E. (2013). Simultaneous removal of cadmium and sulfamethoxazole from aqueous solution by rice straw biochar. *Journal of Zhejiang University. Science. B., 14*(7), 640–649. doi:10.1631/jzus.B1200353 PMID:23825150

Harris, D., Taylor, K. P., Napierkowski, K., & Zechmann, B. (2021). Indoor finish material influence on contamination, transmission, and eradication of methicillin-resistant staphylococcus (MRSA). *HERD, 14*(1), 118–129. doi:10.1177/1937586720952892 PMID:32867539

Hasan, Md. K., Shahriar, A., & Jim, K. U. (2019). Water pollution in Bangladesh and its impact on public health. *Heliyon, 5*(8), e02145. https://doi.org/https://doi.org/10.1016/j.heliyon.2019.e02145

Hawkins, A. T., Pallangyo, A. J., Herman, A. M., Schaumeier, M. J., Smith, A. D., Hevelone, N. D., Crandell, D. M., & Nguyen, L. L. (2016). The effect of social integration on outcomes after major lower extremity amputation. *Journal of Vascular Surgery, 63*(1), 154–162. doi:10.1016/j.jvs.2015.07.100 PMID:26474508

Hazirah, A. R., Zainal, B., & Abdah, M. A. (2013). Total phenolic content, antioxidant and cytotoxic activity of cocoa (Theobroma cacao L.) polyphenols extracts on cancer cell lines. *Malaysian Journal of Nutrition, 19*(2), 223–232.

Heilingloh, C. S., Aufderhorst, U. W., Schipper, L., Dittmer, U., Witzke, O., Yang, D., Zheng, X., Sutter, K., Trilling, M., Alt, M., Steinmann, E., & Krawczyk, A. (2020). *Susceptibility of SARS-CoV-2 to UV irradiation. January.*

Hekmat, D., & Fisk, W. J. (1984). Improving the energy performance of residential clothes dryers. The *35th Annual International Appliance Technical Conference, Ohio State University. History of Millinery Hat styles of the Past.* Millinery Techniques. http://www.millinerytechniques.com/ history-of-millinery.html

Henseler, J., Ringle, C. M., & Sarstedt, M. (2015). A new criterion for assessing discriminant validity in variance-based structural equation modeling. *Journal of the Academy of Marketing Science*, *43*(1), 115–135. doi:10.100711747-014-0403-8

Hizam, M., Noor, M., Lee, W. J., Faiz, M., Azli, M., & Ngadi, N. (2021). Materials Today : Proceedings Moringa oleifera extract as green coagulant for POME Treatment : Preliminary studies and sludge evaluation. *Materials Today: Proceedings*, *2020*(xxxx). doi:10.1016/j.matpr.2021.02.241

Hoefter, A. F. (2001). *The competitiveness of Ghana's industry.* [Unpublished doctoral thesis, Nr. 2494, Difo-Druck Bamberg, Germany].

Hong, J., Hong, T., Kang, H., & Lee, M. (2019). A Framework for Reducing Dust Emissions and Energy Consumption on Construction Sites. *Energy Procedia*, *158*, 5092–5096. doi:10.1016/j.egypro.2019.01.637

Hossain, M. I., Slow, M. L., Ong, S. T., & Said, R. M. (2020). The Role of Environmental Awareness and Green technological Usage to Foster Sustainable Green Practices in Bangladeshi Manufacturing SMEs. *International Journal of Advanced Science and Technology*, *29*(7s), 3115-3124. https://www.researchgate.net/publication/341576178

Hosseini, S. S., Bringas, E., Tan, N. R., Ortiz, I., Ghahramani, M., & Shahmirzadi, M. A. A. (2016). Recent progress in development of high-performance polymeric membranes and materials for metal plating wastewater treatment: A review. *Journal of Water Process Engineering*, *9*, 78–110. doi:10.1016/j.jwpe.2015.11.005

Hrzˇenjak, R., Dolezal, K., & Ujevic, D. (2013). Development of sizing system for girls aged 6 to 12 years in Croatia. *Collegium Antropologicum*, *37*(4), 1095–1103. PMID:24611320

Hrženjak, R., Doležal, K., & Ujević, D. (2015). Sizing system for girls aged 13 - 20 years based on body types. *Textile Research Journal*, *85*(12), 1293–1304. doi:10.1177/0040517514559579

Hsu, C. H. (2009). Developing Accurate Industrial Standards to Facilitate Production in Apparel Manufacturing Based on Anthropometric Data. *Human Factors and Ergonomics in Manufacturing*, *19*(3), 199–211. doi:10.1002/hfm.20148

Hu, X., Gong, J., Renard, E. G., & Parashar, M. (2019). Two-Stage Framework for Big Spatial Data Analytics to Support Disaster Response. *Proceedings - 2019 IEEE International Conference on Big Data, Big Data 2019*. IEEE. 10.1109/BigData47090.2019.9005613

Huang, A., Yan, M., Lin, J., Xu, L., Gong, H., & Gong, H. (2021). A review of processes for removing antibiotics from breeding wastewater. *International Journal of Environmental Research and Public Health*, *18*(9), 4909. doi:10.3390/ijerph18094909 PMID:34062980

Hult, G. T. M., Mena, J. A., Ferrell, O. C., & Ferrell, L. (2011). Stakeholder marketing: A definition and conceptual framework. *AMS Review*, *1*(1), 44–65. doi:10.100713162-011-0002-5

Hu, X., Wu, J., Zhao, J., Yang, Y., & Qiao, H. (2020). Numerical simulation of the surface morphology and residual stress field of IN718 alloy by Gaussian mode laser shock. *Optik (Stuttgart)*, *207*, 164441. doi:10.1016/j.ijleo.2020.164441

Hu, X., Zhao, J., Teng, X., Nie, X., Jiang, Y., & Zhang, Y. (2022). Fatigue resistance improvement on double-sided welded joints of a titanium alloy treated by laser shock peening. *Journal of Materials Engineering and Performance*, *31*(12), 10304–10313. doi:10.100711665-022-07006-w

Hu, Y., Fan, H., Chen, Y., Chang, J., Zhan, X., Wu, H., Zhang, B., Wang, M., Zhang, W., Yang, L., Hou, X., Shen, X., Pan, T., Wu, W., Li, J., Hu, H., & Wei, F. (2021). Spatial patterns and conservation of genetic and phylogenetic diversity of wildlife in China. *Science Advances*, *7*(4), eabd5725. doi:10.1126ciadv.abd5725 PMID:33523945

Hu, Y., Gong, C., Yao, Z., & Hu, J. (2009). Investigation on the non-homogeneity of residual stress field induced by laser shock peening. *Surface and Coatings Technology*, *203*(23), 3503–3508. doi:10.1016/j.surfcoat.2009.04.029

Huyssteen, S. V. (2006). *Development of standardized sizing systems for the South African children's wear market.* [Doctoral thesis, University of Stellenbosch]. https://scholar.sun.ac.za/handle/10019.1/4646

Hwang, D., Stewart, W. P., & Ko, D. (2012). Community Behavior and Sustainable Rural Tourism Development. *Journal of Travel Research*, *51*(3), 328–341. doi:10.1177/0047287511410350

Idris, M. A., Jami, M. S., Jamal, P., & Hammed, A. M. (2016). *Moringa Oleifera Seed Extract : a Ḥ Al Ā L-Friendly Alternative. October.*

Idris, M. A., Jami, M. S., Hammed, A. M., & Jamal, P. (2016). Moringa oleifera seed extract: A review on its environmental applications. *International Journal of Applied Environmental Sciences*, *11*(6), 1469–1486.

Inweregbu, K., Dave, J., & Pittard, A. (2005). Nosocomial infections. *Continuing Education in Anaesthesia, Critical Care & Pain*, *5*(1), 14–17. doi:10.1093/bjaceaccp/mki006

Isberg, R. (2014). *Pathogenic Bacteria: What Distinguishes a Pathogen from a Non-Pathogen?* All Talks in Microbiology. https://www.ibiology.org/microbiology/pathogenic-bacteria-distinguishes-pathogen-non-pathogen/

Iskandar, D. Suliyanto, Dinanti, A., Naufalin, L. R., Krisnaresanti, A., & Chasanah, N. (2023). SME's market development through digital mapping to stimulate business growth. AIP Conference Proceedings, 2482(1), 100003.

ISO/TR 10652. (1991). *Standard Sizing Systems for Clothes*. International Organization for Standardization, Geneva.

Iweriebor, B. C., Iwu, C. J., Obi, L. C., Nwodo, U. U., & Okoh, A. I. (2015). Multiple antibiotic resistances among Shiga toxin producing Escherichia coli O157 in feces of dairy cattle farms in Eastern Cape of South Africa. *BMC Microbiology*, *15*(1), 1–9. doi:10.118612866-015-0553-y PMID:26475706

Jabłońska-Trypuć, A., Makuła, M., Włodarczyk-Makuła, M., Wołejko, E., Wydro, U., Serra-Majem, L., & Wiater, J. (2022). Inanimate surfaces as a source of hospital infections caused by fungi, bacteria and viruses with particular emphasis on SARS-CoV-2. *International Journal of Environmental Research and Public Health, 19*(13), 8121. doi:10.3390/ijerph19138121 PMID:35805776

Janardhanan, K. K., Ravikumar, K. S., & Karuppayil, S. M. (2020). Medicinal mushroom bioactives: Potential sources for anti-cancer drug development. *International Journal of Applied Pharmaceutics, 12*(Special Issue 4), 40–45. doi:10.22159/ijap.2020.v12s4.40103

Järup, L., & Åkesson, A. (2009). Current status of cadmium as an environmental health problem. *Toxicology and Applied Pharmacology, 238*(3), 201–208. doi:10.1016/j.taap.2009.04.020 PMID:19409405

Jaworski, B. J., & Kohli, A. K. (1993). Market Orientation : Antecedents and Consequences. *Journal of Marketing, 57*(3), 53–70. doi:10.1177/002224299305700304

Jayanthi, A., Singh, N., Datta, S., Dey, A., & Chowdhury, A. R. (2015). Antimicrobial activity and cytotoxicity of theobroma cacao extracts. *Der Pharmacia Lettre, 7*(7), 287–294.

Jemal, A., Bray, F., & Ferlay, J. (2011). Global Cancer Statistics: 2011. *CA: a Cancer Journal for Clinicians, 49*(2), 1, 33–64. doi:10.3322/caac.20107.Available

Jen, J.-F., Leu, M.-F., & Yang, T. C. (1998). Determination of hydroxyl radicals in an advanced oxidation process with salicylic acid trapping and liquid chromatography. *Journal of Chromatography. A, 796*(2), 283–288. doi:10.1016/S0021-9673(97)01019-4 PMID:9474786

Jia, W. L., Song, C., He, L. Y., Wang, B., Gao, F. Z., Zhang, M., & Ying, G. G. (2022). Antibiotics in soil and water: Occurrence, fate, and risk. *Current Opinion in Environmental Science & Health*, 100437.

Jin, K. R., & Ji, Z.-G. (2015). An integrated environment model for a constructed wetland – Hydrodynamics and transport processes. *Ecological Engineering, 84*, 416–426. doi:10.1016/j.ecoleng.2015.09.022

Jitt-Aer, K., Wall, G., Jones, D., & Teeuw, R. (2022). Use of GIS and dasymetric mapping for estimating tsunami-affected population to facilitate humanitarian relief logistics: A case study from Phuket, Thailand. *Natural Hazards, 113*(1), 185–211. doi:10.100711069-022-05295-x

Jo, A., Lee, S. K., & Kim, J. (2020). Gender gaps in the use of urban space in seoul: Analyzing spatial patterns of temporary populations using mobile phone data. *Sustainability (Basel), 12*(16), 6481. doi:10.3390u12166481

Jones, J. B. (2017). *Shopping Carts and Restaurant Menus as Community Fomites: A Pilot Study in Henrico County*. AT Still University of Health Sciences.

Jose, J., Pinto, J. S., Kotian, B., Thomas, A. M., & Charyulu, R. N. (2020). Comparison of the regulatory outline of ecopharmacovigilance of pharmaceuticals in Europe, USA, Japan and Australia. *The Science of the Total Environment*, *709*, 134815. doi:10.1016/j.scitotenv.2019.134815 PMID:31887508

Kalyuzhnyi, S., & Gladchenko, M. (2014). Aerobic biodegradation of pharmaceutical compounds. In *Comprehensive Biotechnology* (2nd ed., Vol. 5, pp. 525–538). Academic Press.

Kampf, G., Todt, D., Pfaender, S., & Steinmann, E. (2020). *Persistence of coronaviruses on inanimate surfaces and their inactivation with biocidal agents.*

Kang, Y., Choi, H. S., & Hee, D. W. (2001). A study of the apparel sizing of children's wear - An analysis of the size increments utilized in children's wear based on an anthropometric survey. *Journal of Korean Home Economics Association English Edition*, *2*(1), 95–110. doi:10.1533/9780857096890.1.95

Karagöz, A., Turgut-Kara, N., Çakır, Ö., Demirgan, R., & Arı, Ş. (2007). Cytotoxic activity of crude extracts from astragalus chrysochlorus (leguminosae). *Biotechnology, Biotechnological Equipment*, *21*(2), 220–222. doi:10.1080/13102818.2007.10817449

Karan, E. P., & Irizarry, J. (2014). Developing a spatial data framework for facility management supply chains. *Construction Research Congress 2014: Construction in a Global Network - Proceedings of the 2014 Construction Research Congress*. IEEE. 10.1061/9780784413517.239

Karaouzas, I., Smeti, E., Vourka, A., Vardakas, L., Mentzafou, A., Tornés, E., Sabater, S., Muñoz, I., Skoulikidis, N. Th., & Kalogianni, E. (2018). Assessing the ecological effects of water stress and pollution in a temporary river - Implications for water management. *The Science of the Total Environment*, *618*, 1591–1604. doi:10.1016/j.scitotenv.2017.09.323 PMID:29054662

Kataria, M. V. (2017). Tween consumers: A study of the impact of social media on tween's buying decisions. *International Journal of Marketing & Financial Management*, *5*(1), 2348–3954. doi:10.5281/zenodo.238698

Katzenberger, R. H., Rösel, A., & Vonberg, R. P. (2021). Bacterial survival on inanimate surfaces: A field study. *BMC Research Notes*, *14*(1), 1–10. doi:10.118613104-021-05492-0 PMID:33722269

Kaur, P., Stoltzfus, J., & Yellapu, V. (2018). Descriptive statistics. *International Journal of Academic Medicine*, *4*(1), 60–63. doi:10.4103/IJAM.IJAM_7_18

Kavitha, C., Ramesh, M., Kumaran, S. S., & Lakshmi, S. A. (2012). Toxicity of Moringa oleifera seed extract on some hematological and biochemical profiles in a freshwater fish, Cyprinus carpio. *Experimental and Toxicologic Pathology*, *64*(7–8), 681–687. doi:10.1016/j.etp.2011.01.001 PMID:21282048

Kelly, S. (2010). Qualitative interviewing techniques and styles. In I. Bourgeault, R. Dingwall, & R. de Vries (Eds.), *The Sage handbook of qualitative methods in health research*. SAGE Publications. doi:10.4135/9781446268247.n17

Kenawy, E.-R., Bowlin, G. L., Mansfield, K., Layman, J., Simpson, D. G., Sanders, E. H., & Wnek, G. E. (2002). Release of tetracycline hydrochloride from electrospun poly (ethylene-co-vinylacetate), poly (lactic acid), and a blend. *Journal of Controlled Release*, *81*(1-2), 57–64. doi:10.1016/S0168-3659(02)00041-X PMID:11992678

Kendall, F. P., McCreary, E., Provance, P., Rodgers, M., & Romani, W. (2005). *Muscles: Testing and function with posture and pain* (5th ed.). Lippincott Williams & Wilkins.

Keogh, M. B., Elmusharaf, K., Borde, P., & Mc Guigan, K. G. (2017). Evaluation of the natural coagulant Moringa oleifera as a pretreatment for SODIS in contaminated turbid water. *Solar Energy*, *158*(October), 448–454. doi:10.1016/j.solener.2017.10.010

Khayer, A., Talukder, M. S., Bao, Y., & Hossain, M. N. (2020). Cloud computing adoption and its impact on SMEs' performance for cloud supported operations: A dual-stage analytical approach. *Technology in Society*, *60*(3), 101225. doi:10.1016/j.techsoc.2019.101225

Khozani, M. L., Behzadi, H., Nowkarizi, M., & Neizar, F. S. (2021). We live in cloud computing world, without using it in our libraries. *Library Hi Tech*, *40*(6), 1916–1929. doi:10.1108/LHT-03-2021-0107

Kim, J. H., & Lee, J. W. (2013). Effects of simulation parameters on residual stresses in 3D finite element laser shock peening analysis. *Global Journal of Researches in Engineering Mechanical and Mechanics Engineering*, *13*(9). https://globaljournals.org/GJRE_Volume13/1-Effects-of-Simulation.pdf

Kimuli, S. N. L., Sendawula, K., & Nagujja, S. (2021). Digital technologies in micro and small enterprise: Evidence from Uganda's informal sector during the COVID-19 pandemic. *World Journal of Science, Technology and Sustainable Development*, *18*(2), 93–108.

King, M., López-García, M., Atedoghu, K. P., Zhang, N., Wilson, A. M., Weterings, M., Hiwar, W., Dancer, S. J., Noakes, C. J., & Fletcher, L. A. (2020). Bacterial transfer to fingertips during sequential surface contacts with and without gloves. *Indoor Air*, *30*(5), 993–1004. doi:10.1111/ina.12682 PMID:32329918

Kitsis, G. (2015). Trophic Response to Multiple Stressors Using Species Sensitivity (SSD) Distribution Models. *Scholars Week, 24.* https://cedar.wwu.edu/scholwk/2015/Day_one/24.

Klein, E. Y., Van Boeckel, T. P., Martinez, E. M., Pant, S., Gandra, S., Levin, S. A., Goossens, H., & Laxminarayan, R. (2018). Global increase and geographic convergence in antibiotic consumption between 2000 and 2015. *Proceedings of the National Academy of Sciences of the United States of America*, *115*(15), E3463–E3470. doi:10.1073/pnas.1717295115 PMID:29581252

Kohli, A. K., Jaworski, B. J., & Kumar, A. (1993). MARKOR : A Measure of Market Orientation, *30*(4), 467–477.

Kong, X., Li, C., Zheng, F., & Wang, C. (2020). Improved Deep Belief Network for Short-Term Load Forecasting Considering Demand-Side Management. *IEEE Transactions on Power Systems*, *35*(2), 1531–1538. doi:10.1109/TPWRS.2019.2943972

Korhonen, M., Hellen, L., Hirvonen, J., & Yliruusi, J. (2001). Rheological properties of creams with four different surfactant combinations-effect of storage time and conditions. *International Journal of Pharmaceutics*, *221*(1-2), 187–196. doi:10.1016/S0378-5173(01)00675-5 PMID:11397580

Kosma, C. I., Kapsi, M. G., Konstas, P. G., Trantopoulos, E. P., Boti, V. I., Konstantinou, I. K., & Albanis, T. A. (2020). Assessment of multiclass pharmaceutical active compounds (PhACs) in hospital WWTP influent and effluent samples by UHPLC-Orbitrap MS: Temporal variation, removals and environmental risk assessment. *Environmental Research*, *191*, 110152. doi:10.1016/j.envres.2020.110152 PMID:32877707

Kostich, M. S., Batt, A. L., & Lazorchak, J. M. (2014). Concentrations of prioritized pharmaceuticals in effluents from 50 large wastewater treatment plants in the US and implications for risk estimation. *Environmental Pollution*, *184*, 354–359. doi:10.1016/j.envpol.2013.09.013 PMID:24095705

Kovalova, L., Siegrist, H., Von Gunten, U., Eugster, J., Hagenbuch, M., Wittmer, A., Moser, R., & McArdell, C. S. (2013). Elimination of micropollutants during post-treatment of hospital wastewater with powdered activated carbon, ozone, and UV. *Environmental Science & Technology*, *47*(14), 7899–7908. doi:10.1021/es400708w PMID:23758546

Krakkó, D., Licul-Kucera, V., Záray, G., & Mihucz, V. G. (2019). Single-run ultra-high performance liquid chromatography for quantitative determination of ten popular active pharmaceutical ingredients by quadrupole time-of-flight mass spectrometry after offline preconcentration by solid phase extraction from drinking and river as well as treated wastewater. *Microchemical Journal*, *148*, 108–119. doi:10.1016/j.microc.2019.04.047

Krebs, D. E., Edelstein, J. E., & Thornby, M. A. (1991). Prosthetic management of children with limb deficiencies. *Physical Therapy*, *71*(12), 920–934. doi:10.1093/ptj/71.12.920 PMID:1946626

Krisban, S. (2013). *8 excellent reasons to eat more eggplant*. Care 2 healthy living. www.care2.com/greenliving/8-great-reasons-to-eat-eggplant-html

Krishnamurthy, P. T., Vardarajalu, A., Wadhwani, A., & Patel, V. (2015). Identification and characterization of a potent anticancer fraction from the leaf extracts of moringa oleifera L. *Indian Journal of Experimental Biology*, *53*(2), 98–103. PMID:25757240

Krohn, J., Davis-Manigaulte, J., Fulcher, C., & Tiffany, J. S. (2022). Visualizing Diversity: Spatial Data as a Resource Enabling Extension to Better Engage Communities. *Journal of Human Sciences and Extension*. doi:10.54718/AEKV5040

Krueger, M. C., Harms, H., & Schlosser, D. (2015). Prospects for microbiological solutions to environmental pollution with plastics. *Applied Microbiology and Biotechnology*, *99*(21), 8857–8874. doi:10.100700253-015-6879-4 PMID:26318446

Kuhn, C., & Wehler, S. (2020). A Force-Balanced Fiber Retardation Model to Predict Fiber-Matrix-Separation during Polymer Processing. *Journal of Composites Science*, *4*(4), 165. doi:10.3390/jcs4040165

Kuma-Kpobee, M. A. (2009). *Determination of a Sizing System for Mass Customisation of Ghanaian Women's Traditional Dress: and a Conceptual Framework for Small and Medium Scale Enterprises*. [Unpublished PhD thesis, Manchester Metropolitan University, Manchester].

Kuma-Kpobee, M. A. (2013). The evolution and current manufacturing practice applied to the traditional dress of women in Ghana. *International Journal of Technology and Management Research, 2*, 56–66. www.koforiduapoly.edu.gh

Kumar, K. S., Ganesan, K., & Rao, P. S. (2008). Antioxidant potential of solvent extracts of Kappaphycus alvarezii (Doty) Doty–An edible seaweed. *Food Chemistry, 107*(1), 289–295. doi:10.1016/j.foodchem.2007.08.016

Kümmerer, K. (2010). Pharmaceuticals in the environment. *Annual Review of Environment and Resources, 35*(1), 57–75. doi:10.1146/annurev-environ-052809-161223

Kumssa, D. B., Joy, E. J. M., Young, S. D., Odee, D. W., Ander, E. L., Magare, C., Gitu, J., & Broadley, M. R. (2017). Challenges and opportunities for Moringa growers in southern Ethiopia and Kenya. *PLoS One, 12*(11), e0187651. doi:10.1371/journal.pone.0187651 PMID:29121079

Kwak, Y. K., Colque, P., Byfors, S., Giske, C. G., Möllby, R., & Kühn, I. (2015). Surveillance of antimicrobial resistance among Escherichia coli in wastewater in Stockholm during 1 year: Does it reflect the resistance trends in the society? *International Journal of Antimicrobial Agents, 45*(1), 25–32. doi:10.1016/j.ijantimicag.2014.09.016 PMID:25465520

Kwiecinski, J. M., & Horswill, A. R. (2020). Staphylococcus aureus bloodstream infections: Pathogenesis and regulatory mechanisms. *Current Opinion in Microbiology, 53*, 51–60. doi:10.1016/j.mib.2020.02.005 PMID:32172183

Kwon, Y. I., Apostolidis, E., & Shetty, K. (2008). In vitro studies of eggplant (Solanum Melongena) phenolics as inhibitors of key enzymes relevant for type 2 diabetes and hypertension. *Bioresource Technology, 99*(8), 2981–2988. doi:10.1016/j.biortech.2007.06.035 PMID:17706416

Landrigan, P. J., Fuller, R., Fisher, S., Suk, W. A., Sly, P., Chiles, T. C., & Bose-O'Reilly, S. (2019). Pollution and children's health. *The Science of the Total Environment, 650*, 2389–2394. doi:10.1016/j.scitotenv.2018.09.375 PMID:30292994

Larson, E. A., Adu-Gyamfi, S., Augustine, M., Yamba, P., Kukurah, J., Abdulai, K., Sekyi-Ansah, J., Abdul-Razak, O., Akurugu, E., & Kuzmin, A. (2023). Optimize multiple peening effects on surface integrity and microhardness of aluminum alloy induced by LSP. *Materials Sciences and Applications, 14*(3), 208–221. doi:10.4236/msa.2023.143012

Larson, E. A., Ren, X., Adu-Gyamfi, S., Zhang, H., & Ren, Y. (2019). Effects of scanning path gradient on the residual stress distribution and fatigue life of AA2024-T351 aluminium alloy induced by LSP. *Results in Physics, 13*(February), 102123. doi:10.1016/j.rinp.2019.02.059

Laryea, D. O., Awuah, B., Amoako, Y. A., Osei-Bonsu, E., Dogbe, J., Larsen-Reindorf, R., Ansong, D., Yeboah-Awudzi, K., Oppong, J. K., Konney, T. O., Boadu, K. O., Nguah, S. B., Titiloye, N. A., Frimpong, N. O., Awittor, F. K., & Martin, I. K. (2014). Cancer incidence in Ghana, 2012: Evidence from a population-based cancer registry. *BMC Cancer, 14*(1), 1–8. doi:10.1186/1471-2407-14-362 PMID:24884730

Lawrence, T. B., and Shadnam, M. (2008). Institutional theory. *The international encyclopedia of communication.*

le Roux, A., Khuluse-Makhanya, S., Arnold, K., Engelbrecht, F., Paige-Green, P., & Verhaeghe, B. (2019). A framework for assessing the risks and impacts of rural access roads to a changing climate. *International Journal of Disaster Risk Reduction, 38*, 101175. doi:10.1016/j.ijdrr.2019.101175

Lea, M. (2010). Bioremediation of turbid surface water using seed extract from Moringa oleifera Lam. (drumstick) tree. *Current Protocols in Microbiology, 16*(1), 1G–2. doi:10.1002/9780471729259.mc01g02s16 PMID:20131221

Lee, D., & Choi, K. (2019). Comparison of regulatory frameworks of environmental risk assessment s of human pharmaceuticals in EU, USA and Canada. *The Science of the Total Environment, 671*, 1026–1035. doi:10.1016/j.scitotenv.2019.03.372

Lee, W. C. C., & Zhang, M. (2005). Design of monolimb using finite element modelling and statistics-based Taguchi method. *Clinical Biomechanics (Bristol, Avon), 20*(7), 759–766. doi:10.1016/j.clinbiomech.2005.03.015 PMID:15963612

Liao, D., Zhu, S. P., Correia, J. A. F. O., De Jesus, A. M. P., & Berto, F. (2020). Recent advances on notch effects in metal fatigue: A review. *Fatigue & Fracture of Engineering Materials & Structures, 43*(4), 637–659. Advance online publication. doi:10.1111/ffe.13195

Li, C., Cao, C., Zhang, L., Chen, X., Ren, S., & Zhao, Y. (2017). Effects of corporate environmental responsibility on financial performance: The moderating role of government regulation and organizational slack. *Journal of Cleaner Production, 166*, 1323–1334. doi:10.1016/j.jclepro.2017.08.129

Li, D., Wang, S., & Li, D. (2016). Spatial data mining: Theory and application. In *Spatial Data Mining*. Theory and Application. doi:10.1007/978-3-662-48538-5

Li, H., Chen, H., Zhuang, T., & Chen, J. (2010). Analysis of genetic variation in eggplant and related Solanum species using sequence-related amplified polymorphism markers. *Scientia Horticulturae, 125*(1), 19–24. doi:10.1016/j.scienta.2010.02.023

Li, H., Tang, B., Lu, H., Cheema, M. A., & Jensen, C. S. (2022). Spatial Data Quality in the IoT Era: Management and Exploitation. *Proceedings of the ACM SIGMOD International Conference on Management of Data*. ACM. 10.1145/3514221.3522568

Line, N. D., & Wang, Y. (2017a). A multi-stakeholder market oriented approach to destination marketing. *Journal of Destination Marketing & Management, 6*(1), 84–93. doi:10.1016/j.jdmm.2016.03.003

Linz, M. S., Mattappallil, A., Finkel, D., & Parker, D. (2023). Clinical impact of staphylococcus aureus skin and soft tissue infections. *Antibiotics (Basel, Switzerland)*, *12*(3), 1–27. doi:10.3390/antibiotics12030557 PMID:36978425

Li, R., Wang, J. J., Gaston, L. A., Zhou, B., Li, M., Xiao, R., Wang, Q., Zhang, Z., Huang, H., Liang, W., Huang, H., & Zhang, X. (2018). An overview of carbothermal synthesis of metal–biochar composites for the removal of oxyanion contaminants from aqueous solution. *Carbon*, *129*, 674–687. doi:10.1016/j.carbon.2017.12.070

Li, S., Shi, W., You, M., Zhang, R., Kuang, Y., Dang, C., Sun, W., Zhou, Y., Wang, W., & Ni, J. (2019). Antibiotics in water and sediments of Danjiangkou Reservoir, China: Spatiotemporal distribution and indicator screening. *Environmental Pollution*, *246*, 435–442. doi:10.1016/j.envpol.2018.12.038 PMID:30579212

Li, T., Tang, F., & Zhang, L. (2020). Differential effects of voluntary environmental programs and mandatory regulations on corporate green innovation. *Natural Hazards*, *103*(3), 3437–3456. doi:10.100711069-020-04137-y

Liu, S., & Gao, B. (2022). Econometric Analysis of Leisure Agriculture and Rural Tourism Scenery Based on Spatial Data Analysis. *Computational Intelligence and Neuroscience*, *2022*, 1–8. doi:10.1155/2022/2286803 PMID:35898780

Liu, Y., Liu, X., Dong, W., Zhang, L., Kong, Q., & Wang, W. (2017). Efficient adsorption of sulfamethazine onto modified activated carbon: A plausible adsorption mechanism. *Scientific Reports*, *7*(1), 12437. doi:10.103841598-017-12805-6 PMID:28963547

Li, X., & Kao, C. (2022). Spatial Analysis and Modeling of the Housing Value Changes in the U.S. during the COVID-19 Pandemic. *Journal of Risk and Financial Management*, *15*(3), 139. doi:10.3390/jrfm15030139

Li, Y. H. (2013). *Theory and technology of laser shock processing*. Science Press.

Liyin, S., Hong, Y., & Griffith, A. (2006). Improving Environmental Performance by Means of Empowerment of Contractors. *Management of Environmental Quality*, *17*(3), 242–257. doi:10.1108/14777830610658674

Locey, K. J., Webb, T. A., Khan, J., Antony, A. K., & Hota, B. (2020). An interactive tool to forecast US hospital needs in the coronavirus 2019 pandemic. *JAMIA Open*, *3*(4), 506–512. doi:10.1093/jamiaopen/ooaa045 PMID:33619466

Lompo, P., Heroes, A.-S., Agbobli, E., Kühne, V., Tinto, H., Affolabi, D., & Jacobs, J. (2023). Bacterial contamination of antiseptics, disinfectants and hand hygiene products in healthcare facilities in high-income countries: A scoping review. *Hygiene*, *3*(2), 136–175. doi:10.3390/hygiene3020012

Lorina, I. B. E., Abdulrahman, A. A. S., Hajer, S. A. D., Nouf, K. A. H., Fatimah, A. A. N., & Hani, M. A. F. (2015). Potential pathogenic bacterial contaminants of shared utility devices in a university setting at Al-Hofuf, Saudi Arabia. *African Journal of Microbiological Research, 9*(41), 2139–2144. doi:10.5897/AJMR2015.7577

Lu, J. Z., Luo, K. Y., Yang, D. K., Cheng, X. N., Hu, J. L., Dai, F. Z., Qi, H., Zhang, L., Zhong, J. S., Wang, Q. W., & Zhang, Y. K. (2012). Effects of laser peening on stress corrosion cracking (SCC) of ANSI 304 austenitic stainless steel. *Corrosion Science, 60*, 145–152. doi:10.1016/j.corsci.2012.03.044

Luo, K. Y., Lu, J. Z., Wang, Q. W., Luo, M., Qi, H., & Zhou, J. Z. (2013). Residual stress distribution of Ti-6Al-4V alloy under different ns-LSP processing parameters. *Applied Surface Science, 285*(PARTB), 607–615. doi:10.1016/j.apsusc.2013.08.100

Luo, J., Solimini, N. L., & Elledge, S. J. (2009). Principles of cancer therapy: Oncogene and non-oncogene addiction, the current state of cancer research. *Cell, 136*(5), 823–837. doi:10.1016/j.cell.2009.02.024 PMID:19269363

Luo, X., Dang, N., & Wang, X. (2021). The effect of laser shock peening, shot peening and their combination on the microstructure and fatigue properties of Ti-6Al-4V titanium alloy. *International Journal of Fatigue, 153*, 106465. doi:10.1016/j.ijfatigue.2021.106465

Lu, Y., Zhu, X., & Cui, Q. (2012). Effectiveness and Equity Implications of Carbon Policies in the United States Construction Industry. *Building and Environment, 49*(1), 259–269. doi:10.1016/j.buildenv.2011.10.002

Mackintosh, C. A., & Hoffman, P. N. (1984). An extended model for transfer of micro-organisms via the hands: Differences between organisms and the effect of alcohol disinfection. *Epidemiology and Infection, 92*(3), 345–355. PMID:6429238

Maepa, M., Bodunrin, M. O., Burman, N. W., Croft, J., Engelbrecht, S., Ladenika, A. O., MacGregor, O. S., & Harding, K. G. (2017). Review: Life Cycle Assessments in Nigeria, Ghana, and Ivory Coast. *The International Journal of Life Cycle Assessment, 22*(7), 1159–1164. doi:10.100711367-017-1292-0

Mahmoud, B. & Hinson. (2017). *Market orientation and corporate social responsibility : towards an integrated conceptual framework,* 1–15. doi:10.1186/s40991-017-0020-1

Mahreen, H., Shahid, M., Rajoka, R., Xiong, Y., Cai, H., Muhammad, R., Mahmood, Q., He, Z., & Zhu, Q. (2021). Journal of Environmental Chemical Engineering Green synthesis of a silver nanoparticle using Moringa oleifera seed and its applications for antimicrobial and sun-light mediated photocatalytic water detoxification. *Journal of Environmental Chemical Engineering, 9*(4), 105290. doi:10.1016/j.jece.2021.105290

Majid, N. M., Islam, M. M., Justin, V., Abdu, A., & Ahmadpour, P. (2011). *Evaluation of heavy metal uptake and translocation by Acacia mangium as a phytoremediator of copper contaminated soil. 10*(42), 8373–8379. doi:10.5897/AJB11.392

Majumder, A., Gupta, A. K., Ghosai, P. S., & Varma, M. (2021). A review on hospital wastewater treatment: A special emphasis on occurrence and removal of pharmaceutically active compounds, resistant microorganisms and SARS-CoV-2. *Journal of Environmental Chemical Engineering*, *9*(2), 104812. doi:10.1016/j.jece.2020.104812 PMID:33251108

Maleki, E., Bagherifard, S., Unal, O., Bandini, M., & Guagliano, M. (2022). On the effects of laser shock peening on fatigue behavior of V-notched AlSi10Mg manufactured by laser powder bed fusion. *International Journal of Fatigue*, *163*, 107035. doi:10.1016/j.ijfatigue.2022.107035

Maleki, E., Unal, O., Guagliano, M., & Bagherifard, S. (2021). The effects of shot peening, laser shock peening and ultrasonic nanocrystal surface modification on the fatigue strength of Inconel 718. *Materials Science and Engineering A*, *810*, 141029. doi:10.1016/j.msea.2021.141029

Manna, P. K., De, D., Bera, T. K., Chatterjee, K., & Ghosh, D. (2011). Anthropometric assessment of physical growth and nutritional status among school children of North Bengal. *Anthropologist*, *13*(4), 299–305. doi:10.1080/09720073.2011.11891211

Mann, U. S., Dhingra, A., & Singh, J. (2014). Water pollution. *Causes, Effects and Remedies.*, *02*, 70–74.

Mansouri, F., Chouchene, K., Roche, N., & Ksibi, M. (2021). Removal of pharmaceuticals from water by adsorption and advanced oxidation processes: State of the art and trends. *Applied Sciences (Basel, Switzerland)*, *11*(14), 6659. doi:10.3390/app11146659

Marshall, C. (2010). *A research design for studio-based research in art. Taylor and Francis Group*. L.L.C.

Marsland, S. (2014). *Machine Learning: An Algorithmic Perspective* (2nd ed.). Chapman and Hall/CRC. doi:10.1201/b17476

Martin, M. A., Goya, L., & Ramos, S. (2013). Potential for preventive effects of cocoa and cocoa polyphenols in cancer. *Food and Chemical Toxicology*, *56*, 336–351. doi:10.1016/j.fct.2013.02.020 PMID:23439478

Mastamet Mason, A., De Klerk, H. M., Sommervile, J., & Ashdown, S. P. (2008). Consumers' knowledge on sizing and fit issues : A solution to successful apparel selection in developing countries. *International Journal of Consumer Studies*, *32*(3), 276–284. doi:10.1111/j.1470-6431.2008.00679.x

Mastamet-Mason, A., De Klerk, H. M., & Ashdown, S. (2008). Sizing and fit research at grassroots level: A methodology for the identification of unique body shapes in African developing countries. *Journal of Family Ecology and Consumer Sciences*, *36*, 9–21. doi:10.4314/jfecs.v36i1.47110.x

Matamoros, V., Arias, C. A., & Brix, H. (2019). *Removal of pharmaceutical compounds in constructed wetlands: Mechanisms, performance influencing factors, and implications for treatment design.* Research Gate.

Mathai, E., Allegranzi, B., Kilpatrick, C., & Pittet, D. (2010). Prevention and control of health care-associated infections through improved hand hygiene. *Indian Journal of Medical Microbiology*, *28*(2), 100–106. doi:10.4103/0255-0857.62483 PMID:20404452

Matsubara, K., Kaneyuki, T., Miyake, T., & Mori, M. (2005). Antiangiogenic activity of nasunin, antioxidant anthocyanin, in eggplant peels. *Journal of Agricultural and Food Chemistry*, *53*(16), 6272–6275. doi:10.1021/jf050796r PMID:16076105

Mendoza, A., Aceña, J., Pérez, S., López De Alda, M., Barceló, D., Gil, A., & Valcárcel, Y. (2015). Pharmaceuticals and iodated contrast medium in a hospital wastewater: A case study to analyse their presence and characterize their environmental risk and hazard. *Environmental Research*, *140*, 225–241. doi:10.1016/j.envres.2015.04.003 PMID:25880605

Mensah, S., Nani, G., & Ayarkwa, J. (2021). *Contractors' Adaptation to Environmentally Sustainable Construction: A Micro-Level Implementation Framework*. Smart and Sustainable Built Environment. doi:10.1108/SASBE-11-2020-0174

Merceron, D. L. (2007). *Millinery in Vogue*. New York: Rizzoli International Publications, Inc.

Merovci, F., & Elbatal, I. (2015). Weibull-Rayleigh distribution: Theory and applications. *Applied Mathematics & Information Sciences*, *9*(5), 1–11.

Microbiology, A. S. for. (2005). *Women Better At Hand Hygiene Habits, Hands Down*. ScienceDaily. https://www.sciencedaily.com/releases/2005/09/050922020856.h tm

Mileto, D., Mancon, A., Staurenghi, F., Rizzo, A., Econdi, S., Gismondo, M. R., & Guidotti, M. (2021). Inactivation of SARS-CoV-2 in the Liquid phase: Are aqueous hydrogen peroxide and sodium percarbonate efficient decontamination agents? *Journal of Chemical Health and Safety*, *28*(4), 260–267. doi:10.1021/acs.chas.0c00095 PMID:37556234

Millinery Techniques. (n.d.). *The Art of Millinery - A Guide to making a hat*. Millinery Techniques. http://www.millinerytechniques.com/ history-of millinery.htm l

Minden, V., Deloy, A., Volkert, A. M., Leonhardt, S. D., & Pugal, G. (2017). Antibiotics impact plants traits even at small concentrations. *AoB Plants*, *9*(2), plx010. doi:10.1093/aobpla/plx010 PMID:28439396

Mitchell, G., & Isberg, R. R. (2017). Innate immunity to intracellular pathogens: Balancing microbial elimination and inflammation. *Cell Host & Microbe*, *22*(2), 166–175. doi:10.1016/j.chom.2017.07.005 PMID:28799902

Mitchell, W., Wooliscroft, B., & Higham, J. (2010). Sustainable Market Orientation : A New Approach to Managing Marketing Strategy. *Journal of Macromarketing*, *30*(2), 160–170. doi:10.1177/0276146710361928

Mock, C. N., Quansah, R. E., Addae-Mensah, L., & Donkor, P. (2005). The development of continuing education for trauma care in an African nation. *Injury*, *36*(6), 725–732. doi:10.1016/j.injury.2004.12.044 PMID:15910824

Mohanan, N., Montazer, Z., Sharma, P. K., & Levin, D. B. (2020). Microbial and enzymatic degradation of synthetic plastics. *Frontiers in Microbiology*, *11*, 2837. doi:10.3389/fmicb.2020.580709 PMID:33324366

Montville, R., & Schaffner, D. W. (2003). Inoculum size influences bacterial cross contamination between surfaces. *Applied and Environmental Microbiology*, *69*(12), 7188–7193. doi:10.1128/AEM.69.12.7188-7193.2003 PMID:14660365

Moo, F., & Eyiah, A. (2020). Factors Influencing the Growth of Small and Medium Construction Firms in Northern Ghana. *Journal of African Business*, *21*(3), 416–431. doi:10.1080/15228916.2019.1641306

Morris, D., Harris, S., Morris, C., Commins, E., & Cormican, M. (2008). Hospital effluent: impact on the microbial environment and risk to human health. *Environment protection agency. Ireland*.https://www.epa.ie/publications/research/environment--health/EPA-162-final-web.pdf. Accessed on 01/30/2023.

Mousset, E., Oturan, N., & Oturan, M. A. (2016). Electrochemical advanced oxidation processes: Today and tomorrow. A review. *Environmental Science and Pollution Research International*, *23*(14), 13626–13651. doi:10.100711356-016-6447-y PMID:27300165

Moyo, B., Masika, P. J., Hugo, A., & Muchenje, V. (2011). Nutritional characterization of Moringa (Moringa oleifera Lam.) leaves. *African Journal of Biotechnology*, *10*(60), 12925–12933. doi:10.5897/AJB10.1599

Mpampa, M. L., Azariadis, P. N., & Sapidis, N. S. (2010). A new methodology for the development of sizing systems for the mass customization of garments. *International Journal of Clothing Science and Technology*, *22*(1), 49–68. doi:10.1108/09556221011008802

Mrozik, A., & Piotrowska-Cyplik, A. (2015). Anaerobic microbial degradation of pharmaceuticals—A review. *Environment International*, *79*, 105–118.

Mudholkar, G. S. & Srivastava, D. K. (1993). Exponentiated weibull family for analyzing bathtub failure-rate data. *Reliability, IEEE Transactions on*, *42*(2), 299-302.

Mujumdar, A. S., & Menon, A. S. (1995). *Drying of Solids: Principles*. Classification, and Selection of Dryers.

Mummery, C. L. (1978). The effect of ultrasound on fibroblasts' In Vitro'. King's College London (University of London).

Murdoch, G., Wilson, A. & Bennett, Jr. (1997). *A primer on amputations and artificial limbs*. Charles C Thomas Publisher, Ltd.

Musa, G., Kayat, K., & Thirumoorthi, T. (2010). The experiential aspect of rural home-stay among Chinese and Malay students using diary method. *Tourism and Hospitality Research, 10*(1), 25–41. doi:10.1057/thr.2009.26

Muslim, E., Moch, B. N., Fileinti, N. D., Puspasari, M. A., Sibarani, T. M. L., & Laskana, D. G. N. (2014). The Development of standard size for clothes of Indonesian boys based on anthropometric data as a reference to formulate RSNI 0555. *International Journal of Ergonomics*, *4*(2), 15–23.

Muyandar, A. S., & Devahastin, S. (2014). *Fundamental principles of drying, Handbook of industrial drying* (4th ed.). CRC Press.

Muyibi, S., & Alfugara, A. (2003). Treatment of surface water with Moringa Oleifera seed extract and alum – a comparative study using a pilot scale water treatment plant. *The International Journal of Environmental Studies*, *60*(6), 617–626. doi:10.1080/723032000087925

Muzaffar, H. M., Tahir, A., Ali, A., Ahmad, M., & McArdle, G. (2017). Quality assessment of volunteered geographic information for educational planning. In *Volunteered Geographic Information and the Future of Geospatial Data*. IGI Global. doi:10.4018/978-1-5225-2446-5.ch005

Nadarajah, S., & Kotz, S. (2006b). The exponentiated type distributions. *Acta Applicandae Mathematicae*, *92*(2), 97–111. doi:10.100710440-006-9055-0

Najeeb, A. (2014). Institutional theory and human resource management. *Being Practical with Theory: A Window into Business Research*, 25–30. http://eurekaconnection.files.wordpress.com/2014/02/p-25-30-institutional-theory-theori-ebook_finaljan2014-v3.pdf%0Arese arch-pubs@uow.edu.au

Narver, J. C., & Slater, S. F. (1990). The Effect of a Market Orientation on Business Profitability. *Journal of Marketing*, *54*(4), 20–35. doi:10.1177/002224299005400403

Nasiru, S. (2015). Another weighted weibull distribution from azzalinis family. *European Scientific Journal, 11*(9)

Nasiru, S., Mwita, P. N., & Ngesa, O. (2018). Discussion on generalized modified inverse Rayleigh distribution. *Applied Mathematics & Information Sciences*, *12*(1), 113–124. doi:10.18576/amis/120110

Nawrat, A. (2018). Pharma and the environment: pollution continues despite public pressure. *Pharmaceutical Technology*. https://www.pharmaceutical-technology.com/features/pharma-and-the-environment-pollution-trend/

Nishi, L., Vieira, A. M. S., Vieira, M. F., Bongiovani, M., Camacho, F. P., & Bergamasco, R. (2012). Hybrid process of coagulation/flocculation with Moringa oleifera followed by ultrafiltration to remove Microcystis sp. cells from water supply. *Procedia Engineering*, *42*, 865–872. doi:10.1016/j.proeng.2012.07.479

Nizeyimana, E. (2020). Remote Sensing and GIS Integration. In Managing Human and Social Systems. doi:10.1201/9781003053514-17

Nkamnebe, A. D. (2011). Sustainability marketing in the emerging markets: Imperatives, challenges, and agenda setting. *International Journal of Emerging Markets*, *6*(3), 217–232. doi:10.1108/17468801111144058

Noda, Y., Kneyuki, T., Igarashi, K., Mori, A., & Packer, L. (2000). Antioxidant activity of nasunin, an anthocyanin in eggplant peels. *Toxicology*, *148*(2-3), 119–123. doi:10.1016/S0300-483X(00)00202-X PMID:10962130

Norouzi, N., Jarrott, S., & Chaudhury, H. (2019). Designing intergenerational space through a human-development lens. *Journal of Architectural and Planning Research*, *36*(1).

Nunnally, J. C., & Bernstein, I. H. (1994). *Psychometric Theory* (3rd ed.). McGraw-Hill, Inc.

Nworie, A., Ayeni, J. A., Eze, U. A., & Azi, S. O. (2012). Bacterial contamination of door handles/knobs in selected public conveniences in Abuja metropolis, Nigeria: A public health threat. *Continental Journal of Medical Research*, *6*(1), 7–11.

Nyadanu, D., Aboagye, L. M., Akromah, M. K., Osei, M. K., & Dordoe, M. B. (2014). Agromorphological characterization of gboma eggplant, an indigenous fruit and leafyvegetable in Ghana. *African Crop Science Journal*, *22*, 281–289.

Oakes, J. (1976). Thermally denatured proteins. Nuclear magnetic resonance, binding isotherm and chemical modification studies of thermally denatured bovine serum albumin. *Journal of the Chemical Society, Faraday Transactions 1. Physical Chemistry in Condensed Phases*, *72*, 228–237.

Odoom, R., Anning-Dorson, T., & Acheampong, G. (2017). Antecedents of social media usage and performance benefits in small-and medium-sized enterprises (SMEs). *Journal of Enterprise Information Management*, *30*(3), 383–399. doi:10.1108/JEIM-04-2016-0088

Oduntan, E. B., & Isere, V. O. M. (2022). Digital Technologies for Business Sustainability in Nigeria : An Empirical Analysis. *African Journal of Management and Business Research*, *5*(1), 12–22.

OECD. (2021). *The Frontiers of Digital Learning: bridging the SME digital skills gap*. OECD. https://www.oecd.org/digital/sme/events/Frontiers of Digital Learning - Key Highlights - June 2021.pdf

Oechsle, A. M., Wittmann, X., Gibis, M., Kohlus, R., & Weiss, J. (2014). Collagen entanglement influenced by the addition of acids. *European Polymer Journal*, *58*, 144–156. doi:10.1016/j.eurpolymj.2014.06.015

Oelkers, K. (2020). The accessibility of data on environmental risk assessment of pharmaceuticals – are environmental risk assessments information on emissions with respect to international and European environmental information law? *Regulatory Toxicology and Pharmacology*, *111*, 104571. doi:10.1016/j.yrtph.2019.104571 PMID:31893528

Oelkers, K., & Floeter, C. (2019). The accessibility of data on environmental risk assessment of pharmaceuticals: Is the marketing authorization procedure in conflict with the international right of access to environmental information? *Environmental Sciences Europe*, *31*(1), 58. doi:10.118612302-019-0256-3

Ofosu, J. B., Otchere, F., & Hesse, C. A. (2016). *Intermediate Statistical Methods*. Excellent Publication and Printing.

Oguntunde, P.E., Adejumo, A.O. Okagbue, H.I. & Rastogi, M.K. (2016). Statistical properties and application of a new Lindley exponential distribution. *Gazi University journal of Sciences, 29*(4), 831-838.

Okoro, C. K., Kingsley, R. A., Connor, T. R., Harris, S. R., Parry, C. M., Al-Mashhadani, M. N., Kariuki, S., Msefula, C. L., Gordon, M. A., De Pinna, E., Wain, J., Heyderman, R. S., Obaro, S., Alonso, P. L., Mandomando, I., MacLennan, C. A., Tapia, M. D., Levine, M. M., Tennant, S. M., & Dougan, G. (2012). Intracontinental spread of human invasive Salmonella Typhimurium pathovariants in sub-Saharan Africa. *Nature Genetics*, *44*(11), 1215–1221. doi:10.1038/ng.2423 PMID:23023330

Okuda, T., Baes, A. U., Nishijima, W., & Okada, M. (2001). Coagulation mechanism of salt solution-extracted active component in Moringa oleifera seeds. *Water Research*, *35*(3), 830–834. doi:10.1016/S0043-1354(00)00296-7 PMID:11228982

Olise, C. C., & Simon-Oke, I. (2018). Fomites: Possible vehicle of nosocomial infections. *Journal of Public Health and Nutrition*, *1*(1).

Omali, T. U., & Umoru, K. (2019). A Survey of Remote Sensing and GIS Integration Method. *International Journal of Latest Technology in Engineering*, *VIII*(I).

Omar, S. S., Abdullah, Z. A., Humeid, M. A., & Yamani, M. I. (2012). Optimal composition and heat processing requirements for canning of eggplant dip (Motabbal Al-bathinjan). *Czech Journal of Food Sciences*, *30*(1), 35–44. doi:10.17221/62/2010-CJFS

On, K., & Techapeeraparnich, W. (2021). Barriers to Implementation of Sustainable Construction in Cambodia Construction Industry. *ACM International Conference Proceeding Series*, (pp. 213–219). ACM. 10.1145/3473141.3473251

Onwubiko, N. E., & Chinyeaka, A. H. (2015). Isolation and identification of bacterial contaminants from door handles in a tertiary institution in Umuahia, Abia state, Nigeria. *Nigerian Journal of Microbiology*, *29*, 3139–3147.

Oosterhoff, M., Geertzen, J. H. B., & Dijkstra, P. U. (2022). More than half of persons with lower limb amputation suffer from chronic back pain or residual limb pain: A systematic review with meta-analysis. *Disability and Rehabilitation*, *44*(6), 834–854. doi:10.1080/09638288.2020.1783377 PMID:32603198

Organization for Economic Cooperation and Development. (2018). *Pharmaceutical innovation and access to medicines*. OECD. www.oecd.org health systems.

Orias, F. (2015). Contribution to the evaluation of the ecotoxicological risks of hospital effluents: bioconcentration, bioaccumulation and biomagnification of pharmaceutical residues. *Ecotoxicologie,* École Nationalle des Travaux Publics de l'État [ENTPE], Vaulx-en-Velin France. https://tel.archives-ouvertes.fr/tel-01778521

Orias, F., & Perrodin, Y. (2014). Pharmaceuticals in hospital wastewater: Their ecotoxicity and contribution to the environmental hazard of the effluent. *Chemosphere*, *115*, 31–39. doi:10.1016/j.chemosphere.2014.01.016 PMID:24502927

Ortiz de García, S. A., Pinto Pinto, G., García-Encina, P. A., & Irusta-Mata, R. (2014). Ecotoxicity and environmental risk assessment of pharmaceuticals and personal care products in aquatic environments and wastewater treatment plants. *Ecotoxicology (London, England)*, *23*(8), 1517–1533. doi:10.100710646-014-1293-8 PMID:25064485

Ortiz-de-Mandojana, N., Aguilera-Caracuel, J., & Morales-Raya, M. (2016). Corporate Governance and Environmental Sustainability: The Moderating Role of the National Institutional Context. *Corporate Social Responsibility and Environmental Management*, *23*(3), 150–164. doi:10.1002/csr.1367

Othman, A., Jalil, A. M. M., Weng, K. K., Ismail, A., Ghani, N. A., & Adenan, I. (2010). Epicatechin content and antioxidant capacity of cocoa beans from four different countries. *African Journal of Biotechnology*, *9*(7), 1052–1059. doi:10.5897/AJB09.1219

Otieno, R. B., & Fairhurst, C. (2000). The development of new clothing size charts for female Kenyan children, part 2: Size-range Categorization and Presentation of Final Size Charts. *Journal of the Textile Institute*, *91*(2), 153–162. doi:10.1080/00405000008659535

Owa, F. D. (2013). Water Pollution: Sources, Effects, Control and Management. *Mediterranean Journal of Social Sciences*. doi:10.5901/mjss.2013.v4n8p65

Owens, B. (2015). Pharmaceuticals in the environment: A growing problem. *The Pharmaceutical Journal*, *294*, 7850. doi:10.1211/PJ.2015.20067898

Owusu, D., Ellis, W. O., & Oduro, I. (2008). *Nutritional potential of two leafy vegetables: Moringa oleifera and Ipomoea batatas leaves*.

Pajkos, A., Vickery, K., & Cossart, Y. (2004). Is biofilm accumulation on endoscope tubing a contributor to the failure of cleaning and decontamination? *The Journal of Hospital Infection*, *58*(3), 224–229. doi:10.1016/j.jhin.2004.06.023 PMID:15501338

Pakowski, Z., & Mujumdar, A. S. (1995). Basic process calculations in drying. In A.S. Mujumdar (Ed.), Handbook of Industrial Drying. Marcel Dekker.

Palma, P., Köck-Schulmeyer, M., Alvarenga, P., Ledo, L., Barbosa, I. R., de Alda, M. L., & Barceló, D. (2014). Risk assessment of pesticides detected in surface water of the Alqueva reservoir (Guadiana basin, southern of Portugal). *The Science of the Total Environment*, *488–489*, 208–219. doi:10.1016/j.scitotenv.2014.04.088 PMID:24836129

Panizza, J., & Cerisola, G. (2005). Direct and mediated anodic oxidation of organic pollutants. *Chemical Reviews*, *109*(12), 6541–6569. doi:10.1021/cr9001319 PMID:19658401

Pantelakis, S. G., Petroyiannis, P. V., Bouzakis, K. D., & Mirisidis, I. (2007). Surface hardness increase of 2024 aluminum alloy subjected to cyclic loading. *Theoretical and Applied Fracture Mechanics*, *48*(1), 68–81. doi:10.1016/j.tafmec.2007.04.002

Papadopoulos, T., Baltas, K. N., & Balta, M. E. (2020). The use of digital technologies by small and medium enterprises during COVID-19: Implications for theory and practice. *International Journal of Information Management*, *55*(3), 102–132. doi:10.1016/j.ijinfomgt.2020.102192 PMID:32836646

Papageorgiou, M., Zioris, I., Danis, T., Bikiaris, D., & Lambropoulou, D. (2019). Comprehensive investigation of wide range of pharmaceuticals and personal care products in urban and hospital wastewaters in Greece. *The Science of the Total Environment*, *694*, 133565. doi:10.1016/j.scitotenv.2019.07.371 PMID:31401503

Patel, P., Patel, N., Patel, D., Desai, S., & Meshram, D. (2014). Phytochemical analysis and antifungal activity of moringa oleifera. *International Journal of Pharmacy and Pharmaceutical Sciences*, *6*(5), 144–147.

Pathak, S., Zulic, S., Kaufman, J., Kopecek, J., Stránsky, O., Böhm, M., Brajer, J., Beránek, L., Shukla, A., Ackermann, M., Véle, F., & Mocek, T. (2022). Post-processing of selective laser melting manufactured SS-304L by laser shock peening. *Journal of Materials Research and Technology*, *19*, 4787–4792. doi:10.1016/j.jmrt.2022.07.014

Patist, A., & Bates, D. (2008). Ultrasonic innovations in the food industry: From the laboratory to commercial production. *Innovative Food Science & Emerging Technologies*, *9*(2), 147–154. doi:10.1016/j.ifset.2007.07.004

Paul, A., Rahma, M. M., & Ahmed, T. (2021). Identification of pathogenic bacteria from food handling surfaces (tabletops) from different areas with demonstration of their drug resistance properties. *Journal of Food Safety and Hygiene*, *5*(3). doi:10.18502/jfsh.v5i3.5693

Pawar, T. A. (2016). Use of natural coagulants for the coagulation treatment of water – a review. *International Journal of Research in Engineering* [IJRESTs]. *Science and Technologies*, *2*(1), 156–159.

Peake, B. M., Braund, R., Tong, A. Y. C., & Tremblay, L. A. (2016). Impact of pharmaceuticals on the environment. In book: The Life-Cycle of Pharmaceuticals in the Environmentchapter, 5, 109-152. doi:10.1016/B978-1-907568-25-1.00005-0

Peng, Z., Wang, X., & Wu, Z. (2020). A bundle-based shear-lag model for tensile failure prediction of unidirectional fiber-reinforced polymer composites. *Materials & Design*, *196*, 109103. doi:10.1016/j.matdes.2020.109103

Pereira, A. M. P. T., Silva, L. J. G., Lino, C. M., Meisel, L. M., & Pena, A. (2017). A critical evaluation of different parameters for estimating pharmaceutical exposure seeking an improved environmental risk assessment. *The Science of the Total Environment*, *603-604*, 226–236. doi:10.1016/j.scitotenv.2017.06.022 PMID:28628814

Petersen, H. H., Petersen, T. B., Enemark, H. L., Olsen, A., & Dalsgaard, A. (2016). Removal of Cryptosporidium parvum oocysts in low quality water using Moringa oleifera seed extract as coagulant. *Food and Waterborne Parasitology*, *3*, 1–8. doi:10.1016/j.fawpar.2016.03.002

Peyre, P., Chaieb, I., & Braham, C. (2007). FEM calculation of residual stresses induced by laser shock processing in stainless steels. *Modelling and Simulation in Materials Science and Engineering*, *15*(3), 205–221. doi:10.1088/0965-0393/15/3/002

Peyre, P., & Fabbro, R. (1995). Electromagnetic Gauge study of laser-induced shock waves in aluminum alloys. *Journal de Physique. III*, *5*(12), 1953–1964. doi:10.1051/jp3:1995241

Peyre, P., Fabbro, R., Merrien, P., & Lieurade, H. P. P. (1996). Laser shock processing of aluminium alloys. Application to high cycle fatigue behaviour. *Materials Science and Engineering A*, *210*(1–2), 102–113. doi:10.1016/0921-5093(95)10084-9

Peyre, P., Sollier, A., Chaieb, I., Berthe, L., Bartnicki, E., Braham, C., & Fabbro, R. (2003). FEM simulation of residual stresses induced by laser peening. *The European Physical Journal Applied Physics*, *23*(2), 83–88. doi:10.1051/epjap:2003037

Pham, Q. T., Tran, X. P., Misra, S., Maskeliunas, R., & Damaševičius, R. (2018). Relationship between convenience, perceived value, and repurchase intention in online shopping in Vietnam. Sustainability (Switzerland), 10(1).

Pham, T., & Pham, H. (2021). Improving the Green Performance Of Construction Projects Through Supply Chain Integration: The Role of Environmental Knowledge. *Sustainable Production and Consumption*, *26*, 933–942. doi:10.1016/j.spc.2021.01.004

Pheasant, S. (2003). *Bodyspace anthropometry, ergonomics and the design of work*. Taylor & Francis Ltd.

Physiopedia. (2023). Thomas test. *Physiopedia*. https://www.physio-pedia.com/Thomas_Test

Piccardo, C., Dodoo, A., Gustavsson, L., & Tettey, Y. A. (2019). Comparative Life-Cycle Analysis of Building Materials for the Thermal Upgrade of an Existing Building. In *IOP Conference Series: Earth and Environmental Science*. IOP Science. 10.1088/1755-1315/225/1/012044

Pickering, A. J., Julian, T. R., Marks, S. J., Mattioli, M. C., Boehm, A. B., Schwab, K. J., & Davis, J. (2012). Fecal Contamination and Diarrheal Pathogens on Surfaces and in Soils among Tanzanian Households with and without Improved Sanitation. *Environmental Science & Technology*, *46*(11), 5736–5743. doi:10.1021/es300022c PMID:22545817

Pinto, D. G. L., Loureiro, C. F. G., de Matos Sousa, F. F. L., & Motte-Baumvol, B. (2023). The effects of informality on socio-spatial inequalities in accessibility to job opportunities: Evidence from Fortaleza, Brazil. *Journal of Transport Geography*, *108*, 103577. doi:10.1016/j.jtrangeo.2023.103577

Pirouzi, G., Abu Osman, N. A., Eshraghi, A., Ali, S., Gholizadeh, H., & Wan Abas, W. A. B. (2014). Review of the socket design and interface pressure measurement for transtibial prosthesis. *The Scientific World Journal*, *2014*, 1–9. doi:10.1155/2014/849073 PMID:25197716

Podbevek, T. (2014). For a good fitted skirt, the waist-to-hip distance should be measured. *Anthropological Notebooks, 20*(2), 77-88.

Pollock, A. (2015, May 20). *Amputees control bionic legs with their thoughts.* Reuters. https://www.reuters.com/article/us-iceland-mind-controlled-limb-idUSKBN0O51EQ20150520

Polo Pena, A. I., Frias Jamilena, D. M., & Rodriguez Molina, M. A. (2013). Market orientation as a strategy for the rural tourism sector: Its effect on tourist behavior and the performance of enterprises. *Journal of Travel Research*, *52*(2), 225–239. doi:10.1177/0047287512459108

Poonekar, P. (1992). *Prosthetics and orthotics in India, In prosthetic and orthotic research for the twenty-first century*. National Institute of Child Health and Human Development.

Prabhasnkar, V. P., Joshua, D. I., Balakrishna, K., Siddiqui, I. F., Taniyasu, S., Yamashita, N., Kannan, K., Akiba, M., Praveenkumarreddy, Y., & Guruge, K. S. (2016). Removal rates of antibiotics in four sewage treatment plants in South India. *Environmental Science and Pollution Research International*, *23*(9), 8679–8685. doi:10.100711356-015-5968-3 PMID:26797959

Prasetyo, P. E. (2019). Role of entrepreneurial culture as the driver of economic growth. *International Journal of Economics and Financial Issues*, *9*(3), 237–243. doi:10.32479/ijefi.7882

Priyadarshinee, P., Raut, R. D., Jha, M. K., & Kamble, S. S. (2017). A cloud computing adoption in Indian SMEs: Scale development and validation approach. *The Journal of High Technology Management Research*, *28*(2), 221–245. doi:10.1016/j.hitech.2017.10.010

Procházka, R., Stehlík, A., Kotous, J., Salvetr, P., Bucki, T., Stránský, O., & Zulić, S. (2023). Fatigue properties of spring steels after advanced processing. *Materials (Basel)*, *16*(9), 3327. doi:10.3390/ma16093327 PMID:37176208

Puri, A. (2013). Efficacy of Pattern Making Software in Product Development. *International Journal of Advanced Quality Management*, *1*(1), 21–39.

Putra, G. (2011). Nutrition is a key to health. *Nurtient.* www.nutrient.net

Qin, Z., Li, B., Huang, X., Zhang, H., Chen, R., Adeel, M., & Xue, H. (2022). The effect of laser shock peening on surface integrity and high and very high cycle fatigue properties of 2024-T351 aluminum alloy. *Optics & Laser Technology*, *149*, 107897. doi:10.1016/j.optlastec.2022.107897

Qiu, T., Ma, X., Ye, M., Yuan, R., & Wu, Y. (2013). Purification, structure, lipid lowering and liver protecting effects of polysaccharide from Lachnum YM281. *Carbohydrate Polymers*, *98*(1), 922–930. doi:10.1016/j.carbpol.2013.07.014 PMID:23987429

Quartey, P. (2006). The textile and clothing industry in Ghana. In H. Jauch & R. Traub- Merz (Eds.), the future of the textile and clothing industry in Sub-Saharan Africa, (pp. 35-146). Bonn: Friedrich-Ebert-Stiftung.

Quitmann, C., Sauerborn, R., & Herrmann, A. (2021). *Gaps in Reporting Greenhouse Gas Emissions by German Hospitals: A Systematic Grey Literature Review.*

R'enyi, A. (1961). On measures of entropy and information. In *Proceedings of the Fourth Berkeley Symposium on Mathematical Statistics and Probability*, (pp. 547-561). University of California Press, Berkeley, CA.

Rabiei-Dastjerdi, H., McArdle, G., Matthews, S. A., & Keenan, P. (2021). Gap analysis in decision support systems for real-estate in the era of the digital earth. *International Journal of Digital Earth*, *14*(1), 121–138. doi:10.1080/17538947.2020.1808719

Raghavan, V., Wani, M., & Abraham, D. M. (2018). Exploring E-Business in Indian SMEs: Adoption, Trends and the Way Forward. In *Emerging Markets from a Multidisciplinary Perspective* (pp. 95–106). Springer. doi:10.1007/978-3-319-75013-2_9

Rahman, M., Aziz, S., & Huges, M. (2020). The product-market performance benefits of environmental policy. Why customer awareness? *Business Strategy and Environment.* WILEY. https://doi.org/DOI:10.1002/bse.2484

Rajabifard, A., & Williamson, I. P. (2001). Spatial data infrastructures: concept, SDI hierarchy and future directions. *Proceedings of Geomatics*, 10. https://doi.org/10.1.1.9.1919

Raj, R., Tripathi, A., Das, S., & Ghangrekar, M. M. (2021). Removal of caffeine from wastewater using electrochemical advanced oxidation process: A mini review. *Case Studies in Chemical and Environmental Engineering*, *4*, 100129. doi:10.1016/j.cscee.2021.100129

Ramachandran, K. M., & Tsokos, C. P. (2015). *Mathematical Statistics with Applications in R* (2nd ed.). Elsevier.

Ramachandra, T. V., Bharath, A. H., & Sowmyashree, M. V. (2015). Monitoring urbanization and its implications in a mega city from space: Spatiotemporal patterns and its indicators. *Journal of Environmental Management*, *148*, 67–81. doi:10.1016/j.jenvman.2014.02.015 PMID:24768450

Randall, K., Ewing, E. T., Marr, L. C., Jimenez, J. L., & Bourouiba, L. (2021). How did we get here: What are droplets and aerosols and how far do they go? A historical perspective on the transmission of respiratory infectious diseases. *Interface Focus*, *11*(6), 20210049. doi:10.1098/rsfs.2021.0049 PMID:34956601

Rand-Weaver, M., Margiotta-Casaluci, L., Patel, A., Panter, G. H., Owen, S. F., & Sumpter, J. P. (2013). The read-across hypothesis and environmental risk assessment of pharmaceuticals. *Environmental Science & Technology*, *47*(20), 11384–11395. doi:10.1021/es402065a PMID:24006913

Ranjbari, M., Morales-Alonso, G., Esfandabadi, Z. S., & Carrasco-Gallego, R. (2019). Sustainability and the sharing economy: Modelling the interconnections. *Direccion y Organizacion*, *68*, 33–40. doi:10.37610/dyo.v0i68.549

Rathod, M., Haldar, S., & Basha, S. (2015). Nanocrystalline cellulose for removal of tetracycline hydrochloride from water via biosorption: Equilibrium, kinetic and thermodynamic studies. *Ecological Engineering*, *84*, 240–249. doi:10.1016/j.ecoleng.2015.09.031

Ratoarinoro, N., Wilhelm, A., Berlan, J., & Delmas, H. (1992). Effects of ultrasound emitter type and power on a heterogeneous reaction. *The Chemical Engineering Journal*, *50*(1), 27–31. doi:10.1016/0300-9467(92)80003-S

Ravikumar, K., & Udayakumar, J. (2020). *Moringa oleifera seed composite a novel material for hazardous heavy metals (Cd, Cr and Pb) removal from aqueous systems. 11*(1), 123–138.

Ravikumar, K., & Sheeja, K. A. (2014). Removal of Fluoride from Aqueous System Using Moringa oleifera Seed Cake and its Composite Coagulants, I. *International Journal of Current Engineering and Technology*, *4*(3), 1356–1360.

Redovniković, I. R., Delonga, K., Mazor, S., Dragović-Uzelac, V., Carić, M., & Vorkapić-Furač, J. (2009). Polyphenolic content and composition and antioxidative activity of different Cocoa liquors. *Czech Journal of Food Sciences*, *27*(5), 330–337. doi:10.17221/119/2008-CJFS

Reinkensmeyer, D. J. (2009). Robotic assistance for upper extremity training after stroke. *Studies in Health Technology and Informatics*, *145*, 25–39. PMID:19592784

Renjith Kumar, R., Devasena, T., & Abeens, M. (2023). Improved mechanical properties of nano structured 316 L stainless steel by statistically optimized low pulsed laser shock peening process for aeronautical application. *Optik (Stuttgart)*, *276*, 170654. doi:10.1016/j.ijleo.2023.170654

Report, F. (2010). Rapid Urbanization and Mega Cities: The Need for Spatial Information Management. In The International Federation of Surveyors (FIG) Kalvebod Brygge 31–33, DK-1780 Copenhagen V DENMARK (48).

Reuter, C., Goebel, P., & Foerstl, K. (2012). The impact of stakeholder orientation on sustainability and cost prevalence in supplier selection decisions. *Journal of Purchasing and Supply Management*, *18*(4), 270–281. doi:10.1016/j.pursup.2012.06.004

Reynolds, K. A., Watt, P. M., Boone, S. A., & Gerba, C. P. (2005). Occurrence of bacteria and biochemical markers on public surfaces. *International Journal of Environmental Health Research*, *15*(3), 225–234. doi:10.1080/09603120500115298 PMID:16134485

Rezakazemi, M., Khajeh, A., & Mesbah, M. (2018). Membrane filtration of wastewater from gas and oil production. *Environmental Chemistry Letters*, *16*(2), 367–388. doi:10.100710311-017-0693-4

Rikalovic, A., Soares, G. A., & Ignjatic, J. (2018). Spatial analysis of logistics center location: A comprehensive approach. *Decision Making: Applications in Management and Engineering*, *1*(1), 38–50. doi:10.31181/dmame180138r

Ritchie, H., & Roser, M. (2020). CO_2 and Greenhouse Gas Emissions. *OurWorldInData.Org*.

Roberts, S., & Tribe, J. (2008). Sustainability indicators for small tourism enterprises — An exploratory perspective. *Journal of Sustainable Tourism*, *16*(5), 575–594. doi:10.1080/09669580802159644

Rohini, M. V., & Padmini, E. (2016). Preliminary phytochemical screening of selected medicinal plants of polyherbal formulation. *Journal of Pharmacognosy and Phytochemistry*, *5*(5), 277.

Romm, S. (1989). Arms by design. *Plastic and Reconstructive Surgery*, *84*(1), 158–163. doi:10.1097/00006534-198907000-00029 PMID:2660173

Ruixia, Z. (2018). Effects of laser shock peening on the corrosion behavior and biocompatibility of nickel-titanium alloy. *Journal of Biomedical Materials Research. Part B, Applied Biomaterials*, *107*(6), 1854–1863. PMID:30550636

Rusin, P., Maxwell, S., & Gerba, C. (2002). Comparative surface-to-hand and fingertip-to-mouth transfer efficiency of gram-positive bacteria, gram-negative bacteria, and phage. *Journal of Applied Microbiology*, *93*(4), 585–592. doi:10.1046/j.1365-2672.2002.01734.x PMID:12234341

Russotto, V., Cortegiani, A., Raineri, S. M., & Giarratano, A. (2015). Bacterial contamination of inanimate surfaces and equipment in the intensive care unit. *Journal of Intensive Care*, *3*(1), 54. doi:10.118640560-015-0120-5 PMID:26693023

Rutala, W. A., & Weber, D. J. (2013b). Disinfection and sterilization in healthcare facilities. Bennett & Brachman's Hospital Infections: Sixth Edition, May. doi:10.1016/j.idc.2021.04.004

Rutala, W. A., & Weber, D. J. (2018). Disinfection Chapter. *Practical Healthcare Epidemiology*, 58.

Rutala, W. A., & Weber, D. J. (2013a). Disinfectants used for environmental disinfection and new room decontamination technology. *American Journal of Infection Control*, *41*(5), S36–S41. doi:10.1016/j.ajic.2012.11.006 PMID:23622746

Rutala, W. A., & Weber, D. J. (2019a). Best practices for disinfection of noncritical environmental surfaces and equipment in health care facilities: A bundle approach. *American Journal of Infection Control*, *47*, A96–A105. doi:10.1016/j.ajic.2019.01.014 PMID:31146858

Rutala, W. A., & Weber, D. J. (2019b). Disinfection, sterilization, and antisepsis: An overview. *American Journal of Infection Control*, *47*, A3–A9. doi:10.1016/j.ajic.2019.01.018 PMID:31146848

Safo-adu, G., & Hanson, R. (2019). PRACTICES AND KNOWLEDGE ABOUT ENVIRONMENTAL SANITATION AND HYGIENE AMONG URBAN DWELLERS IN GOMOA EAST DISTRICT. *GHANA*, *6*(1), 1–11.

Saghir, A. M., Tazeem, S., & Ahad, I. (2017). The weighted exponentiated inverted Weibull distribution: Properties and application. *Journal of Information and Mathematical Sciences*, *9*(1), 137–151.

Sahar, E., David, I., Gelman, Y., Chikurel, H., Aharoni, A., Messalem, R., & Brenner, A. (2011). The use of RO to remove emerging micropollutants following CAS/UF or MBR treatment of municipal wastewater. *Desalination*, *273*(1), 142–147. doi:10.1016/j.desal.2010.11.004

Sakr, D. A., Sherif, A., & El-Haggar, S. M. (2010). Environmental Management Systems' Awareness: An Investigation of Top 50 Contractors In Egypt. *Journal of Cleaner Production*, *18*(3), 210–218. doi:10.1016/j.jclepro.2009.09.021

Salehi, M., & Shahrabi, J. (2012). Upgrading fitness in the production of garments. *Research Journal of Textile and Apparel*, *16*(4), 100–104. doi:10.1108/RJTA-16-04-2012-B011.x

Salgot, M., & Folch, M. (2018). Wastewater treatment and water reuse. *Current Opinion in Environmental Science & Health*, *2*, 64–74. doi:10.1016/j.coesh.2018.03.005

Salkind, N. J. (Ed.). (2010). *Encyclopedia of research design* (Vol. 1). SAGE Publications, Inc. doi:10.4135/9781412961288

Sandanasamy, J., Hamid Nour, A., Nizam Tajuddin, S., Hamid Nour, A., Abd Hamid, H., & Ziad Sulaiman, A. (2015). Bioassay guided isolation and identification of a cytotoxic compound form azadirachta indica leaves. *Journal of the Chinese Chemical Society (Taipei)*, *62*(8), 680–688. doi:10.1002/jccs.201400461

Sano, Y., Mukai, N., Okazaki, K., & Obata, M. (1997). Residual stress improvement in metal surface by underwater laser irradiation. *Nuclear Instruments & Methods in Physics Research. Section B, Beam Interactions with Materials and Atoms*, *121*(1–4), 432–436. doi:10.1016/S0168-583X(96)00551-4

Santacatalina, J. V., Ozuna, C., Cárcel, J. A., García-Pérez, J. V., & Mulet, A. (2011) Quality assessment of dried eggplant using different drying methods: hot air drying, vacuum freeze drying and atmospheric freeze drying. *International Congress on Engineering and Food Conference* (pp. 1-6). ICEF.

Santos, L. H. M. L. M., Gros, M., Rodriguez-Mozaz, S., Delerue-Matos, C., Pena, A., Barceló, D., & Montenegro, M. C. B. S. M. (2013). Contribution of hospital effluents to the load of pharmaceuticals in urban wastewaters: Identification of ecologically relevant pharmaceuticals. *The Science of the Total Environment*, *461–462*, 302–316. doi:10.1016/j.scitotenv.2013.04.077 PMID:23732224

Sarker, S., Rahman, M. M., Yadav, A. K., & Islam, M. M. (2019). Zoning of marine protected areas for biodiversity conservation in Bangladesh through socio-spatial data. *Ocean and Coastal Management*, *173*, 114–122. doi:10.1016/j.ocecoaman.2019.03.002

Sarstedt, M., Ringle, C. M., & Hair, J. F. (2021). Partial least squares structural equation modeling. In *Handbook of market research* (pp. 587–632). Springer International Publishing. doi:10.1007/978-3-319-05542-8_15-2

Sattar, S. A., Springthorpe, S., Mani, S., Gallant, M., Nair, R. C., Scott, E., & Kain, J. (2001). Transfer of bacteria from fabrics to hands and other fabrics: Development and application of a quantitative method using Staphylococcus aureus as a model. *Journal of Applied Microbiology*, *90*(6), 962–970. doi:10.1046/j.1365-2672.2001.01347.x PMID:11412326

Saunders, M. N. K., Lewis, P., & Thornhill, A. (2019). *Research Methods for Business Students* (Eighth Edi). Pearson Education Limited. . doi:10.1108/qmr.2000.3.4.215.2

Schoenfuss, H. L., Furlong, E. T., Phillips, P. J., Scott, T. M., Kolpin, D. W., Cetkovic-Cvrlje, M., Lesteberg, K. E., & Rearick, D. C. (2016). Complex mixtures, complex responses: Assessing pharmaceutical mixtures using field and laboratory approaches. *Environmental Toxicology and Chemistry*, *35*(4), 953–965. doi:10.1002/etc.3147 PMID:26561986

Scott, E., & Bloomfield, S. F. (1990). The survival and transfer of microbial contamination via cloths, hands and utensils. *The Journal of Applied Bacteriology*, *68*(3), 271–278. doi:10.1111/j.1365-2672.1990.tb02574.x PMID:2111304

Sękara, A., Cebula, S., & Kunicki, E. (2007). Cultivated eggplants – origin, breeding objectives and genetic resources, A Review'. *Folia Horticulture Ann*, *19*, 97–114.

Semedo, M. C., Karmali, A., & Fonseca, L. (2015). A high throughput colorimetric assay of β-1, 3-d-glucans by Congo red dye. *Journal of Microbiological Methods*, *109*, 140–148. doi:10.1016/j.mimet.2014.12.020 PMID:25555819

Seow, T. W., Lim, C. K., Nor, M. H. M., Mubarak, M. F. M., Lam, C. Y., & Yahya, A., & Zaharah Ibrahim. (2012). REVIEW ON WASTEWATER TREATMENT TECHNOLOGIES Jayashree Dhote. [IJERT]. *International Journal of Engineering Research & Technology (Ahmedabad)*, *1*(5), 1–10.

Severo, E. A., De Guimarães, J. C. F., Dellarmelin, M. L., & Ribeiro, R. P. (2019). The Influence of Social Networks on Environmental Awareness and the Social Responsibility Of Generations. *Brazilian Business Review*, *16*(5), 500–518. doi:10.15728/bbr.2019.16.5.5

Sewell, P., Noroozi, S., Vinney, J., & Andrews, S. (2000). Developments in the transtibial prosthetic socket fitting process: A review of past and present research. *Prosthetics and Orthotics International*, *24*(2), 97–107. doi:10.1080/03093640008726532 PMID:11061196

Shabanpour, N., Razavi-Termeh, S. V., Sadeghi-Niaraki, A., Choi, S. M., & Abuhmed, T. (2022). Integration of machine learning algorithms and GIS-based approaches to cutaneous leishmaniasis prevalence risk mapping. *International Journal of Applied Earth Observation and Geoinformation*, *112*, 102854. doi:10.1016/j.jag.2022.102854

Shah, M. P., Yao, K. M., & Seshadri, R. (2015). Aerobic microbial degradation of organic pollutants: Revisiting the biochemistry behind the pathway. *Applied Microbiology and Biotechnology*, *99*(10), 4155–4173.

Shang, S., Du, S., Du, S., & Zhu, S. (2021). Estimating building-scale population using multi-source spatial data. *Cities (London, England)*, *111*, 103002. doi:10.1016/j.cities.2020.103002

Shan, T. C., Al Matar, M., Makky, E. A., & Ali, E. N. (2017). The use of Moringa oleifera seed as a natural coagulant for wastewater treatment and heavy metals removal. *Applied Water Science*, *7*(3), 1369–1376. doi:10.100713201-016-0499-8

Sharma, D. K., Sharma, N., & Rana, S. (2013). Seed-borne diseases of brinjal (Solanum Melongena L.) and their control measures: A Review. *Review Article International Journal of Bioassays*, *2*, 1428–1433.

Shengju, Z., Rui, W., & Xingang, W. (2019). Study the Influencing Factors of Carbon Emission in the Production Stage of Building Materials and the Countermeasures for Reducing Carbon Emission. *IOP Conference Series. Earth and Environmental Science*, *371*(3), 032082. doi:10.1088/1755-1315/371/3/032082

Shmueli, G., Sarstedt, M., Hair, J. F., Cheah, J.-H., Ting, H., Vaithilingam, S., & Ringle, C. M. (2019). Predictive model assessment in PLS-SEM: Guidelines for using PLSpredict. *European Journal of Marketing*, *53*(11), 2322–2347. doi:10.1108/EJM-02-2019-0189

Shoben, M. M., & Taylor, P. J. (2004). *Grading for the fashion industry, the theory and practice* (3rd ed.). LCFS Fashion Media.

Simmonds, R., Lee, D., & Hayhurst, E. (2020). Mobile phones as fomites for potential pathogens in hospitals: Microbiome analysis reveals hidden contaminants. *The Journal of Hospital Infection*, *104*(2), 207–213. doi:10.1016/j.jhin.2019.09.010 PMID:31585142

Singh, A., Mehrotra, R., Rajput, V. D., Dmitriev, P., Singh, A. K., Kumar, P., Tomar, R. S., Singh, O., & Singh, A. K. (2022). Geoinformatics, artificial intelligence, sensor technology, big data: emerging modern tools for sustainable agriculture. *Sustainable Agriculture Systems and Technologies*, 295–313.

Singh, D., Joshi, K., Samuel, A., Patra, J., & Mahindroo, N. (2020). *Alcohol-based hand sanitisers as first line of defence against SARS-CoV-2 : a review of biology, chemistry and formulations.*

Singh, A. (2019). Remote sensing and GIS applications for municipal waste management. *Journal of Environmental Management*, *243*, 22–29. doi:10.1016/j.jenvman.2019.05.017 PMID:31077867

Sinha, S. N. (2012). Phytochemical analysis and antibacterial potential of Moringa oleifera Lam. *International Journal of Science Innovations and Discoveries*, *2*(4), 401–407.

Sinha, V., & Kumria, R. (2001). Polysaccharides in colon-specific drug delivery. *International Journal of Pharmaceutics, 224*(1-2), 19–38. doi:10.1016/S0378-5173(01)00720-7 PMID:11472812

Sirés, I., Brillas, E., & Oturan, M. A. (2014). Electrochemical advanced oxidation processes: today and tomorrow. In *Electrochemical Water Treatment Methods* (pp. 261–304). Royal Society of Chemistry. doi:10.1039/9781782620113-00261

Skip The Pie. (2015) USDA Record"11210" Nutritional Comparison: Kale, raw vs Eggplant, cooked, boiled. *Skip the Pie.*

Small, E. (2009). *Top 100 food plants; the world's most important culinary crops.* NRC Research Press.

Smith, M. W. (2015). The 10 most-prescribed and top-selling medications. *Drug & Medications.* https://www.webmd.com/drug-medication/news/20150508.

Spoden, M., Nimptsch, U., & Mansky, T. (2019). Amputation rates of the lower limb by amputation level – observational study using German national hospital discharge data from 2005 to 2015. *BMC Health Services Research, 19*(1), 8. doi:10.118612913-018-3759-5 PMID:30612550

Sreshtaa, V. S., & Geetha, R. V. (2021). Role of handwashing and hand sanitising in preventing infection. *Annals of the Romanian Society for Cell Biology*, 724–737.

Ssekitoleko, R. T., Oshabaheebwa, S., Munabi, I. G., Tusabe, M. S., Namayega, C., Ngabirano, B. A., Matovu, B., Mugaga, J., Reichert, W. M., & Joloba, M. L. (2020). The role of medical equipment in the spread of nosocomial infections: A cross-sectional study in four tertiary public health facilities in Uganda. *BMC Public Health, 20*(1), 1–11. doi:10.118612889-020-09662-w PMID:33066745

Stephens, B., Azimi, P., Thoemmes, M. S., Heidarinejad, M., Allen, J. G., & Gilbert, J. A. (2019). Microbial exchange via fomites and implications for human health. *Current Pollution Reports, 5*(4), 198–213. doi:10.100740726-019-00123-6 PMID:34171005

Stevens, P. M., DePalma, R. R., & Wurdeman, S. R. (2019). Transtibial socket design, interface, and suspension: A clinical practice guideline. *Journal of Prosthetics and Orthotics, 31*(3), 172–178. doi:10.1097/JPO.0000000000000219

Stewart Fotheringham, A., & Rogerson, P. A. (1993). GIS and spatial analytical problems. *International Journal of Geographical Information Systems, 7*(1), 3–19. doi:10.1080/02693799308901936

Sulbarán Rangel, B. C., Madrigal Olveira, A. E., Romero Arellano, V. H., & Guzmán González, C. A. (2019). Cellulosic nanomaterials for adsorption of emerging contaminants. *Tecnura, 23*(62), 13–20. doi:10.14483/22487638.15451

Suleyman, G., Alangaden, G., & Bardossy, A. C. (2018). The role of environmental contamination in the transmission of nosocomial pathogens and health-associated infections. *Current Infectious Disease Reports, 20*(6), 12. Advance online publication. doi:10.100711908-018-0620-2 PMID:29704133

Sun, P., Li, Y., Meng, T., Zhang, R., Song, M., & Ren, J. (2018). Removal of sulfonamide antibiotics and human metabolite by biochar and biochar/H2O2 in synthetic urine. *Water Research, 147*, 91–100. doi:10.1016/j.watres.2018.09.051 PMID:30300785

Sun, Y., Wang, S., Li, T., Li, X., Jiao, L., & Zhang, L. (2008). Purification, structure and immunobiological activity of a new water-soluble polysaccharide from the mycelium of Polyporus albicans (Imaz.) Teng. *Bioresource Technology, 99*(4), 900–904. doi:10.1016/j.biortech.2007.01.029 PMID:17368889

Sutanapong, C. & Louangrath, P.I. (2015). Descriptive and Inferential Statistics. *Inter. J. Res. Methodol. Soc. Sci.*

Su, Y., Yrastorza, J. T., Matis, M., Cusick, J., Zhao, S., Wang, G., & Xie, J. (2022). Biofilms: Formation, research models, potential targets, and methods for prevention and treatment. *Advancement of Science, 9*(29), 2203291. doi:10.1002/advs.202203291 PMID:36031384

Swelam, A. A., Sherif, S. S., & Ibrahim, A. (2018). THE ADSORPTION KINETICS AND MODELING FOR Pb (II). *REMOVAL FROM SYNTHETIC AND REAL WASTEWATER BY MORINGA OLEIFERA SEEDS., 29*(2), 105–124.

Tabur, S., Eren, M. A., Celik, Y., Dag, O. F., Sabunca, Z. A., & Savas, E. (2015). The major predictors of amputation and length of stay in diabetic patients with acute foot ulceration. *Wiener Klinische Wochenschrift, 127*(1-2), 45–50. doi:10.100700508-014-0630-5 PMID:25398288

Takht Ravanchi, M., Kaghazchi, T., & Kargari, A. (2009). Application of membrane separation processes in petrochemical industry: A review. *Desalination, 235*(1–3), 199–244. doi:10.1016/j.desal.2007.10.042

Talmadge, J., Chavez, J., Jacobs, L., Munger, C., Chinnah, T., Chow, J. T., Williamson, D., & Yates, K. (2004). Fractionation of Aloe vera L. inner gel, purification and molecular profiling of activity. *International Immunopharmacology, 4*(14), 1757–1773. doi:10.1016/j.intimp.2004.07.013 PMID:15531292

Tang, Y., Ge, M. Z., Zhang, Y., Wang, T., & Zhou, W. (2020). Improvement of fatigue life of GH3039 superalloy by laser shock peening. *Materials (Basel), 13*(17), 3849. doi:10.3390/ma13173849 PMID:32878304

Tan, S., Xu, Q., Luo, Z., Liu, Z., Yang, H., & Yang, L. (2011). Inquiry of water-soluble polysaccharide extraction conditions from grapefruit skin. *Engineering (London), 3*(11), 1090.

Taylor, P., Tie, J., Li, P., Xu, Z., Zhou, Y., Li, C., & Zhang, X. (2014). *Desalination and Water Treatment Removal of Congo red from aqueous solution using Moringa oleifera seed cake as natural coagulant.* Taylor and Francis. doi:10.1080/19443994.2014.905980

Teixeira, S., Delerue-Matos, C., & Santos, L. (2012). Removal of sulfamethoxazole from solution by raw and chemically treated walnut shells. *Environmental Science and Pollution Research International, 19*(8), 3096–3106. doi:10.100711356-012-0853-9 PMID:22744161

Tevatia, A., & Srivastava, S. K. (2015). Modified shear lag theory-based fatigue crack growth life prediction model for short-fiber reinforced metal matrix composites. *International Journal of Fatigue*, *70*, 123–129. doi:10.1016/j.ijfatigue.2014.09.004

Todd, E. C. D., Greig, J. D., Bartleson, C. A., & Michaels, B. S. (2009). Outbreaks where food workers have been implicated in the spread of foodborne disease. Part 6. Transmission and survival of pathogens in the food processing and preparation environment. *Journal of Food Protection*, *72*(1), 202–219. doi:10.4315/0362-028X-72.1.202 PMID:19205488

Todd, E. C. D., Michaels, B. S., Greig, J. D., Smith, D., Holah, J., & Bartleson, C. A. (2010). Outbreaks where food workers have been implicated in the spread of foodborne disease. Part 7. Barriers to reduce contamination of food by workers. *Journal of Food Protection*, *73*(8), 1552–1565. doi:10.4315/0362-028X-73.8.1552 PMID:20819372

Torb & Reiner. (2014). *A Brief History of Millinery*. Torb & Reiner. http://www.torbandreiner. com/Millinery Materials/history-of -millinery

Torburn, L., Powers, C. M., Guiterrez, R., & Perry, J. (1995). Energy expenditure during ambulation in dysvascular and traumatic below-knee amputees: A comparison of five prosthetic feet. *Journal of Rehabilitation Research and Development*, *32*(2), 111–119. PMID:7562650

Traverso, N., Ricciarelli, R., Nitti, M., Marengo, B., Furfaro, A. L., Pronzato, M. A., Marinari, U. M., & Domenicotti, C. (2013). Role of glutathione in cancer progression and chemoresistance. *Oxidative Medicine and Cellular Longevity*, *2013*, 1–10. doi:10.1155/2013/972913 PMID:23766865

Trösch, W. (2009). Water treatment. In *Technology Guide*. Principles - Applications - Trends., doi:10.1007/978 3 540 88546 7_73

Tweneboah-Koduah, E. Y., & Coffie, I. S. (2022). Social Distancing, Hand Washing and Handshaking Behaviour During and Beyond Coronavirus Pandemic: A Social Marketing Perspective. *Social Marketing Quarterly*, *28*(4), 288–306. doi:10.1177/15245004221134208

Twilley, D., Rademan, S., & Lall, N. (2020). A review on traditionally used South African medicinal plants, their secondary metabolites and their potential development into anticancer agents. *Journal of Ethnopharmacology*, *261*, 113101. doi:10.1016/j.jep.2020.113101 PMID:32562876

Ueda Yamaguchi, N., Cusioli, L. F., Quesada, H. B., Camargo Ferreira, M. E., Fagundes-Klen, M. R., Salcedo Vieira, A. M., Gomes, R. G., Vieira, M. F., & Bergamasco, R. (2021). A review of Moringa oleifera seeds in water treatment: Trends and future challenges. *Process Safety and Environmental Protection*, *147*, 405–420. doi:10.1016/j.psep.2020.09.044

Umayal Sundari, A. R., Neelamegam, P., & Subramanian, C. V. (2013). An experimental study and analysis on solar drying of bitter gourd using an evacuated tube air collector in Thanjavur, Tamil Nadu, India. *Conference Papers in Energy*, *2*, 125–128. doi:10.1155/2013/125628

UNCTAD. (2021). *COVID-19 and E-commerce : findings from a survey of online consumers in 9 countries*. UNCTAD.

UNICEF, W. H. O. (2015). *Progress on Sanitation and Drinking Water: 2015 Update and MDG Assessment.* UN.

Urbanek, A. K., Rymowicz, W., & Mirończuk, A. M. (2018). Degradation of plastics and plastic-degrading bacteria in cold marine habitats. *Applied Microbiology and Biotechnology, 102*(18), 7669–7678. doi:10.100700253-018-9195-y PMID:29992436

Urevig, A. (2019). 5 Things we can do now to reduce the threat of drugs in our environment. *Ensia.*. https://ensia.com/notable/pharmaceuticals-drugs-medications-antiobiotics-endocrine-water-pollution/

USDA database. (2012). *USDA National Nutrient Database for Standard Reference Release,* 28, 1028-1035. USDA.

van Aswegen, D. C., & Polese, C. (2021). Experimental and analytical investigation of the effects of laser shock peening processing strategy on fatigue crack growth in thin 2024 aluminium alloy panels. *International Journal of Fatigue, 142*, 105969. doi:10.1016/j.ijfatigue.2020.105969

Van Leeuwen, T., & Jewitt, C. (2004). *The handbook of visual analysis.* SAGE Publications., doi:10.4135/9780857020062

Van Seventer, J. M., & Hochberg, N. S. (2017). Principles of infectious diseases: transmission, diagnosis, prevention, and control. International Encyclopedia of Public Health, 22.

Velmurugan, A., Swarnam, P., Subramani, T., Meena, B., & Kaledhonkar, M. J. (2020). *Water demand and salinity.* Desalination-Challenges and Opportunities. doi:10.5772/intechopen.88095

Vestel, J., Cadwell, D. J., Constantine, L., D'Aco, V. J., Davidson, T., Dolan, D. G., Millard, S. P., Murray-Smith, R., Parke, N. J., Ryan, J. J., Straub, J. O., & Wilson, P. (2016). Use of acute and chronic ecotoxicity data in risk assessment of pharmaceuticals. *Environmental Toxicology and Chemistry, 35*(5), 1201–1212. doi:10.1002/etc.3260 PMID:26403382

Vigneswaran, S., & Sundaravadivel, M. (2009). Traditional and household water purification methods of rural communities in developing countries. *Wastewater Recycle. Reuse and Reclamation, 2*, 84–96.

Villares, A., Mateo-Vivaracho, L., & Guillamón, E. (2012). Structural Features and Healthy Properties of Polysaccharides Occurring in Mushrooms. *Revista de Agricultura (Piracicaba), 2*(4), 452–471.

Villaseñor-Basulto, D. L., Astudillo-Sánchez, P. D., del Real-Olvera, J., & Bandala, E. R. (2018). Wastewater treatment using Moringa oleifera Lam seeds: A review. *Journal of Water Process Engineering, 23*(February), 151–164. doi:10.1016/j.jwpe.2018.03.017

Vuong, Q. V., Thanh, D. T., Bhuyan, D. J., Goldsmith, C. D., Sadeqzadeh, E., Scarlett, C. J., & Bowyer, M. C. (2015). Optimization of ultrasound-assisted extraction conditions for euphol from the medicinal plant, Euphorbia tirucalli, using response surface methodology. *Industrial Crops and Products, 63*, 197–202. doi:10.1016/j.indcrop.2014.09.057

Vyas, G. S., & Jha, K. N. (2016). Benchmarking Green Building Attributes to Achieve Cost Effectiveness Using a Data Envelopment Analysis. *Sustainable Cities and Society*. doi:10.1016/j.scs.2016.08.028

Vymazal, J. (2005). Constructed wetlands for wastewater treatment. *Ecological Engineering*, *25*(5), 475–477. doi:10.1016/j.ecoleng.2005.07.002

Walliman, N. S. R. (2001). *Your research project: A step-by-step guide for the first-time researcher*. SAGE Publications.

Walter, S., & Mitkidis, K. (2018). The risk assessment of pharmaceuticals in environment: EU and US Regulatory Approach. *European Journal of Risk Regulation*, *9*(3), 527–547. doi:10.1017/err.2018.33

Wang, J., Lian, P., Yu, Q., Wei, J., & Kang, W.-y. (2017). Purification, characterization and procoagulant activity of polysaccharides from Angelica dahurice roots. *Chemistry Central Journal*, *11*(1), 17. doi:10.118613065-017-0243-y PMID:28246546

Wang, L., Wang, L., Xue, Y., Zhang, H., & Fu, H. (2015). Nanoindentation response of laser shock peened Ti-based bulk metallic glass. *AIP Advances*, *5*(5), 57156. doi:10.1063/1.4921898

Wang, L., Zhou, L., Liu, L., He, W., Pan, X., Nie, X., & Luo, S. (2022). Fatigue strength improvement in Ti-6Al-4V subjected to foreign object damage by combined treatment of laser shock peening and shot peening. *International Journal of Fatigue*, *155*, 106581. doi:10.1016/j.ijfatigue.2021.106581

Wang, X., Rye, D., Houtkooper, R. H., & Auwerx, J. (2015). Antibiotic use and abuse: A threat to mitochondria and chloroplasts with impact on research, health, and environment. *BioEssays*, *37*(10), 1045–1053. doi:10.1002/bies.201500071 PMID:26347282

Wan, S., Chen, Y., Xiao, Y., Zhao, Q., Li, M., & Wu, S. (2021). Spatial analysis and evaluation of medical resource allocation in China based on geographic big data. *BMC Health Services Research*, *21*(1), 1084. doi:10.118612913-021-07119-3 PMID:34641850

Water, U. N. (2017). *The United Nations world water development report 2017: wastewater the untapped resource*. UN Water.

Weibull, W. A. (1951). A statistical distribution function of wide applicability. *Journal of Applied Mechanics*, *18*(3), 293–297. doi:10.1115/1.4010337

Wei, X., Filleter, T., & Espinosa, H. D. (2015). Statistical shear lag model–Unraveling the size effect in hierarchical composites. *Acta Biomaterialia*, *18*, 206–212. doi:10.1016/j.actbio.2015.01.040 PMID:25684701

Whitaker, B. D., & Stommel, J. R. (2003). Distribution of Hydroxycinnamic acid conjugates in fruit of commercial eggplant (Solanum melongena L.) cultivars. *Journal of Agricultural and Food Chemistry*, *51*(11), 3448–3454. doi:10.1021/jf026250b PMID:12744682

Widaningrum, D. L., Surjandari, I., & Sudiana, D. (2020). Discovering spatial patterns of fast-food restaurants in Jakarta, Indonesia. *Journal of Industrial and Production Engineering, 37*(8), 403–421. doi:10.1080/21681015.2020.1823495

Widyanti, A., Mahachandra, M., Soetisna, H. R., & Sutalaksan, I. Z. (2017). Anthropometry of Indonesian Sundanese children and the development of clothing size system for Indonesian Sundanese children aged 6 to10 years. *International Journal of Industrial Ergonomics, 6*, 37–46. doi:10.1016/j.ergon.2017.05.008.x

Wieland, T. (2017). Market area analysis for retail and service locations with MCI. *The R Journal, 9*(1), 298. doi:10.32614/RJ-2017-020

Wigley, M. L. (1991). Prosthetic theory: The disciplining of architecture. *Assemblage*, (15), 6–29. doi:10.2307/3171122

Wild, P. C., Wilederpass, E., & Sterwart, W. B. (2020). *World Cancer Report : Cancer research for cancer prevention.* https://doi.org/Licence: CC BY-NC-ND 3.0 IGO

WillacyH. (2020). Antibiotics. In *Patient.* https://patient.info/infections/antibiotics-leaflet

Wire, C. (2003). Creative metal clay jewelry: Techniques, projects, inspirations. Lark Books, Division of Sterling Publishing Co., Inc.

Wißmann, J. E., Kirchhoff, L., Brüggemann, Y., Todt, D., Steinmann, J., & Steinmann, E. (2021). Persistence of pathogens on inanimate surfaces: A narrative review. *Microorganisms, 9*(2), 1–37. doi:10.3390/microorganisms9020343 PMID:33572303

Wollgast, J., & Anklam, E. (2000). Review on polyphenols in theobroma cacao: Changes in composition during the manufacture of chocolate and methodology for identification and quantification. *Food Research International, 33*(6), 423–447. doi:10.1016/S0963-9969(00)00068-5

World Bank. (2021). *Unmasking the Impact of COVID-19 on Businesses : Firm Level Evidence from Across the World.* World Bank. https://openknowledge.worldbank.org/handle/10986/34626

World Health Organisation. (2018). *Resource mobilisation for antimicrobial resistance (AMR): Getting AMR into plans and budgets of government and development partners.* WHO. www.who.int. Accessed on 03/23/2023.

WTTC. (2018). *Travel & Tourism Economic Impact 2018 World.* London: WTTC. https://www.wttc.org/-/media/files/reports/economic-impactresearch/regions-2018/world2018.pdf

WTTC. (2019). *Home.* World Travel & Tourism Council. http://www.wttc.org

Wuana, R., & Okieimen, F. (2011). Heavy Metals in Contaminated Soils: A Review of Sources, Chemistry, Risks and Best Available Strategies for Remediation. *ISRN Ecology, 2011*, 1–20. doi:10.5402/2011/402647

Wu, B., & Shin, Y. C. (2005). A self-closed thermal model for laser shock peening under the water confinement regime configuration and comparisons to experiments. *Journal of Applied Physics*, *97*(11), 113517. doi:10.1063/1.1915537

Xiaowei, J., Zha, J., Xu, Y., Giesy, J. P., Richardson, K. L., & Wang, Z. (2011). Derivation of predicted no effect concentrations (PNEC) for 2,4,6-throchlorophenol base on Chinese resident species. *Chemosphere*, *86*, 17–23. https://www.usask.ca/toxicology/jgiesy/pdf/publications/JA-6 24 PMID:21955353

Yadav, V., Sherly, M. A., Ranjan, P., Tinoco, R. O., Boldrin, A., Damgaard, A., & Laurent, A. (2020). Framework for quantifying environmental losses of plastics from landfills. *Resources, Conservation and Recycling*, *161*, 104914. doi:10.1016/j.resconrec.2020.104914

Yahya, M., Ginting, B., & Saidi, N. (2022). Anticancer and antiretroviral activities of methanolic extract from theobroma cacao L pod husk: Focusing on the ethyl acetate partition. *F1000 Research*, *11*, 1395. doi:10.12688/f1000research.128048.1

Yang, X., Flowers, R. C., Weinberg, H. S., & Singer, P. C. (2011). Occurrence and removal of pharmaceuticals and personal care products (PPCPs) in an advanced wastewater reclamation plant. *Water Research*, *45*(16), 5218–5228. doi:10.1016/j.watres.2011.07.026 PMID:21864879

Yeboah-Boateng, E. O., & Essandoh, K. A. (2014). Factors Influencing the Adoption of Cloud Computing by Small and Medium Enterprises in Developing Economies. *International Journal of Emerging Science and Engineering*, *2*(4), 631–642.

Ye, C., Liao, Y., & Cheng, G. J. (2010). Warm laser shock peening driven nanostructures and their effects on fatigue performance in Aluminum Alloy 6160. *Advanced Engineering Materials*, *12*(4), 291–297. doi:10.1002/adem.200900290

Yepiz-Gomez, M. S., Bright, K. R., & Gerba, C. P. (2006). Identity and numbers of bacteria present on tabletops and in dishcloths used to wipe down tabletops in public restaurants and bars. *Food Protection Trends*, *26*(11), 786–792.

Yertutan, C. (2001). Problems of young people related to anthropometric measures of ready-made clothes. *Journal of Qafqaz University, Fall(8)*, 27–32.

Ying, Z., Han, X., & Li, J. (2011). Ultrasound-assisted extraction of polysaccharides from mulberry leaves. *Food Chemistry*, *127*(3), 1273–1279. doi:10.1016/j.foodchem.2011.01.083 PMID:25214126

Younis, K. M. (2015). Views on Potential Methods For Raising Environmental Awareness in Developing Countries: A Study on Social Responsibility Engagement in Liberia. *Global Bioethics*, *26*(2), 128–144. doi:10.1080/11287462.2015.1031469

Yusianto, R., Marimin, Suprihatin, & Hardjomidjojo, H. (2020). Intelligent spatial decision support system concept in the potato agro-industry supply chain. *2020 International Conference on Computer Science and Its Application in Agriculture, ICOSICA 2020*. IEEE. 10.1109/ICOSICA49951.2020.9243233

Zabeen, S., Preuss, M., & Withers, P. J. (2015). Evolution of a laser shock peened residual stress field locally with foreign object damage and subsequent fatigue crack growth. *Acta Materialia*, *83*, 216–226. doi:10.1016/j.actamat.2014.09.032

Zabulis, X., Meghini, C., Partarakis, N., Beisswenger, C., Dubois, A., Fasoula, M., Nitti, V., Ntoa, S., Adami, I., Chatziantoniou, A., Bartalesi, V., Metilli, D., Stivaktakis, N., Patsiouras, N., Doulgeraki, P., Karuzaki, E., Stefanidi, E., Qammaz, A., Kaplanidi, D., & Galanakis, G. (2020). Representation and preservation of heritage crafts. *Sustainability (Basel)*, *12*(4), 1461. doi:10.3390u12041461

Zahra, T., Thamboo, J., & Asad, M. (2021). Compressive strength and deformation characteristics of concrete block masonry made with different mortars, blocks and mortar beddings types. *Journal of Building Engineering*, *38*, 102213. doi:10.1016/j.jobe.2021.102213

Zakaria, N. (2014). Body shape analysis and identification of key dimensions for apparel sizing systems. Woodhead Publishing Limited. https://dx.doi.org/10.1533/9780857096890.1.95.

Zakaria, N. (2011). Sizing system for functional clothing- Uniforms for school children. *Indian Journal of Fibre and Textile Research*, *36*, 348–357.

Zakaria, N. (2016). Size and fit of clothing for children and teenagers. In *Clothing for children and teenagers* (pp. 15–32). Elsevier Ltd. doi:10.1016/B978-0-08-100226-1.00002-3

Zakaria, N. (2016b). Conducting anthropometric surveys of children and teenagers. In *Clothing or children and teenagers* (pp. 33–51). Elsevier Ltd. doi:10.1016/B978-0-08-100226-1.00003-5

Zakaria, N., & Gupta, D. (2014). *Apparel sizing: existing sizing systems and the development of new sizing systems*. Woodhead Publishing Limited. doi:10.1533/9780857096890.1.3.x

Zhang, C., Dong, Y., & Ye, C. (2021). Recent developments and novel applications of laser shock peening: A review. In Advanced Engineering Materials, 23(7). doi:10.1002/adem.202001216

Zhang, H., Cai, Z., Wan, Z., Peng, P., Zhang, H., Sun, R., Che, Z., Guo, C., Li, B., & Guo, W. (2020). Microstructure and mechanical properties of laser shock peened 38CrSi steel. *Materials Science and Engineering A*, *788*, 139486. doi:10.1016/j.msea.2020.139486

Zhang, R., Zhou, X., Gao, H., Mankoci, S., Liu, Y., Sang, X., Qin, H., Hou, X., Ren, Z., Doll, G. L., Martini, A., Dong, Y., Sahai, N., & Ye, C. (2018). The effects of laser shock peening on the mechanical properties and biomedical behavior of AZ31B magnesium alloy. *Surface and Coatings Technology*, *339*, 48–56. doi:10.1016/j.surfcoat.2018.02.009

Zhang, T., & Xie, M. (2007). Failure data analysis with extended Weibull distribution. *Communications in Statistics. Simulation and Computation*, *36*(3), 579–592. doi:10.1080/03610910701236081

Zhang, Y. N., Niu, Q., Gu, X., Yang, N., & Zhao, G. (2019). Recent progress on carbon nanomaterials for the electrochemical detection and removal of environmental pollutants. *Nanoscale*, *11*(25), 11992–12014. doi:10.1039/C9NR02935D PMID:31140537

Zhao, Z., Wang, L., Lu, K., Li, Y., Chen, W., & Liu, L. (2020). Effect of foreign object damage on high-cycle fatigue strength of titanium alloy for aero-engine blade. *Engineering Failure Analysis*, *118*, 104842. doi:10.1016/j.engfailanal.2020.104842

Zhou, C., Li, Y., Yu, X., Yang, H., Ma, H., Yagoub, A. E. A., Cheng, Y., Hu, J., & Otu, P. N. Y. (2016). Extraction and characterization of chicken feet soluble collagen. *Lebensmittel-Wissenschaft + Technologie*, *74*, 145–153. doi:10.1016/j.lwt.2016.07.024

Zhou, H.-C., Lin, Y.-M., Li, Y.-Y., Li, M., Wei, S.-D., Chai, W.-M., & Tam, N. F.-y. (2011). Antioxidant properties of polymeric proanthocyanidins from fruit stones and pericarps of Litchi chinensis Sonn. *Food Research International*, *44*(2), 613–620. doi:10.1016/j.foodres.2010.12.016

Zhou, J. Z., Huang, S., Sheng, J., Lu, J. Z., Wang, C. D., Chen, K. M., Ruan, H. Y., & Chen, H. S. (2012). Effect of repeated impacts on mechanical properties and fatigue fracture morphologies of 6061-T6 aluminum subject to laser peening. *Materials Science and Engineering A*, *539*, 360–368. doi:10.1016/j.msea.2012.01.125

Zhou, J., Zhou, X., Li, H., Hu, J., Han, X., & Liu, S. (2022). In-situ laser shock peening for improved surface quality and mechanical properties of laser-directed energy-deposited AlSi10Mg alloy. *Additive Manufacturing*, *60*, 103177. Advance online publication. doi:10.1016/j.addma.2022.103177

Zhou, S., Paolo, C. D., Wu, X., Shao, Y., Seiler, T. B., & Hollert, H. (2019). Optimization of screening-level risk assessment and priority selection of emerging pollutants – The case of pharmaceuticals in European surface waters. *Environment International*, *128*, 1–10. doi:10.1016/j.envint.2019.04.034 PMID:31029973

Zhou, Z., Ruan, D., Jiang, L.-M., Yang, Y., Ge, H., & Wang, L. (2019). Comparison on treatment strategy for chemical cleaning wastewater: Pollutants removal, process design and techno-economic analysis. *Journal of Environmental Management*, *235*, 161–168. doi:10.1016/j.jenvman.2019.01.053 PMID:30682668

Zhu, Q., & Sarkis, J. (2007). The moderating effects of institutional pressures on emergent green supply chain practices and performance. *International Journal of Production Research*, *45*(18–19), 4333–4355. doi:10.1080/00207540701440345

Zhu, Y., Fu, J., Zheng, C., & Ji, Z. (2015). Effect of laser shock peening without absorbent coating on the mechanical properties of Zr-based bulk metallic glass. *Optics & Laser Technology*, *75*, 157–163. doi:10.1016/j.optlastec.2015.06.029

Zou, P., Yang, X., Huang, W.-W., Zhao, H.-T., Wang, J., Xu, R.-B., Hu, X.-L., Shen, S.-Y., & Qin, D. (2013). Characterization and bioactivity of polysaccharides obtained from pine cones of Pinus koraiensis by graded ethanol precipitation. *Molecules (Basel, Switzerland)*, *18*(8), 9933–9948. doi:10.3390/molecules18089933 PMID:23966080

Zumanu, V. O. (2019). Performance of SMEs in Ghana: Do mobile money payments matter? University of Ghana.

About the Contributors

Nana Yaw Asabere is currently an Associate Professor in the Computer Science Department, and Director of Research, Innovation, Publication, and Technology Transfer (DRIPTT), in Accra Technical University (ATU), Ghana. He received his Bachelor of Science (B.Sc.) Degree in Computer Science (CS) from Kwame Nkrumah University of Science and Technology, Kumasi, Ghana in 2005 and his Master of Science (M.Sc.) Degree in ICT from Aalborg University, Aalborg, Denmark in 2010. Furthermore, through his academic and professional accomplishments, he was awarded a Full Scholarship by the Chinese Government through the Chinese Scholarship Council (CSC) in May, 2011 to pursue his Doctor of Philosophy (Ph.D.) Degree in Computer Science. Consequently, he pursued his Ph.D. from September, 2011 at the School of Software, Dalian University of Technology, Dalian, China, and successfully completed it in September, 2014 under the supervision of Prof. Feng Xia (Ph.D.). During his Ph.D., he developed and evaluated five (5) different algorithms (pseudocodes) to solve three (3) major scientific problems related to Socially-Aware Recommendations in Smart Conferences. Prof. Asabere has approximately seventeen (17) years of teaching (graduate and undergraduate levels) and research experience as well as research project supervision experience with five (5) years of academic leadership at the tertiary level of education in Ghana. In ATU, he has held positions such as Coordinator/Head of Department for Non-Tertiary and Professional Programmes as well as the Examinations Officer in the Computer Science Department and Head of Computer Science Department. Currently, he is a member of the current Academic Board. Additionally, he is a member of the Faculty of Applied Sciences Board as well as a sub-committee of the Faculty Board in ATU (Appointments and Promotions Review Committee). As Head of Computer Science Department (November, 2018 – December, 2020), he has successfully spearheaded the accreditation and reaccreditation of undergraduate computer science programmes and is currently facilitating the introduction of new programmes in his Department. He has also facilitated the improvement of ICT facilities, computing resources, and his Departmental Staff in ATU. Prof. Asabere has several publications in International Journals and Conference Proceedings such as IEEE/ACM Transactions /Journals/

Conferences (Web of Science: SCI, Scopus and EI Indexed). Currently, he has a total of 746 research article citations in Google Scholar (h-index = 15, i10 index = 16) and 359 citations in Scopus (h-index = 11). In Research Gate, he currently has an 18.54 RG Score (Impact Points), 710 citations and h-index = 15. Additionally, his h-index is 11 in his Web of Science and Scopus Researcher Profiles. His current teaching and technological research interests include Mobile and Social Recommender Systems, Artificial Intelligence, Data Science, Big Data, ICT in Education (e-learning and m-learning), and Software Engineering. He received the Best Paper Award in the10th IEEE International Conference on Ubiquitous Intelligence and Computing (2013) which constituted part of his Ph.D. research and thesis. Additionally, he also received the Best Paper Award in the 7th IEEE International Conference on Adaptive Science & Technology (2018). Furthermore, he is a Reviewer as well as an Editor of several Computer Science Journals and Conference Proceedings. Among these include IEEE Transactions/Systems Journal, Proceedings of WWW International Conference (ACM), ACM Transactions on Multimedia Computing, Communications and Applications, and Frontiers in Big Data. He has been a Technical Program Committee (TPC) Member of the All Workshops on Big Scholarly Data since its inception. He is currently a member of IEEE, IEEE Computational Intelligence Society (CIS), and the Institute of ICT Professionals Ghana (IIPGh). Prof. Asabere currently leads the Digital Capacity (DigiCAP) Training Programme in ATU, which financially mobilized €200,000 funded by AFOS Foundation for Entrepreneurial Development Cooperation. Prof. Asabere also consults/has consulted internationally for the United Nations (UN) Global Pulse (Report on AI Ecosystem Mapping in Ghana) and locally for the Ghana Tertiary Education Commission (GTEC), and Commission for Technical and Vocational Education and Training (CTVET), Ghana.

Amevi Acakpovi is a Professor in the Electrical/Electronic Engineering, the Pro Vice-Chancellor and currently Acting Vice-Chancellor of Accra Technical University. He is also a Fellow Professional Engineer from the Institution of Engineering and Technology-Ghana and a Senior Member of the Institute of Electrical and Electronics Engineers (IEEE). He received a Bachelor of Science Degree in Computer and Electrical Engineering from the Lokossa Institute of Technology (Republic of Benin) in 2006 and a Certificate of Designer Engineer (Eq. MSc.) in Electrical Engineering from the Abomey-Calavi University (Republic of Benin) in 2009 and a PhD degree in Energy Systems Engineering from the Open University Malaysia in 2017. Prof Acakpovi is originally from the field of Electrical, Electronic and Energy Systems Engineering. His areas of research include Hybrid Energy Systems, Mini and Microgrid Optimization, Smart Grid Modelling and Optimization, Solar and Wind Energy Modelling and Optimization, Artificial Intelligence in Energy Systems and Recommender Systems. Prof. Acakpovi was a recipient of many awards and laurels including recently the award of the decade researcher at the Applied Research Conference in Africa (ARCA) in August 2021 and the Stairs Award of Distinguished Professor in Electrical Engineering, Pune, India in December, 2021. He also won the “Excellence in Research & Development Leadership” award by the SOLARQUARTER GRANDMASTERS AWARDS in September 2022. He has also served a number of national and international committees including chairing of the Electromagnetic Compatibility (EMC) committee (TC77), of the African Electrotechnical Standardization Commission (AFSEC) under the African Union (AU).

***

Doreen Adi is a senior lecturer and research scientist.

Francisca Adomako Serwaa holds first degree in medical laboratory science and currently working at Incas diagnostic services in Kumasi metropolis. She also is in her final year in postgraduate studies in molecular medicine at the department of molecular medicine in KNUST

Abenaa Owusuwaa Adu is an Assistant Lecturer at the Department of Pharmaceutical Sciences, KsTU. She is a pharmacist and holds a B.Pharm (Pharmacy) and an MPhil (Pharmaceutical Chemistry), all from KNUST. She is quick to transform every space into a “classroom” as she is passionate about inspiring and mentoring people through imparting knowledge and developing potential. Abenaa’s love to challenge and solve real problems inspires her to formulate and produce various pharmaceutical products. Currently, she is a PhD candidate in Pharmaceutical Chemistry at KNUST. Her research interests include Pharmaceuticals and the Environment, Quality Assessment of Pharmaceuticals, Analysis of Environmental Pollutants, Formulation Science and any research that affects Public Health.

Nina Afriyie is a lecturer at the Department of Entrepreneurship and Agribusiness, CCTU

Emmanuel Agbesi is a practical educationist and marketing professional who is working at deepening the relevance of industrial knowledge in the classroom. He is an adjunct lecturer in the Department of Marketing and Supply Chain Management of the Evangelical Presbyterian University College, along with other distance learning centers in Ho, Ghana. He has taught courses in the area of Integrated Marketing Communications, Marketing Analysis and Decision Making, Tourism Marketing; as well as assisted to teach Principles of Marketing, Retail and Brands Management, Marketing Planning and Control, Operations Management, International Business and International Business Management. His research interests are in branding and brand architecture, consumer behaviour, tourism marketing, sustainability development, and educational development. His research interests are in branding and brand architecture, consumer behaviour, sales and strategic marketing, tourism marketing, sustainability development, and educational development.

Stephen Agyefi-Mensah is a Senior Lecturer in the Department of Construction Technology and Management, and the Dean of the School of Graduate Studies, Cape Coast Technical University, CCTU.

Munir Ahmad is a Ph.D. in Computer Science with over 23 years of extensive experience in spatial data development, management, processing, visualization, and quality control. He has dedicated expertise in open data, crowdsourced data, volunteered geographic information, and spatial data infrastructure. A seasoned professional with extensive knowledge in the field, he has served as a trainer for the latest spatial technologies. With a passion for research and over 25 publications in the same field, he obtained his Ph.D. degree in Computer Science from Preston University Pakistan, in 2022. He is committed to advancing the industry and sharing knowledge through his expertise and experience. #SpatialData #GIS #GeoTech.

Isaac Aidoo (PhD) is a Building Technologist and Lecturer in the Building Technology Department of Accra Technical University, Accra Ghana

Rex Ifelaja Akinruntan is a Senior Lecturer at the Department of Sculpture Technology, Takoradi Technical University. He is a PhD candidate pursuing Painting and Sculpture at the Kwame Nkrumah University of Science and Technology, Kumasi.

Nene Kwabla Amoatey is currently an Assistant Lecturer at the Kumasi Technical University, Ghana. He holds a Master of Science Degree in Chemical Engineering from Kwame Nkrumah University of Science and Technology, Kumasi, Ghana, and a BSc. in Chemical Engineering from the from Kwame Nkrumah University of Science and Technology, Kumasi, Ghana. His current research include estimation of the methane generation from various landfills, plasma reactors for wastewater treatment, treatment of wastewater using waste materials, waste management, pyrolysis-gasification and bioconversion of wastes and biomass to produce valuable products.

Nana Benyi Ansah is a civil engineer and a lecturer at Accra Technical University. He lectures in several engineering subjects, including the strength of materials, Construction Technology, International Construction and Building Economics. He is an adjunct lecturer at Pentecost University. Nana Benyi Ansah's research interest is in Construction sustainability, Construction Life Cycle Assessment, Construction Health and Safety and Circularity. He is a member of the Building Technical Committee (TC 11) of the Ghana Standard Authority and a member of the committee on the Ghana Building Code (GS1207) 2018. He has fifteen years of construction site practice experience on concrete structures.

Maxwell Hubert Antwi is an Assistant Lecturer at the Department of Medical laboratory Science at Koforidua Technical University. He hold BSc in Medical laboratory Science and an MPhil in Immunology.

Gretel Rejoice Mawunyo Asigbe-Tsriku is a senior assistant registrar at Accra Technical University. She holds a Bachelor of Arts Degree in Integrated Development Studies from the University for Development Studies and a masters degree in Human Resource Development from the University of Cape Coast, Ghana. Gretel has publications in other international journals and presented papers a couple of international conferences. She has interest in women and development, politics, entertainment and trends in tertiary education.

Benjamin Asinyo is a lecturer at knust Kumasi He lecture in the department of industrial Art The researcher is the coordinator for post graduates students of industrial Art.

Milku Augustine joined the Department of Mechanical Engineering of the Bolgatanga Technical University, Bolgatanga, Ghana in October 2020 where he is currently acting as the Head of Department. He holds BTech (Hons) Automobile Engineering and M.Sc. in Engineering and Management from the Kwame Nkrumah University of Science and Technology, Ghana and Coventry University, UK respectively. Milku is a registered Professional Engineer with the Ghana Institution of Engineering and has more than 25 years of experience in aircraft maintenance engineering with the Ghana Air Force. He is an author of several other journal papers. He is in his final stage of his PhD studies where is working on the contribution of motor vehicle exhaust emissions to air pollution in the city of Accra, Ghana.

Jacob Ayembilla is a lecturer and a research scientist.

David Azanu is currently a Lecturer at the Kwame Nkrumah University of Science and Technology, Ghana. He holds a PhD in analytical chemistry and Master of Science Degree in analytical chemistry from Kwame Nkrumah University of Science and Technology, Kumasi, Ghana, and a BSc. in Chemistry from the Kwame

Nkrumah University of Science and Technology, Ghana. His current research interests include Toxicology, analytical instrumentation and Engineering.

Salifu Tahiru Azeko obtained his PhD in Materials Science and Engineering in 2015. He then had his Postdoctoral Research Studies in the US and France between 2017-2019. He is currently a Senior Lecturer in the Mechanical Engineering Department of Tamale Technical University, Tamale, Ghana. His research area focuses on Structural and Multifunctional Materials as well as Materials Selection and Design, Recycling of Polymers, Metals and Natural Wastes into Composite Materials for different applications. He is a Professional Member of the Ghana Academy of Arts and Science, Ghana Institution of Engineering (Ghie), Institution of Engineering and Technology (IET), Ghana, and a Corporate Member of the American Society of Civil Engineering (ASCE). Dr Azeko held several positions in Tamale Technical University, including: Vice Dean, Faculty of Engineering; Head of Department Mechanical Engineering; Head of Department Water and Environmental Engineering; as well as Head of the Amatrol Laboratory. He is also an External Moderator/ Examiner, Commission for Technical Vocational Education and Training (CTVET), External Assesor, Ghana Tertiary Education Commission (GTEC) as well as a National Short Term Scholar for Curriculum Development in Industrial Mechanics as well as Agro-Machinery Manufacturing.

Nana Esi Bedua-Taylor is a Lecturer in the Hospitality Management Department, CCTU.

Joseph Boadu is a Snr. Engineer in the Works and Physical Development Directorate of the Tamale Technical University and Head of the Estates Section of the directorate. A professional member of the Institution of Engineering and Technology. I have more than a decade worth of experience in the Building & Construction industry, an astute draughtsman and a professional cost engineer in the industry. I hold a BSc. Building Technology from the Kwame Nkrumah University of Science and Technology, Kumasi ; MSc. Development Management, from the University for Development Studies, Tamale.

Fredrick Boakye-Yiadom forges a new theoretical approach to the study of African Art that is thoroughly grounded in thoughts and actions of the users. This approach emphasizes on the beliefs the users of the Arts attach to the individual Arts from different African Cultures, their functionality as well as Africans in the diaspora. Rich description of the Arts is rendered more vivid by a generous selection of Art works. Boakye-Yiadom is a holder of Ph.D of Art and Culture at University of Education Winneba. As an Associate Professor, Art Historian, Conservation and

Restoration artist, Boakye-Yiadom specialises in conserving and restoring outdoor sculptures of all kinds. Having been associated with teaching for more than twenty-three years, He has gained the understanding of how Africans treat their Art pieces. He taught at the basic level for eight years and subsequently at the tertiary level for the last fifteen years teaching history of art (History of Ancient Art and History of African Art), Assemblage and Construction, Interior Design, and others at Takoradi Technical University, Takoradi. Besides, teaching and supervising students' projects, He serve as the University's Teacher Union Secretary and a Hall Warden for the past eight years.

Isaac Sewornu Coffie obtained his Doctor of Philosophy Degree from the University of Ghana with a specialization in social marketing. He is currently an adjunct lecturer at Accra Technical University. His teaching and research interests are social marketing, corporate social responsibility, entrepreneurship and SME development.

Daniel Danso is a lecturer at Akenten Appiah-Menka University of Skills Training and Entrepreneurial Development (AAMUSTED), Ghana, He lectures in textiles designing and fashion illustration. He lives in Kumasi and is married.

James Darmey is currently an Assistant Lecturer at the Kumasi Technical University, Ghana. He holds a Master of Science Degree in Chemical Engineering from Kwame Nkrumah University of Science and Technology, Kumasi, Ghana, and a BSc. in Minerals Engineering from the University of Mines and Technology, Tarkwa, Ghana. His current research interests include mineral processing and extractive metallurgy (ferrous metallurgy), waste management, pyrolysis-gasification and bioconversion of wastes and biomass to produce valuable products, including activated carbon for gold adsorption, energy, and other useful materials.

Evans Kwadwo Donkor is a an Associate Professor at the Sculpture Technology Department, Takoradi Technical University. He is best known for his composite sculptures made from scrap metals and discarded automobile parts and other industrial detritus. Donkor has a PhD in Arts & Culture from University of Education, Winneba. As an artist, he strongly believes in material culture by bringing every material to life. He reshapes and attaches the various components, giving old and new materials a new purpose as well as a bit of soul.

Andrews Dodzi Kobla Dzikunu is currently reading Masters in Software Engineering at the Zhejiang Normal University in China. He received his MSc in Computer Science and Technology from the same University in 2021. He writes and presents widely on system administration and Computer Hardware and Networking.

Frances Betty Fraikue is a Senior Lecturer and the Head of Department for Hospitality Management in Takoradi Technical University (TTU) from 2020-2023. She started teaching in 1993 after she completed Teacher Training at Holy Child Teacher Training College. She attained her Bachelor of Education in Home Economics at the University of Education, a Master of Education in Home Economics, a Master of Philosophy (M.Phil.) in Hospitality Management, and a Ph.D. in Hospitality Management and Catering Technology at M. S. Ramaiah University of Applied Sciences, Bangalore, India as well as being a certified Curriculum Based Training Educator. She loves research and her areas of interest are recipe development, food production and service as well as blended research by collaboration.

Victor Worlanyo Gbedawo is a PhD candidate reading computer science at the C.k Tedam University of Technology and Applied Sciences in Ghana. He received his MSc. in Information technology from the Kwame Nkrumah University of Science and Technology in 2017 and holds a BSc. in Computer Science from the University for Development Studies, Ghana. In 2012, he was hired as the University's Webmaster by Accra Technical University and is currently a Lecturer and a Researcher at the Computer science department of Ho Technical University since 2020. He writes and presents widely on computer security and machine learning.

Glenn Kwabena Gyimah (PhD) is a Lecturer in the Department of Mechanical Engineering, at Accra Technical University, and Ag. Head of Department, Research, Innovation and Grant (DRIPTT). He holds a PhD in Mechanical Design and Theory (South China University of Technology, 2014), Masters in Mechanical Design and Theory (South China University of Technology, 2010), MBA in Human Resource Management (University of Ghana) and an HND in Mechanical Engineering, Production option (Accra Polytechnic now Accra Technical University). He is a certified Member of the American Society of Mechanical Engineers (ASME), the Ghana Institute of Engineering (GhIE) of Ghana, a registered member of the Institute of Mechanical Engineers (IMechE) of UK, and a Member of the International Association of Engineers (IAENG), Hong Kong. Glenn has over 19 years of experience in the engineering industry, academic training and research, management, industrial research and innovation in government institutions and private businesses in China, Hong Kong, Europe and Ghana. Glenn was the Chair of the Auto-Tribology of the 4th International Tribology Symposium of IFToMM & 1st Korea Tribology International Symposium, 2017, Jeju, Korea. He serves on the National Mirror Committee (Ghana Standard Authority), a member of ISO/TC 323 Circular Economy – Framework, Principles, Terminology, and Management System Standard (Active in WG1, WG2 and WG4) – International Standard Organization (ISO) and a member of International Telecommunication Union (ITU) United for

Smart Sustainable City (U4SSC). Thematic Group: Developing ICT Procurement Guidelines for Smart Sustainable Cities. Glenn's research focuses on Tribology: Friction, Wear and Lubrication; Conventional and Non-conventional manufacturing systems; Materials Engineering; Production planning and control; Wastewater and faecal treatment systems; Solid waste treatment systems and Quality control and assurance. Dr. Glenn is well-read with over seventeen (17) referred journals, book chapter and conference publications. He has rich consulting experience, both local and international. He has worked extensively with the Liquid and Solid waste sectors in Ghana, Austria, Netherlands and China on Engineering Design, Procurement, Construction, Installation and Commissioning.

Alhassan Abdul Latif Husein is a Lecturer at Tamale Technical University in the Department of Water and Environmental Engineering.

Mansuur Husein is a Lecturer at Tamale Technical University in the Department of Water and Environmental Engineering.

Hamdu Ibrahim is HOD, Agricultural Engineering Department, Tamale Technical University.

Zakaria Issaka (PhD) Associate Professor, Department of Agricultural Engineering, Tamale Technical University

Peter Kaba is a PhD candidate, United Graduate School of Agricultural Sciences, Ehime University (Japan).

Niib Konwuruk is a Lecturer at Tamale Technical University in the Department of Water and Environmental Engineering.

Enoch Asuako Larson is a Mechanical Engineer with research interest in Material characterization, Laser surface treatment, Design and manufacturing of mechanical systems, Material testing, and failure analysis. He has two decades of teaching, supervison and research in higher education. He received his Ph.D. in Mechanical Engineering from Jiangsu University (JU) in China in 2018. In 2015 and 2009, he obtained his MSc. in Mechanical Engineering and Bachelor of Technology in Automobile Engineering respectively from KNUST and the Ho Technical University/ KNUST. He was employed as an assistant instructor at Tamale Polytechnic in 2003 having completed his Higher National Diploma (HND) in Mechanical Engineering from then Tamale Polytechnic. He subsequently rose to the rank of lecturer in 2015. At Tamale Technical University, he served on various committees. He was later appointed as the Deputy Research Director for Research and Innovation and later the Head of the Department for Civil Engineering in 2020 and 2021, respectively. In 2022, he was employed as a lecturer in the Department of Mechanical and Industrial Engineering (DMIE) of the University for Development Studies (UDS). Tamale Ghana. He is currently a senior lecturer and the head of the Department of Mechanical and Industrial Engineering at UDS. He has trained several undergraduate and postgraduate students and supervised many research theses. Dr. Asuako has attended numerous conferences locally and internationally and is credited with 18 peer review publications. He is a reviewer of some prestigious journals. He is a member of the Ghana Institution of Engineering (GhIE) and a student member of the American Society of Mechanical Engineers (ASME). He has developed several curricula for CTVET, and policy documents among others.

Mahmoud Abdulai Mahmoud holds a faculty position in the Department of Marketing and Entrepreneurship, University of Ghana Business School, University of Ghana, Legon. He is an Associate Professor in Marketing. He obtained his PhD, Master of Philosophy and Bachelor of Science in Business Administration Degrees from the University of Ghana Business School, University of Ghana. Professor Mahmoud's area of research interest is Strategic Marketing with a specific focus on market orientation and other strategic orientations (learning, innovation and corporate social responsibility). Professor Mahmoud has a number of publications to his credit in a number of international peer review journals including; Journal of Financial Services Marketing, Corporate Social Responsibility Journal, International Journal of Public Sector Management, International Journal of Marketing Studies, International Journal of Productivity and Performance Management. He has participated in a number of local and international conferences.

Donne Komla Muddey is a Lecturer and head of Department of Mathematics and Statistics, Ho Technical University. He holds Ph.D in Applied Statistics from University for Development Studies. He has about eighteen years of teaching experience in the field of statistics in the Polytechnic/University with specialization in data management and analysis (Computing Statistics), and Probability distributions. His research areas are; probability distributions, informal sector and Social Protection. He is a member of Ghana Statistical Association, professional body. He also has strong technical skills as well as excellent entrepreneurial skills, hence, establishment of the following companies; Zendo Farms, Laststop Computer Ventures, and Health Links Homoeopathic Clinic conscientious and self-motivated individual with great enthusiasm and determination to succeed in all fields. He attends international and local conferences; conferences on Employment Promotion and Economic

Development for African English-Speaking Countries, Sponsored by the Ministry of Commerce, P. R. China, and, a conference paper presented at the University for Studies, on Modification of the Weighted Weibull Distribution.

Alhassan Mumuni is a Senior Lecturer and the head of Electrical/Electronics Department, CCTU

Beauty Awusi Nyavor Akporyo staff of Accra Technical University since 2007 till date A Technical staff of Fashion Department working in the production also she holds MHIL in Fashions design Technology from KNUST and (MT)in Fashion design and Textiles from UEW. Currently pursuing her Phd in fashion design from Goenka University .

Elikem Chosniel Ocloo is a Senior Lecturer in the Marketing Department and currently, the Head of the Marketing Department. He holds a Bachelor of Arts Degree in Geography and Resource Development (Major) with First Class Honours from the University of Ghana and a Master's degree in Business Administration (Marketing option) from the University of Leicester, United Kingdom, and a Doctor of Philosophy (Ph.D.) degree from Jiangsu University, China in Management Science and Engineering (Marketing-focused). He has acquired experience in both academia and industry and is an ardent seeker of innovation and creative ways to solve societal and business marketing problems. He is an Associate Member of the Chartered Institute of Marketing Ghana (CIMG). Dr. Ocloo has publications in International Journals and other journals with Scopus indexing, SSCI-Indexed, and ESCI-Indexed and presented papers in both local and international conferences. He has a special interest in current business trends in Digitalisation and 4th Industrial Revolution concepts. His current research interests include E-commerce/Digitalisation, Health Marketing, SMEs Development, Social Enterprise, Social Marketing, and Marketing Management.

Frank Frimpong Opuni (PhD, MBA, B.ed) is a Senior Lecturer in the Marketing Department within the Faculty of Business of Accra Technical University, Ghana. He has successfully defended his PhD from the University of Bolton, UK, has an MBA in Marketing from KNUST, Ghana and a Bachelor of Education from the University of Cape Coast, Ghana. Frank has published a good number of articles in high ranking peer reviewed journals across the globe. His special research interests are: Social Marketing, Digital Marketing, Health Marketing and Pro-environmental/ Sustainability Marketing.

Dorcas Serwaa is a lecturer, researcher and African Union Scholar currently doing her PhD with University of Melbourne. She has either led or co-authored over 20 peer-reviewed articles and hope to contribute more to the scientific community through evidence-based studies

Philip Yamba is an esteemed senior lecturer in the field of mechanical engineering at the Tamale Technical University. He is passionate about teaching and sharing his knowledge with students and colleagues. Dr. Philip earned his bachelor's degree at Kwame Nkrumah University of Science and Technology. He later went on to earn his Master's degree in Engineering Project Management at the Ghana Technology University College and obtained his PhD in Mechanical Engineering at Jiangsu University, China. Dr. Philip has been teaching at Tamale Technical University since 2005. He has taught a wide variety of courses in the areas of mechanical engineering, and his research focuses on the area of computer vision systems in non-destructive testing and evaluation in laser and conventional welding and materials technology. Dr. Philip has authored or co-authored numerous papers on the subject and is a well-known figure in the field. Dr. Philip is also highly active in the university community and is a strong advocate for diversity and inclusion. He is the head of the mechanical engineering department and a member of several university committees and advisory boards.

Index

E

F

G

H

I

L

M

N

O

P

R

S

T

V

W

www.ingramcontent.com/pod-product-compliance
Lightning Source LLC
LaVergne TN
LVHW080611300926
849162LV00005B/1123